JPEG
STILL IMAGE
DATA COMPRESSION
STANDARD

JPEG
STILL IMAGE
DATA COMPRESSION
STANDARD

William B. Pennebaker
Joan L. Mitchell

VAN NOSTRAND REINHOLD
New York

Library of Congress Catalog Card Number 92-32860
ISBN 0-442-01272-1

I(T)P Van Nostrand Reinhold is a division of International Thomson Publishing.
       ITP logo is a trademark under license.

Printed in the United States of America

Van Nostrand Reinhold
115 Fifth Avenue
New York, New York 10003

International Thomson Publishing
Berkshire House
168-173 High Holborn
London, WC1V 7AA, England

Thomas Nelson Australia
102 Dodds Street
South Melbourne 3205
Victoria, Australia

Nelson Canada
1120 Birchmount Road
Scarborough, Ontario M1K 5G4, Canada

16 15 14 13 12 11 10 9 8 7 6 5 4 3

Library of Congress Cataloging-in-Publication Data

Pennebaker, William B.
    JPEG still image data compression standard / William B.
Pennebaker, Joan L. Mitchell.
        p. cm.
    Includes bibliographical references and index.
    ISBN 0-442-01272-1
    1. Image processing—Digital techniques—Standards. 2. Data
compression (Telecommunication)—Standards. 3. Algorithms.
I. Mitchell, Joan L. II. Title.
TA1632.P45 1992
621.36'7—dc20   92-32860
                                                      CIP

To Margaret, Betsy, Patty,

Doris, and Bill

CONTENTS

FOREWORD xiii

ACKNOWLEDGMENTS xv

TRADEMARKS xvii

CHAPTER 1. INTRODUCTION 1
1.1 Examples of JPEG image compression ○ 4
1.2 Organization of the book ○ 4
1.3 An architecture for image compression ○ 6
1.4 JPEG baseline and extended systems ○ 6
1.5 An evolving standard ○ 7
1.6 An international collaboration ○ 7

CHAPTER 2. IMAGE CONCEPTS AND VOCABULARY 9
2.1 Digital images ○ 10
2.2 Sampling ○ 10
2.3 Two-dimensional arrays of samples ○ 12
2.4 Digital image data types ○ 13

2.5 Large amounts of data ○   21

CHAPTER 3. ASPECTS OF THE HUMAN VISUAL SYSTEM   23
3.1 Luminance sampling ○   23
3.2 Sample precision ○   25
3.3 Chrominance sampling ○   25
3.4 Linearity ○   25

CHAPTER 4. THE DISCRETE COSINE TRANSFORM (DCT)   29
4.1 Basic DCT concepts ○   29
4.2 Mathematical definition of the FDCT and IDCT ◑   39
4.3 Fast DCTs ◑   41

CHAPTER 5. IMAGE COMPRESSION SYSTEMS   65
5.1 Basic structure of image compression systems ○   65
5.2 Image compression models ○   67
5.3 JPEG entropy encoder and entropy decoder structures ○   73
5.4 Transcoding ○   77
5.5 JPEG lossless and lossy compression ○   78
5.6 Sequential and progressive coding ○   79
5.7 Hierarchical coding ○   79
5.8 Compression measures ○   79

CHAPTER 6. JPEG MODES OF OPERATION   81
6.1 Sequential DCT-based mode of operation ○   81
6.2 Progressive DCT-based mode of operation ○   86
6.3 Sequential lossless mode of operation ○   92
6.4 Hierarchical mode of operation ○   93

CHAPTER 7. JPEG SYNTAX AND DATA ORGANIZATION   97
7.1 Control procedures and compressed data structure ○   97
7.2 Interchange and abbreviated compressed data formats ○   99
7.3 Image data ordering ○   99
7.4 Marker definitions ◑   105
7.5 Frame header ◑   110
7.6 Scan header ◑   113
7.7 Limit on the number of data units in an MCU ◑   116
7.8 Marker segments for tables and parameters ◑   117
7.9 Hierarchical progression marker segments ◑   120
7.10 Examples of JPEG data streams ◑   122
7.11 Backus-Naur Form ●   127
7.12 JPEG BNF ●   130

CHAPTER 8. ENTROPY CODING CONCEPTS   135
8.1 Entropy and information ○   135
8.2 An example to illustrate entropy coding ○   137
8.3 Variable-length code words ○   137
8.4 Statistical modeling ○   143
8.5 Adaptive coding ○   147

CHAPTER 9. JPEG BINARY ARITHMETIC CODING   149
9.1 The QM-encoder ◑   151
9.2 The QM-decoder ◑   162
9.3 More about the QM-coder ◑   166

CHAPTER 10. JPEG CODING MODELS   169
10.1 JPEG sequential DCT-based coding models ◑   170
10.2 Models for progressive DCT-based coding ◑   173
10.3 Coding model for lossless coding ◑   182
10.4 Models for hierarchical coding ◑   185

CHAPTER 11. JPEG HUFFMAN ENTROPY CODING   189
11.1 Statistical models for the Huffman DCT-based sequential
  mode ◑   190
11.2 Statistical models for progressive DCT-based coding ◑   194
11.3 Statistical models for lossless coding and hierarchical mode spatial
  corrections ◑   198
11.4 Generation of Huffman tables ◑   198

CHAPTER 12. ARITHMETIC CODING STATISTICAL
MODELS   203
12.1 Overview of JPEG binary arithmetic-coding procedures ◑   203
12.2 Decision trees and notation ◑   204
12.3 Statistical models for the DCT-based sequential mode with
  arithmetic coding ◑   206
12.4 Statistical models for progressive DCT-based coding ◑   214
12.5 Statistical models for lossless coding and hierarchical-mode spatial
  corrections ◑   216
12.6 Arithmetic coding conditioning tables ◑   218

CHAPTER 13. MORE ON ARITHMETIC CODING   219
13.1 Optimal procedures for hardware and software ●   220
13.2 Fast software encoder implementations ●   225
13.3 Fast software decoder implementations ●   228
13.4 Conditional exchange ●   229
13.5 QM-coder versus Q-coder ●   229
13.6 Resynchronization of decoders ●   230
13.7 Speedup mode ●   231

CHAPTER 14. PROBABILITY ESTIMATION   233
14.1 Bayesian estimation ◑   234
14.2 Renormalization-driven estimation ◑   235
14.3 Markov-chain modelling of the probability estimation ●   236
14.4 Approximate model ●   238
14.5 Single- and mixed-context models ●   240
14.6 Single-context model ●   241
14.7 Mixed-context model ●   243
14.8 Application of the estimation models to the QM-coder ◑   244
14.9 Initial learning ◑   247

14.10 Robustness of estimators versus refinement of models ◑   249
14.11 Other estimation tables ◑   249

CHAPTER 15. COMPRESSION PERFORMANCE   253
15.1 Results for baseline sequential DCT ◑   254
15.2 Results for sequential DCT with arithmetic coding ◑   255
15.3 Results for sequential DCT with restart capability ◑   255
15.4 Results for progressive DCT with arithmetic coding ◑   256
15.5 Results for lossless mode with arithmetic coding ◑   257
15.6 Summary of Results ○   258

CHAPTER 16. JPEG ENHANCEMENTS   261
16.1 Removing blocking artifacts with AC prediction ◑   261
16.2 Low bitrate VQ enhanced decoding ◑   264
16.3 An approximate form of adaptive quantization ◑   264
16.4 Display-adjusted decoding ○   266

CHAPTER 17. JPEG APPLICATIONS AND VENDORS   267
17.1 Adobe Systems Incorporated ○   269
17.2 AT&T Microelectronics ○   270
17.3 AutoGraph International ApS ○   271
17.4 AutoView ○   272
17.5 Bulletin board systems ○   272
17.6 California Department of Motor Vehicles ○   273
17.7 C-Cube Microsystems, Inc. ○   273
17.8 Data Link ○   275
17.9 Discovery Technologies, Inc. ○   275
17.10 DSP ○   276
17.11 Eastman Kodak Company ○   276
17.12 Handmade Software ○   277
17.13 IBM ○   278
17.14 Identix ○   279
17.15 IIT ○   279
17.16 Independent JPEG Group ○   280
17.17 ITR ○   281
17.18 Lewis Siwell, Inc. ○   281
17.19 LSI Logic ○   282
17.20 Moore Data Management Services ○   283
17.21 NBS Imaging ○   283
17.22 NTT Electronics Technology Ltd. ○   284
17.23 OPTIBASE® ○   284
17.24 Optivision, Inc. ○   284
17.25 Philips Kommunikations Industrie ○   285
17.26 PRISM ○   286
17.27 Storm Technology ○   286
17.28 Telephoto Communications ○   287
17.29 Tribune Publishing Co. ○   288
17.30 VideoTelecom ○   288
17.31 XImage ○   289

17.32 Xing ○   289
17.33 Zoran Corporation ○   290
17.34 3M ○   291
17.35 File formats ○   292

CHAPTER 18. OVERVIEW OF CCITT, ISO, AND IEC   295
18.1 ISO ○   296
18.2 CCITT ○   297
18.3 IEC ○   298
18.4 Joint coordination ○   299

CHAPTER 19. HISTORY OF JPEG   301
19.1 Formation of JPEG ○   301
19.2 Original JPEG Goals ○   302
19.3 Selecting an approach ○   302
19.4 Functional requirements ○   303
19.5 Refining the ADCT technique ○   306
19.6 Technical specifications ◑   307
19.7 ISO 10918 Part 1 ○   309
19.8 JPEG Part 1 DIS ballot results ○   311
19.9 CCITT Recommendation T.81 ○   311
19.10 ISO 10918 Part 2 ○   311
19.11 JPEG Goals Achieved ○   313

CHAPTER 20. OTHER IMAGE COMPRESSION STANDARDS   317
20.1 CCITT G3 and G4 ○   317
20.2 H.261 ○   318
20.3 JBIG ○   318
20.4 MPEG ○   325

CHAPTER 21. POSSIBLE FUTURE JPEG DIRECTIONS   331
21.1 Adaptive quantization ○   331
21.2 Improvements to lossless coding ◑   332
21.3 Other possible addenda ○   333
21.4 Backwards compatibility ○   333

APPENDIX A. ISO DIS 10918-1 REQUIREMENTS AND GUIDELINES   335

APPENDIX B. DRAFT ISO DIS 10918-2 COMPLIANCE TESTING   545

REFERENCES   627

INDEX   632

FOREWORD

In the latter part of 1985, I began looking for a group in a position to establish an international standard for compression of continuous-tone still images. At the time, my department at Digital was beginning to build bi-level image archiving systems for insurance companies and the like, based on the CCITT G3/G4 compression methods used by fax machines. For anyone close to this work, it was readily apparent that compression was an essential enabling technology, and that any compression method employed needed to be an accepted international standard. Even then customers were shying away from technologies that threatened to lock them into a single vendor.

It seemed obvious that future applications would need to store and transmit grayscale and color continuous-tone images, yet there were no such standard compression methods. From the literature survey and experimentation that my colleagues and I at Digital had conducted, I felt that current research had reached diminishing returns in image-quality/compression performance. Moreover, no major industrial concerns appeared to have vested interests in any particular solution. The timing and the technology seemed ripe for standardization.

In March of 1986, I attended my first ISO/TC97/SC2/WG8 meeting in Boston. A small subgroup of WG8—Graham Hudson, Alain Leger, Hiroshi Yasuda, Istvan Sebestyen, and a few others—formed the precursor to JPEG. As I participated over the following months and the group grew, I became convinced that it had the expertise, the international representation, and the proper organizational vehicles (ISO and CCITT) to define the accepted world standard.

Unlike the traditional model of standardization in which committees are relegated to "harmonizing" conflicting preexisting standards, JPEG grew into a true world-wide applied research collaboration, that extended the practical state of the art in still image compression. Indeed, the widely-acclaimed MPEG (Motion Picture Experts Group) committee modeled itself after the best of JPEG's experiences. Naturally, JPEG was a most rewarding technical experience, but it also provided a greatly rewarding human experience: the chance to watch many bright and culturally diverse engineers transform from a combative collection of individual researchers into a cooperative team of collaborators.

Two of the most insightful, energetic, and prolific members of this JPEG team were Joan Mitchell and Bill Pennebaker. It is a distinct privilege to introduce their comprehensive book on JPEG.

This book should be especially welcome to the technical community. At the present time, the Baseline JPEG method is being widely implemented and discussed (though not always so widely understood), but JPEG's many powerful extended features are little known. As key contributors to all aspects of JPEG, Mitchell and Pennebaker are eminently qualified to cover this subject matter.

I highly recommend this book to anyone seeking a greater understanding of the technology and applications for the ISO/CCITT JPEG standard.

Gregory K. Wallace
Digital Equipment Corporation

ACKNOWLEDGMENTS

The ISO 10918-1 JPEG Draft International Standard | CCITT Recommendation T.81 is the result of the dedicated work and collaborative efforts of a team of technical experts from many countries. It has been a privilege and honor to work with these people. In particular, L. Conte (Italy), A. Gill (Israel), E. Hamilton (USA), A. Leger (France), A. Ligtenberg (USA/The Netherlands), H. Lohscheller (Germany), T. Omachi (Japan), H. Poulsen (Denmark), and J. Vaaben (Denmark) have been long-standing contributors and deserve special recognition. Many others have contributed, but the list is so long that it is impractical to mention specific names.

We would also like to express our appreciation to G. Hudson (former JPEG chair, UK), G. Wallace (present JPEG chair, USA, and author of the foreword to this book), H. Yasuda (Convenor of SC29, Japan), and I. Sebestyen (Special Rapporteur for CCITT SGVIII, Germany) for their guidance and encouragement to the JPEG committee.

We are extremely grateful to the ISO Central Secretariat for permission to reproduce both Part 1 and Part 2 of the JPEG Draft International Standard (DIS) in our book. Although Part 2 is still in unapproved form, these documents will make the book far more useful to our readers. We are also grateful to E. Hamilton, editor of Part 2, for the extra effort he made to get the latest draft of this document to us.

Most of the images that we have used for the examples in this book are JPEG test images. The original source images were digitized for JPEG by the Independent Broadcasting Authority (IBA, UK). These images are in the public domain, but IBA requests that users acknowledge the source of the images.

Our IBM colleagues I. R. Finlay, J. H. Morgan, and K. L. Anderson deserve special thanks for their continuing involvement in JPEG activities. K. L. Anderson was deeply involved in much of the testing and validation work, and her name is well known to the JPEG committee.

Much appreciation is due to others at IBM for their help during the past five years. We particularly want to thank L. A. Allman, T. P. Dowd, C. A. Gonzales, T. McCarthy, R. J. Moorhead, H. A. Peterson, and D. S. Thornton. In addition, R. B. Arps, G. Goertzel, G. G. Langdon, R. C. Pasco, J. J. Rissanen, and others in the IBM Research Division made significant contributions to the arithmetic coding technology that became part of JPEG.

K. S. Pennington and D. F. Liddell, our managers during most of our participation in JPEG, have been very supportive of our JPEG involvement. C. F. Touchton, the IBM image standards project authority, educated us about standards work, attended many meetings with us, started the US X3L2.8 committee, and has continued to encourage us during our involvement with JPEG and the writing of this book.

Finally, we want to express our appreciation to S. Chapman of Van Nostrand Reinhold, E. A. McAuliffe and J. W. Dunkin of IBM Research, who helped us create this book with BookMaster™, R. L. Lucchese, who assisted us with the chapter on JPEG vendors and applications, and S. Pendak, who did all the final formatting and graphics in the JPEG DIS Part 1. We also want to thank the vendors who supplied information for this book, G. R. Thompson, who helped us prepare the images for publication, D. P. de Garrido, who supplied the images illustrating AC prediction, K. L. Anderson and J. H. Morgan, who spent many hours proofreading, E. Viscito, who checked our MPEG sections for accuracy, W. Equitz and R. B. Arps, who reviewed the JBIG sections, and E. L. Pennebaker, who helped us to make the tutorial chapters more readable.

TRADEMARKS

Adobe PostScript Language, Adobe Photoshop, and Adobe Display PostScript are trademarks of Adobe Systems Incorporated, which may be registered in certain jurisdictions.

Aldus is a registered trademark of Aldus Corporation. TIFF is a trademark of Aldus Corporation.

Apple is a registered trademark of Apple Computer, Inc. Macintosh is a trademark licensed to Apple Computer, Inc.

AT&T and VCOS are registered trademarks of American Telephone and Telegraph Company.

EasyTech/Codec, EasyCopy/X, and EasyCopy/PC+ are trademarks of AutoGraph International ApS.

CL550 is a trademark of C-Cube Microsystems, Inc.

CompuServe is a registered trademark of CompuServe Inc.

EPASS is a registered trademark of Data Link Information Solutions, Inc.

FilmFax is a registered trademark of Discovery Technologies, Inc.

Kodak is a registered trademark of the Eastman Kodak Company. Kodak Professional Digital Camera System, Kodak 35mm Rapid Film

1

INTRODUCTION

This book is about JPEG, the new international standard for color image compression.

JPEG is an acronym for "Joint Photographic Experts Group." This group has been working under the auspices of three major international standards organizations—the International Organization for Standardization (ISO), the International Telegraph and Telephone Consultative Committee (CCITT), and the International Electrotechnical Commission (IEC)—for the purpose of developing a standard for color image compression. As the standard evolved and implementers started deploying it, the name quite naturally became attached in an informal way to the standard itself.

JPEG now stands at the threshold of widespread use in diverse applications. Many new technologies are converging to help make this happen. High-quality continuous-tone color displays are now a part of most personal computing systems. Most of these systems measure their storage in megabytes, and the processing power at the desk is approaching that of mainframes of just a few years ago. Communication over telephone lines is now routinely at 9,600 baud, and with each year modem capabilities improve. LANs are now in widespread use. CD-ROM and other

(a)

(b)

Figure   1-1. Examples of **JPEG** reconstructed images. (a) Original (720
         pixels horizontally by 576 pixels vertically for a total of
         414,720 bytes). (b) Compressed to 29,471 bytes.

(c)

(d)

Figure 1-1 continued. (c) Compressed to 9,416 bytes. (d) Compressed to 3,729 bytes.

mass-storage devices are opening up the era of electronic books. Multimedia applications promise to use vast numbers of images and digital cameras are already commercially available.

These technology trends are opening up both a capability and a need for digital continuous-tone color images. However, until JPEG compression came upon the scene, the massive storage requirement for large numbers of high-quality images was a technical impediment to widespread use of images. The problem was not so much the lack of algorithms for image compression (as there is a long history of technical work in this area), but, rather, the lack of a standard algorithm—one which would allow an interchange of images between diverse applications. JPEG has provided a high-quality yet very practical and simple solution to this problem.

1.1 Examples of JPEG image compression ○

The purpose of image compression is to represent images with less data in order to save storage costs or transmission time and costs. Obviously, the less data required to represent an image, the better, provided there is no penalty in obtaining a greater reduction. However, the most effective compression is achieved by approximating the original image (rather than reproducing it exactly), and the greater the compression, the more approximate ("lossy") the rendition is likely to be.

The user of JPEG compression can adjust the compression parameters so as to achieve an increase in compression in return for a decrease in image quality. Figure 1-1 shows a series of JPEG reconstructed grayscale images that illustrate some of the possible choices. Figure 1-1a is the 414,720-byte original image. This image can be compressed to 219,599 bytes without any loss or change in the image. Figures 1-1b, 1-1c, and 1-1d show reconstructed images after lossy compression to 29,471 bytes, 9,416 bytes, and 3,729 bytes, respectively. These same images in color (small sections of which are shown on the cover) compress to 35,037 bytes (0.68 bits/pixel) 11,741 bytes (0.23 bits/pixel), and 5,915 bytes (0.11 bits/pixel). Note that it takes only a modest additional amount of compressed data to support full color.

1.2 Organization of the book ○

This book is intended for a rather diverse audience. As the use of JPEG grows, we expect that many people without technical backgrounds will need to understand what JPEG can do for them and how to use JPEG intelligently. Others will need to understand more technical aspects of JPEG as they integrate JPEG capabilities into their applications. Still others, such as implementers and students of data compression, will need in-depth technical information about JPEG.

We have made an estimate of the level of difficulty of each section, and have marked each section heading with one of three indicators:

○ indicates sections suitable for the nontechnical reader. These sections should be useful for those who want to understand the functions and capabilities of JPEG, but have only a passing interest in "how it actually works." Mathematical equations are almost nonexistent in these sections and many of the concepts are illustrated with photographs and graphics. A reader can bypass all of the more advanced sections, as ○ sections are not dependent on a knowledge of the more advanced sections.

◐ indicates a section of intermediate technical difficulty. These sections build upon the material in the ○ sections and usually have some technical content. Some college-level training in mathematics, engineering, or science is needed to fully understand many of these sections.

● marks sections that involve details of interest only to a JPEG implementer, or that require much deeper mathematical skills.

The first three chapters of this book are tutorials for people unfamiliar with the concepts and vocabulary relating to images and image compression. The chapters which follow cover material more specific to JPEG.

Chapter 4 covers the discrete cosine transform (DCT), a very important aspect of JPEG. The DCT allows users to select a level of image quality, and is a crucial element in the compression-versus-quality trade-off that can be made in JPEG. Although the beginning part of Chapter 4 is tutorial, the later parts are a selective and increasingly sophisticated review of modern algorithms for computing fast DCTs. Chapter 5 contains a tutorial on image compression systems.

At that point the foundation is laid for a proper understanding of the JPEG standard. Chapter 6 on JPEG modes of operation should provide an understanding of the functional capabilities of JPEG. The six chapters that follow get into more technical details, and should be of interest to those who need to understand the structure of the JPEG compressed data streams and the inner workings of the compression algorithms. Some of the function and flexibility in JPEG may be truly appreciated and understood only when the material in these chapters is mastered.

Because arithmetic coding is relatively new, we have devoted a chapter to additional information about efficient implementations of the JPEG arithmetic coder and another chapter to the underlying structure of the probability estimation used in the arithmetic coder.

Compression performance is described in Chapter 15, and Chapter 16 discusses some optional ways of improving the quality of images compressed using a standard JPEG compressed data stream. These chapters are followed by chapters on applications and vendors, standards committees, and the history of JPEG.

Chapter 20 provides a brief review of the systems defined by two other image compression standards committees, JBIG and MPEG. The Joint Bi-level Image Experts Group (JBIG) has defined a new technique for bi-level or black/white image compression that will improve the compression of facsimile images. The Moving Picture Experts Group (MPEG) has defined a technique for motion sequences. In their respective domains, these techniques are more appropriate than JPEG.

Appendix A contains a complete reproduction of the JPEG DIS Part 1, and Appendix B contains a draft of JPEG DIS Part 2. Of course, if the reader should discover any conflict between our descriptions and the JPEG DIS, the JPEG DIS must take precedence. There have been minor editorial changes to Part 1 since the JPEG DIS was issued, but we know of no significant technical changes or corrections to the DIS in its transition to International Standard (IS) status. Although JPEG has reached consensus on all basic aspects of Part 2, some details remain to be completed.

We shall refer frequently to the JPEG DIS, and unless we explicitly state Part 2, we mean Part 1. We encourage the reader to examine the DIS and use it as a cross-reference for the rest of this book. We would like to call attention to the good introductory section (Clause 3), the definitions of terms (Clause 4) that are specific to the DIS, Annex K, which contains informational and guidance material, Annex L, which lists patents that may be relevant, and Annex M, which contains additional references. The index for this book covers the material in the DIS, and, because the DIS itself has no index, this may be helpful to the reader.

1.3 An architecture for image compression ○

JPEG is more than an algorithm for compressing images. Rather, it is an architecture for a set of image compression functions. It contains a rich set of capabilities that make it suitable for a wide range of applications involving image compression.

In one respect, however, JPEG is not a complete architecture for image exchange. The JPEG data streams are defined only in terms of what a JPEG decoder needs to decompress the data stream. Major elements are lacking that are needed to define the meaning and format of the resulting image. The JPEG committee recognized that these aspects are quite controversial and would probably have delayed the decision-making process needed to complete JPEG. They decided that, necessary as these parameters and constraints are, they are more properly the domain of application standards. The committee therefore deliberately did not include them in JPEG.

1.4 JPEG baseline and extended systems ○

JPEG has defined a "baseline" capability which must be present in all of the JPEG modes of operation which use the DCT. In the authors' experience many people who consider themselves experts on JPEG know only about the baseline system. There is much more to JPEG than the baseline system, and one of the goals of this book is to make these extended system functions better known.*

Indeed, JPEG has already become a standard of comparison in judging new compression algorithms, and the "name of the game" is to show how one's new algorithm is superior to JPEG. For that reason, another of the goals of this book is to educate users so that these comparisons can be made properly and with an understanding of the trade-offs between

---

* For example, the JPEG "progressive" mode with an alternate coding method (arithmetic coding) was used to obtain the set of images shown in Figure 1-1 and also the set on the cover. With custom tables the baseline system requires 38,955 bytes to code the example in Figure 1-1b.

complexity, compression performance, and function that were made by the JPEG committee.

1.5 An evolving standard ○

One of the problems in preparing a book of this type is that the standard is evolving even as we complete this book. Although JPEG is now officially a CCITT recommendation, the document is still undergoing minor editorial changes in ISO. Therefore, a note of caution is in order: The reader should recognize that JPEG is not yet an ISO International Standard, and, until it is, is subject to change. In addition, although we have made a concerted effort to present an accurate description of JPEG in this book, what we present is our interpretation. The official documents for ISO International Standards and CCITT Recommendations are distributed by the various standards organizations.

1.6 An international collaboration ○

As is clearly stated in the acknowledgments, the JPEG standard is the result of the international technical community's collaboration to create a standard that applies to a broad range of applications. Although we certainly contributed to this work, we are really reporting here on the work of this large group. As the reader will find in going through the DIS, standards documents can be relatively terse and often do not contain the background information that would help one to understand the underlying technologies and how to apply them. This book is intended to fill in some of the missing information.

2

IMAGE CONCEPTS AND VOCABULARY

In this chapter we develop some of the basic concepts and vocabulary needed for an understanding of the image compression functions contained in JPEG. Our approach in this chapter will be qualitative rather than mathematical.

The reader undoubtedly has an intuitive grasp of what the term "image" means. We are surrounded by images: photographs, television, printed material, etc., all aimed at the incredibly powerful visual input channel into the human brain. We are used to thinking of images as two-dimensional, but in fact, they often have several more dimensions. Color, time (motion), and depth (three-dimensional and stereo pairs) can add additional dimensions to the information in an image, and, if we want to consider the image as a fractal, the question of dimensionality becomes yet more complex. By definition, however, JPEG is concerned primarily with images that have two spatial dimensions, contain grayscale or color information, and possess no temporal dependence (hence the word "still" in the title of the JPEG standard).

2.1 Digital images ○

The images compressed with the JPEG techniques are digital images. Many readers are already familiar with the way in which television images are scanned. Starting at the top of the screen, a television field is displayed on a CRT (cathode-ray tube) by the scanning of an electron beam one horizontal line at a time in rapid sequence. An image created in this way is already segmented into a set of discrete scan lines. To turn this into a true digital image we do two more things: take samples along each scan line at regular intervals and convert these samples into binary numbers which can be stored in a computer.

2.2 Sampling ○

The sampling process used to create a digital image necessarily discards an enormous amount of information: the fine spatial detail and tiny amplitude variations that are, by definition, not important enough to preserve. The term *lossless compression*, which we shall discuss later in this chapter implies preservation of all information, but in fact such compression preserves only the information in the sampled data, totally ignoring the loss of information that is a necessary part of capturing any digital image.

There are three important parameters in sampling a signal: precision, sampling interval, and sampling aperture.

Suppose we have the one-dimensional analog signal, perhaps the output of a television camera, shown in Figure 2-1a. When we take measurements of this signal at discrete points, marked by dots, we are sampling the signal. When we convert to a digital representation of the signal we are expressing the sample with some fixed number of bits per sample, which is the precision of the sample. In the process we are assuming that this fixed number of bits is sufficient to represent the signal, much as we restrict the number of decimal places when we write down a number. The only difference is that we use binary digits or "bits," the traditional way of expressing numbers in computers.

The interval between samples determines the resolution of the sampled signal. For the signal in Figure 2-1a resolution can be defined as the number of samples taken per unit time. For a given precision in our samples, the higher the resolution (that is, the more samples we take in a given time interval), the more accurately we can represent the variations in the signal. The resolution thus provides an upper limit on the frequencies that can be represented in the sampled signal. Just how we sample a signal is very important, however, because we can get very misleading measurements if we do it improperly.

Figure 2-1b illustrates a signal that varies rapidly between samples. If we connect the dots as shown in Figure 2-1b, the sampled output appears to have a much lower frequency than the original analog signal. This distortion is termed *aliasing*, and is closely related to an effect seen in western-movie scenes, in which wagon wheels appear to turn backwards. Figure 2-1b illustrates an important point. You must have enough samples

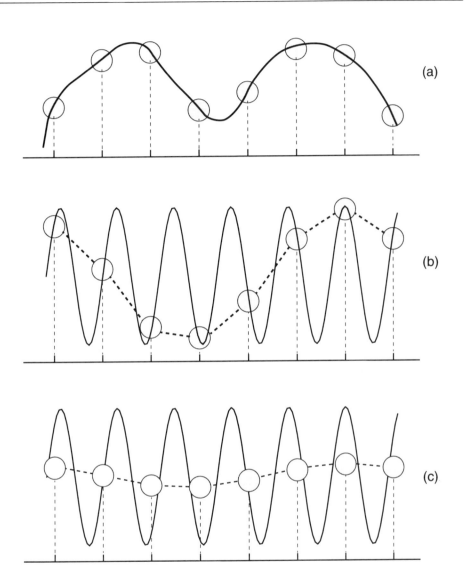

Figure   2-1. **Sampling illustrations.** (a) Illustrates sampling of a one-
          dimensional analog signal. (b) Illustrates how aliasing can
          occur when a high-frequency signal is sampled too infre-
          quently. (c) Shows the suppression of aliasing when the sig-
          nal is averaged over a window equal to the sampling interval.

to represent properly a signal that is changing.[1] When you do not have
enough samples, you may get aliasing effects.

Aliasing can be largely suppressed if the input signal is averaged over
an interval (the aperture) that is roughly equal to the interval between

samples. Figure 2-1c illustrates this. Optical scanning input devices often do this averaging automatically in the process of scanning an image, usually because the effective aperture (opening) through which the light is acquired by each sensor is nearly as wide as the spatial interval between light sensors. However, even the best image input devices will exhibit minor aliasing artifacts.

One of the common space- and cost-saving techniques, especially in working with images, is to reduce the spatial resolution of the data (i.e., the number of samples taken in a given space). This can be done in a number of ways, including a simple method known as "subsampling." In subsampling by a factor of two, for example, every other sample is discarded, thereby halving the number of samples representing the data. Note that this effectively makes the sampling interval twice as large, while leaving the aperture unchanged. The net result is equivalent to sampling with too small an aperture. Sampling with too small an aperture or too large a sampling interval, as was done in Figure 2-1b, will provide an insufficient number of samples for a truly accurate representation of the original image, and can result in severe aliasing effects.

The solution to this problem is to use an average of two or more samples rather than a single sample in the process of reducing the resolution. We call this process filtering, and the weight we give to each sample as we average them is called the filter coefficient. The only restriction is that these weights should sum to one. Figure 2-2 gives two simple weighted averages (filters) that might be used when subsampling by a factor of two in one dimension. Note that the center of each subsampled point labelled Y coincides with the center of one of the original sampling points labelled X; the center of each subsampled point labelled Z coincides with a boundary between sample points labelled X. These are examples of two different sample registrations.

Taking a weighted average is also known as *low-pass filtering*, because the effect of the averaging is to remove high spatial frequencies. These filters are also called *downsampling filters*. With different weights one can also *high-pass filter*, i.e., selectively enhance the high spatial frequencies. High-pass filters are also known as *upsampling* filters.

2.3 Two-dimensional arrays of samples ○

For simplicity let us consider a digital image obtained by scanning a monochrome (grayscale) photograph. We create a two-dimensional array of samples, each sample with 8-bit precision. Although sampling grids other than rectangular are possible, JPEG has chosen to use only simple rectangular grids; we shall therefore limit our discussion to such grids. We shall have more to say about precision requirements in following sections, but one sample per byte (eight bits) is a convenient precision for computer storage of images. This precision is usually sufficient to give the illusion of a continuous range of grayscale values when the image is observed on a typical CRT display. Some applications, notably those dealing with medical images, may require higher precision (12 bits/sample or more).

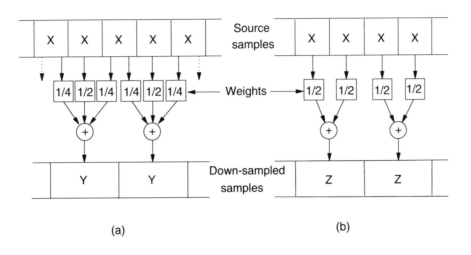

Figure   2-2. Two examples of low-pass filtering

Each sample in a monochrome image is called a pixel or pel. *Pixel* is a contraction of "picture element" and is used in the display industry. *Pel* is a contraction for "print element" and is used in the printing industry. In modern usage the two are almost interchangeable, and the distinction between them has been lost. This is probably fortunate, since in many applications there is a need both to display and to print images. Note that in color images a set of samples is usually required to represent a single pixel.*

Precision and aspect ratio are two attributes of a sample. Precision determines how many levels of intensity can be represented and is expressed as the number of bits/sample. Aspect ratio describes the shape of the sample, and this is determined by the relationship between the physical size of the image and the rectangular grid of samples. If the shape of the sample is not the same between input and output devices, geometric distortion will result.

Figure 2-3 illustrates the appearance of an image at three different precisions. Figure 2-3*a* shows an image at 8 bits/sample, Figure 2-3*b* shows the same image at one bit/sample and Figure 2-3*c* shows this image at four bits/sample.

2.4 Digital image data types ○

The examples in Figure 2-3 illustrate some of the different digital image data types. These are all examples of monochrome still images, where the

---

* If an appropriate color-encoding format is used for the analog signal, single samples can be used to represent color information. However, JPEG does not apply to color images sampled in this format.

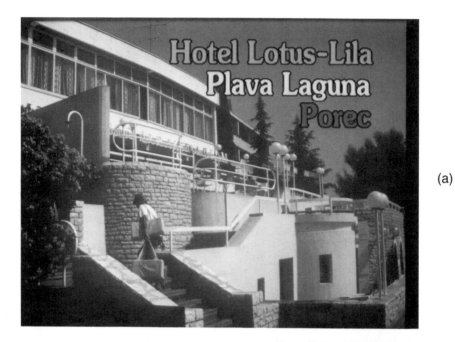

(a)

(b)

Figure   2-3. Examples of still-image precisions. (a) Eight bits/sample. (b)
        One bit/sample. (a) is the luminance component of the JPEG
        test image, "hotel".

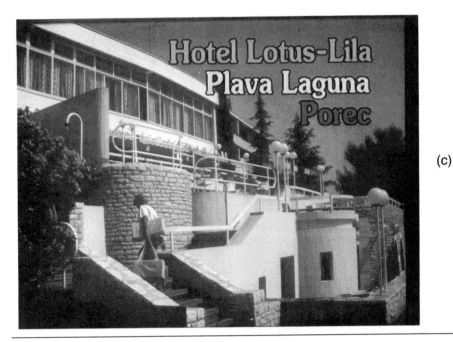

(c)

Figure   2-3   continued.   (c) Four bits/sample.

term "still" is used to distinguish from the set of images that make up a motion sequence. Unless otherwise stated, the images should be assumed to be "still" in this book.

2.4.1 Grayscale ○

At eight bits/sample, the example shown in Figure 2-3a has sufficient precision to be considered a grayscale or continuous-tone image. JPEG is designed for this class of data—data that are of high enough precision to give the appearance of continuity to an observer. JPEG can be used to compress limited-precision data, but may not be as efficient as alternative approaches designed for such data. A discontinuous representation such as a color-palette image must be remapped to a continuous representation before JPEG can be used effectively.

2.4.2 Binary image ○

Figure 2-3b shows a binary or bi-level image that could be transmitted by a facsimile machine. These images have only one bit per sample and are usually images of text or line drawings in which the original has only two tones, black and white. The resolution of these images is usually very high, such that the eye would have trouble discerning subtle changes in grayscale value from sample to sample. At these higher resolutions, a process called *digital halftoning* creates the effect of continuous tones in a binary image by alternating between closely spaced black and white samples. The effect is similar to the technique used to reproduce photographs in newspapers.

Indeed, photographs can be regarded as the most limiting form of halftoning, in that the photographic grains are either white or black. However, the effective resolution required to resolve single grains is on the order of 5,000 pels/inch (2,000 pels/cm). Therefore, even at the highest resolutions currently used for binary facsimile images (400 pels/inch), digital halftoning is an imperfect substitute for true grayscale capability. The JBIG Draft International Standard discussed in Chapter 20 is designed for these binary images.

2.4.3 Limited bits/sample ○

Figure 2-3c is an example of a lower-precision or limited bits/sample format that is typical of computer graphics. The bound between limited bits/sample and continuous tone is not rigorously defined, but can be taken to be about four bits/sample.

At intermediate resolutions, images of black/white documents can benefit from additional precision. A limited number of bits per sample can provide for grayscale in transition regions between black and white. This greatly improves the apparent quality of the images. The use of grayscale in this manner is sometimes called anti-aliasing and can be an effective tool for removing the "jaggies" in graphics images.

2.4.4 Color ○

According to the trichromatic theory, the sensation of color is produced by selectively exciting three classes of receptors in the eye. Certain frequencies in the visible light spectrum will excite certain receptors, producing the effect of color. Furthermore, if we provide the same stimulus to these receptors from two different sources, the two sources will appear to have the same color.

In color imaging there are two basic ways of producing this selective excitation: additive color and subtractive color. Additive color is used with active light-emitting systems in which the light from sources of different colors is added together to produce the perceived color. Subtractive color is used with passive systems, in which light from a given source is selectively absorbed at different wavelengths, leaving only the wavelengths that will be perceived as the desired colors. Color can also be produced by other physical processes, such as refraction and interference, but these are simply alternative ways of creating additive or subtractive color.

In an additive color device such as a display CRT, the light is produced by three primary phosphors, red, green, and blue (RGB). These phosphors are excited separately by the electron beam in the CRT, and the light emitted from the three phosphors stimulates the three types of receptors in the eye to produce the perception of color. When excited in the right proportion, the three phosphors produce the perception of white light; the absence of excitation produces black.

Subtractive color is used in the printing industry. Three colors (and sometimes a fourth) are superimposed on a white reflective surface such as paper. The inks, typically cyan (a blue-green), magenta, and yellow (CMY), selectively absorb certain ranges of wavelengths of light. The eye

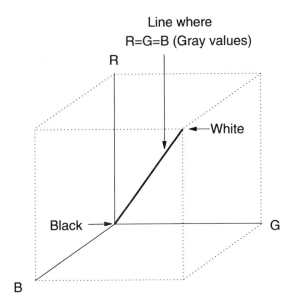

Figure   2-4. RGB (red-green-blue) color coordinate system. On the line
            labelled R=G=B are shades of gray ranging from black to
            white.

perceives the reflected light, which has not been absorbed; hence the term
"subtractive." Where no ink is on the paper the light reflected is white;
where all three inks are present, the light is (in principle) absorbed and the
appearance is black. In practice, complete absorption is difficult to achieve
in printing inks and a fourth ink, black (thus, CMYK, in which K stands
for *blacK*), is used as well.

2.4.4.1 *Color spaces/coordinates* ○

Many representations of color images are possible. The trichromatic theory tells us that, ideally, three arrays of samples (three components) should
be sufficient to represent a color image.[2] However, output devices are
limited, so not all colors can be obtained.
      RGB is one example of a color representation requiring three inde-
pendent values to describe the colors. Each of the values can be varied
independently, and we can therefore create a three-dimensional space with
the three components, R, G, and B, as independent coordinates. Colors
are represented as points in this space, as shown in Figure 2-4. Note that
shades of gray from black to white are found on the diagonal line in this
plot. In general pixels in a color image have information from the samples
of each component, and the color image is comprised of the two-
dimensional arrays of the component samples.
      Color representations such as RGB or CMY are not always the most
convenient. Other representations are available that use color components

that are closely related to the criteria used to describe color perception: brightness, hue, and saturation. Brightness describes the intensity of the light (revealing whether it is white, gray, or black) and this can be related to the luminance of the source. Hue describes what color is present (red, green, yellow, etc.) and this can be related to the dominant wavelength of the light source. Saturation describes how vivid the color is (very strong, pastel, nearly white) and this can be related to the purity or narrowness of the spectral distribution of the source.

Color spaces or color coordinate systems in which one component is the luminance and the other two components are related to hue and saturation are called luminance-chrominance representations. The luminance provides a grayscale version of the image (such as the image on a monochrome television receiver), and the chrominance components provide the extra information that converts the grayscale image to a color image. Luminance-chrominance representations are particularly important for good image compression.

2.4.4.2 *Linear color transformations* ○

The luminance of a display system is the sum of the luminances of the red, green, and blue phosphors. For convenience, the values of the three colors, R, G, and B, are expressed by a relative scale from 0 to 1, where 0 indicates no excitation of the phosphor and 1 indicates maximum excitation. In addition, the display is adjusted so that a gray value between black and white results whenever the three primary signals, R (red), G (green) and B (blue), are equal. For a particular definition of red, green and blue, the luminance (Y) of any color can be calculated from the following weighted sum:[3]

$$Y = 0.3R + 0.6G + 0.1B \qquad\qquad [2\text{-}1]$$

The scaling is chosen such that the luminance is also expressed by a relative scale from 0 to 1 and the weights reflect the contributions of the individual primaries to the total luminance.

The term *chrominance* is defined as the difference between a color and a reference white at the same luminance. The chrominance information can therefore be expressed by a set of color differences, V and U, where V and U are defined by:

$$V = R - Y \qquad\qquad [2\text{-}2]$$

$$U = B - Y \qquad\qquad [2\text{-}3]$$

These color differences are zero whenever $R = G = B$, as this condition produces gray, which has no chrominance. The V component controls colors ranging from red ($V > 0$) to blue-green ($V < 0$), whereas the U component controls colors ranging from blue ($U > 0$) to yellow ($U < 0$). Together with the luminance, these chrominance coordinates make up the color coordinate system known as YUV. Figure 2-5 illustrates the relationship between this YUV color space and the RGB color space.

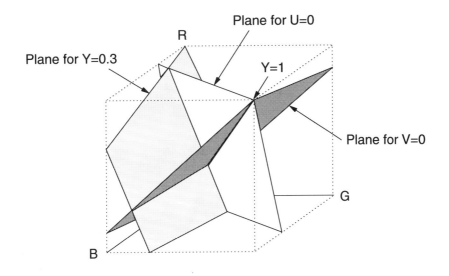

Figure   2-5. Relationship between the RGB and YUV coordinate systems

Chrominance values of zero are located on the diagonal where $R = G = B$, and the realizable range of YUV values is bounded by the RGB cube. The YUV color space is used in the European television systems.

Another color space, YIQ, is used in the North American television systems. The YIQ space is related to the YUV space as follows: Y is the same for both spaces whereas I and Q are related only to U and V. I and Q are given by:

$$I = 0.74V - 0.27U \qquad\qquad\qquad\qquad\qquad [2\text{-}4]$$

$$Q = 0.48V + 0.41U \qquad\qquad\qquad\qquad\qquad [2\text{-}5]$$

Still another color coordinate system, YCbCr, was used extensively in the development of the JPEG standard. This color coordinate system is closely related to YUV. It uses the same Y coordinate as the YUV system, whereas U and V are scaled and zero-shifted to produce the variables Cb and Cr, respectively. The equations are:

$$Cb = (U/2) + 0.5 \qquad\qquad\qquad\qquad\qquad [2\text{-}6]$$

$$Cr = (V/1.6) + 0.5 \qquad\qquad\qquad\qquad\qquad [2\text{-}7]$$

With this scaling and zero shifting the chrominance values are always in the range 0 to 1. The values of Cb and Cr are sometimes multiplied by 255, so that they can be represented by an 8-bit integer.

In principle, a number of other color spaces, including the CIE (Commission Internationale de l'Eclairage) tristimulus values, can be related to these spaces by simple linear equations.

2.4.4.3 Uniform perceptual color spaces ○

Although luminance-chrominance coordinates such as YUV are very convenient from the standpoint of ease of conversion to and from RGB, they suffer from one drawback: the amount of perceived color change produced by a fixed small change in these coordinates is quite nonuniform, and depends on the particular values of luminance and chrominance. Two color spaces, CIELUV and CIELAB, which provide a relatively uniform perceptual space for describing colors, have been defined. A uniform perceived change for given input change is a very desirable property, in that the precision required to express colors to a particular degree of fidelity can be more readily specified in a space with uniform perceptual characteristics. Unfortunately, the transformations between the linear spaces described in section 2.4.4.2 and these uniform perceptual spaces are relatively complex. Nonetheless, these spaces are expected to be used extensively in applications requiring precise color reproduction. Accurate color reproduction is a very complex subject and a complete discussion of it is well beyond the scope of this book.

2.4.4.4 Colorblindness of JPEG ○

From the perspective of JPEG, the question of which coordinate system to use for color images is important only in the sense that a poor choice of color coordinates can adversely affect the compression. In terms of the structure of the compression algorithms, JPEG is "colorblind"; however, applications that use JPEG may not be.

2.4.4.5 Additional details about color ○

Illumination has a very significant effect on the perception of color, as is well known to anyone who has tried to match colors under two different lighting conditions. The "white" used in the equations above assumes a particular illumination known as Standard Illuminant C.[3]

The range of colors that can be represented by any output device is limited. A display system can display colors only within a range dictated by the phosphors in the display. Since phosphors always emit a band of wavelengths, pure spectral colors can never be produced by a CRT.

Broadcast TV systems use a modified form of RGB known as gamma-corrected RGB. The "gamma correction" is made at the studio in order to correct for nonlinearities in television receiver CRTs. This is done for purposes of economy in manufacturing the receivers. Many modern display systems are relatively linear—that is, a set of equal-intensity steps will produce on the screen a corresponding set of grayscale steps that have a relatively uniform set of intensity changes. In critical applications the display response should be measured and corrections should be applied to the data to linearize the displayed output.

In low-cost computer display systems the display storage is sometimes limited to eight bits/pixel. This restricts the possible set of colors to 256 distinct values—the "palette." These colors do not form a continuous set; rather, the entries in the palette can be arbitrarily assigned. The JPEG algorithms assume continuity, and are not appropriate for direct compression of palette representations. However, JPEG is suitable for any of the continuous representations of color. It can be used for palette representations if the palette entries are converted to one of the continuous three-component representations before compression is performed.

2.4.5 Image sequences ○

Sometimes images come in correlated sets of two-dimensional arrays. One of the most common of these is moving pictures. Each frame of a movie consists of a two-dimensional array of pixels, but in areas in which there is no motion, the correlation (similarity) between images is very high. MPEG has a defined a new standard that is designed to take advantage of this interframe correlation.* MPEG can achieve about three times the compression of JPEG, but requires several frame buffers. JPEG can be implemented without any frame buffers. JPEG compresses each frame independently, and is almost symmetric in terms of complexity between encoders and decoders. For these reasons, some applications requiring motion sequences use JPEG in spite of the lower compression performance.

Some medical applications such as CAT (computer-aided tomography) scans use sequences of images. However, in unpublished research Peng[4] found that the correlation between images is not high enough to improve the compression significantly. This occurs because the sampling interval between images is typically much larger than the sampling intervals within the images.

2.5 Large amounts of data ○

The amount of data in a digital image can be extremely large—even millions of bytes. Assuming one byte per pixel, a 320×240 pixel image (which is relatively small) occupies 76,800 bytes of storage. This is equivalent to about 25 pages of dense text at one byte per character. An IBM® PC VGA display has 640×480 pixels, and images for this display occupy 307,200 bytes of storage. Displays are now available exceeding 2000×2000 pixels and images for these exceed four million bytes. Color images for display can require as much as three times more data, and color images for printing may require even more data. Motion sequences may be still more demanding.

This is one of the motivating factors behind image compression. Modern workstations and personal computers are capable of displaying

---

* A brief review of MPEG is found in Chapter 20.

high-resolution images, and the eye can sometimes analyze one image in a fraction of a second. It is very easy to construct applications that quickly exceed the capacity of hard-disk and even optical-disk storage. Properly used, JPEG compression can reduce the storage requirements by more than an order of magnitude, and may improve the system response time in the process.

3

ASPECTS OF THE HUMAN VISUAL SYSTEM

In this chapter we review some of the basic properties of the human visual system. The initial purpose in doing this is to see what the ideal image-sampling parameters should be. However, we shall also use these same data in later chapters to understand how to minimize the visual impact of distortions introduced by JPEG compression.

3.1 Luminance sampling ○

The upper bound on the number of samples needed in an image is determined by the response of the human visual system to changes in intensity. The sensitivity of the eye to varying changes in luminance intensity is shown in Figure 3-1, where the curve labelled "Luminance" is a replot of Van Nes and Bouman's measurements at higher intensities.[5] (We shall consider the other two curves in Figure 3-1 in section 3.3.) This plot gives us the relative response of the eye to threshold intensity changes at different angular frequencies (where the angle in question is measured relative to the eye of the observer). Any point above the curve represents a condition in which the eye is unable to detect any intensity change; points below the curve represent conditions in which the eye can see the variation in intensity.

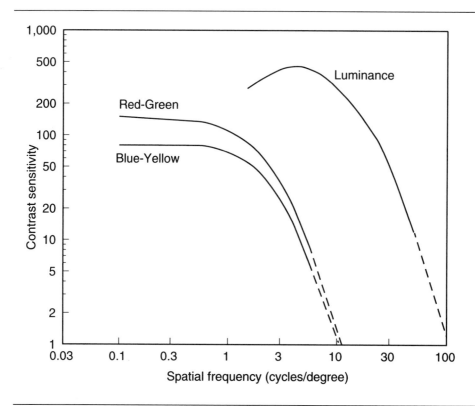

Figure 3-1. Sensitivity of the eye to luminance and chrominance intensity changes

There are two things to note about the luminance data in Figure 3-1. First, the peak in contrast sensitivity occurs at about 5 cycles/degree. Second, the sensitivity goes to zero at frequencies above about 100 cycles/degree. To see how this relates to what we see in a display, we translate these to a distance across the face of the display for a particular viewing distance. A quick calculation reveals that the sensitivity of the eye to intensity change is greatest for objects with dimensions of about 0.2 cm., if viewed at a distance of 1 meter. At that same viewing distance the eye has trouble resolving objects smaller than about 0.01 cm. Therefore, at a 1-meter viewing distance an ideal digital image 40 cm. wide should have about 4,000 pixels per line, which is about twice the performance achieved with the best monochrome displays commercially available today.

Current-generation computer displays typically have about 1,000 or fewer pixels per line, yet individual pixels are not always easily discerned. The limitation is certainly not in the human visual system; therefore, it must be in the display system. For example, some commercial television receivers use high-pass filtering to improve the quality of the displayed image.

The frequency response of the eye for vertical patterns is similar to the response for horizontal patterns. On the diagonals, however, the response is significantly reduced.[6] Some compression systems, JPEG among them, are able to take advantage of this property of the human visual system.[7]

3.2 Sample precision ○

The ideal precision needed for samples can also be inferred from the contrast sensitivity function plotted in Figure 3-1. The data in this figure can be used to infer that the eye can distinguish about 1,000 uniformly spaced gray levels under ideal conditions, which suggests a limiting precision of about 10 bits/pixel. However, this is for linear color spaces. If the display is calibrated in terms of one of the uniform-perceptual spaces, the just-noticeable difference gives only about 100 distinguishable luminance steps.[3] In practical systems, however, conditions are not ideal and eight bits/pixel is usually required to give good results. The ability to perceive subtle change in grayscale intensity is degraded in the presence of noise and near edges (large intensity changes). This latter effect is called *spatial masking*.[8]

3.3 Chrominance sampling ○

The contrast sensitivity function for luminance shown in Figure 3-1 is very different from the other two curves in the same figure. These other curves, plotted from measurements by Mullen,[9] are for constant-luminance red–green and blue–yellow gratings. Because color vision depends on relative response of three sets of physical receptors, it should come as no surprise that the contrast sensitivity function for chrominance is quite different from the contrast sensitivity function for luminance.* The contrast sensitivity for chrominance reaches a maximum for angular frequencies below about 1 cycle per degree, and the sensitivity to contrast change is negligible at frequencies above about 12 cycles/degree.

The much-reduced sensitivity to spatial variations in the two chrominance coordinates relative to the sensitivity in the luminance coordinate is a very important property of color vision from the standpoint of data compression. It is the basic reason why a well-designed color compression scheme can often compress the color information in a small fraction of the bits needed to compress the luminance. In common with most compression systems, the JPEG color compression algorithms perform best when luminance/chrominance coordinate systems are used.

3.4 Linearity ○

An idealized system for capturing, displaying, and viewing is shown in Figure 3-2. Nonlinearity can (and does) occur in each of the three major components of this system.

---

* The contrast sensitivities shown in Figure 3-1 are appropriate for the Cb and Cr chrominance components, and these happen to be widely used with JPEG.

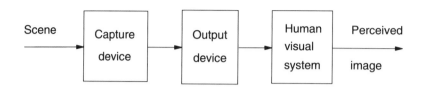

Figure   3-2. Idealized system for capturing, displaying and viewing

The capture device is typically the most linear part of the system, in that the quantum efficiency (the number of electrons emitted for each photon arriving from the image) is quite constant in a well-designed and well-calibrated imaging device. (Quantum efficiency is usually dependent on the color of the light.)

The output device, especially if it is a CRT display, is often quite nonlinear. A gamma correction is often applied to displays to correct for the nonlinear characteristics of the electron gun and to make the visually perceived output fairly linear. In critical applications the device characteristics should be measured and the input corrected to achieve a linear response.

If a printing system is the output device in Figure 3-2, a different set of nonlinearities must be handled. In some applications images must be both displayed and printed, and finding a representation of the image that is suitable for both media can be a difficult problem.

The eye has a nonlinear response to changes in intensity that is inversely proportional to the average intensity. This characteristic, known as Weber's law, means that the eye has high sensitivity at low intensity levels and greatly reduced sensitivity at high intensity levels. Superimposed on this is a dependence on the intensity of the region surrounding the region under test. As the surrounding-region intensity is increased, the relative sensitivity in dark areas is reduced and the sensitivity in light areas is increased. The calculated dependence of subjective grayscale value on reflectance from a printed page is shown for various background reflectances in Figure 3-3.[10] This general behavior also holds for display systems, as there is in general a large shift in sensitivity to luminance as the eye adapts to different average luminance levels.[11]

Many of the representations of chrominance are also perceptually nonuniform, as has already been discussed in section 2.4.4.3. The uniform-color spaces CIELAB and CIELUV have been defined as one solution to this problem.

In practical image display systems the environment is not always controlled, the equipment is not always calibrated, and the user will not always do what is expected with the images. Because the perception of an image may be quite different, depending on the details of equipment, surrounding conditions, and use, some caution in applying these measured

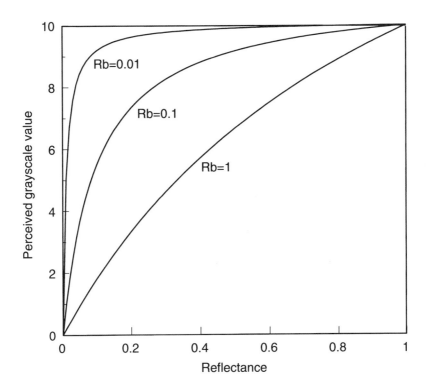

Figure 3-3. Calculated dependence of subjective grayscale value on reflectance from a printed page. The three curves are for different background reflectance values Rb. $Rb = 1$ is appropriate for viewing the images printed in this book.

response characteristics is necessary. With that caveat, we shall continue to use the measured response characteristics of the eye as a guide in determining visibility of distortion.

4

THE DISCRETE COSINE
TRANSFORM (DCT)

This chapter is devoted to the discrete cosine transform, one of the basic building blocks for JPEG. The discrete cosine transform was first applied to image compression in Ahmed, Natarajan, and Rao's pioneering work,[12] in which they showed that this particular transform was very close to the KLH (Karhunen-Loeve-Hotelling) transform, a transform that produces uncorrelated coefficients. Decorrelation of the coefficients is very important for compression, because each coefficient can then be treated independently without loss of compression efficiency. Another important aspect of the DCT is the ability to quantize the DCT coefficients using visually-weighted quantization values, as will be discussed in this chapter.

The first third of this chapter is devoted to basic concepts that the nontechnical reader should find useful. The latter two-thirds treats topics that are progressively more advanced, culminating in the final sections, which contain a discussion of the most efficient fast two-dimensional (2-D) DCT currently known.

4.1 Basic DCT concepts ○

We have seen that the human visual system response is very dependent on spatial frequency. If we could somehow decompose the image into a set

29

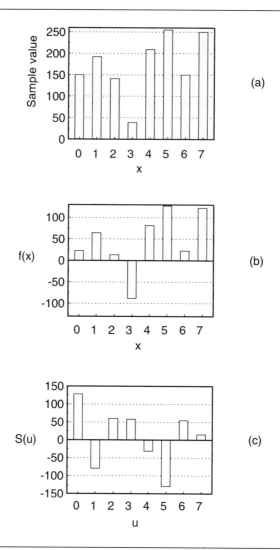

Figure   4-1. 1-D DCT decomposition. (a) Eight arbitrary grayscale samples.   (b) Level shift of (a) by 128.   (c) Coefficients for
decomposition into waveforms.

of waveforms, each with a particular spatial frequency, we might be able
to separate the image structure the eye can see from the structure that is
imperceptible.   The DCT can provide a good approximation to this
decomposition.

4.1.1 The one-dimensional DCT ○

To understand how an image can be decomposed into its underlying spatial
frequencies, we first consider a one-dimensional (1-D) case. We start with
a set of eight arbitrary grayscale samples such as is shown in Figure 4-1a.

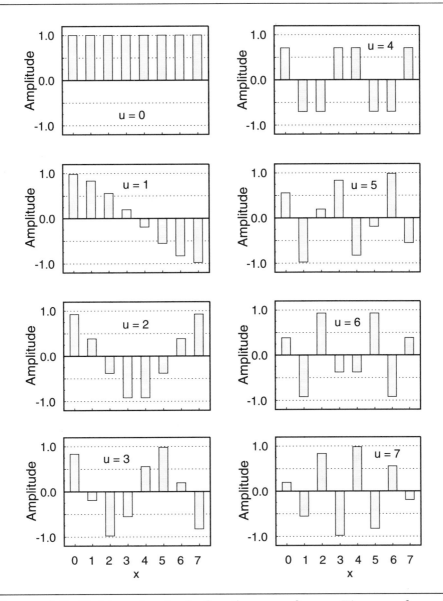

Figure   4-2. Eight cosine basis function waveforms. The waveform for
$u = 0$ is a constant. The other waveforms show an alternating
behavior at progressively higher frequencies.

The samples have values in the range 0 to 255, but after a level shift by
128 (as is done by JPEG), we get the values $f(x)$ in Figure 4-1b. We want
to decompose these eight sample values into a set of waveforms of different
spatial frequencies.

    Figure 4-2 shows a set of eight different cosine waveforms of uniform
amplitude, each sampled at eight points. The top-left waveform ($u = 0$) is

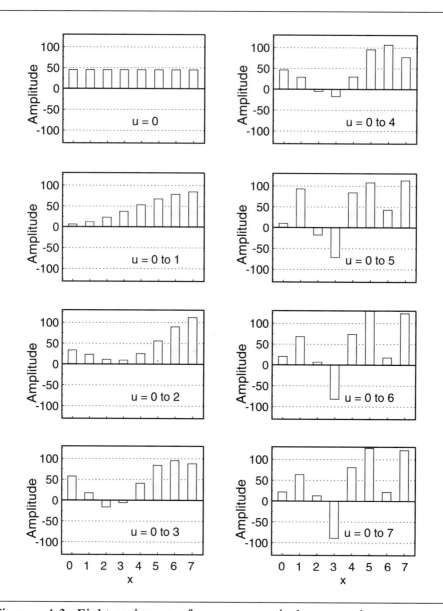

Figure   4-3. Eight cosine waveforms progressively summed

simply a constant, whereas the other seven waveforms ($u = 1, \ldots, 7$) show an alternating behavior at progressively higher frequencies.

These waveforms (which are called cosine basis functions) are said to be orthogonal. A set of waveforms is orthogonal if it has the following interesting property: if we take the product of any two waveforms in the set at each sampling point, and sum these products over all sampling points, the result is zero. If the waveform is multiplied by itself and summed, the result is a constant. For example, if we take the product of waveform 0 and waveform 1, and sum over all sample points, the result is zero. On the other hand, if we take the product of waveform 1 with itself, the product at each sample point is the square of the waveform value; therefore, the sum of the products over all sample points is a positive constant (which is used to define a scale factor for the waveforms).

Orthogonal waveforms are independent. That is, there is no way that a given waveform can be represented by any combination of the other waveforms. However, the complete set of eight waveforms, when scaled by numbers called coefficients and added together, can be used to represent any eight sample values such as those in Figure 4-1b. The coefficients $S(u)$ are plotted in Figure 4-1c. Figure 4-3 shows a sequence in which the eight scaled waveforms are progressively summed, starting with the lowest frequency (adding one more each time), until finally the original set of samples is reconstructed. The coefficients plotted in Figure 4-1c are the output of an 8-point DCT for the eight sample values in Figure 4-1b.

The coefficient that scales the constant basis function ($u = 0$) is called the DC coefficient. The other coefficients are called AC coefficients. These names are derived from the historical use of the DCT for analyzing electrical currents that had both direct- and alternating-current (DC and AC) terms. Note that the DC term gives the average over the set of samples.

The process of decomposing a set of samples into a scaled set of cosine basis functions is called the forward discrete cosine transform (FDCT). The process of reconstructing the set of samples from the scaled set of cosine basis functions is called the inverse discrete cosine transform (IDCT). If the sample sequence is longer than eight samples, it can be divided into eight-sample groups and the DCT can be computed independently for each group. Because the cosine basis functions always have the same set of values at each of the discrete sampling points, only the coefficient values change from one group of samples to the next.

4.1.2 The two-dimensional DCT ○

The 1-D DCT can be extended to apply to 2-D image arrays. Figure 4-4 shows a set of 64 2-D cosine basis functions that are created by multiplying a horizontally oriented set of 1-D 8-point basis functions (shown in Figure 4-2) by a vertically oriented set of the same functions. The horizontally oriented set of basis functions represents horizontal frequencies and the other set of basis functions represents vertical frequencies. By convention, the DC term of the horizontal basis functions is to the left, and the DC term for the vertical basis functions is at the top. Consequently, the top row and left column have 1-D intensity variations, which, if plotted, would be the same as in Figure 4-2. (For purposes of illustration, neutral gray represents zero in these figures, white represents positive amplitudes, and black represents negative amplitudes.)

When scaled by an appropriate set of 64 coefficients, these 64 basis functions can be used to represent any 64 sample values such as the 8×8 block of samples shown at the upper left corner of Figure 4-5. Figure 4-5 also shows the sequence in which these 64 basis functions are progressively summed following the zigzag sequence shown by the white connecting

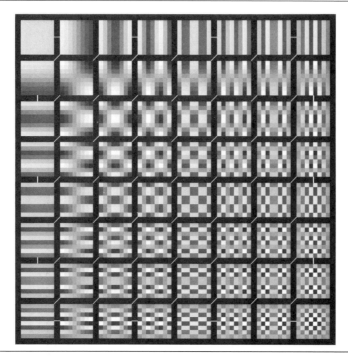

Figure 4-4. **DCT basis functions.** Each 8×8 array of samples shows a different 2-D basis function. The horizontal DCT frequency of the basis function increases from left to right and the vertical DCT frequency of the basis function increases from top to bottom. The DC basis function is therefore in the upper left corner. The white lines connecting the 8×8 arrays show the zigzag sequence.

lines. This zigzag pattern, which is used in the JPEG algorithms, approximately orders the basis functions from low to high spatial frequencies. Figure A.6 of the JPEG DIS shows the relationship between zigzag indices and DCT coefficient positions.

Because the 2-D DCT basis functions are products of two 1-D DCT basis functions, the only constant basis function is in the upper left corner of the array; the coefficient for this basis function is called the DC coefficient, whereas the rest of the coefficients are called AC coefficients.

4.1.3 DCT integer representations and quantization ○

Quantization allows us to reduce the accuracy with which the DCT coefficients are represented when converting the DCT to an integer representation. This can be very important in image compression, as it tends to make many coefficients zero—especially those for high spatial frequencies.

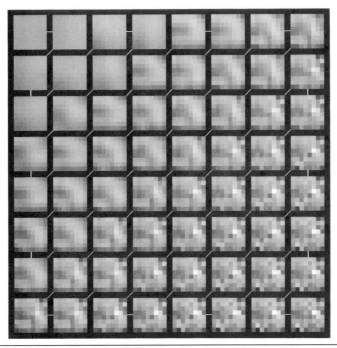

Figure 4-5. Example of reconstruction of an 8×8 array of samples. The 8×8 array of samples which is to be reconstructed is shown at the upper left of this figure. The sequence of 8×8 sample arrays beneath show how the reconstruction proceeds as basis functions are added in zigzag order.

The quantization values can be set individually for each DCT coefficient, using criteria based on visibility of the basis functions.* If we measure the threshold for visibility of a given basis function—the coefficient amplitude that is just detectable by the human eye—we can divide (quantize) the coefficients by that value (with appropriate rounding to integer values). If we multiply (dequantize) the scaled-down coefficients by that value before reconstructing, we create a condition in which the eye should not be able to detect any difference between quantized and unquantized DCT coefficients. If we are willing to tolerate some visible artifacts in the reconstructed image, we might divide by a value larger than the visibility threshold value.

---

* It is a curious fact that although visually weighed quantization was suggested by Wintz[13] in 1972 for the KLH transform, a close relative of the DCT, not until the work of Lohscheller[14] in 1984 was such quantization applied to the DCT.

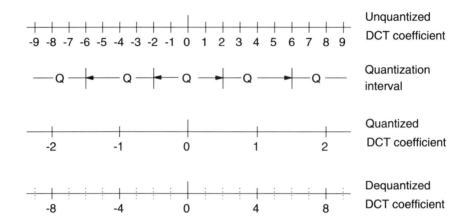

Figure 4-6. Quantization and dequantization

This process of scaling the DCT coefficients and truncating them to integer values is called quantization, and the rescaling to restore approximately the original DCT coefficient magnitude is called dequantization. Quantization and dequantization are illustrated in Figure 4-6.

The quantization value chosen for the example in Figure 4-6 is 4. Note that half of this value (i.e., 2 in this example) is added to the coefficient magnitude before the magnitude is divided by the quantization value and truncated. This is the rounding convention adopted by JPEG. To dequantize the quantized DCT coefficients, we multiply them by the quantization value. Note that the quantization used by JPEG is linear—that is, for a given quantization value the steps are uniform in size.

In the JPEG system the quantized DCT coefficients are always integers, as are the quantization values. The integer representation is chosen such that 8-bit precision image samples transform to 11-bit precision quantized DCT coefficients for quantization values of 1. Consequently, for 8-bit precision input samples, a quantization value of 16 produces a quantized DC coefficient with 7-bit precision (128 levels).

The visibility of the 8×8 DCT basis functions has been measured for an image with luminance resolution of 720 samples per horizontal line × 576 lines and chrominance resolution of 360 samples per horizontal line × 576 lines, and a viewing distance equal to six times the screen width.[14] Table 4-1 gives the threshold quantization values measured for luminance; Table 4-2 gives similar data for the two chrominance components. In the authors' experience, although these values are supposed to be at the threshold of visibility, images reconstructed using these tables usually show noticeable artifacts when viewed on high-quality displays. If the values are all divided by two, however, the reconstructed images are usually indistinguishable from the source images. This is reassuring, because one

Table   4-1. Luminance quantization table

16	11	10	16	24	40	51	61
12	12	14	19	26	58	60	55
14	13	16	24	40	57	69	56
14	17	22	29	51	87	80	62
18	22	37	56	68	109	103	77
24	35	55	64	81	104	113	92
49	64	78	87	103	121	120	101
72	92	95	98	112	100	103	99

of the assumptions made in these measurements is that the distortions introduced by the quantization of the individual coefficients are independent. Little is known about the visibility thresholds when more than one AC coefficient is non-zero in a block.

The general character of these tables can be understood from the contrast sensitivity functions for luminance and chrominance (see Figure 3-1). Since the lowest-frequency basis function in the DCT block contains a half cycle, the lowest-frequency basis function has one cycle per 16 pixels. At a viewing distance of six times screen width we calculate approximately 5 cycles/degree for this lowest frequency, which is just below the value at which the luminance contrast sensitivity function peaks. The dip in the luminance quantization values therefore occurs just about where we would expect. The chrominance quantization values show no minimum, and, again, this is what we would expect from the chrominance contrast sensitivity function. Note that if the viewing distance is not six times the screen

Table   4-2. Chrominance quantization table

17	18	24	47	99	99	99	99
18	21	26	66	99	99	99	99
24	26	56	99	99	99	99	99
47	66	99	99	99	99	99	99
99	99	99	99	99	99	99	99
99	99	99	99	99	99	99	99
99	99	99	99	99	99	99	99
99	99	99	99	99	99	99	99

width (and the viewer is usually closer than this to a computer display), the threshold quantization values will be lower at intermediate frequencies.

Measurements of quantization tables for the RGB coordinate system can be found in Peterson.[15] Ahumada and Peterson[16] have recently proposed a model for DCT quantization that goes well beyond the qualitative discussion above, potentially providing quantitative prediction of the quantization tables needed for a given display and viewing condition.

4.1.4 Zero-shift ○

JPEG uses a zero-shift in the input samples to convert 8-bit image data from the range 0 to 255 to the range −128 to +127. This is done by subtracting 128 from the data before the DCT is calculated. (For 12-bit data the zero-shift is accomplished by subtracting 2,048, converting the samples from the range 0 to 4,095 to the range −2,048 to +2,047.) This zero-shift reduces the internal precision requirements in the DCT calculations.

4.1.5 DCT blocks ○

In principle any size array of samples can be decomposed into a 2-D DCT. The example given here is 8×8, and we chose that size because it is the DCT used by JPEG. Of course, most images are much larger than 8×8 and must be divided into 8×8 blocks of samples. Each 8×8 block is transformed independently, and this segmentation into blocks leads to some problems.

Because the image is divided into 8×8 blocks, the spatial frequencies in the image and the spatial frequencies of the cosine basis functions are not precisely equivalent. A single cosine basis function of a given frequency can represent only a particular phase of a spatial oscillation within the block; representing an arbitrary phase also requires either a sine function or a cosine harmonic of the fundamental frequency. Nonetheless, the relationship between DCT frequency and spatial frequency is close enough that the contrast sensitivity function can still be a good qualitative guide to the relative visibility of the DCT basis functions.

Segmentation into 8×8 blocks can also lead to "blocking artifacts"—visible discontinuities between adjacent blocks. If we were to set all AC coefficients to zero, the DC coefficient would then have to represent all the samples in a block. Wherever the DC terms differed significantly from block to block, the discontinuity would be visible. These blocking artifacts appear as edges in the image, and abrupt edges imply high spatial frequencies. However, these high spatial frequencies occur at the transitions between blocks and are caused by the absence of AC coefficients. In this case non-zero AC coefficients are required to suppress the blocking.

In Annex K of the JPEG DIS a decoder option is discussed in which the DC values in each block and its eight nearest neighbors are used to predict the low-frequency AC coefficients. A quadratic surface is fitted to the nine DC values in such a way that the average within the block is preserved. AC coefficients are then computed from the parameters of the quadratic surface, giving the equations listed in the DIS.   "AC prediction" can very effectively suppress blocking artifacts in smooth regions. Some illustrations of AC prediction can be found in Chapter 16.

4.1.6 Computational complexity of DCTs ○

A straightforward calculation for the 8×8 DCT blocks would suggest that each coefficient needs 64 multiplies and 63 additions. The quantization step would introduce one more addition and a division. However, this gives a misleading impression about complexity. Cost-effective hardware and software implementations are already commercially available, both for the DCT alone and for complete DCT-based JPEG compression modes.

The most efficient integer arithmetic implementations of the DCT and quantization require less than one multiplication and nine additions per coefficient. Other new approaches are available that take good advantage of the highly-pipelined architecture and parallelism in modern computers and signal processors. These new machines even make possible very efficient floating point implementations.[17]

In the following sections we shall summarize a few of the fast DCT approaches and discuss very recent work on scaled DCTs, DCT implementations in which the quantization and dequantization are integrated into the DCT computations in order to reduce the total number of operations.

4.2 Mathematical definition of the FDCT and IDCT ◑

Until now we have avoided the mathematical definitions of the FDCT and IDCT because they tend to intimidate nonmathematically-oriented readers. However, anyone who wants to understand thoroughly and/or implement JPEG will have to master the mathematics.

4.2.1 One-dimensional FDCT and IDCT ◑

The 1-D DCT appropriate for JPEG is defined in Rao and Yip's book:[18]

FDCT:*

$$S(u) = \frac{C(u)}{2} \sum_{x=0}^{7} s(x) \cos\left[(2x+1)u\pi/16\right] \qquad [4\text{-}1]$$

IDCT:

$$s(x) = \sum_{u=0}^{7} \frac{C(u)}{2} S(u) \cos\left[(2x+1)u\pi/16\right] \qquad [4\text{-}2]$$

---

* We shall follow a notation here that is a compromise between the notation used by Rao and Yip and the notation used in the JPEG documentation.

where

$$C(u) = 1/\sqrt{2} \quad \text{for } u = 0$$
$$C(u) = 1 \qquad \text{for } u > 0$$
$$s(x) = \text{1-D sample value}$$
$$S(u) = \text{1-D DCT coefficient}$$

This transformation is orthonormal, meaning that the following relationship involving the sum exists:

$$\sum_{x=0}^{7} \frac{C(u)}{2} \cos\left[(2x+1)u\pi/16\right] \frac{C(u')}{2}\cos\left[(2x+1)u'\pi/16\right] = \delta(u,u') \quad [4\text{-}3]$$

where

$$\delta(u,u') = 1 \text{ if } u = u'$$
$$\delta(u,u') = 0 \text{ if } u \neq u'$$

It is this property of orthonormality that allows us to decompose any sequence of eight sample values into the set of weighted cosine functions.

4.2.2 Two-dimensional reference FDCT and IDCT ◑

The 2-D FDCT and IDCT can be constructed from products of the terms of a horizontal 1-D DCT (using u and x) and a vertical 1-D DCT (using v and y, where v represents vertical frequencies and y represents vertical displacements). This leads us to the reference 2-D FDCT and IDCT as defined by JPEG, but cast in a slightly different form to emphasize the connection with the 1-D DCT.

FDCT:

$$S(v,u) = \qquad\qquad\qquad\qquad\qquad\qquad\qquad\qquad\qquad\qquad [4\text{-}4a]$$

$$\frac{C(v)}{2} \frac{C(u)}{2} \sum_{y=0}^{7} \sum_{x=0}^{7} s(y,x) \cos\left[(2x+1)u\pi/16\right] \cos\left[(2y+1)v\pi/16\right]$$

IDCT:

$$s(y,x) = \qquad\qquad\qquad\qquad\qquad\qquad\qquad\qquad\qquad\qquad [4\text{-}4b]$$

$$\sum_{v=0}^{7} \frac{C(v)}{2} \sum_{u=0}^{7} \frac{C(u)}{2} S(v,u) \cos\left[(2x+1)u\pi/16\right] \cos\left[(2y+1)v\pi/16\right]$$

where

$$C(u) = 1/\sqrt{2} \quad \text{for } u = 0$$
$$C(u) = 1 \qquad \text{for } u > 0$$

$$C(v) = 1/\sqrt{2} \quad \text{for } v = 0$$
$$C(v) = 1 \quad \text{for } v > 0$$
$$s(y,x) = \text{2-D sample value}$$
$$S(v,u) = \text{2-D DCT coefficient}$$

These equations, when implemented with 64-bit precision floating-point arithmetic, define the reference FDCT and IDCT used in the JPEG compliance tests in Appendix B. Note that overflow and underflow of the IDCT output beyond the range determined by the source image data precision must be appropriately clamped.

There are quite a few FDCT and IDCT implementations that use algorithms designed to reduce the number of multiplications and additions in the transformation. These practical, fast DCT algorithms use fixed-precision integer arithmetic or take advantage of new high-speed floating-point hardware. Some aspects of these fast DCT algorithms, including very recent work on fast scaled DCTs, will be reviewed in the next section. The reader should be aware of the following important point, however: because round-off and truncation effects depend on the way the calculations are done, different FDCT and IDCT implementations will probably give slightly different results.

4.2.3 Compliance tests for the FDCT and IDCT ◑

Although there is no formal specification of either the FDCT or IDCT, the techniques for computing the FDCT and IDCT should have sufficient accuracy relative to the reference calculation to pass either the JPEG generic compliance tests or (if appropriate) application-specific compliance tests. Appendix B contains a draft of ISO DIS 10918 Part 2 (Compliance Testing).

4.3 Fast DCTs ◑

In this section we shall explore some of the principles and techniques used in fast DCT algorithms, the origins of which go back to the Cooley and Tukey algorithm for the fast Fourier transform[19] implementation of the discrete Fourier transform (DFT). An extensive review of fast DFT techniques published recently by Duhamel and Vetterli[20] contains an interesting section on the history of this field of research. Fast DCT techniques as a separate field started with the paper by Chen, Smith and Fralick,[21] but the linkage between the DCT and the DFT continues to be important. We shall not attempt to carry out an exhaustive survey of the field, as a text[18] is available that serves that purpose well. What we shall do is to develop some of the ideas—mostly at an intermediate level—that will give the reader a feeling for the concepts involved in the most efficient algorithms currently known. In the process we shall develop the equations in enough detail to permit relatively easy translation into software implementations. We shall first investigate the 1-D DCT and then apply what we have learned to the more difficult 2-D case.

The most straightforward way to implement the DCT is to follow the theoretical equations. When we do this, we get an upper limit of 64 multiplications and 56 additions for each 1-D 8-point DCT. (Note that the cosine terms can be combined with the constants before the computation because the cosines become discrete numbers at each position.) Therefore, a full 8×8 DCT done in this way in separable 1-D format—eight rows and then eight columns—would require 1,024 multiplications and 896 additions plus additional operations to quantize the coefficients.

If the DCT really required this many operations to compute, JPEG would have chosen a different algorithm. In fact, the most efficient algorithm for the 8×8 DCT, a true 2-D method developed by Feig,[22] requires only 54 multiplications, 464 additions, and 6 arithmetic shifts to produce a form suitable for quantization.

4.3.1 A simple example of a fast 1-D DCT ❶

The fast DCT techniques take advantage of the symmetries in the DCT equations. An examination of the basis functions in Figure 4-2 shows that the 1-D basis functions have symmetries that can be exploited to reduce the number of operations. Amplitudes can be combined with appropriate signs before the multiplication. To illustrate the combination of terms, the eight equations are presented below, using sums and differences of the samples. We define the following:

$$C_k = \cos(k\pi/16)$$
$$S_k = \sin(k\pi/16)$$

$$C_1 = S_7 = 0.9808$$
$$C_2 = S_6 = 0.9239$$
$$C_3 = S_5 = 0.8315$$
$$C_4 = S_4 = 0.7071$$
$$C_5 = S_3 = 0.5556$$
$$C_6 = S_2 = 0.3827$$
$$C_7 = S_1 = 0.1951$$

Sums:

$$s_{jk} = s(j) + s(k)$$
$$s_{07} = s(0) + s(7)$$
$$s_{16} = s(1) + s(6)$$
$$s_{25} = s(2) + s(5)$$
$$s_{34} = s(3) + s(4)$$
$$s_{0734} = s_{07} + s_{34}$$
$$s_{1625} = s_{16} + s_{25}$$

Differences:

$$d_{jk} = s(j) - s(k)$$
$$d_{07} = s(0) - s(7)$$
$$d_{16} = s(1) - s(6)$$
$$d_{25} = s(2) - s(5)$$
$$d_{34} = s(3) - s(4)$$
$$d_{0734} = s_{07} - s_{34}$$
$$d_{1625} = s_{16} - s_{25}$$

Then, the DCT equations can be written as:

$$2S(0) = C_4(s_{0734} + s_{1625})$$ [4-5a]

$$2S(1) = C_1 d_{07} + C_3 d_{16} + C_5 d_{25} + C_7 d_{34}$$ [4-5b]

$$2S(2) = C_2 d_{0734} + C_6 d_{1625}$$ [4-5c]

$$2S(3) = C_3 d_{07} - C_7 d_{16} - C_1 d_{25} - C_5 d_{34}$$ [4-5d]

$$2S(4) = C_4(s_{0734} - s_{1625})$$ [4-5e]

$$2S(5) = C_5 d_{07} - C_1 d_{16} + C_7 d_{25} + C_3 d_{34}$$ [4-5f]

$$2S(6) = C_6 d_{0734} - C_2 d_{1625}$$ [4-5g]

$$2S(7) = C_7 d_{07} - C_5 d_{16} + C_3 d_{25} - C_1 d_{34}$$ [4-5h]

The equations above have been arranged to make it clear that the samples appear as sum and difference pairs that are used more than once. Note that the total number of operations is 22 multiplications and 28 additions. Furthermore, if the DCT coefficients are scaled such that one multiplication is combined with the quantization, we have only 14 multiplications and 28 additions to produce a form suitable for quantization. Thus, even this very simple attempt to exploit the symmetries in the equations has given us a fast DCT.

4.3.2 A more sophisticated fast DCT ❶

Ligtenberg and Vetterli[23] have derived a fast DCT that serves as an excellent vehicle for explaining how the symmetries are exploited in high-performance fast DCTs.

We again write down the basic equations for the 8×8 DCT in terms of a set of sums and differences, but using slightly different groupings than we did for the first example.

$$2S(0) = C_4((s_{07} + s_{12}) + (s_{34} + s_{56}))$$ [4-6a]

$$2S(1) = C_1 d_{07} + C_3 d_{16} + C_5 d_{25} + C_7 d_{34}$$ [4-6b]

$$2S(2) = C_2(s_{07} - s_{34}) + C_6(d_{12} - d_{56})$$ [4-6c]

$$2S(3) = C_3 d_{07} - (C_5 d_{34} + C_7 d_{16} + C_1 d_{25})$$ [4-6d]

$$2S(4) = C_4((s_{07} + s_{34}) - (s_{12} + s_{56}))$$ [4-6e]

$$2S(5) = C_5 d_{07} + C_7 d_{25} + C_3 d_{34} - C_1 d_{16} \qquad\qquad \text{[4-6f]}$$

$$2S(6) = C_6(s_{07} - s_{34}) + C_2(s_{25} - s_{16}) \qquad\qquad \text{[4-6g]}$$

$$2S(7) = C_7 d_{07} + C_3 d_{25} - (C_5 d_{16} + C_1 d_{34}) \qquad\qquad \text{[4-6h]}$$

We shall also need the following standard trigonometric identities:

$$C_i - C_j = -2 \sin((i+j)\pi/32) \sin((i-j)\pi/32) \qquad\qquad \text{[4-7a]}$$

$$C_i + C_j = 2 \cos((i+j)\pi/32) \cos((i-j)\pi/32) \qquad\qquad \text{[4-7b]}$$

The fast DCT algorithms make use of a basic operation called a rotation. The rotation of two inputs x and y by an angle Q to produce a new representation X and Y is given by:

$$X = x \cos(Q) + y \sin(Q) \qquad\qquad \text{[4-8a]}$$

$$Y = -x \sin(Q) + y \cos(Q) \qquad\qquad \text{[4-8b]}$$

If the angle of rotation is $k\pi/16$ we get:

$$X = C_k x + S_k y \qquad\qquad \text{[4-9a]}$$

$$Y = -S_k x + C_k y \qquad\qquad \text{[4-9b]}$$

which is a form we shall encounter frequently.

As written above, the rotation requires four multiplications and two additions. It can be recast as three multiplications and three additions if multiplications are computationally more costly (they often are). We do this by the following:

$$X = C_k x + S_k y + C_k y - C_k y \qquad\qquad \text{[4-10a]}$$

$$Y = -S_k x + C_k y + C_k x - C_k x \qquad\qquad \text{[4-10b]}$$

which, with a little algebra, can be put into the form:

$$X = C_k(x + y) + (S_k - C_k)y \qquad\qquad \text{[4-11a]}$$

$$Y = -(S_k + C_k)x + C_k(x + y) \qquad\qquad \text{[4-11b]}$$

which requires three multiplications and three additions. (Note that $S_k - C_k$ and $S_k + C_k$ are constants.)

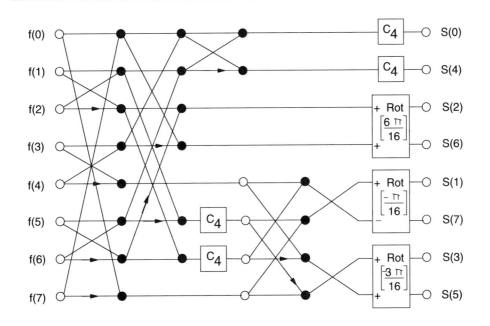

Figure   4-7. Flowgraph for Vetterli and Ligtenberg fast 1-D DCT

Rotations can be by negative as well as positive angles. Note that a negation of an input to a rotation can be absorbed into the constants in the rotation and does not constitute an extra operation.

A fast DCT is a sequence of operations that efficiently computes the set of weighted sums and differences making up the DCT coefficients. Of course, efficiency is determined in part by the architecture of the computing engine used, so there is no "best" algorithm. There are some very good ones, however.

The flowgraph for the Ligtenberg and Vetterli fast 1-D DCT is shown in Figure 4-7. In this graph the flow of operations is from left to right, and lines are summed where they merge at a node. If a line contains an arrow, the signal is negated before the addition to (i.e., subtracted from) the other signal fed to the node. Thus, if we look at input lines 0 and 7 (the top and bottom lines at the left side of the diagram), the signals are summed as we move to the right along the top line, and subtracted $(s(0) - s(7))$ as we move to the right along the bottom line. Multiplication of a signal (in this case by C_4) is indicated by C_4 in a box.

Rotation is performed in the boxes labelled "Rot." If the input to the rotation is positive, the input line is labelled "+"; if the input is negative, the line is labelled "−". Figure 4-7 is a modification of a flowgraph in Rao and Yip[18] describing the Ligtenberg and Vetterli algorithm. The modifications were necessary to correct some errors and to allow the use of a standard rotation.

Let us analyze the DCT equations to see how this flowgraph is derived. (There are more sophisticated derivations for those familiar with matrix algebra; see Rao and Yip's book for details.)

The equations for $S(0)$ and $S(4)$ are straightforward:

$$2S(0) = C_4[(s_{07} + s_{12}) + (s_{34} + s_{56})] \qquad [4\text{-}12a]$$

$$2S(4) = C_4[(s_{07} + s_{34}) - (s_{12} + s_{56})] \qquad [4\text{-}12b]$$

Referring to the flowgraph, we see that the sum in $S(0)$ is created by repeated summation of terms, followed by multiplication by C_4. (The factor of 2 is often ignored in flowgraphs.)

Similarly, $S(4)$ is calculated by a sequence of additions, a subtraction, and a multiplication by C_4.

Equations 4-6c and 4-6g can be recast as:

$$2S(2) = C_2(s_{07} - s_{34}) + C_6(d_{12} - d_{56}) \qquad [4\text{-}13a]$$

$$2S(6) = C_6(s_{07} - s_{34}) + C_2(s_{25} - s_{16}) \qquad [4\text{-}13b]$$

Since $s_{25} - s_{16} = - (d_{12} - d_{56})$,

$$2S(2) = C_6(d_{12} - d_{56}) + S_6(s_{07} - s_{34}) \qquad [4\text{-}13c]$$

$$2S(6) = - S_6(d_{12} - d_{56}) + C_6(s_{07} - s_{34}) \qquad [4\text{-}13d]$$

which is a rotation of inputs x and y by $6\pi/16$, where x and y are given by:

$$x = d_{12} - d_{56} \qquad [4\text{-}14a]$$

$$y = s_{07} - s_{34} \qquad [4\text{-}14b]$$

Referring to Figure 4-7, $d_{12} - d_{56}$ and $s_{07} - s_{34}$ are created from the inputs by appropriate sums and differences. The rotation then creates coefficients $S(2)$ and $S(6)$.

Coefficients $S(3)$ and $S(5)$ are given by:

$$2S(3) = C_3 d_{07} - [S_3 d_{34} + S_1 d_{16} + C_1 d_{25}] \qquad [4\text{-}15a]$$

$$2S(5) = S_3 d_{07} + S_1 d_{25} + C_3 d_{34} - C_1 d_{16} \qquad [4\text{-}15b]$$

Rearranging terms,

$$2S(3) = [C_3 d_{07} - S_3 d_{34}] - [S_1 d_{16} + C_1 d_{25}] \qquad [4\text{-}16]$$

Let us recast the second term in brackets in terms of s_{12}, d_{12}, s_{56}, and d_{56}. To do this we define a linear sum with unknown coefficients A, B, C, and D:

$$S_1 d_{16} + C_1 d_{25} = As_{12} + Bd_{12} + Cs_{56} + Dd_{56} \qquad [4\text{-}17]$$

We solve for the coefficients by equating terms for $s(1)$, $s(2)$, $s(5)$, and $s(6)$ independently to get the following four equations:

For $s(1)$: $S_1 = A + B$
For $s(2)$: $C_1 = A - B$
For $s(6)$: $-S_1 = C - D$
For $s(5)$: $-C_1 = C + D$

which gives:

$A = \ \ C_4 C_3$
$B = -C_4 S_3$
$C = -C_4 C_3$
$D = -C_4 S_3$

The second term in brackets in the equation for $S(3)$ becomes:

$$S_1 d_{16} + C_1 d_{25} = C_4 [C_3 (s_{12} - s_{56}) - S_3 (d_{12} + d_{56})] \qquad [4\text{-}18]$$

and, therefore,

$$2S(3) = [C_3 d_{07} - S_3 d_{34}] - C_4 [C_3 (s_{12} - s_{56}) - S_3 (d_{12} + d_{56})] \qquad [4\text{-}19]$$

We follow a similar scheme for $S(5)$:

$$2S(5) = S_3 d_{07} + S_1 d_{25} + C_3 d_{34} - C_1 d_{16} \qquad [4\text{-}20a]$$

$$2S(5) = [S_3 d_{07} + C_3 d_{34}] + [S_1 d_{25} - C_1 d_{16}] \qquad [4\text{-}20b]$$

$$S_1 d_{25} - C_1 d_{16} = As_{12} + Bd_{12} + Cs_{56} + Dd_{56} \qquad [4\text{-}20c]$$

For $s(1)$: $-C_1 = A + B$
For $s(2)$: $S_1 = A - B$
For $s(5)$: $-S_1 = C + D$
For $s(6)$: $C_1 = C - D$

which gives:

$A = -C_4 S_3$
$B = -C_4 C_3$
$C = \ \ C_4 S_3$
$D = -C_4 C_3$

The second term in brackets in $S(5)$ becomes:

$$S_1 d_{25} - C_1 d_{16} = C_4[-S_3 s_{12} - C_3 d_{12} + S_3 s_{56} - C_3 d_{56}] \qquad [4\text{-}21]$$

and, therefore,

$$2S(5) = [S_3 d_{07} + C_3 d_{34}] + C_4[-S_3(s_{12} - s_{56}) - C_3(d_{12} + d_{56})] \qquad [4\text{-}22]$$

The equations for $S(3)$ and $S(5)$ can now be rearranged to give:

$$2S(3) = C_3[d_{07} - C_4(s_{12} - s_{56})] - S_3[d_{34} - C_4(d_{12} + d_{56})] \qquad [4\text{-}23a]$$

$$2S(5) = S_3[d_{07} - C_4(s_{12} - s_{56})] + C_3[d_{34} - C_4(d_{12} + d_{56})] \qquad [4\text{-}23b]$$

Referring to our equation for the rotation, we see that this is a rotation of x and y, where:

$$Q = -3\pi/16 \qquad [4\text{-}24a]$$

$$x = d_{07} - C_4(s_{12} - s_{56}) \qquad [4\text{-}24b]$$

$$y = d_{34} - C_4(d_{12} + d_{56}) \qquad [4\text{-}24c]$$

The sums and differences that make up x and y are readily created from the starting input sums and differences.

The same algebraic manipulation can be applied to coefficients $S(1)$ and $S(7)$. Skipping the details (the procedure is identical to that shown above), we get:

$$2S(1) = C_1[d_{07} + C_4(s_{12} - s_{56})] - S_1[-d_{34} - C_4(d_{12} + d_{56})] \qquad [4\text{-}25a]$$

$$2S(7) = S_1[d_{07} + C_4(s_{12} - s_{56})] + C_1[-d_{34} - C_4(d_{12} + d_{56})] \qquad [4\text{-}25b]$$

which is a rotation in which:

$$Q = -\pi/16 \qquad [4\text{-}26a]$$

$$x = d_{07} + C_4(s_{12} - s_{56}) \qquad [4\text{-}26b]$$

$$y = -[d_{34} + C_4(d_{12} + d_{56})] \qquad [4\text{-}26c]$$

Because x and y contain the same sets of sums and differences of pairs that occurred in the equations for $S(3)$ and $S(5)$, many of the calculations can be reused. Note that this rotation is by a negative angle.

We now have terms involving sums and differences, some of which are scaled by C_4. We leave it as an exercise for the reader to compare these equations to the structure of the flowgraph in Figure 4-7. When we count the number of operations in this flowgraph, we get 13 multiplications and

29 additions. Ligtenberg, Wright and O'Neill[24] have explored the entire family of algorithms that have this basic structure.

4.3.3 Fast DCTs based on the discrete Fourier transform ●

There are a number of relationships between the DFT on real inputs and the DCT. For example, Vetterli and Nussbaumer[25] showed that the N-point DCT can be expressed in terms of the real and imaginary parts of an N-point DFT and rotations of the DFT outputs. (That algorithm is quite similar to the Ligtenberg and Vetterli algorithm described in section 4.3.2.) Following a very different line of reasoning, Haralick[26] showed that the first N coefficients of a $2N$-point DFT with appropriate symmetry of input values can be used to compute an N-point DCT. Haralick's analysis is of particular interest, as it will lead us to a very efficient scaled DCT structure.

Defining $W_K = \exp(-j2\pi/K)$, the K-point DFT is defined by

$$F(u) = \sum_{x=0}^{K-1} s(x) W_K^{ux} \qquad [4\text{-}27]$$

We first extend an N-point sequence $s(x)$, $x = 0, \ldots, N-1$, by defining another N points with symmetry about the point $(2N-1)/2$, i.e.:

$$s(x) = s(2N - x - 1) \qquad\qquad x = N, \ldots, 2N-1 \qquad [4\text{-}28]$$

Then, the terms in the portion of the DFT sum from $x = N, \ldots, 2N - 1$ become:

$$F(u) = \sum_{x=0}^{N-1} s(x) W_K^{ux} + \sum_{x=N}^{2N-1} s(2N - x - 1) W_K^{ux} \qquad [4\text{-}29]$$

If we define a new index $k = 2N - x - 1$ for the portion of the sum from $x = N$ to $x = 2N - 1$ and note that $W_K^{2N} = 1$, this becomes:

$$F(u) = \sum_{x=0}^{N-1} s(x) W_K^{ux} + \sum_{k=0}^{N-1} s(k) W_K^{-u(k+1)} \qquad [4\text{-}30]$$

Then, replacing the index k by x and multiplying by $(1/2)W_K^{u/2}$,

$$(1/2)F(u)W_K^{u/2} = \sum_{x=0}^{N-1} s(x) \cos((2x+1)u\pi/2N) \qquad [4\text{-}31]$$

which proves that the first eight DFT coefficients, when multiplied by a complex scaling factor, give the 8-point DCT coefficients.

Tseng and Miller[27] took this a bit further, showing that the DCT can be obtained solely from the real part of $F(u)$ when the symmetry above is defined. Basically, since the right-hand side of Equation 4-31 is real, the left-hand side must also be real. Therefore, denoting the real and imaginary parts of $F(u)$ by A_u and B_u respectively,

$$F(u)W_K^{u/2} = (A_u + jB_u)(\cos(\pi u/2N) - j\sin(\pi u/2N)) \qquad [4\text{-}32]$$

Expanding this equation into separate real and imaginary parts and setting the imaginary part to zero (the DCT is real), we get:

$$B_u = A_u \sin(\pi u/2N)/\cos(\pi u/2N) \qquad [4\text{-}33]$$

When this is substituted back into Equation 4-32, it gives:

$$F(u)W_K^{u/2} = A_u \sec(\pi u/2N) = Re(F(u))\sec(\pi u/2N) \qquad [4\text{-}34]$$

Therefore, from Equation 4-31 we get:

$$\sum_{x=0}^{N-1} s(x)\cos((2x+1)u\pi/2N) = (1/2)Re(F(u))\sec(\pi u/2N) \qquad [4\text{-}35]$$

Therefore, the DCT coefficients can be obtained by a simple scaling of the real part of the DFT coefficients.

4.3.4 The 16-point DFT for real inputs ●

If we are to use the result in Equation 4-35 to calculate an 8-point DCT, we need the equations for the 16-point DFT.

An optimum form for the 16-point DFT has been developed by Winograd[28] (see also Silverman[29]). In its most general form the 16-point DFT requires 18 multiplications and 74 additions. However, as noted by Arai, Agui, and Nakajima,[30] we reduce the Winograd solution to a much simpler form when we force the symmetry of input values needed to derive an 8-point DCT from the 16-point DFT (Equation 4-28). The Winograd form for the DFT computes real and imaginary terms separately. Therefore, since we are concerned only with the real part of the final solution, we can simply discard all imaginary terms. We can then cast the Winograd solution in the following form, which we state without proof. (The reader may find it interesting to derive these results from the equations provided by Winograd.[28])

Using the definitions of s_{jk} and d_{jk} defined in section 4.3.1, we define the following intermediate products:

$$m_1 = 2[(s_{07} + s_{34}) + (s_{25} + s_{16})] \qquad [4\text{-}36a]$$

$$m_2 = (s_{07} + s_{34}) - (s_{25} + s_{16}) \qquad \text{[4-36b]}$$

$$m_3 = (s_{07} - s_{34}) \qquad \text{[4-36c]}$$

$$m_4 = d_{07} \qquad \text{[4-36d]}$$

$$m_5 = \cos(2u)[(s_{16} - s_{25}) + (s_{07} - s_{34})] \qquad \text{[4-36e]}$$

$$m_6 = \cos(2u)(d_{25} + d_{16}) \qquad \text{[4-36f]}$$

$$m_7 = \cos(3u)[(d_{16} + d_{07}) - (d_{25} + d_{34})] \qquad \text{[4-36g]}$$

$$m_8 = [\cos(u) + \cos(3u)](d_{16} + d_{07}) \qquad \text{[4-36h]}$$

$$m_9 = [\cos(u) - \cos(3u)](d_{25} + d_{34}) \qquad \text{[4-36i]}$$

where $u = \pi/8$. Then from the above and the following partial sums:

$$s_5 = m_4 + m_6 \qquad \text{[4-37a]}$$

$$s_6 = m_4 - m_6 \qquad \text{[4-37b]}$$

$$s_7 = m_8 - m_7 \qquad \text{[4-37c]}$$

$$s_8 = m_9 - m_7 \qquad \text{[4-37d]}$$

we construct the first eight coefficients of the DFT for the assumed input symmetry:

$$16F(0) = m_1 \qquad \text{[4-38a]}$$

$$16F(4) = m_2 \qquad \text{[4-38b]}$$

$$16F(2) = m_3 + m_5 \qquad \text{[4-38c]}$$

$$16F(6) = m_3 - m_5 \qquad \text{[4-38d]}$$

$$16F(3) = s_6 - s_8 \qquad \text{[4-38e]}$$

$$16F(5) = s_6 + s_8 \qquad \text{[4-38f]}$$

$$16F(1) = s_5 + s_7 \qquad \text{[4-38g]}$$

$$16F(7) = s_5 - s_7 \qquad \text{[4-38h]}$$

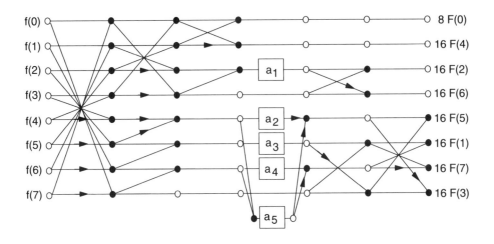

Figure 4-8. Flowgraph for 1-D DCT adapted from Arai, Agui, and Nakajima. $a_1 = 0.707$, $a_2 = 0.541$, $a_3 = 0.707$, $a_4 = 1.307$, and $a_5 = 0.383$.

4.3.5 Scaled 1-D DCT ●

To get the DCT from the equations for the 16-point DFT with real inputs, we multiply each DFT output by a scaling factor derived from Equation 4-35 and the normalization required by the DCT definition. This simplified form is the basis for the flowgraph shown in Figure 4-8, which was adapted from the flowgraph by Arai, Agui, and Nakajima.[30]

This version of the DCT has 13 multiplications and 29 additions, which is competitive with the best that has been achieved by other techniques. What makes it special, however, are the properties it has for a scaled DCT implementation. Although 13 multiplications are required, eight of the multiplies are to scale the final output to the correct range. If the output is to be quantized—as in the case of JPEG—the outputs can be left in this scaled form and the scaling factors can be absorbed into the divisions needed to quantize the outputs.[30] Only 5 multiplications are actually needed before the quantization, making this the most efficient 1-D quantized DCT known.

4.3.6 The fast IDCT ●

So far we have only been considering the forward DCT, yet an equally important part of the system, the decoder, requires the inverse DCT. It turns out that the IDCT can be obtained by simply reversing the direction of the flowgraph—i.e., going from right to left. This results from a basic property of orthogonal transforms:[31] the inverse of the transform is equal to the transpose, and the transpose can be obtained by transposing the flowgraph. Duhamel and Vetterli[20] have shown that the number of branches in a flowgraph is equal to the number of merges, and, therefore, the computational complexity of the IDCT is identical to that of the FDCT for any given fast DCT algorithm. Note that any scaling or normalization must be done properly for this approach to the IDCT to work correctly.

4.3.7 Fast 2-D DCTs ●

Because the 2-D DCT is separable, the summations can be done as eight 1-D DCTs on all rows, and then eight 1-D DCTs on the eight columns formed by the coefficients for the rows. In general, however, this is not as efficient as a true 2-D approach. Vetterli,[31] for example, describes a 2-D DCT that requires 104 multiplications and 460 additions. When the rotations are denormalized, the number of multiplications is reduced to only 81.

Note, however, that if we use the scaled 1-D DCT we can delay the row output scaling until after the column 1-D DCTs are complete.[30] Consequently, we can imbed the output scaling for both the row DCTs and the column DCTs in the final quantization step. To get to the final scaled DCT output we therefore need 16 1-D scaled DCTs, and therefore a total of $16 \times 5 = 80$ multiplications and $16 \times 29 = 464$ additions. This is the best approach known for a separable 2-D scaled DCT.

Feig[22] has extended this scaled DCT approach to the 2-D 8×8 DCT with significant further computational savings. We have already recognized that as a result of the DCT separability, the output scalings for 1-D DCTs on both rows and columns can be imbedded in the final quantization step. Note that when the output scalings are passed through to the quantization step, what is left is a 2-D DFT over the range $0 \leq u < 8$ and $0 \leq v < 8$. Therefore, the problem is reduced to finding a good solution to the 2-D DFT. Feig's solution to this problem requires 54 multiplications, 462 additions, and 6 multiplications by 1/2 (which can be done by arithmetic shifts).

We shall now review Feig's approach to the scaled 2-D DCT. A knowledge of matrix algebra and tensor products is assumed for the rest of this section. Reviews of properties of tensor products can be found in Granata, Conner, and Tolimieri[32] and in the introductory sections of An, et al.[33]

We first cast the 1-D N-point DCT based on the $2N$-point DFT in terms of matrix equations:

$$S = DF \qquad\qquad [4\text{-}39]$$

where S is the DCT matrix, F is the DFT matrix, and D is a diagonal matrix that contains the constant scaling terms. Following Feig's approach, we first note that the 1-D flowgraph can be broken into four parts, which represent a factoring of the matrix F into:

$$F = PR_1MR_2 \qquad\qquad [4\text{-}40]$$

R_2 is a matrix that creates input sums and differences. It in turn is factored into three matrices:

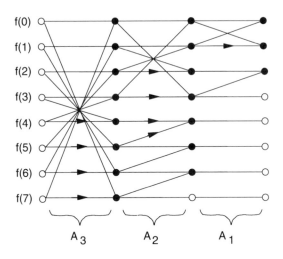

Figure   4-9. Flowgraph for R_2 (A_1, A_2, and A_3)

$$R_2 = A_1 A_2 A_3 \qquad\qquad\qquad [4\text{-}41]$$

where

$$A_1 = \begin{bmatrix}
1 & 1 & 0 & 0 & 0 & 0 & 0 & 0 \\
1 & -1 & 0 & 0 & 0 & 0 & 0 & 0 \\
0 & 0 & 1 & 1 & 0 & 0 & 0 & 0 \\
0 & 0 & 0 & 1 & 0 & 0 & 0 & 0 \\
0 & 0 & 0 & 0 & 1 & 0 & 0 & 0 \\
0 & 0 & 0 & 0 & 0 & 1 & 0 & 0 \\
0 & 0 & 0 & 0 & 0 & 0 & 1 & 0 \\
0 & 0 & 0 & 0 & 0 & 0 & 0 & 1
\end{bmatrix} \qquad [4\text{-}41a]$$

$$A_2 = \begin{bmatrix}
1 & 0 & 0 & 1 & 0 & 0 & 0 & 0 \\
0 & 1 & 1 & 0 & 0 & 0 & 0 & 0 \\
0 & 1 & -1 & 0 & 0 & 0 & 0 & 0 \\
1 & 0 & 0 & -1 & 0 & 0 & 0 & 0 \\
0 & 0 & 0 & 0 & -1 & -1 & 0 & 0 \\
0 & 0 & 0 & 0 & 0 & 1 & 1 & 0 \\
0 & 0 & 0 & 0 & 0 & 0 & 1 & 1 \\
0 & 0 & 0 & 0 & 0 & 0 & 0 & 1
\end{bmatrix} \qquad [4\text{-}41b]$$

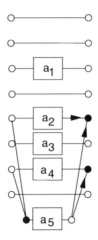

Figure   4-10. Flowgraph for M

$$A_3 = \begin{bmatrix} 1 & 0 & 0 & 0 & 0 & 0 & 0 & 1 \\ 0 & 1 & 0 & 0 & 0 & 0 & 1 & 0 \\ 0 & 0 & 1 & 0 & 0 & 1 & 0 & 0 \\ 0 & 0 & 0 & 1 & 1 & 0 & 0 & 0 \\ 0 & 0 & 0 & 1 & -1 & 0 & 0 & 0 \\ 0 & 0 & 1 & 0 & 0 & -1 & 0 & 0 \\ 0 & 1 & 0 & 0 & 0 & 0 & -1 & 0 \\ 1 & 0 & 0 & 0 & 0 & 0 & 0 & -1 \end{bmatrix} \qquad \text{[4-41c]}$$

The flowgraph for R_2 (A_1, A_2, and A_3) is shown in Figure 4-9. Comparing this figure with Figure 4-8, we see that A_1, A_2, and A_3 are the three stages of the 1-D DFT that create the sums and differences from the input values.

   M is given by:

$$M = \begin{bmatrix} 1 & 0 & 0 & 0 & 0 & 0 & 0 & 0 \\ 0 & 1 & 0 & 0 & 0 & 0 & 0 & 0 \\ 0 & 0 & C_4 & 0 & 0 & 0 & 0 & 0 \\ 0 & 0 & 0 & 1 & 0 & 0 & 0 & 0 \\ 0 & 0 & 0 & 0 & -C_2 & 0 & -C_6 & 0 \\ 0 & 0 & 0 & 0 & 0 & C_4 & 0 & 0 \\ 0 & 0 & 0 & 0 & -C_6 & 0 & C_2 & 0 \\ 0 & 0 & 0 & 0 & 0 & 0 & 0 & 1 \end{bmatrix} \qquad \text{[4-42]}$$

where C_2, C_4 and C_6 are defined in section 4.3.1.

   The flowgraph for M is shown in Figure 4-10. Comparing this to Figure 4-8, we see that M contains the multiplications in the center. It is interesting to note that the sub-array of elements in M

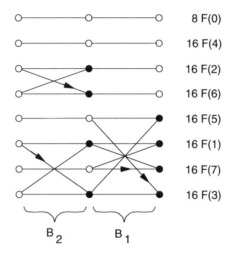

Figure   4-11. Flowgraph for R_1 (B_1 and B_2)

$$\begin{array}{ccc} -C_2 & . & -C_6 \\ . & . & . \\ -C_6 & . & C_2 \end{array} \qquad \text{[4-43]}$$

is a permutation and rotation of inputs to rows 5 and 7.
    The matrix R_1 is factored into two matrices:

$$R_1 = B_1 B_2 \qquad\qquad\qquad\qquad\qquad \text{[4-44a]}$$

where

$$B_1 = \begin{bmatrix}
1 & 0 & 0 & 0 & 0 & 0 & 0 & 0 \\
0 & 1 & 0 & 0 & 0 & 0 & 0 & 0 \\
0 & 0 & 1 & 0 & 0 & 0 & 0 & 0 \\
0 & 0 & 0 & 1 & 0 & 0 & 0 & 0 \\
0 & 0 & 0 & 0 & 1 & 0 & 0 & 1 \\
0 & 0 & 0 & 0 & 0 & 1 & 1 & 0 \\
0 & 0 & 0 & 0 & 0 & 1 & -1 & 0 \\
0 & 0 & 0 & 0 & -1 & 0 & 0 & 1
\end{bmatrix} \qquad \text{[4-44b]}$$

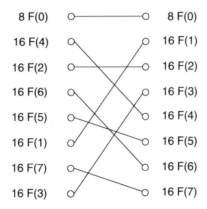

Figure   4-12. Flowgraph for P which permutes the rows

$$B_2 = \begin{bmatrix} 1 & 0 & 0 & 0 & 0 & 0 & 0 & 0 \\ 0 & 1 & 0 & 0 & 0 & 0 & 0 & 0 \\ 0 & 0 & 1 & 1 & 0 & 0 & 0 & 0 \\ 0 & 0 & -1 & 1 & 0 & 0 & 0 & 0 \\ 0 & 0 & 0 & 0 & 1 & 0 & 0 & 0 \\ 0 & 0 & 0 & 0 & 0 & 1 & 0 & 1 \\ 0 & 0 & 0 & 0 & 0 & 0 & 1 & 0 \\ 0 & 0 & 0 & 0 & 0 & -1 & 0 & 1 \end{bmatrix}$$ [4-44c]

The flowgraph for R_1 is shown in Figure 4-11, and, comparing this to Figure 4-8, we see that R_1 corresponds to the operations to the right of the multiplications in that figure.

P is a permutation matrix given by:

$$P = \begin{bmatrix} 1 & 0 & 0 & 0 & 0 & 0 & 0 & 0 \\ 0 & 0 & 0 & 0 & 0 & 1 & 0 & 0 \\ 0 & 0 & 1 & 0 & 0 & 0 & 0 & 0 \\ 0 & 0 & 0 & 0 & 0 & 0 & 0 & 1 \\ 0 & 1 & 0 & 0 & 0 & 0 & 0 & 0 \\ 0 & 0 & 0 & 0 & 1 & 0 & 0 & 0 \\ 0 & 0 & 0 & 1 & 0 & 0 & 0 & 0 \\ 0 & 0 & 0 & 0 & 0 & 0 & 1 & 0 \end{bmatrix}$$ [4-45]

The flowgraph for P is shown in Figure 4-12, from which it is seen that the sole function of this matrix is to permute the output rows to "natural order."

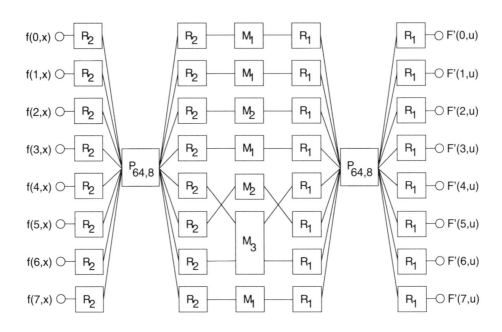

Figure   4-13. Flowgraph for fast 2-D DCT

Having noted the factors implicit in the 1-D flowgraph, we can then proceed to develop the 2-D DCT. The 2-D DCT is given by the tensor product of the separable 1-D DCTs:

$$S \otimes S = DF \otimes DF \tag{4-46}$$

where $\otimes$ denotes the Kroenecker, or tensor, product. From the distributive property of the tensor product we get:

$$S \otimes S = (D \otimes D)(F \otimes F) \tag{4-47}$$

The matrix $(D \otimes D)$ is diagonal and contains the scaling factors that will be absorbed into the quantization. What is left is the 2-D DFT matrix $(F \otimes F)$.

Using the factors developed for the 1-D DFT, we have:

$$F \otimes F = PR_1 MR_2 \otimes PR_1 MR_2 \tag{4-48}$$

which again, from the distributive property of tensor products, can be written as:

$$F \otimes F = (P \otimes P)(R_1 \otimes R_1)(M \otimes M)(R_2 \otimes R_2) \tag{4-49}$$

This equation is the key to Feig's approach. For each of the separate tensor products we have a choice: we can either evaluate the terms as a separable row–column sequence or we can take the tensor product first.

$(P \otimes P)$ is a permutation matrix of size 64×64, which simply describes the permutation that must be made to the output to get to what is sometimes termed natural order. Permutation matrices may give some complexity of data ordering, but are of no consequence from a computational standpoint.

The $(R_1 \otimes R_1)$ and $(R_2 \otimes R_2)$ terms are handled in row–column form, and the $(M \otimes M)$ processing is separated into three parts. Figure 4-13 illustrates the sequence of processing steps.[22]

The first part of the flowgraph involving matrix R_2 uses row–column processing of the tensor product. In general, if A_s and B_r are square matrices of order s and r, and $n = rs$, the commutative property of tensor products gives us:

$$A_s \otimes B_r = P[n,s](B_r \otimes A_s)P[n,r] \qquad [4\text{-}50]$$

Therefore, if the 64×64 matrix $R_2 \otimes R_2$ operates on inputs X to produce outputs Y, Y can be expressed as a tensor product involving the 8×8 matrix R_2, i.e.:

$$Y = (R_2 \otimes R_2)X = (I_8 R_2 \otimes R_2 I_8)X = (I_8 \otimes R_2)(R_2 \otimes I_8)X \qquad [4\text{-}51]$$

where I_8 is the 8×8 identity matrix. From the commutative property of tensor products:

$$(R_2 \otimes I_8) = P[64,8](I_8 \otimes R_2)P[64,8] \qquad [4\text{-}52]$$

where $P[64,8]$ is a special form of permutation matrix called a stride-by-8 matrix.[32] Therefore,

$P[64,8]$ permutes the row and column elements of the original 8×8 input array in X.

When we expand $(I_8 \otimes R_2)$, we get:

$$I_8 \otimes R_2 = \begin{bmatrix} R_2 & & & & & & & \\ & R_2 & & & & & 0 & \\ & & R_2 & & & & & \\ & & & R_2 & & & & \\ & & & & R_2 & & & \\ & & & & & R_2 & & \\ & 0 & & & & & R_2 & \\ & & & & & & & R_2 \end{bmatrix} \qquad [4\text{-}53]$$

This produces the flowgraph for $(R_2 \otimes R_2)$ shown in Figure 4-13. Note that the inputs to the flowgraph are rows of input samples, and thus include the effect of the first permutation matrix. A_1, A_2, and A_3 require 3, 7, and

8 additions, respectively, and, because R_2 is invoked a total of 16 times, this part of the flowgraph requires 288 additions.

$$Y = (I_8 \otimes R_2)P[64,8](I_8 \otimes R_2)P[64,8]X \qquad [4\text{-}54]$$

A similar structure holds for the $(R_1 \otimes R_1)$ section of the flowgraph. In this case we use a slight variation of Equation 4-54:

$$Y = P[64,8](I_8 \otimes R_1)P[64,8](I_8 \otimes R_1)X \qquad [4\text{-}55]$$

Note that the output is missing the final $P[64,8]$ permutation matrix. This can be incorporated into the $P \otimes P$ permutation matrix in Equation 4-49. Since B_1 and B_2 require 4 additions each and R_1 is invoked 16 times, this part of the flowgraph requires 128 additions.

$(M \otimes M)$ contains all of the multiplications, and therefore must be treated carefully. Going back to the definition of M in Equation 4-42 and the definition of the tensor product, one can write:

$$M \otimes M = \qquad [4\text{-}56]$$

$$\begin{bmatrix}
M & 0 & 0 & 0 & 0 & 0 & 0 & 0 \\
0 & M & 0 & 0 & 0 & 0 & 0 & 0 \\
0 & 0 & C_4M & 0 & 0 & 0 & 0 & 0 \\
0 & 0 & 0 & M & 0 & 0 & 0 & 0 \\
0 & 0 & 0 & 0 & -C_2M & 0 & -C_6M & 0 \\
0 & 0 & 0 & 0 & 0 & C_4M & 0 & 0 \\
0 & 0 & 0 & 0 & -C_6M & 0 & C_2M & 0 \\
0 & 0 & 0 & 0 & 0 & 0 & 0 & M
\end{bmatrix}$$

where each element in Equation 4-56 is an 8×8 array, and each row corresponds to eight elements of the input vector.

Four of the rows of Equation 4-56 represent M operating on a segment of the input vector of length 8, and, as such, require 5 multiplications and 3 additions per row, for a total of 20 multiplications and 12 additions. These four rows go through a box labelled M_1 in the flowgraph in Figure 4-13. The elements of M_1 are shown in Figure 4-10.

Two of the rows of Equation 4-56 represent M scaled by C_4 operating on the input vector segments. These two rows go through boxes labelled M_2 in the flowgraph. In principle each row requires 9 multiplications. Note, however, that $C_4 = 1/\sqrt{2}$, and therefore, $C_4^2 = 1/2$. Therefore, two of the multiplications for each row are by a factor of $1/2$, which can be implemented by a shift. This gives a total of 14 multiplications, 6 additions, and 4 shifts for these two rows. The flowgraph in Figure 4-14 shows the operations in box M_2.

The remaining two rows, rows 5 and 7 of Equation 4-56 (representing 16 input vector elements), go through the box labeled M_3 in Figure 4-13. In this box we perform the tensor product $N \otimes M$, where N is defined by the elements from rows 5 and 7 and columns 5 and 7 of the matrix M. Thus,

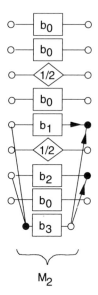

Figure   4-14. Flowgraph for M_2 operations

$$N = \begin{bmatrix} -C_2 & -C_6 \\ -C_6 & C_2 \end{bmatrix}$$ [4-57]

Using the relationship in Equation 4-50, the tensor product $N \otimes M$ can be rewritten as:

$$N \otimes M = P[16,2](M \otimes N)P[16,8]$$ [4-58a]

where

$$M \otimes N =$$ [4-58b]

$$\begin{bmatrix} N & 0 & 0 & 0 & 0 & 0 & 0 & 0 \\ 0 & N & 0 & 0 & 0 & 0 & 0 & 0 \\ 0 & 0 & C_4N & 0 & 0 & 0 & 0 & 0 \\ 0 & 0 & 0 & N & 0 & 0 & 0 & 0 \\ 0 & 0 & 0 & 0 & -C_2N & 0 & -C_6N & 0 \\ 0 & 0 & 0 & 0 & 0 & C_4N & 0 & 0 \\ 0 & 0 & 0 & 0 & -C_6N & 0 & C_2N & 0 \\ 0 & 0 & 0 & 0 & 0 & 0 & 0 & N \end{bmatrix}$$

Thus, the first pair, second pair, fourth pair, and eighth pair of lines are matrix-multiplied by N and the third and sixth pairs are multiplied by N

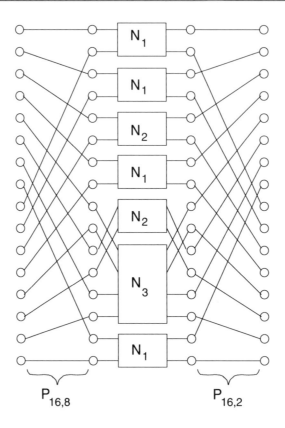

$$P_{16,8} \qquad\qquad\qquad P_{16,2}$$

Figure   4-15. Flowgraph for $M \otimes N$ tensor product

scaled by C_4. The matrix multiplication by N, either scaled or unscaled, requires 3 multiplications and 3 additions.

What is left is the tensor product $N \otimes N$, which, after substitution for the trigonometric identities

$$C_2 C_6 = C_4/2$$
$$C_2 C_2 = (1 + C_4)/2$$
$$C_6 C_6 = (1 - C_4)/2$$

becomes:

$$N \otimes N = \qquad\qquad\qquad\qquad\qquad\qquad\qquad\qquad [4\text{-}59]$$

$$(1/2) \begin{bmatrix} (1 + C_4) & C_4 & C_4 & (1 - C_4) \\ C_4 & -(1 + C_4) & (1 - C_4) & -C_4 \\ C_4 & (1 - C_4) & -(1 + C_4) & -C_4 \\ (1 - C_4) & -C_4 & -C_4 & (1 + C_4) \end{bmatrix}$$

Feig factors this matrix into the following:

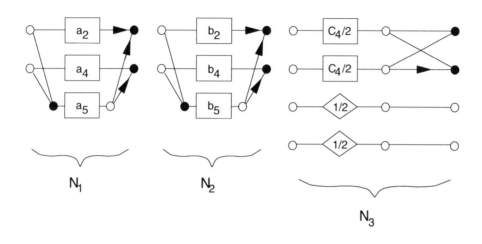

Figure   4-16. Flowgraphs for N_1, N_2, and N_3

$$N \otimes N = \tag{4-60}$$

$$
\begin{bmatrix}
1 & 0 & 1 & 0 \\
0 & 1 & 0 & 1 \\
0 & 1 & 0 & -1 \\
-1 & 0 & 1 & 0
\end{bmatrix}
\begin{bmatrix}
C_4/2 & C_4/2 & 0 & 0 \\
C_4/2 & -C_4/2 & 0 & 0 \\
0 & 0 & 1/2 & 0 \\
0 & 0 & 0 & -1/2
\end{bmatrix}
\begin{bmatrix}
1 & 0 & 0 & -1 \\
0 & 1 & 1 & 0 \\
1 & 0 & 0 & 1 \\
0 & 1 & -1 & 0
\end{bmatrix}
$$

which can be implemented with 2 multiplications, 10 additions, and 2 shifts.

Figure 4-15 shows the flowgraph for the $M \otimes N$ tensor product. The three boxes, N_1, N_2, and N_3, are shown in Figure 4-16. From this flowgraph and the other flowgraphs in this section, one can write down the equations for the 2-D DCT.

Table 4-3 summarizes the number of operations required to implement this DCT. From this we see that the net complexity is 54 multiplications, 462 additions, and 6 arithmetic shifts, not including the calculations to quantize the output.

4.3.8 Other approaches to fast DCT algorithms ◑

The scaled DCT algorithms described in the sections above are optimized assuming that multiplications and additions are separate operations. It is also assumed that multiplications are relatively expensive compared to shifts and additions. This is not always the case. Some computer and signal processor architectures use a fused multiply/add instruction (i.e., multiplication and addition are done in one instruction), and the fast algorithms should be modified to take advantage of this instruction. Linzer and Feig[17] investigated this problem and found a solution in which

Table   4-3. Operations for Feig's scaled 2-D DCT

Stage	Mult.	Add.	Shifts
Pre-addition stage ($R_1 \otimes R_1$)	0	16×18	0
Post-addition stage ($R_2 \otimes R_2$)	0	16×8	0
($M \otimes M$)			
Rows 1, 2, 4, 8	4×5	4×3	0
Rows 3, 6	2×7	2×3	2×2
Rows 5, 7 ($M \otimes N$)			
Rows 1, 2, 4, 8	4×3	4×3	0
Rows 3, 6	2×3	2×3	0
Rows 5, 7 ($N \otimes N$)	2	10	2
Totals:	54	462	6

both the FDCT and IDCT require 416 operations, not including the quantization and dequantization.

5

IMAGE COMPRESSION SYSTEMS

5.1 Basic structure of image compression systems ○

Image compression is the art and science of reducing (compressing) the number of bits required to describe an image. The two basic components of an image compression system are illustrated in Figure 5-1. The device that compresses the "source" image (the original digital image) is called an encoder and the output of this encoder is called compressed data (or coded data). The compressed data may be either stored or transmitted, but are at some point fed to a decoder. The decoder is a device that recreates or "reconstructs" an image from the compressed data.

5.1.1 Encoder structure ○

In general, a data compression encoding system can be broken into the following basic parts: an encoder model, an encoder statistical model, and an entropy encoder. The encoder model generates a sequence of "descriptors" that is an abstract representation of the image. The statistical model converts these descriptors into symbols and passes them on to the entropy encoder. The entropy encoder compresses the symbols to form

65

Figure   5-1. Image compression system

the compressed data. Figure 5-2 shows a sketch of this general encoder structure.

The two main functional blocks in Figure 5-2 divide the system in a way that is slightly different from the classical division found in many image compression books. Figure 5-2 follows the convention used in the JPEG documentation, where the division of the system is made at a point that cleanly separates the modeling from parts that are dependent on the particular form of entropy coding used. The traditional "symbols"— the items that are assigned codes—are generated by the statistical model inside the JPEG entropy encoder. We use the term "descriptors" here so that there is no confusion with the more traditional terminology.

The encoder may require external tables—that is, tables specified externally when the encoder is invoked. We define two classes of these tables. Model tables are those tables that are needed in the procedures that generate the descriptors. Entropy-coding tables are those tables needed by the JPEG entropy-coding procedures.

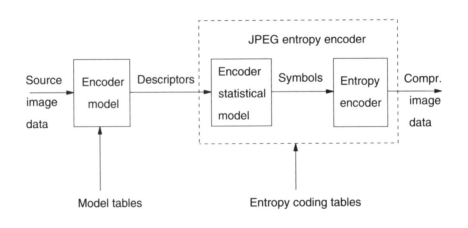

Figure   5-2. The basic parts of an encoder

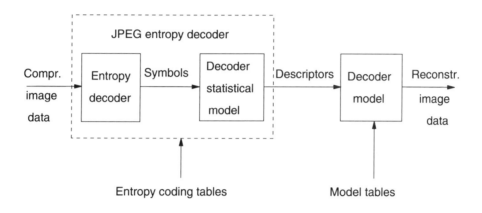

Figure   5-3. The basic parts of a decoder

5.1.2 Decoder structure ○

A data compression decoding system can be broken into basic parts that have an inverse function relative to the parts of the encoding system. As shown in Figure 5-3, the compressed data are fed to the JPEG entropy decoder. The entropy decoder decodes a sequence of descriptors that exactly matches the sequence that was compressed by the entropy encoder. The sequence of descriptors is converted to a reconstructed image by the decoder model. The decoder requires the same model tables and entropy-coding tables as the encoder. If they are not known to both *a priori*, the encoder must transmit the tables as part of the compressed data.

Unless the compressed image data have been corrupted (by errors introduced during storage or transmission), the descriptor input to the entropy encoder and the output from the entropy decoder are identical. Any loss or distortion, any difference between the source and reconstructed image data, is introduced only by the encoder and decoder models. One of the goals of image compression is to minimize the amount of distortion, especially the amount of visible distortion, for a given amount of compression.

Models that introduce distortion are termed "lossy" models, whereas models that allow no distortion are termed "lossless." The entropy encoding and decoding systems are also "lossless," as this is a term applied to any coding system or subsystem in which the input to the encoder is exactly matched by the output from the decoder.

5.2 Image compression models ○

The encoder model is defined as a unit that generates a sequence of descriptors. In this section we shall give some examples that should make this definition a little less abstract. Most of the examples given here are actually used in JPEG.

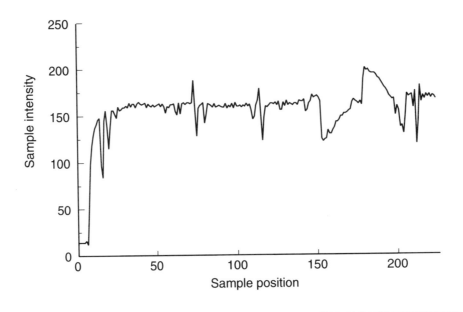

Figure   5-4. Plot of a line of image sample intensities

Suppose we have some image data that we need to code. We could simply feed the sample values to the entropy encoder, in which case the encoder model in Figure 5-2 is essentially an empty box. This is the simplest possible encoder model; it is also one of the poorest from the standpoint of compression of grayscale images. It is known as "pulse code modulation," or PCM. For PCM the set of descriptors is all possible values of the samples.

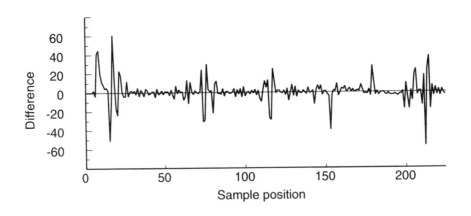

Figure   5-5. Plot of intensity differences between each sample and the nearest neighbor sample to left

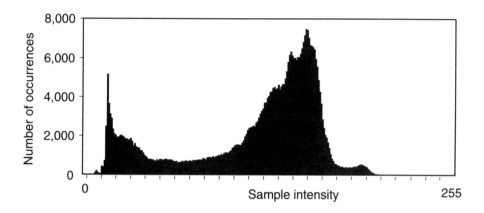

Figure   5-6. Histogram of image intensities

5.2.1 DPCM model ○

A much better model is realized if one attempts to predict the sample value from a combination of samples already coded in the image (the decoder must know those values too, so it can make the same prediction). For example, assuming we are coding from left to right, we might use the sample to the left as an estimate for the current sample. Then, the descriptor fed to the entropy encoder can be the difference between the sample being coded and the sample to the left. Since in a continuous-tone image the differences between one sample and the next are likely to be small, it is more efficient to encode the difference values than to encode each sample independently. This class of coding models is known as "differential pulse code modulation," or DPCM. For DPCM the set of descriptors is all possible difference values.

Figure 5-4 shows a plot of a typical line of image data. Figure 5-5 shows the corresponding differences between each sample and the sample immediately to the left. Note that the difference values are usually close to zero. This may be seen more clearly from the histogram of intensities for an entire image in Figure 5-6 and the histogram of the differences in Figure 5-7. The histogram of differences is tightly clustered around zero, indicating that small differences are very probable.

Given a known starting value, differences from one sample to the next can be used to exactly reconstruct the original intensity values. Therefore the set of differences is a representation that is entirely equivalent to the original intensity values. However, creating a representation in which a few events or symbols are very probable is important for data compression. The essence of data compression is the assignment of shorter code words to the more probable symbols, and longer code words to the less probable symbols. We might suspect, therefore, that DPCM gives better compression than does PCM, and indeed, it does. Phrased another way, the better performance of DPCM relative to PCM is due to the high degree of

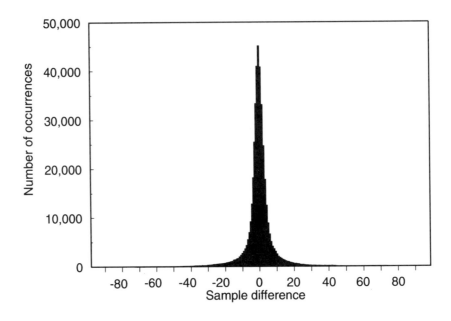

Figure   5-7. Histogram of differences between each sample and the nearest
           neighbor sample to left

correlation found in most images. Only if there is strong correlation will
the small differences be very probable.

A schematic of a DPCM encoder model that uses the neighboring
sample to the left as the prediction is shown in Figure 5-8. The corresponding decoder model is shown in Figure 5-9.

In these figures subtraction is indicated by "−" and addition by "+."
There is an implicit storage of at least one sample in both schematics, as
differences are always calculated relative to the previous sample.*

In two places JPEG uses forms of DPCM similar to that shown in
Figure 5-8: in the lossless coding mode and in the coding of the DCT DC
coefficient. Note that in lossless coding other predictors besides the value
of the neighboring sample to the left are allowed. For many images the
most effective predictor is an average of the samples immediately above
and to the left.

---

*   Note that as shown here, this DPCM model is lossless. Many forms of DPCM
quantize the difference values, scaling them in a manner that is quite similar to the
way DCT coefficients are quantized. The decoder must rescale the differences, and
the net effect is to introduce distortion between encoder input and decoder output.

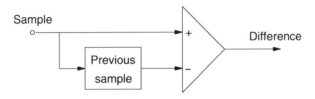

Figure   5-8. DPCM encoder model

5.2.2 DCT model ○

The DCT (FDCT and IDCT) and the associated quantization and dequantization of the coefficients are part of the encoder and decoder models used in the lossy JPEG modes. This encoder model is sketched in Figure 5-10.

Quantization is done in the box labelled "Q." Note that the DC term is fed separately to an additional stage containing a DPCM model. The DC differences and AC coefficients are the descriptors that are fed to the entropy-coding block. The entropy-coding block codes these two classes of descriptors differently.

The corresponding decoder model for the lossy JPEG modes is shown in Figure 5-11. The dequantization is done in the box labelled "DQ."

The FDCT, quantization, dequantization, and IDCT are the cause of the distortion in the images reconstructed by a JPEG lossy decoder. Arithmetic approximations in the integer arithmetic typically used in computing the FDCT and IDCT introduce a small amount of this distortion. The principal source of loss or distortion, however, is the quantization and dequantization of the coefficients.

The quantization of each coefficient is done independently, and can therefore be matched to the response of the human visual system. It is this aspect that makes the JPEG lossy modes so effective. However, the reader should recognize that the quantization rules are defined for a particular set of viewing conditions, image content, and perhaps even application

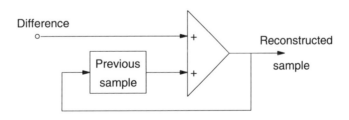

Figure   5-9. DPCM decoder model

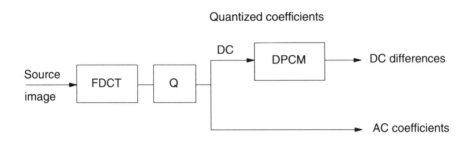

Figure   5-10. Encoder model for DCT systems

demands. If these change, the distortion introduced by quantization may become a problem. An example of this can be seen by comparing the image in Figure 1-1*b* to the enlarged section in Figure 5-12. Distortion that is negligible for normal viewing conditions becomes objectionable in a close-up.

There are additional attributes of DCT models that help to simplify the statistical modeling, one of them being the almost ideal decorrelation between the DCT basis functions.

5.2.3 Other models ○

The DPCM model is one particular instance of a general class of coding known as *predictive coding*. In predictive coding information already sent is used to predict future values and the differences are coded. The DCT model is also a particular form of a general class of coding known as *transform coding*. There are many other classes of coding models.

Because our goal in this book is to provide a complete review of the technologies that apply specifically to JPEG, we shall not discuss the many other models that have been devised for image coding. We note, however, that excellent implementations of block truncation coding, vector

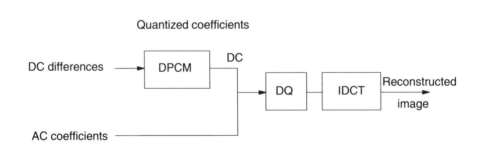

Figure   5-11. Decoder model for DCT systems

Figure   5-12. Visible distortion in a close-up. When a portion of Figure
1-1 *b* is enlarged 4× horizontally and vertically, the distortion is far more visible.

quantization, other transform coding schemes, sub-band coding, and
several predictive coding schemes were considered by JPEG as candidates
for the standard.   JPEG's selection in January, 1988, of a DCT-based
approach for lossy compression reflects the fact that of all the proposals
submitted, the DCT provided the best image quality for a given bit rate.
This is discussed in more detail in the chapter on JPEG history. For a
comprehensive review of image compression models we refer the reader to
two excellent texts on image compression.[1, 34]

5.3 JPEG entropy encoder and entropy decoder structures ○

JPEG uses two techniques for entropy coding: Huffman coding[35] and
arithmetic coding. Huffman coding, devised about 40 years ago, is the
more familiar of the two, is computationally simpler, and is simpler to
implement. However, it requires the code tables to be known at the start
of entropy coding. Arithmetic coding provides systematically higher compression performance (typically 10% or more) and one-pass adaptive coding in which the "code book" adapts dynamically to the data being coded.
In this section we shall only introduce the major functional blocks in these
entropy coders. The principles of entropy coding and the actual coding
procedures will be described in a later chapter.
      Inside an entropy coder there are four basic building blocks: a "sta-
tistical model," an "adaptor," a storage area, and an encoder. The way

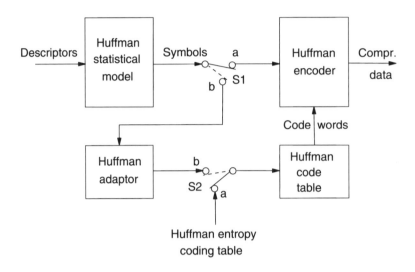

Figure   5-13. Huffman entropy encoder

the building blocks are linked together is a function of the particular
entropy coding.   The statistical model translates the descriptors into
"symbols," each symbol being assigned a particular code word or probability.*   The adapter is responsible for the assignment of code words
(Huffman coding) or probability estimates (arithmetic coding) needed by
the encoder.   The storage area contains the Huffman code table or the
arithmetic-coding probability estimates.   With the help of the code words
or probability estimates in the storage area, the encoder converts the symbols to bits of compressed data.

The symbols created by the statistical model are members of an
"alphabet" that contains all possible symbols that the model can generate.
Usually, some of these symbols are quite probable and some are very
improbable.   The objective in entropy coding is to assign short code words
to the very probable symbols and longer code words to the less probable
symbols.   Samuel Morse did this in an intuitive way more than a century
ago when he invented Morse code.   When the code assignment is done in
an optimal way, the result is entropy coding.

5.3.1 Huffman entropy encoder and decoder structures ○

The four basic building blocks of the Huffman entropy encoder are illustrated in Figure 5-13.   The Huffman statistical model converts the
descriptors into the actual symbols that are coded.   For example, the JPEG

---

* Although statistical considerations are also important in the design of the model
that produces the descriptor sequences, such considerations are not a part of the
statistical model.

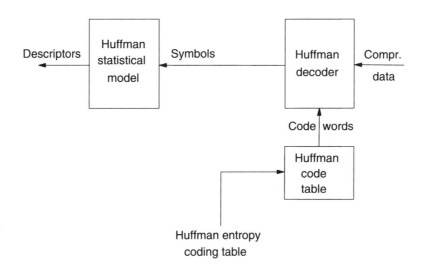

Figure 5-14. Huffman entropy decoder

Huffman statistical model converts 16 contiguous zero AC coefficients into a single symbol meaning a run of 16 zeros. This type of conversion is usually done to improve the efficiency of the Huffman codes. The symbols created by the statistical model are directed by switch S1 to either the Huffman encoder or the Huffman adaptor.

Whenever the adaptor is used, two passes through the data are required. In the first pass the symbols are fed to the adaptor, in which the data is analyzed and used to create a "custom" Huffman code table. (Huffman code tables are the "entropy coding tables" for a Huffman entropy coder.) The second pass then encodes the data. Since two passes through the data are not always convenient, switch S2 allows "fixed" Huffman code tables from an external source to be used instead. Custom tables typically improve the coding efficiency by a few percent relative to the efficiency achieved with fixed tables.

The Huffman entropy decoder is illustrated in Figure 5-14. The Huffman compressed data are decoded into symbols by the Huffman decoder block. The code words always come from a fixed Huffman code table that is loaded into the Huffman code table storage area before decoding begins. The data needed to generate this table may be incorporated into the compressed data. The symbols are translated back into the descriptors by the Huffman statistical model block.

5.3.2 Arithmetic entropy encoder and decoder structures ○

The four basic building blocks of the arithmetic-coding entropy encoder are illustrated in Figure 5-15.

JPEG uses an adaptive binary arithmetic coder that can code only two symbols, 0 and 1. (This may seem restrictive until you realize that com-

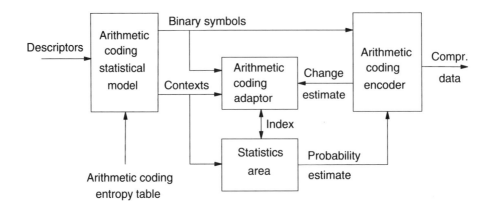

Figure 5-15. Arithmetic entropy encoder

puters also use only binary symbols.)   Therefore, the arithmetic coding
statistical model must translate the descriptors into a set of binary decisions. The statistical model also generates a "context" that selects a particular probability estimate to be used in coding the binary decision. The
binary decisions and contexts are fed in parallel to both the arithmetic-
coding adaptor and the arithmetic coder. The arithmetic-coding "conditioning table" (the arithmetic coding version of the entropy-coding table)
provides parameters needed in generating the contexts.

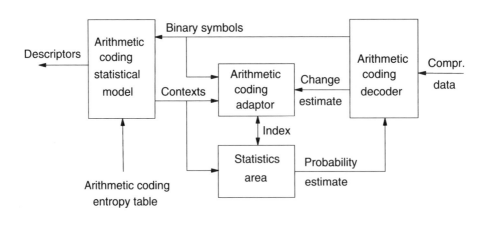

Figure 5-16. Arithmetic entropy decoder

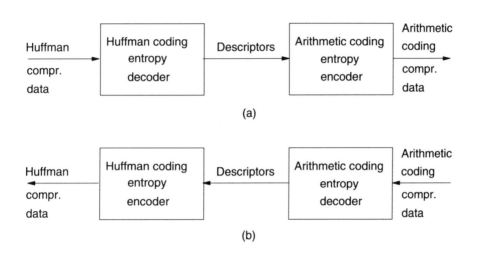

Figure   5-17. Transcoding between entropy coders

The arithmetic encoder encodes the symbol using the probability estimate supplied to it. In the process of coding, it also keeps an approximate count of the 0's and 1's, and occasionally signals the adaptor to tell it to make the probability estimate for a particular context larger or smaller.

The arithmetic coder is identical to that adopted by JBIG (see Chapter 20). Consequently, everything but the statistical model and perhaps the size of the probability storage area are unchanged when used by JBIG. Note that the JPEG/JBIG arithmetic entropy coder is a single-pass adaptive coder.

The arithmetic entropy decoder is illustrated in Figure 5-16. The compressed data are fed to the decoder, which, with the benefit of the same probability estimates used for encoding, determines the sequence of binary symbols. The binary symbols are in turn translated back into descriptors by the arithmetic-coding statistical model. Note that the decoder and encoder statistical models generate the same context for a given binary decision. In general, the contexts, adaptation, and probability estimates must be identical in the entropy encoder and decoder.

5.4 Transcoding ○

Because the Huffman and arithmetic entropy coders encode and decode the same set of descriptors, it is possible to "transcode" between these systems. Figure 5-17 illustrates this procedure for a conversion from Huffman compressed data to arithmetic-coding compressed data. Of course, transcoding is a reversible process and compressed data generated by an arithmetic coder can be transcoded to Huffman coded compressed data.

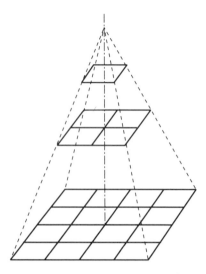

Figure   5-18. Pyramidal coding using the JPEG hierarchical mode

5.5 JPEG lossless and lossy compression ○

JPEG defines both lossy and lossless compression processes. The lossy JPEG processes are based on the Discrete Cosine Transform (DCT), whereas the lossless processes are based on forms of DPCM.

Lossy JPEG DCT-based compression is particularly effective in minimizing visible distortion. The improvement in compression that results when a small amount of visible distortion is allowed can be quite significant. Color images that can be compressed by a ratio of about 2:1 with the JPEG lossless techniques can often be compressed with JPEG lossy techniques by more than 20:1 yet have nearly imperceptible levels of visible distortion in the reconstructed images.*

---

* When judging compression for a given degree of visual distortion, compression ratios can be misleading. Compression ratios can be artificially inflated by using excessively high precisions or too many samples in the chrominance components. It would be better to state that nearly imperceptible distortion is achieved at compressions of a little more than 1 bit/pixel with the JPEG lossy techniques. Of course, bits/pixel can also be misleading, as it can be radically affected by changes in viewing conditions and picture resolution. Really meaningful comparison between compression systems requires a fixed set of viewing conditions, a carefully calibrated display system, and a set of images with scene content and resolution that are representative of the application. The real cost in storage and transmission comes from the amount of compressed data required to represent the image at some particular quality.

5.6 Sequential and progressive coding ○

A sequential encoder codes the image in a single scan or pass through the data. The decoder necessarily follows the same order in decoding the data, reconstructing the image at full quality in a single scan in the same order as it was encoded.

A progressive encoder encodes the image in two or more scans or passes through the data, first coding an approximation of the entire image and then coding finer details for the whole image with each succeeding scan. The decoder follows the same order in decoding, first reconstructing the approximation of the image, and then adding finer detail with each succeeding scan.

The particular form of progression defined by JPEG can provide a final image that is identical to that of the JPEG sequential mode. Furthermore, under the right conditions, slightly better compression is achieved with the JPEG progressive mode than with the JPEG sequential mode. Examples of JPEG sequential and progressive encoding will be given in Chapter 6.

5.7 Hierarchical coding ○

An alternative form of progressive coding known as hierarchical coding uses a set of successively smaller images that are created by "downsampling" (low-pass filtering and subsampling) the preceding larger image in the set. Then, starting with the smallest image in the set the image set is coded with increasing resolution. After the first stage, each lower-resolution image is scaled up to the next resolution (upsampled) and used as a prediction for the following stage. When the set of images is stacked, it sometimes resembles a pyramid (Figure 5-18). Consequently, this form of coding is also called pyramidal coding.

Just as smaller images generated as part of a hierarchical progression can be scaled up to full resolution, full-resolution images generated as part of a nonhierarchical progression can be scaled down to smaller sizes. These scalings are done in the decoder, and are not restricted to powers of 2.

5.8 Compression measures ○

The amount of compression that an encoder achieves can be measured in two different ways. Sometimes the parameter of interest is compression ratio—the ratio between the original source data and the compressed data sizes. However, for continuous-tone images another measure, the average number of compressed bits/pixel, is sometimes a more useful parameter for judging the performance of an encoding system. For a given image, however, the two are simply different ways of expressing the same compression.

6

JPEG MODES
OF OPERATION

The JPEG standard describes a family of image compression techniques rather than a single compression technique. It provides a "tool kit" of compression techniques from which applications can select elements that satisfy their particular requirements.

The four JPEG modes of operation are the sequential DCT-based mode, the progressive DCT-based mode, the sequential lossless mode, and the hierarchical mode. The DCT-based modes provide lossy compression, whereas a form of DPCM predictive coding is used for the sequential lossless mode. The hierarchical mode uses extensions of either DCT-based or predictive coding. Figure 6-1 gives a schematic of these four modes and some of the couplings between them. Figure 6-2 is the original image that will be used to demonstrate these four modes of operation.

6.1 Sequential DCT-based mode of operation ○

In the sequential DCT-based mode the image components are compressed either individually or in groups, a single "scan" coding a component or group of components completely in one pass. If groups of components are coded, the data are interleaved in a manner that will be described in Chapter 7. However, from a conceptual standpoint each component is still

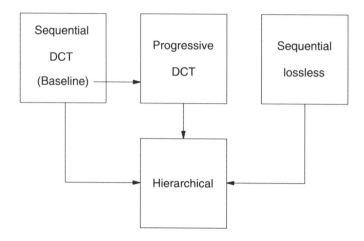

Figure 6-1. The four JPEG modes of operation

coded separately, even though the data and coding operations are inter-leaved. The interleaving allows color images to be compressed and decompressed with a minimum of buffering.

Figure 6-2. Original image used for all compression examples in this chapter

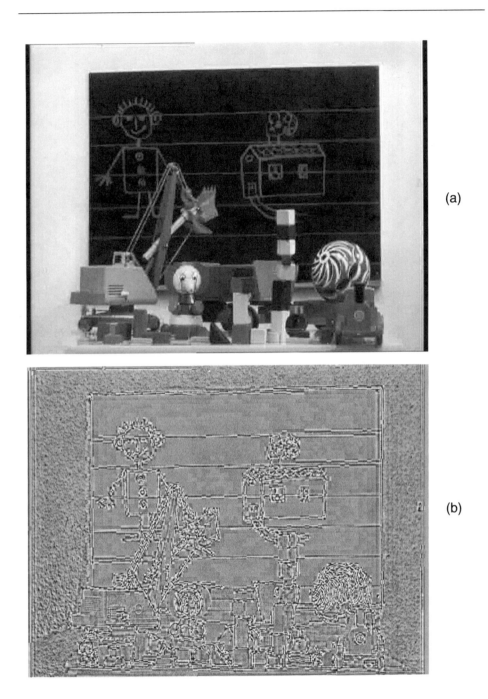

(a)

(b)

Figure   6-3. Sequential DCT mode of operation. (a) 0.158 bit/pixel (AC
quantization values multiplied by 6.08). (b) 8× difference
between original and (a).

(c)

(d)

Figure 6-3 continued. (c) 0.243 bit/pixel (AC quantization values multiplied by 2.68). (d) 8× difference between original and (c).

Figure 6-3 continued. (e) 0.440 bit/pixel (suggested table). (f) 8× difference between original and (e).

A particular restricted form of this sequential DCT-based mode of operation is called the "baseline system." It represents a minimum capability that must be present in all DCT-based decoder systems. Sequential DCT-based systems that have capabilities beyond the baseline requirements are called "extended sequential systems."

Figure 6-3 illustrates the image quality which can be achieved by the JPEG sequential modes. Reconstructions at three different bit rates are shown. These bit rates were obtained by multiplying the AC coefficient quantization values listed in Table 4-1 in Chapter 4 by the values given in the figure captions. The DC quantization value was unmodified.

These examples were created using the arithmetic coding entropy coder option (as were all the examples in this chapter). The same image quality can be obtained with the JPEG baseline system, but with somewhat poorer compression. Chapter 15 provides data on the performance of many of the various JPEG modes and options.

Many of the reconstructed images in this chapter are accompanied by a differential image created by subtracting the reconstructed image from the original image. These differential images are multiplied by 8 to make the error more visible; they are then level-shifted such that zero differences are mapped to a neutral gray.

6.2 Progressive DCT-based mode of operation ○

The progressive DCT-based mode is achieved by a sequence of "scans," each of which codes a part of the quantized DCT coefficient information. Two complementary ways of doing this are used: spectral selection and successive approximation. In spectral selection the DCT coefficients are grouped into "spectral" bands of related spatial frequencies and the lower-frequency bands are (usually) sent first. In successive approximation the information is first sent with lower precision and then refined in later scans.

The reconstructed images from a typical spectral-selection sequence are shown in Figure 6-4. The DC coefficients are sent first, giving the rather blocky and crude reconstruction in Figure 6-4a. The first AC scan, which includes only the two lowest-frequency AC coefficients, removes much of the blocking, as can be seen in Figure 6-4c. The image quality is significantly improved after five AC coefficients are coded, as can be seen in Figure 6-4e. When all of the DCT coefficients for each block have been coded, the image quality is exactly the same as in Figure 6-3e. This final stage is completed at 0.448 bit/pixel for this spectral-selection progression.

Successive approximation gives better quality at low bit-rates. Figure 6-5a. shows the quality obtained when the AC coefficients are divided by 4, and Figure 6-5c. shows the quality when the AC coefficients are divided by 2. The reconstructions in these two figures should be compared to Figure 6-4c and Figure 6-4e respectively. In successive approximation the AC coefficients are not represented with sufficient accuracy to eliminate

(a)

(b)

Figure   6-4. Spectral-selection progression. (a) 0.065 bit/pixel (DC coefficients only). (b) 8× difference between original and (a).

(c)

(d)

Figure 6-4 continued. (c) 0.157 bit/pixel (first two AC coefficients). (d)
8× difference between original and (c).

Figure 6-4 continued. (e) 0.256 bit/pixel (first five AC coefficients). (f) 8× difference between original and (e).

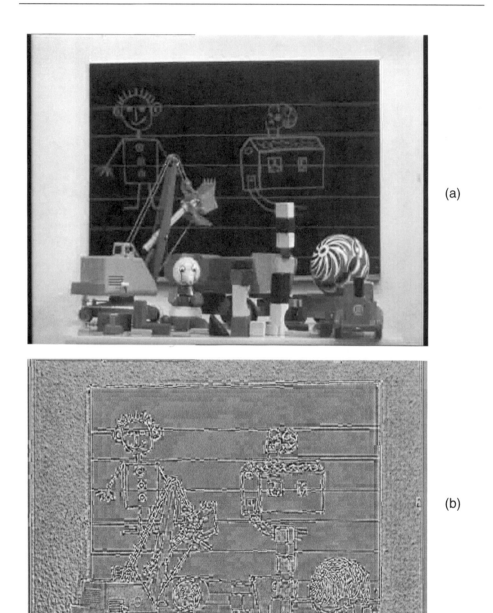

(a)

(b)

Figure   6-5. Successive approximation progression. (a) 0.158 bit/pixel (AC
coefficients divided by 4). (b) 8× difference between original
and (a).

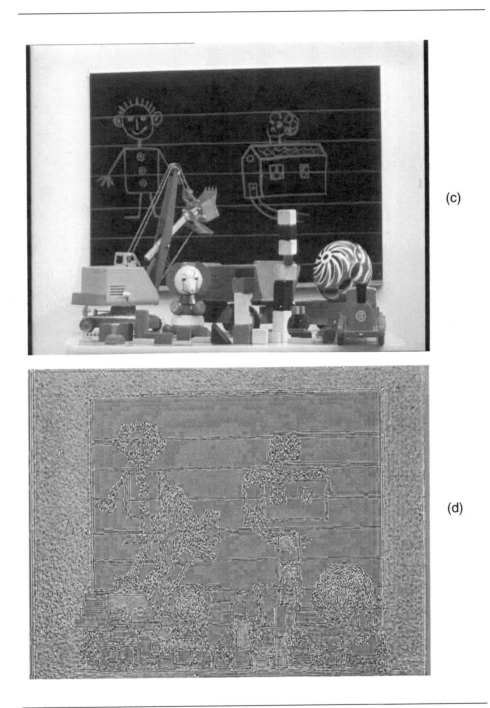

Figure 6-5 continued. (c) 0.243 bit/pixel (AC coefficients divided by 2). (d) 8× difference between original and (c).

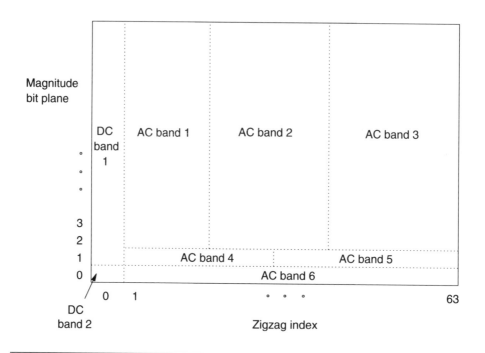

Figure 6-6. Example of segmentation of the DCT for progressive coding

blocking artifacts.* In other respects, however, the images are of significantly higher quality for a given bit-rate. The full precision DCT is completed at 0.428 bit/pixel for this successive-approximation progression.

These two progressive DCT processes may be intermixed to provide a very graceful progression. Figure 6-6 illustrates a possible mixture of spectral-selection and successive-approximation scans. The DCT coefficients for each block are ordered from DC to highest frequency AC coefficient according to the zigzag sequence described in section 4.1.2. The first scan compresses all but the least significant bit of the DC coefficients, thereby providing a rather blocky and crude rendition of the image such as in Figure 6-5 a Then the AC coefficients, except for the two least significant magnitude bits, are sent in three spectral bands (in three separate scans). The next magnitude bit of the AC coefficients is split into two bands, and the final bit is sent as AC band 6. Finally, the least significant bit of each DC coefficient is sent.

6.3 Sequential lossless mode of operation ○

An exact match between encoder input and decoder output cannot be guaranteed between two different implementations of the JPEG

---

* A technique for suppressing blocking artifacts is discussed in Chapter 16.

Table   6-1. Predictors for lossless coding

Selection-value	Prediction
0	no prediction (differential coding)
1	a
2	b
3	c
4	$a + b - c$
5	$a + (b - c)/2$
6	$b + (a - c)/2$
7	$(a + b)/2$

DCT-based systems because the IDCT is not rigorously defined for the DCT-based modes of operation.

The lossless mode of operation uses a predictive coding technique that is an extension of the coding model used for the DC coefficients in the DCT modes. The reconstructed neighboring samples (a, b, and c) illustrated in Figure 6-7 are used to predict the current sample x. The prediction equation is selected (for a given scan through a component or group of components) from the choices shown in from Table 6-1. The first line of an image component is always coded one-dimensionally, using the predictor defined for selection value 1 from Table 6-1.

A point transform may be used to reduce the precision of the source data. This point transform divides each sample value by 2^{Pt}, where Pt is an integer value defined in the compressed data header. Note that the predictions are always calculated from the point transformed values.

6.4 Hierarchical mode of operation ○

The hierarchical mode of operation provides for progressive coding with increasing spatial resolution between progressive stages. The first stage (lowest resolution) is coded using one of the sequential or progressive JPEG modes. The output of each hierarchical stage is then upsampled if neces-

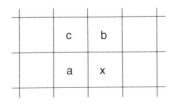

Figure   6-7. Neighboring samples for lossless predictors

(a)

(b)

Figure   6-8. Hierarchical progression. (a) 0.053 bit/pixel (expanded 4:1 by pixel replication in each dimension). (b) 0.224 bit/pixel (expanded 2:1 by pixel replication in each dimension).

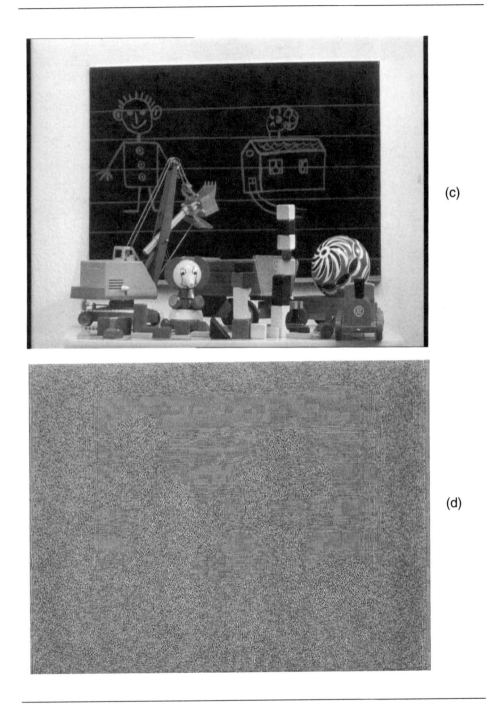

(c)

(d)

Figure   6-8   continued.   (c) 0.568 bit/pixel (full resolution).   (d) 8× difference between original and (c).

sary (interpolated to increase the spatial resolution), and becomes the prediction for the next stage. The upsampling filters match the spatial resolution of source image (possibly downsampled) and double the resolution
horizontally, vertically, or in both dimensions. The image quality for a
three stage DCT-based hierarchical progression is illustrated in Figure 6-8
a, *b*, and *c*. The image quality at extremely low bit rates surpasses any of
the other JPEG modes, but this is achieved at the expense of a higher bit
rate at the completion of the progression. In this example the final image
required about 33% more bits than a successive-approximation progression
with comparable final image quality.

Refinement in image quality at a given spatial resolution is also possible. The differences are coded using modifications of either the
DCT-based or the lossless modes of operation (the modifications allow
them to work with differential images). JPEG defines constraints on how
these different coding modes may be mixed.*

---

* Although we usually use the term "lossless" in describing the final differential
refinement stages, the term really applies to the coding process after a point
transform of the differences. The point transform is used when some distortion
can be allowed between the source and reconstructed image, and can substantially
improve the compression.

7

JPEG SYNTAX AND DATA ORGANIZATION

This chapter describes the organization JPEG has adopted for image data and compressed data, and the conventions adopted for interleaving data from multiple components. It also discusses the JPEG headers, the special codes used to segment the compressed data, and the control procedures that are common to all the JPEG processes.

Certain parameters that are needed by many applications are not part of the JPEG compressed data format and may therefore be needed in application-specific "wrappers" surrounding the JPEG data. Among these parameters are image aspect ratio, pixel shape, meaning of the image components, orientation of the image (vertical, horizontal, upside down, etc.), and sample registration between the components. JPEG found, however, that other standards committees regarded this aspect of the data sets as their domain. JPEG therefore decided to restrict the JPEG compressed data almost entirely to those parameters needed to decompress the data stream.

7.1 Control procedures and compressed data structure ○

JPEG compressed data contains two classes of segments: entropy-coded segments and marker segments. Entropy-coded segments contain the

```
| start-of-image marker |
   | table/misc. marker segment(s) | frame header |
      | table/misc. marker segment(s) | scan header |
         | entropy-coded segment |
   | end-of-image marker |
```

Figure 7-1. Structure of typical compressed image data stream

entropy-coded data, whereas marker segments contain header information, tables, and other information required to interpret and decode the compressed image data. Marker segments always begin with a "marker," a unique two-byte code that identifies the function of the segment. Detailed information about markers is found in section 7.4.

The JPEG processes use a common control structure that closely parallels the compressed data structure. A JPEG compression process consists of a single "frame" (nonhierarchical process) or a sequence of frames (hierarchical process). Each frame is composed of one or more "scans" through the data, where a scan is a single pass through the data for one or more components of the image.

At the beginning of each frame, the control procedures generate a frame header that contains parameter values needed for decoding the frame. Similarly, at the beginning of each scan, the control procedures generate a scan header that contains parameter values needed for decoding the scan. Each frame and scan header starts with a marker that can be located in the compressed data without decoding the entropy-coded segments. Marker segments defining quantization tables, entropy-coding tables, and other parameters may precede the frame and scan headers.

Scan headers are always followed by one or more entropy-coded segments. The input data for the scan may be divided into fixed intervals called "restart intervals," and the data in each restart interval are compressed into a single entropy-coded segment. When a scan contains more than one entropy-coded segment, the segments are separated by special two-byte codes called a "restart markers" that can be located in the compressed data without decoding. The encoding and decoding of each entropy-coded segment is done independently of the other entropy-coded segments in the scan.

This compressed image data structure is reflected directly in the control procedures that apply to all of the JPEG modes of operation. In Annex E of Part 1 of the JPEG DIS control procedures are described that are common to the encoding and decoding of frames, scans, and restart intervals for all nonhierarchical JPEG modes of operation. Additional higher-level control procedures are described in Annex J for the hierarchical mode.

These common control procedures make use of a number of markers that we shall define in this chapter. Section 7.3 of this chapter describes the image data structures used in the lowest level of the control structure.

| start-of-image marker |
  | table/misc. marker segment(s) |
| end-of-image marker |

Figure   7-2. Structure of abbreviated format for table specification data

7.2 Interchange and abbreviated compressed data formats ○

A JPEG compressed image data stream always contains two isolated markers that start and end the compressed image data. Between these two markers there are typically two or more marker segments that contain frame headers, scan headers, tables, other miscellaneous parameters needed for decoding, and at least one entropy-coded segment. Figure 7-1 illustrates this structure.

There are three formats for JPEG compressed data:

1. Interchange format for compressed image data

2. Abbreviated format for compressed image data

3. Abbreviated format for table specification data

The interchange format for compressed image data includes all tables that are required by the decoder. If compressed image data are intended for general use, the interchange format should be used.

The abbreviated format for compressed image data may omit some or all of the tables needed for decoding. These tables must have been inherited from a previous compressed data stream or must be known through some other mechanism (such as defaults defined for a particular application).

The abbreviated format for table specification data (Figure 7-2) contains no frames, scans, or entropy-coded segments. This abbreviated format is used to establish tables that may be inherited by the abbreviated format for compressed image data. If these abbreviated formats are used, the application is responsible for coordinating table specifications and compressed image data.

7.3 Image data ordering ○

In this section we shall review the conventions JPEG has defined for ordering of interleaved and non-interleaved data. One of the more complex aspects of JPEG, at least from an implementer's point of view, is the interleaving of data from image components during the compression process. This interleaving makes it possible, however, to decompress the image, convert from a luminance–chrominance representation to a red–green–blue display output, and store the result directly in a display buffer in a relatively "seamless" manner and with a minimum of intermediate buffering.

7.3.1 Internal representation ○

A JPEG compressed image data stream contains a single image that can have up to 255 unique components. Each component of the image is represented internally in the compression/decompression system as a rectangular array of samples. All parameters and conventions for processing the data are defined with respect to the internal representation,* as this is the representation for which the order of encoding and decoding operations is defined. The rectangular arrays of samples are processed from left to right along rows, and from top row to bottom row; the first sample coded is, by definition, in the upper-left corner of the array.

The internal representation is merely a convenience. Applications are free to choose other representations if they wish; in all cases the data are decoded and returned to the application in the same order in which they were presented to the encoder. The restriction to rectangular arrays is a significant limitation, however, as certain nonrectangular sampling patterns are not readily adapted to the JPEG format.

7.3.2 Sampling factors and component dimensions ◑

JPEG defines horizontal and vertical sampling factors, H_i and V_i, that specify the number of samples in the ith component relative to the other components in the frame. Therefore, if the overall horizontal and vertical dimensions of the image are X and Y, the dimensions of the ith component (the number of horizontal samples, X_i, and vertical samples, Y_i) are given by:

$$X_i = \text{ceiling} \left[X \frac{H_i}{H_{\max}} \right] \qquad\qquad \text{[7-1a]}$$

$$Y_i = \text{ceiling} \left[Y \frac{V_i}{V_{\max}} \right] \qquad\qquad \text{[7-1b]}$$

$H_{\max}$ and $V_{\max}$ are the maximum horizontal and vertical sampling factors among the components in the frame, and "ceiling" indicates the greatest integer value (i.e., round up). Therefore, X and Y are equal to the dimensions of the components with the maximum horizontal and vertical sampling factors, respectively. The sampling factors and image dimensions are specified in the frame header (see section 7.5).

---

* The word "internal" in describing the image representation refers to the fact that JPEG defines a compressed data syntax that is sufficient for decoding, but that is missing certain attributes and is lacking some constraints that may be needed to define completely the meaning of the image for a particular application. The omission of these additional attributes and constraints in defining the JPEG compressed data syntax was by choice, as other standards committees are responsible for this aspect of the system.

7.3.3 Data units for DCT-based and lossless processes ○

A data unit is the smallest logical unit of source data that can be processed in a given JPEG mode of operation. In lossless modes the arrays are processed one sample at a time; therefore, the data unit for lossless processes is one sample. By definition, the samples are processed from left to right along rows, and from top row to bottom row of each component.

In DCT-based modes, however, the component arrays are divided into 8×8 blocks for purposes of computing the DCT. Therefore, the data unit for DCT-based processes is a single 8×8 block of samples. By definition, the upper-left 8×8 block is aligned with the upper-left 8×8 group of samples in the component and the blocks in the component are processed from left to right along block rows, and from top block row to bottom block row. Figure A.4 in the JPEG DIS illustrates the blocking pattern.

7.3.4 Minimum coded units ◑

For the coding procedures, data units are assembled into groups called minimum coded units (MCU). In scans with only one component, the data are non-interleaved and the MCU is one data unit. In scans with more than one component, the data are interleaved, and the MCU defines the repeating pattern of interleaved data units.

The order of data units and the number of data units in the MCU for interleaved data are determined from the horizontal and vertical sampling factors and the components in the scan (see section 7.6).

When the data are interleaved, the sampling factors define a two-dimensional array of data units, $H_i \times V_i$ in size, for the ith component of the frame. If the 1, 2, ..., Ns components in the scan correspond to frame components $i(1)$, $i(2)$, ..., $i(Ns)$, each MCU in the scan is constructed by taking $H_{i(1)} \times V_{i(1)}$ data units from frame component $i(1)$, $H_{i(2)} \times V_{i(2)}$ data units from frame component $i(2)$, ..., and $H_{i(Ns)} \times V_{i(Ns)}$ data units from frame component $i(Ns)$. Within each $H_i \times V_i$ array of data units, the data units are ordered from left to right and from top row to bottom row. Processing always starts at the upper left corner of each component array.

Entropy coding is always done in complete MCU. Therefore, in the encoder any incomplete MCU must be completed by replication of the far-right column and/or the bottom row of each component (or by any other suitable padding technique). Any extra columns and rows added by the encoder are discarded by the decoder. For example, if the luminance component is 14 samples wide, the 8×8 block segmentation for the DCT algorithm scan requires two MCU horizontally. The missing two columns can be supplied by replication of the 14th column of samples.

7.3.5 Examples of interleaved and non-interleaved MCU ◑

In the sequential DCT algorithm the DCT for each 8×8 block is coded as a complete unit, independent of whether the data are interleaved or non-interleaved. If the data are non-interleaved, then the MCU is just one data unit—i.e., an 8×8 block. A single quantization table is used in each scan,

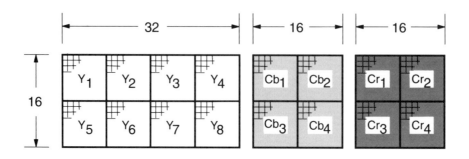

Figure   7-3. Three-component image with chrominance subsampled

Table   7-1. Non-interleaved data units

Component block	Tq	PRED	MCU
Scan1:			
Y1	00	0	1
Y2	00	DC_{Y_1}	2
Y3	00	DC_{Y_2}	3
Y4	00	DC_{Y_3}	4
Y5	00	DC_{Y_4}	5
Y6	00	DC_{Y_5}	6
Y7	00	DC_{Y_6}	7
Y8	00	DC_{Y_7}	8
Scan2:			
Cb1	01	0	1
Cb2	01	DC_{Cb_1}	2
Cb3	01	DC_{Cb_2}	3
Cb4	01	DC_{Cb_3}	4
Scan3:			
Cr1	01	0	1
Cr2	01	DC_{Cr_1}	2
Cr3	01	DC_{Cr_2}	3
Cr4	01	DC_{Cr_3}	4

and the preceding DC value is always the predictor, *PRED*, for the current DC value.

For interleaved data, however, the DCT blocks are ordered according to the parameters in the frame and scan headers, as described in section 7.3.4. The frame header also specifies a parameter Tq_i that identifies the quantization table to use with the ith component in the frame. We shall discuss this in more detail in section 7.5.6. For interleaved data, particular attention must be paid to using the correct quantization table when quantizing the DCT for a block, and to using the correct *PRED* value when coding the DC coefficient of a block. (*PRED* is always the preceding DC value coded for the same component.)

Figure 7-3 shows a small image with twice as many luminance blocks horizontally as chrominance component blocks. The sampling factors are therefore ($H = 2$, $V = 1$) for the luminance Y, and ($H = 1$, $V = 1$) for the Cb and Cr chrominance components.

When the data units are coded in non-interleaved order, three separate scans are made. Table 7-1 lists the order of the DCT blocks (data units) for non-interleaved data, each component of which is coded in a separate scan. If the data units for all three components are interleaved in a single scan, the ordering would be as shown in Table 7-2.

Table 7-2. Horizontally-interleaved data units

Component block	Tq	*PRED*	MCU
Scan1:			
Y1	00	0	1
Y2	00	DC_{Y_1}	1
Cb1	01	0	1
Cr1	01	0	1
Y3	00	DC_{Y_2}	2
Y4	00	DC_{Y_3}	2
Cb2	01	DC_{Cb_1}	2
Cr2	01	DC_{Cr_1}	2
Y5	00	DC_{Y_4}	3
Y6	00	DC_{Y_5}	3
Cb3	01	DC_{Cb_2}	3
Cr3	01	DC_{Cr_2}	3
Y7	00	DC_{Y_6}	4
Y8	00	DC_{Y_7}	4
Cb4	01	DC_{Cb_3}	4
Cr4	01	DC_{Cr_3}	4

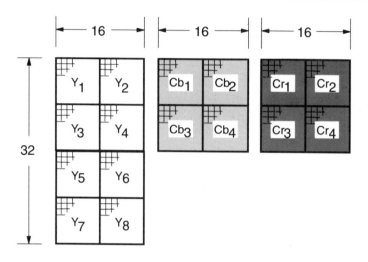

Figure   7-4. Three-component image with twice as many luminance sam-
         ples vertically

As a second example consider an image in which the chrominance
components are subsampled 2:1 in the vertical dimension, as shown in
Figure 7-4. In this example the sampling factors are ($H = 1$, $V = 2$) for Y,
and ($H = 1$, $V = 1$) for Cb and Cr. The data-unit sequence when all three
components are interleaved in one scan is given in Table 7-3. Note that
the different ordering of blocks in the MCU makes the descriptors
produced by the DPCM coding of the luminance different from those
produced by the previous two examples. The possibility of changes in the
predictions must be considered when transcoding from one format to
another. If the block order changes, as is sometimes the case when vertical
sampling factors are greater than one, care must be taken to go back to the
original DC coefficient whenever the predictions change.[*]

7.3.6 A cautionary note about interleaves in the MCU ❶

The number of data units of each component in the MCU is defined in the
frame header parameters. Unless there is only one component in the scan
(non-interleaved), the number of data units of a given component in the
MCU is not affected by other components in the scan. In the example
given in section A.2.3 of Part 1 of the JPEG DIS the frame has four components in which the first component has $H_1 = 2$, $V_1 = 2$, the second
component has $H_2 = 2$, $V_2 = 1$, the third component has $H_3 = 1$, $V_3 = 2$,
and the fourth component has $H_4 = 1$, $V_4 = 1$. If only the first and third

---

[*] For the lossless algorithms this is not a concern. The prediction is always calculated in the same manner, independent of the interleave.

Table   7-3. Vertically-interleaved data units

Component block	Tq	$PRED$	MCU
Scan1:			
Y1	00	0	1
Y3	00	DC_{Y_1}	1
Cb1	01	0	1
Cr1	01	0	1
Y2	00	DC_{Y_3}	2
Y4	00	DC_{Y_2}	2
Cb2	01	DC_{Cb_1}	2
Cr2	01	DC_{Cr_1}	2
Y5	00	DC_{Y_4}	3
Y7	00	DC_{Y_5}	3
Cb3	01	DC_{Cb_2}	3
Cr3	01	DC_{Cr_2}	3
Y6	00	DC_{Y_7}	4
Y8	00	DC_{Y_6}	4
Cb4	01	DC_{Cb_3}	4
Cr4	01	DC_{Cr_3}	4

components were present in a scan, it might be tempting to renormalize and interleave as if $H_1 = 2$, $V_1 = 1$ and $H_3 = 1, V_3 = 1$. This is not allowed, however.

7.4 Marker definitions ❶

Each marker segment begins with one or more byte-aligned X'FF' (hexadecimal FF) bytes and a non-zero one-byte "marker code" that identifies the function of the segment. JPEG terms the marker code and its X'FF' prefix a marker. Marker segments and entropy-coded segments always contain an integer number of bytes.

In generating the entropy-coded segments, X'FF' bytes are occasionally created. To prevent accidental generation of markers in the entropy-coded segments, each occurrence of a X'FF' byte is followed by a "stuffed" zero byte. Marker segments have known lengths, and therefore may contain accidental markers. Stuffing of zero bytes is not done in marker segments.

All marker segments and entropy-coded segments are followed by another marker. In Huffman coding one-bits are used to pad the entropy-coded data to achieve byte alignment for the next marker. Because a leading one-bit sequence is always a prefix for a longer code, padding with

Table 7-4. Definition of SOF_n markers

Name	Code	Length	Category
Nondifferential Huffman-coding frames:			
SOF_0	X'FFC0'	V	Baseline DCT
SOF_1	X'FFC1'	V	Extended sequential DCT
SOF_2	X'FFC2'	V	Progressive DCT
SOF_3	X'FFC3'	V	Lossless (sequential)
Differential Huffman-coding frames:			
SOF_5	X'FFC5'	V	Differential sequential DCT
SOF_6	X'FFC6'	V	Differential progressive DCT
SOF_7	X'FFC7'	V	Differential lossless
Nondifferential arithmetic-coding frames:			
SOF_9	X'FFC9'	V	Extended sequential DCT
SOF_{10}	X'FFCA'	V	Progressive DCT
SOF_{11}	X'FFCB'	V	Lossless (sequential)
Differential arithmetic-coding frames:			
SOF_{13}	X'FFCD'	V	Differential sequential DCT
SOF_{14}	X'FFCE'	V	Differential progressive DCT
SOF_{15}	X'FFCF'	V	Differential lossless

V – Variable length with known structure.

---

one-bits cannot create valid Huffman codes that might be decoded before the marker is identified. In arithmetic coding the entropy-coded segment is terminated at a byte boundary in a way that guarantees that the next marker will be detected before decoding is complete. For either entropy coder, a marker immediately following the entropy-coded segment can provide information needed to terminate decoding.

The markers fall into two categories: those without parameters and those followed by a variable-length sequence of parameters of known structure. For those markers with parameters, the first parameter is a two-byte parameter giving the length (in bytes) of the sequence of parameters. This length includes the length parameter itself, but excludes the marker that defines the segment.

The rest of this section contains a brief summary of markers and their function. Table 7-4 lists the start of frame (SOF_n) markers. Differential frames referred to there are used only within a hierarchical progression. All other markers are listed in Table 7-5. Additional information about markers can be found in Annex B of the DIS.

Table 7-5. Definition of non-SOF$_n$ markers

Name	Code	Length	Category
APP$_n$	X'FFE0'		
	– X'FFEF'	V	Reserved for application use
COM	X'FFFE'	V	Comment
DAC	X'FFCC'	V	Define arithmetic conditioning table(s)
DHP	X'FFDE'	V	Define hierarchical progression
DHT	X'FFC4'	V	Define Huffman table(s)
DNL	X'FFDC'	4	Define number of lines
DQT	X'FFDB'	V	Define quantization table(s)
DRI	X'FFDD'	4	Define restart interval
EOI	X'FFD9'	N	End of image
EXP	X'FFDF'	3	Expand reference image(s)
JPG	X'FFC8'	U	Reserved for JPEG extensions
JPG$_n$	X'FFF0'		
	– X'FFFD'	U	Reserved for JPEG extensions
RES	X'FF02'		
	– X'FFBF'	U	Reserved
RST$_m$	X'FFD0'		
	– X'FFD7'	N	Restart with modulo 8 counter m
SOI	X'FFD8'	N	Start of image
SOS	X'FFDA'	V	Start of scan
TEM	X'FF01'	N	For temporary use in arithmetic coding

N – No length or parameter sequence follows.
U – Undefined at this time.
V – Variable length with known structure.

APP$_n$: The APP$_n$ (Application) marker segments are reserved for application use. Since these segments may be defined differently for different applications, the JPEG DIS states that they should be removed when the data are transferred between application environments. However, the reader should be aware that JPEG did not make the removal of these APP$_n$ marker segments mandatory. Therefore, if two applications require conflicting usage of an APP$_n$ marker segment, checking of the data may be needed when JPEG compressed data is imported or exported from one application to the other.

COM: The COM (Comment) marker segment is intended for text fields. It may not contain information that could affect decoding. JPEG has not specified the character set to be used in this segment.

DAC: The DAC (Define Arithmetic coding Conditioning) marker segment defines one or more tables of parameters for the conditioning of the probability estimates used in the arithmetic coding procedures.

DHP: The DHP (Define Hierarchical Progression) marker segment is used to signal the parameters identifying the dimensions, components, and relative sampling of the final image of the hierarchical progression. It uses the same syntax as the frame header (see section 7.5) and must precede the first frame in the progression.

DHT: The DHT (Define Huffman Table) marker segment defines one or more Huffman tables.

DNL: The DNL (Define Number of Lines) marker segment provides a mechanism for defining or redefining the number of lines in the image at the end of the first scan. The two-byte length (always 4) is followed by a two-byte value giving the number of lines in the frame. The number of lines specified in the DNL marker segment must be consistent with the number of lines and the sampling factors already specified in the frame header and with the number of lines already decoded. Thus, unless the value specified in the frame header is zero, the number of lines specified in the DNL marker segment must be less than or equal to the frame header value, and in all cases must be a value somewhere within the range of the last row of MCU decoded. Note that if the data are directly written to an image buffer, spurious data may be written to the buffer if rows of samples are padded in this last MCU row.

DQT: The DQT (Define Quantization Table) marker segment defines one or more quantization tables. Quantization tables used in a progression should not be changed between scans of that progression. The DQT marker has meaning only for DCT-based algorithms.

DRI: The DRI (Define Restart Interval) marker segment provides the mechanism for setting and resetting the restart interval. The two-byte length (always 4) is followed by a two-byte value giving the number of MCU in the restart interval. The restart interval is set to zero at the start of an image. When the restart interval is zero, the restart function is disabled.

EOI: The EOI (End Of Image) marker terminates the JPEG compressed data stream.

EXP: The EXP (Expand) marker segment signals the expansion of the spatial resolution of the reference data needed by the hierarchical differential frame that follows. It is followed by a two-byte length (always 3) and an eight-bit integer that has one of the values X'10', X'01' or X'11'. X'10' signals a horizontal expansion by a factor of 2, X'01' signals a vertical expansion by a factor of 2, and X'11' signals both horizontal and vertical expansion by a factor of 2.

JPG; JPG$_n$: The JPG and JPG$_n$ marker segments are reserved for future JPEG extensions.

RES: The RES markers are reserved for future JPEG extensions.

RST$_m$: The RST$_m$ (Restart) marker is appended to the compressed data between restart intervals. This provides a unique byte-aligned code that can be located by scanning the compressed data. A three-bit field (m) in this marker code provides a modulo-8 restart interval count. The modulo count of the first RST$_m$ code in a scan is zero.

SOF$_n$: An SOF$_n$ (Start Of Frame) marker begins a frame header. The variable-length SOF$_n$ marker segment gives the frame parameters that apply to all scans within the frame. The particular value of n in SOF$_n$ identifies the mode of compression and the entropy coder used within the frame.

SOI: The SOI (Start Of Image) marker begins the compressed data stream. Note that the SOI marker can be used to detect problems with bit order, bit sense, and bit alignment of the data.

SOS: The SOS (Start Of Scan) marker begins a scan header. The scan header is always followed immediately by entropy-coded data for the scan. The two-byte length gives the number of bytes of scan parameters.

TEM: The TEM marker is reserved for temporary private use by arithmetic encoders. It is used to signal temporarily an unresolved carry-over, so that postprocessing can carry out the resolution off-line. It may not occur in JPEG compressed data streams.

The DHT, DAC, DRI, DQT, COM, and APP$_n$ marker segments may be inserted into the compressed data before the frame and scan headers. They may be used in any order and as many times as desired. If the DHT and DQT marker segments follow the start-of-image marker (SOI), they may be followed immediately by the end-of-image (EOI) marker. This is the JPEG abbreviated format for table specification data. COM and APP$_n$ marker segments may also be used in this abbreviated sequence. The DAC and DRI may appear but have no function, because the next SOI will set default values.

7.4.1 Structure of the compressed data stream ❶

From the definitions of the segments in the previous section the marker and segment structure of a nonhierarchical encoder output might be as shown in Figure 7-5. Note:

1. The marker terminating the final restart interval of the first scan may be preceded by the DNL marker.

2. The last entropy-coded segment in a scan is not followed by an RST$_m$ marker.

3. If arithmetic coding is used, the DAC marker segment is used instead of the DHT marker segment.

SOI
   DQT, length, quantization table definition(s).
   DRI, length, restart interval
      SOF_n, length, frame parameters
        DHT, length, Huffman table definition(s)
        SOS, length, scan parameters
            compressed data for restart interval, RST_0
            ... etc. ...
            compressed data for restart interval, RST_m
            ... etc. ...
            compressed data for final restart interval
        DHT, length, Huffman table definition(s)
        SOS, length, scan parameters
            ... etc. ...
EOI

Figure   7-5. Example of nonhierarchical compressed data structure

7.4.2 Bit and byte ordering conventions in the compressed data ◗

The bit ordering and byte ordering conventions in the compressed data are as follows:

1. The coded data are eight-bit byte aligned.

2. Integers are sent with the most significant byte first. If an integer is split between two or more bytes, the byte containing the most significant bit (MSB) of the integer is sent first.

3. The root of a Huffman code is placed toward the MSB of the byte, and successive bits are placed in the direction MSB to least significant bit (LSB) of the byte. Remaining bits, if any, go into the next byte, following the same rules.

4. Integers associated with Huffman codes are appended with the MSB adjacent to the leaf of the Huffman code.

7.5 Frame header ◗

As noted in section 7.1, JPEG divides the compression sequence into frames and scans. For nonhierarchical processes the frame defines the basic attributes of the image, including size, number of components, precision, and entropy-coding technique. For nonhierarchical encoding, only a single frame is allowed, whereas in the hierarchical mode a sequence of frames may be used.

    The parameters following the SOF_n marker (the frame header) are given in Table 7-6. All integer parameters in the frame header are unsigned.

Table   7-6. Frame header structure

Parameter	Symbol	Size (bits)
Marker (X'FFC0–3, 5–7, 9–B, D–F')	SOF_n	16
Frame header length	Lf	16
Sample precision	P	8
Number of lines	Y	16
Number of samples/line	X	16
Number of components in frame	Nf	8
Frame component specification ($i = 1, ..., Nf$)		
Component identifier	C_i	8
Horizontal sampling factor	H_i	4
Vertical sampling factor	V_i	4
Quantization table destination selector	Tq_i	8

7.5.1 Frame header length ❶

A 16-bit integer, Lf, gives the length in bytes of the frame parameters. The length value excludes the bytes allocated to the frame marker, but includes the two bytes in the length field.

7.5.2 Sample precision ❶

An eight-bit integer, P, specifies the sample precision in bits. A precision of either eight or 12 bits is allowed for the DCT compression algorithms. Precisions of from two to 16 bits are allowed for the lossless DPCM coding algorithms. All image components in a frame must have the same sample precision. Both source input and reconstructed output samples are defined to have an unsigned representation.

The bounds on precision of the quantized DCT are determined by the sample precision, the smallest quantization value, and the normalization defined for the DCT calculations. The largest quantized DCT coefficient precision possible is 15 bits, corresponding to an input precision of 12 bits and quantization table values of one. In a DCT-based hierarchical progression the differential samples can have a precision of nine bits or 13 bits, giving a maximum DCT precision of 12 bits or 16 bits. In the baseline system the sample precision is eight bits and the quantized DCT coefficient precision is limited to a maximum of 11 bits.

Output samples have the same precision as input samples. However, because the DCT is quantized, it is possible for the the IDCT output to either underflow or overflow the range allowed by the sample precision. Provisions should be made for properly clamping the output data when this occurs.

7.5.3 Number of lines ◑

A 16-bit integer, Y, specifies the number of raster lines in the internal representation of the frame. The value excludes any lines added to complete the bottom row of MCU in the frame.

If the number of lines is set to zero, it is unspecified at the start of compression. In this case the DNL segment must be used at the end of the first scan of the frame to signal the number of lines.

7.5.4 Number of samples per line ◑

A 16-bit integer, X, specifies the number of samples per raster line in the internal representation of the frame. The value excludes any columns added to complete the MCU at the right edge of the frame.

7.5.5 Number of components in frame ◑

An eight-bit integer, Nf, specifies the number of components in a frame. Nf can have any value from 1 to 255 in sequential DCT-based frames and lossless frames. In progressive frames Nf may have only values from 1 to 4. A value of zero is undefined and is forbidden.

7.5.6 Frame component specification parameters ◑

For each component in a frame a component identifier, the horizontal and vertical sampling factors, and the quantization table selector must be specified.

The frame component specification parameters are given by:

$$C_1,(H_1, V_1), Tq_1; \quad C_2,(H_2, V_2), Tq_2; \quad ...; \quad C_{Nf},(H_{Nf}, V_{Nf}), Tq_{Nf}$$

where,

C_i = component identifier assigned to ith component

H_i = horizontal sampling factor of ith component

V_i = vertical sampling factor of ith component

Tq_i = quantization table selector for ith component

The component identifiers, C_i, are assigned by the application; any eight-bit value is allowed and each component must be assigned a unique number that is used for that component in the scans in the image.

The sampling factors have already been discussed in section 7.3.4. The (H_i, V_i) pair of parameters makes up an eight-bit integer in which the H_i value is in the high order four bits and the V_i value is in the low order four bits. The allowed values of H_i and V_i are 1, 2, 3, and 4, all other values being forbidden.

For the DCT algorithms the Tq_i parameter may have values of 0, 1, 2, or 3; it selects the quantization table to be used for the ith component.

Table 7-7. Scan header structure

Parameter	Symbol	Size (bits)
Marker (X'FFDA')	SOS	16
Scan header length	Ls	16
Number of components in scan	Ns	8
Scan component specification ($k = 1,...,Ns$)		
Scan component selector	Cs_k	8
DC entropy coding table selector	Td_k	4
AC entropy coding table selector	Ta_k	4
Start of spectral selection or predictor selection	Ss	8
End of spectral selection	Se	8
Successive approximation bit position high	Ah	4
Successive approximation bit position low		
or point transform	Al	4

A maximum of four quantization tables may be defined. For the lossless algorithms Tq_i serves no function and must be set to zero.

As an example, consider a three-component system, YCrCb, with identical sampling for all components in the vertical direction and twice as many Y samples as Cr or Cb samples in the horizontal direction. Let the first component, Y, use quantization table 0 and the rest use quantization table 1. If Y, Cr, and Cb are numbered 0, 1, and 2 respectively, the component specification is:

$$\begin{array}{cccccccccc} Nf, & C_1, & (H_1 \; V_1), & Tq_1; & C_2, & (H_2 \; V_2), & Tq_2; & C_3, & (H_3 \; V_3), & Tq_3 \\ X'03 & 00 & 21 & 00 & 01 & 11 & 01 & 02 & 11 & 01' \end{array}$$

Note that a quantization table may be specified by a DQT marker segment after the table selection has been made.

7.6 Scan header ◑

Many scans may occur within a frame, especially in the JPEG progressive modes. The number of components in the scan dictates the type of data ordering used in the scan. If the scan has only one component the data are non-interleaved and each MCU contains only one data unit. If the scan contains more than one component the data are interleaved and the number of data units in the MCU is determined by the sampling factors of the components in the scan. Note that even when the data are interleaved, the components are still coded independently. Only the order of processing and the entropy-coded bit stream are interleaved.[*]

The parameters following the SOS marker (the scan header) are given in Table 7-7. These parameters must be present for all scans, even if a parameter is always set to zero.

7.6.1 Scan header length ◑

A 16-bit integer, Ls, gives the length in bytes of the scan header. The length value excludes the bytes allocated to the SOS marker, but includes the two bytes in the length field.

7.6.2 Number of components in scan ◑

An eight-bit integer, Ns, specifies the number of components in a scan. Ns may have only values from 1 to 4 (or 1 to Nf if $Nf < 4$).

7.6.3 Scan component specification parameters ◑

For each component in a scan, a scan component selector and entropy-coding table selectors must be specified. The components selected for a scan may be a subset of the components specified for the frame, but must follow the same order as in the frame component specification.

The scan component specification parameters are given by:

$$Cs_1,(Td_1, Ta_1); \ \ Cs_2,(Td_2, Ta_2); \ \ ...; \ \ Cs_{Ns},(Td_{Ns}, Ta_{Ns})$$

where

    Cs_k = scan component selector for the kth scan component

    Td_k = DC entropy-coding table selector for the kth scan component

    Ta_k = AC entropy-coding table selector for the kth scan component

The identification numbers $Cs_1, ..., Cs_{Ns}$ must match one of the component identification numbers specified in the component specification defined for the frame.

The parameters (Td_k, Ta_k) select the entropy-coding table(s) for component Cs_k. Each (Td_k, Ta_k) pair is an 8-bit integer in which Td_k is in the high order 4 bits and Ta_k is in the low order 4 bits. In extended systems Ta_k and Td_k may have values from 0 to 3, whereas for the baseline system, Ta_k and Td_k may have only the values 0 and 1. If a table selection field is not applicable to the scan, it is set to zero. Note that in the progressive DCT modes DC and AC coefficients may not be coded in the same scan.

---

[*] JPEG makes no attempt to exploit correlations between the components of an image; such effects are assumed to be too small to justify the additional complexity that intercomponent correlation would introduce.

For Huffman entropy coding, the parameter values must be consistent with the Huffman table specifications already made. For arithmetic entropy coding the four statistics areas are always available. Default conditioning is used with a statistics area if the conditioning table for that area has not been specified by a DAC marker segment.

7.6.4 Start of spectral selection or predictor selection ◑

In spectral selection a band of coefficients, in zigzag order and contiguous, is coded in the scan. The start of spectral selection, Ss, identifies the starting index of this band. For a DCT sequential process Ss is zero, whereas for a progressive DCT scan Ss may have any value from 0 to 63. A value of 0 signals the coding of DC coefficients. Note that in the progressive DCT scan, DC coefficients are always coded separately from AC coefficients. For the lossless nondifferential algorithm Ss selects the predictor to be used for all components in the scan.

7.6.5 End of spectral selection ◑

The end of spectral selection, Se, identifies the index of the last coefficient in the spectral band. For a progressive DCT scan, Se may have any value from Ss to 63, whereas for a DCT sequential process the end of spectral selection, Se, is set to 63. Note that bands sent in multiple scans do not have to be contiguous.

If the start of spectral selection parameter is set to zero in the progressive DCT mode, only DC coefficients may be coded. Therefore, if Ss is zero, Se must also be zero. For the lossless algorithms, Se is always set to zero.

7.6.6 Successive approximation bit position high ◑

If successive approximation is used in the first scan of a spectral-selection band the coefficients are coded at reduced precision. The successive approximation bit position high parameter, Ah, is set to the value of Al of the preceding scan of the same band (see the following section). If there was no preceding scan of the same band (i.e., for the first successive approximation scan, spectral-selection-only scans, and sequential DCT processes), Ah is set to zero. Ah is also set to zero for lossless modes.

7.6.7 Successive approximation bit position low or point transform ◑

A four-bit integer, Al, specifies the point transform used to reduce the precision of the DCT coefficients. The AC coefficients in the band are scaled by dividing them by 2^{Al}, whereas the DC coefficients are scaled by an arithmetic-right-shift by Al.* Effectively, Al gives the magnitude least significant bit position after scaling. For subsequent (successive approxi-

---

* The arithmetic shift is the equivalent of a divide if the input data are not zero-shifted. Once the data are zero-shifted, a divide introduces nonuniform quantization and the arithmetic shift must be used instead.

Table 7-8. DHT marker segment structure

Parameter	Symbol	Size (bits)
Marker (X'FFC4')	DHT	16
Huffman table definition length	Lh	16
For each Huffman table:		
Table class	Tc	4
Huffman table identifier	Th	4
Number of Huffman codes of length i		
for $i = 1, ..., 16$	L_i	8
Value associated with each Huffman code		
for $i = 1, ..., 16; j = 1, ..., L_i$	V_{ij}	8

mation) scans, Al gives the bit position of the magnitude bit that will be coded in the scan. All coefficients in a band must be coded to the same precision before invoking the successive approximation algorithm.

In the first scan for a given spectral selection band Ah is set to zero and Al is set to the desired point transform value. For scans after the first for a given band, Al must be one unit smaller than Ah. Any other combinations of values for Ah and Al are forbidden.

For the lossless DPCM process, Al defines the point transformation to be used on the input data and Ah is set to zero. The point transformation is a division by 2^{Al}. However, the division should be unsigned for unsigned nondifferential inputs and signed for signed differential inputs.*

7.7 Limit on the number of data units in an MCU ◑

If more than one component is specified for the scan, the components in the scan and their sampling factors (specified in the frame header) define the ordering of components and the number of data units (blocks or samples, depending on the mode) of each component in the MCU. The total number of data units in the MCU, Nb, is the sum of the data units for all the components in the scan.

$$Nb = \sum_{s=1}^{Ns} H_{i(s)} V_{i(s)} \qquad\qquad [7\text{-}2]$$

---

* The JPEG DIS does not make this distinction between unsigned and signed division. However, if signed division is used with unsigned 16-bit data or unsigned division is used with any signed input data, the point-transformed data will have a discontinuity between sample values of X'7FFF' and X'8000'.

The total number of data units in the MCU may not exceed 10. Therefore, any combinations of components and sampling factors that would give $Nb > 10$ are forbidden.

7.8 Marker segments for tables and parameters ◗

In addition to the frame and scan headers a number of other marker segments are needed to define entropy-coding tables, quantization tables and parameters.

7.8.1 Huffman table specification ◗

The efficiency of Huffman coding can be significantly improved by the use of custom Huffman code tables, and the DHT marker segment provides a mechanism for specifying these custom tables. The DHT segment may precede either the SOF_n or the SOS marker segments. One or more Huffman table specifications follow the marker and 16-bit length. The last table specification in the DHT marker segment is always followed by the X'FF' prefix of the next marker.

The parameters following the DHT marker are given in Table 7-8. The table class, Tc, is 0 for DC and lossless coding, whereas Tc is 1 for AC code tables. The code table identifier, Th, may have values from 0 to 3, except in the baseline DCT system, in which the value is restricted to 0 and 1.

The Huffman table is specified by identifying the number of codes of each length from 1 to 16, by means of the 16 L_i parameters. Values, V_{ij}, are then listed for each code of each length. Note that the code table is constrained such that 16 bits is the maximum code length. Note also that the maximum value is 255.

Once a Huffman table has been specified, it may be used for subsequent images. If the table has never been specified, the results are unpredictable. If the abbreviated format for compressed data is used and tables are inherited from a previous image, care must be taken to make sure the correct Huffman tables are used.

7.8.2 Arithmetic coding conditioning table specification ◗

One of the important aspects of the statistical models for arithmetic coding is the use of conditional probabilities in the coding decisions. The DAC marker provides a mechanism for setting the parameters that control the conditioning of the probability estimates, thereby improving the coding performance.

Conditioning tables may not be inherited from a previous image. Instead, default values for the conditioning parameters used in the arithmetic-coding procedures are installed in each conditioning table at the start of the image. They may be overruled by the DAC (Define Arithmetic coding Conditioning) marker segment. This marker segment may precede either the SOF_n or the SOS marker segments.

Table 7-9. DAC marker segment structure

Parameter	Symbol	Size (bits)
Marker (X'FFCC')	DAC	16
Arithmetic coding conditioning table length	La	16
For each arithmetic-coding table:		
Table class	Tc	4
Arithmetic coding conditioning identifier	Ta	4
Conditioning table value	Cs	8

The parameters following the DAC marker are given in Table 7-9. The table class, Tc, is 0 for DC and lossless coding, whereas Tc is 1 for AC tables. The code table identifier, Ta, may have values from 0 to 3.[*]

For DC and lossless tables Cs contains two four-bit fields L and U. These two parameters control the conditioning of some of the probabilities used in coding the DPCM differences. L and U may have values from 0 to 15. U is in the high order four bits of Cs, and L is in the low order four bits. For AC tables Cs contains the parameter Kx, which has a value from 1 to 63. Kx controls the conditioning of some of the probabilities used in coding the AC coefficients. Chapter 12 gives a complete description of these conditioning parameters. The default values established by the SOI marker code are $L = 0$, $U = 1$, and $Kx = 5$.

7.8.3 Quantization table specification ❶

The amount of distortion in a lossy JPEG image is controlled almost entirely by the quantization tables. Quantization tables may be inherited from one image to the next, and may also be established for a given application by mechanisms independent of the JPEG compressed image data syntax. However, the tables must be included in the interchange format for compressed image data, and DQT marker segments provide the means to do this.

The parameters following the DQT marker are given in Table 7-10. Pq specifies the precision of the quantization table elements (0 for 8-bit precision, 1 for 16-bit precision). The quantization table identifier, Tq, may have values from 0 to 3, and quantization table elements, Q_k, may have values from 1 to the maximum value permitted for the precision specified. However, for input precisions of eight bits, Pq must be zero. The table elements are specified in zigzag order.

Multiple tables may be specified in one DQT marker segment. Note that the last table definition will always be followed by the X'FF' prefix of the next marker.

---

[*] The symbol for the code table identifier has been changed to Tb in the JPEG IS.

Table 7-10. DQT marker segment structure

Parameter	Symbol	Size (bits)
Marker (X'FFDE')	DQT	16
Quantization table definition length	Lq	16
For each quantization table:		
Quantization table element precision	Pq	4
Quantization table identifier	Tq	4
Quantization table element ($k = 0, ..., 63$)	Q_k	8 or 16

Once a quantization table has been defined, it may be used for subsequent images. This inheritance property allows applications to define defaults and to use the abbreviated format for the compressed data. However, the disassociation of the quantization table from the compressed image data places a responsibility on the applications to keep the correct quantization table associated with each image. Clearly, if a table has never been defined, the results are unpredictable.

7.8.4 Restart interval ◑

As noted in section 7.1, JPEG has defined a mechanism for partitioning the compressed data stream into independently decodable segments called restart intervals. Except for the final restart interval in a scan, restart intervals contain a fixed number of MCU that is established by a DRI marker segment. The final interval contains only the number of MCU required to complete the scan.

The restart capability is disabled by the SOI (Start Of Image) marker. It is enabled for an image if the DRI (Define Restart Interval) marker sets the restart interval to a non-zero value. If restart is enabled, an RST_m marker must be placed in the compressed data stream between the entropy-coded segments for each restart interval. The parameters following the DRI marker are listed in Table 7-11.

The restart interval, Ri, is specified in MCU. Note that the MCU is dependent on the number of components specified in the scan interleave and on the sampling factor specified in the frame parameters. The number of data units in the restart interval may therefore change from one scan to the next.

For DCT-based modes the restart interval is defined to be an integer multiple of MCU. Because of the two-dimensional prediction used in the spatial algorithms, the restart interval for the spatial algorithms must be an integer multiple of the number of MCU in a row of MCU in the scan.

7.8.5 Restart marker ◑

The RST_m markers separate the entropy-coded segments for each restart interval and allow each segment to be encoded and decoded independently of other intervals in the scan. These markers provide some very interesting capabilities for error recovery, localized decoding of sections of an image, and parallelism in encoding and decoding. The error recovery procedures are optional, and are not defined by JPEG. The decoders must, however, be able to handle the segmented compressed data stream.

The RST_m markers incorporate a modulo-8 count, m, of the restart intervals. This count is started at zero for each scan and is incremented by one after each RST_m marker is added to the code stream.

Encoders and decoders are reinitialized at the beginning of each restart interval. All DC predictions are reset to 0 for the DCT algorithms. For the spatial algorithms the prediction is reset to $2^{(P - Pt - 1)}$, where P is the precision and Pt is the shift specified for the point transform.   If arithmetic coding is used, all statistics areas must be re-initialized.   If Huffman coding is used, all run counts must be reset to zero.

7.9 Hierarchical progression marker segments ◑

The JPEG hierarchical mode provides a means to create a progression involving resolution changes between stages of the progression. For closed systems in which encoders and decoders use the same IDCT, hierarchical progressions can also be used to provide a lossy first stage followed by a correction to a lossless output.*

In the JPEG hierarchical mode of operation the progression is achieved by a sequence of frames. The first frame must be preceded by the DHP marker segment. This segment defines the image parameters for the completed progression. The syntax for the DHP segment is exactly the same as for the frame header, except for the substitution of the DHP marker for the SOF_n marker. Although not strictly needed for decoding,

---

Table   7-11. DRI marker segment structure

Parameter	Symbol	Size (bits)
Marker (X'FFDD')	DRI	16
Define restart interval segment length	Lr	16
Restart interval	Ri	16

---

* If images are interchanged with other systems, the differences between IDCT implementations may introduce distortions comparable to the distortion of the IDCT output relative to the original source image.

```
FF D8
FF DB  00 43 00 10 0B 0C 0E 0C 0A 10 0E 0D 0E 12 11 10
       13 18 28 1A 18 16 16 18 31 23 25 1D 28 3A 33 3D
       3C 39 33 38 37 40 48 5C 4E 40 44 57 45 37 38 50
       6D 51 57 5F 62 67 68 67 3E 4D 71 79 70 64 78 5C
       65 67 63
FF C0  00 0B 02 40 02 DB 80 01 01 11 00
FF C4  00 D2 00 00 01 05 01 01 01 01 01 01 00 00 00 00
       00 00 00 00 01 02 03 04 05 06 07 08 09 0A 0B 10
       00 02 01 03 03 02 04 03 05 05 04 04 00 00 01 7D
       01 02 03 00 04 11 05 12 21 31 41 06 13 51 61 07
       22 71 14 32 81 91 A1 08 23 42 B1 C1 15 52 D1 F0
       24 33 62 72 82 09 0A 16 17 18 19 1A 25 26 27 28
       29 2A 34 35 36 37 38 39 3A 43 44 45 46 47 48 49
       4A 53 54 55 56 57 58 59 5A 63 64 65 66 67 68 69
       6A 73 74 75 76 77 78 79 7A 83 84 85 86 87 88 89
       8A 92 93 94 95 96 97 98 99 9A A2 A3 A4 A5 A6 A7
       A8 A9 AA B2 B3 B4 B5 B6 B7 B8 B9 BA C2 C3 C4 C5
       C6 C7 C8 C9 CA D2 D3 D4 D5 D6 D7 D8 D9 DA E1 E2
       E3 E4 E5 E6 E7 E8 E9 EA F1 F2 F3 F4 F5 F6 F7 F8
       F9 FA
FF DA  00 08 01 01 00 00 3F 00 ...
FF D9
```

Figure   7-6. Baseline compressed data stream for Y

this segment allows the decoder to determine, before starting the decoding, the resources necessary to handle the final completed image.

The EXP segment is also needed in hierarchical progressions that change the spatial resolution of the image. This segment must precede any differential hierarchical frame in which the reference components decoded in the preceding frames must be upsampled. The upsampling must be done with the bilinear interpolation filter defined in Annex J of Part 1 of the JPEG DIS. Note that the EXP segment may not be used to signal the upsampling of a reference component that does not exist. Note also that only one EXP segment is allowed before a differential frame.

There are restrictions on how the DCT and lossless frames may be intermixed. Basically, if a hierarchical process uses DCT-based processes, only the final frame for each component may be lossless. Alternatively, if a hierarchical progression uses a lossless process for the first frame, DCT-based processes may not be used.

Note that in differential frames the input data to the process are signed two's complement differences between source and reference samples. Therefore, when the point transform is applied, it is a signed integer divide by 2^{Pt} that cannot be approximated by an arithmetic-shift-right by Pt.

```
FF D8
FF DB  00 84 00 10 0B 0C 0E 0C 0A 10 0E 0D 0E 12 11 10
       13 18 28 1A 18 16 16 18 31 23 25 1D 28 3A 33 3D
       3C 39 33 38 37 40 48 5C 4E 40 44 57 45 37 38 50
       6D 51 57 5F 62 67 68 67 3E 4D 71 79 70 64 78 5C
       65 67 63 01 11 12 12 18 15 18 2F 1A 1A 2F 63 42
       38 42 63 63 63 63 63 63 63 63 63 63 63 63 63 63
       63 63 63 63 63 63 63 63 63 63 63 63 63 63 63 63
       63 63 63 63 63 63 63 63 63 63 63 63 63 63 63 63
       63 63 63 63
FF C0  00 11 08 02 40 02 D0 03 01 21 00 02 11 01 03 11
       01
FF C4  01 A2 00 00 01 05 01 01 01 01 01 01 00 00 00 00
       00 00 00 00 01 02 03 04 05 06 07 08 09 0A 0B 10
       00 02 01 03 03 02 04 03 05 05 04 04 00 00 01 7D
       01 02 03 00 04 11 05 12 21 31 41 06 13 51 61 07
       22 71 14 32 81 91 A1 08 23 42 B1 C1 15 52 D1 F0
       24 33 62 72 82 09 0A 16 17 18 19 1A 25 26 27 28
       29 2A 34 35 36 37 38 39 3A 43 44 45 46 47 48 49
       4A 53 54 55 56 57 58 59 5A 63 64 65 66 67 68 69
       6A 73 74 75 76 77 78 79 7A 83 84 85 86 87 88 89
       8A 92 93 94 95 96 97 98 99 9A A2 A3 A4 A5 A6 A7
       A8 A9 AA B2 B3 B4 B5 B6 B7 B8 B9 BA C2 C3 C4 C5
       C6 C7 C8 C9 CA D2 D3 D4 D5 D6 D7 D8 D9 DA E1 E2
       E3 E4 E5 E6 E7 E8 E9 EA F1 F2 F3 F4 F5 F6 F7 F8
       F9 FA 01 00 03 01 01 01 01 01 01 01 01 01 00 00
       00 00 00 00 01 02 03 04 05 06 07 08 09 0A 0B 11
       00 02 01 02 04 04 03 04 07 05 04 04 00 01 02 77
       00 01 02 03 11 04 05 21 31 06 12 41 51 07 61 71
       13 22 32 81 08 14 42 91 A1 B1 C1 09 23 33 52 F0
       15 62 72 D1 0A 16 24 34 E1 25 F1 17 18 19 1A 26
       27 28 29 2A 35 36 37 38 39 3A 43 44 45 46 47 48
       49 4A 53 54 55 56 57 58 59 5A 63 64 65 66 67 68
       69 6A 73 74 75 76 77 78 79 7A 82 83 84 85 86 87
       88 89 8A 92 93 94 95 96 97 98 99 9A A2 A3 A4 A5
       A6 A7 A8 A9 AA B2 B3 B4 B5 B6 B7 B8 B9 BA C2 C3
       C4 C5 C6 C7 C8 C9 CA D2 D3 D4 D5 D6 D7 D8 D9 DA
       E2 E3 E4 E5 E6 E7 E8 E9 EA F2 F3 F4 F5 F6 F7 F8
       F9 FA
FF DA  00 0C 03 01 00 02 11 03 11 00 3F 00 ...
FF D9
```

Figure   7-7. Baseline compressed data for interleaved YYCbCr

7.10 Examples of JPEG data streams ◑

Figures 7-6 through 7-15 provide examples of JPEG data streams expressed
as hexadecimal bytes. Ellipses (...) indicate where the entropy-coded seg-

```
FFD8
FFDB 00 43 00 08 06 06 07 06 05 08 07 07 07 09 09 08
     0A 0C 14 0D 0C 0B 0B 0C 19 12 13 0F 14 1D 1A 1F
     1E 1D 1A 1C 1C 20 24 2E 27 20 22 2C 23 1C 1C 28
     37 29 2C 30 31 34 34 34 1F 27 39 3D 38 32 3C 2E
     33 34 32
FFC9 00 0B 08 00 00 02 D0 01 01 11 00
FFCC 00 06 00 10 10 05
FFDA 00 08 01 01 00 00 3F 00 ...
FFDC 00 04 02 40
FFD9
```

Figure   7-8. Sequential DCT-based arithmetic-coding compressed data for
          Y

ments have been left out. All markers except RST_m markers are left-
justified. Table 7-4 can be used to look up the SOF_n markers. Table 7-5
gives the hexadecimal codes for the rest of the markers. The first and last
markers in these figures are the SOI (X'FFD8') and EOI (X'FFD9')
markers that begin and end every JPEG compressed data stream.

Figure 7-6 gives an example of a baseline-system compressed image
data stream for a single grayscale component. The DQT (X'FFDB')
marker segment defines the quantization table for the single luminance (Y)
component. This table is the example luminance table given in Annex K
of the JPEG DIS (Part 1). The SOF_0 (X'FFC0') marker segment defines
this as a baseline frame. The DHT (X'FFC4') defines the DC and AC
example luminance Huffman-coding tables given in Annex K of the JPEG

```
FFD8
FFDB 00 84 00 10 0B 0C 0E 0C 0A 10 0E 0D 0E 12 11 10
     13 18 28 1A 18 16 16 18 31 23 25 1D 28 3A 33 3D
     3C 39 33 38 37 40 48 5C 4E 40 44 57 45 37 38 50
     6D 51 57 5F 62 67 68 67 3E 4D 71 79 70 64 78 5C
     65 67 63 01 11 12 12 18 15 18 2F 1A 1A 2F 63 42
     38 42 63 63 63 63 63 63 63 63 63 63 63 63 63 63
     63 63 63 63 63 63 63 63 63 63 63 63 63 63 63 63
     63 63 63 63 63 63 63 63 63 63 63 63 63 63 63 63
     63 63 63 63
FFC9 00 11 08 02 40 02 D0 03 01 21 00 02 11 01 03 11
     01
FFCC 00 0A 00 10 10 05 01 10 11 05
FFDA 00 0C 03 01 00 02 11 03 11 00 3F 00 ...
FFD9
```

Figure   7-9. Sequential DCT-based arithmetic-coding compressed data for
          interleaved YYCbCr

```
FFD8
FFDB   00 84 00 10 0B 0C 0E 0C 0A 10 0E 0D 0E 12 11 10
       13 18 28 1A 18 16 16 18 31 23 25 1D 28 3A 33 3D
       3C 39 33 38 37 40 48 5C 4E 40 44 57 45 37 38 50
       6D 51 57 5F 62 67 68 67 3E 4D 71 79 70 64 78 5C
       65 67 63 01 11 12 12 18 15 18 2F 1A 1A 2F 63 42
       38 42 63 63 63 63 63 63 63 63 63 63 63 63 63 63
       63 63 63 63 63 63 63 63 63 63 63 63 63 63 63 63
       63 63 63 63 63 63 63 63 63 63 63 63 63 63 63 63
       63 63 63 63
FFD9
```

Figure   7-10. Abbreviated format compressed data for table specification

DIS. The SOS (X'FFDA') marker segment would be followed by the entropy-coded data in a real data stream.

Figure 7-7 shows the compressed image data for a color image compressed with the baseline system. In this case the DQT defines both quantization tables given in Annex K of the JPEG DIS. The SOF_0 marker segment defines $Y = 576$, $X = 720$, and the first component as having twice as many horizontal samples as the other two components. These values would be appropriate for the JPEG test images reproduced in this book. The first component uses quantization table 0 and Huffman coding DC and AC tables 0. The other two components use quantization table 1 and Huffman coding DC and AC tables 1. The DHT marker defines all four Huffman tables given in Annex K of the JPEG DIS. The three components are all interleaved in one scan.

Figure 7-8 provides an illustration of compressed image data for a single component compressed with the extended-sequential DCT mode with arithmetic coding. The DAC (X'FFCC') marker segment parameters are actually the default values established by the SOI. The quantization table values are half those given in Figure 7-6. The SOF_9 (X'FFC9') marker parameters could have been identical to those of SOF_0, but instead,

```
FFD8
FFC9   00 11 08 00 80 00 80 03 01 11 00 02 11 00 03 11
       00
FFCC   00 06 00 10 10 05
FFDA   00 08 01 01 00 00 3F 00 ...
FFCC   00 06 01 10 11 05
FFDA   00 08 01 02 11 00 3F 00 ...
FFDA   00 08 01 03 11 00 3F 00 ...
FFD9
```

Figure   7-11. Abbreviated-format sequential arithmetic-coding compressed
         data for non-interleaved YCbCr

```
FFD8
FFCA  00 11 08 00 80 00 80 03 01 11 00 02 11 01 03 11
      01
FFCC  00 06 00 10 10 05
FFDA  00 08 01 01 00 00 00 01 ...
FFDA  00 08 01 01 00 01 02 01 ...
FFDA  00 08 01 01 00 03 05 01 ...
FFDA  00 08 01 01 00 06 09 01 ...
FFDA  00 08 01 01 00 0A 14 01 ...
FFDA  00 08 01 01 00 15 23 01 ...
FFDA  00 08 01 01 00 24 3F 01 ...
FFDA  00 08 01 02 11 00 00 01 ...
FFDA  00 08 01 02 11 01 05 01 ...
FFDA  00 08 01 02 11 06 09 01 ...
FFDA  00 08 01 02 11 0A 3F 01 ...
FFDA  00 08 01 03 11 00 00 01 ...
FFDA  00 08 01 03 11 01 05 01 ...
FFDA  00 08 01 03 11 06 09 01 ...
FFDA  00 08 01 03 11 0A 3F 01 ...
FFDA  00 08 01 01 00 00 00 10 ...
FFDA  00 08 01 01 00 01 14 10 ...
FFDA  00 08 01 01 00 15 3F 10 ...
FFDA  00 08 01 02 11 00 00 10 ...
FFDA  00 08 01 02 11 01 09 10 ...
FFDA  00 08 01 02 11 0A 3F 10 ...
FFDA  00 08 01 03 11 00 00 10 ...
FFDA  00 08 01 03 11 01 09 10 ...
FFDA  00 08 01 03 11 0A 3F 10 ...
FFD9
```

Figure 7-12. Abbreviated-format progressive arithmetic-coding compressed data for non-interleaved YCbCr

Y has been set to zero and the entropy-coded segment is followed by the DNL (X'FFDC') marker segment defining Y=576 (X'0240').

The compressed data stream in Figure 7-9 is similar to that shown in Figure 7-7, except that arithmetic-entropy coding is used in Figure 7-9.

Figure 7-10 is an example of the abbreviated format for table specification. Only quantization tables are present in this data stream. Applications can download tables with this format. For the following figures we will assume that these quantization tables will be inherited.

Figure 7-11 illustrates a compressed data stream describing an image whose three components have equal dimensions, as in a YCbCr formatted image. The three components are coded independently in this non-interleaved sequential arithmetic-coding data stream. Because the quantization tables are missing, this is an example of abbreviated format compressed image data.

```
FFD8
FFFE 00 06 49 42 4D 01
FFDD 00 04 01 00
FFCA 00 11 08 01 48 00 E4 03 01 11 00 02 11 01 03 11 01
FFCC 00 0A 00 10 10 05 01 10 11 05
FFDA 00 0C 03 01 00 02 11 03 11 00 00 02 ...
      FFD0 ... FFD1 ... FFD2 ... FFD3 ...
FFDA 00 08 01 01 00 01 05 02 ...
      FFD0 ... FFD1 ... FFD2 ... FFD3 ...
FFDA 00 08 01 01 00 06 3F 02 ...
      FFD0 ... FFD1 ... FFD2 ... FFD3 ...
FFDA 00 08 01 01 00 00 00 21 ...
      FFD0 ... FFD1 ... FFD2 ... FFD3 ...
FFDA 00 08 01 01 00 01 09 21 ...
      FFD0 ... FFD1 ... FFD2 ... FFD3 ...
FFDA 00 08 01 01 00 0A 3F 21 ...
      FFD0 ... FFD1 ... FFD2 ... FFD3 ...
FFDA 00 08 01 02 11 00 00 21 ...
      FFD0 ... FFD1 ... FFD2 ... FFD3 ...
FFDA 00 08 01 02 11 01 3F 01 ...
      FFD0 ... FFD1 ... FFD2 ... FFD3 ...
FFDA 00 08 01 03 11 00 00 21 ...
      FFD0 ... FFD1 ... FFD2 ... FFD3 ...
FFDA 00 08 01 03 11 01 3F 01 ...
      FFD0 ... FFD1 ... FFD2 ... FFD3 ...
FFDA 00 08 01 01 00 00 00 10 ...
      FFD0 ... FFD1 ... FFD2 ... FFD3 ...
FFDA 00 08 01 01 00 01 09 10 ...
      FFD0 ... FFD1 ... FFD2 ... FFD3 ...
FFDA 00 08 01 01 00 0A 3F 10 ...
      FFD0 ... FFD1 ... FFD2 ... FFD3 ...
FFDA 00 08 01 02 11 00 00 10 ...
      FFD0 ... FFD1 ... FFD2 ... FFD3 ...
FFDA 00 08 01 02 11 01 3F 10 ...
      FFD0 ... FFD1 ... FFD2 ... FFD3 ...
FFDA 00 08 01 03 11 00 00 10 ...
      FFD0 ... FFD1 ... FFD2 ... FFD3 ...
FFDA 00 08 01 03 11 01 3F 10 ...
      FFD0 ... FFD1 ... FFD2 ... FFD3 ...
FFD9
```

Figure   7-13. Abbreviated-format   progressive   arithmetic-coding   compressed data with restarts for interleaved YCbCr

Figure 7-12 contains an example of abbreviated format for a full-progressive arithmetic entropy coder. (The quantization table is not present, and is assumed to be provided to the decoder by some other mechanism, such as inheritance.) The components are compressed with a mix-

```
FFD8
FFC4  00 1F 00 00 01 05 01 01 01 01 01 01 00 00 00 00
      00 00 00 00 01 02 03 04 05 06 07 08 09 0A 0B
FFC3  00 11 08 00 80 00 80 03 01 11 00 02 11 00 03 11
      00
FFDA  00 08 01 01 00 07 00 00 ...
FFDA  00 08 01 02 00 07 00 00 ...
FFDA  00 08 01 03 00 07 00 00 ...
FFD9
```

Figure 7-14. Lossless Huffman-coding compressed data for non-interleaved YCbCr

ture of successive approximation and spectral selection. The DC values of the first component are coded except for the LSB; then, the AC coefficients of the same component are coded at reduced precision (Al=1) in bands of 1–2, 3–5, 6–9, 10–20, 21–35, and 36–63. Components 2 and 3 are then coded in similar fashion except that the final spectral band is from 10–63. Then, for each component, three scans—one for the DC coefficients and two for the AC coefficients—complete the coding of the DC and AC coefficients.

Figure 7-13 gives an arithmetic-coded, progressive DCT-based data stream with restart intervals. Note that the first restart marker follows some entropy-coded data and is always RST_0.

Figure 7-14 gives an example of a lossless data stream for Huffman coding and non-interleaved data. All three components have equal dimensions. Figure 7-15 gives the data stream for lossless coding of the same image with arithmetic coding and interleaved data. Both examples use predictor selector 7. Both are examples of interchange format compressed image data, because all of the tables required for decoding are present in the data stream.

7.11 Backus-Naur Form ●

This section presents a description of the JPEG compressed data stream in the Backus-Naur Form (BNF). It builds upon Thornton and Anderson's

```
FFD8
FFCB  00 11 08 00 80 00 80 03 01 11 00 02 11 00 03 11
      00
FFDA  00 0C 03 01 00 02 11 03 11 07 00 00 ...
FFD9
```

Figure 7-15. Lossless arithmetic-coding compressed data for interleaved YCbCr

application of BNF to the nonhierarchical JPEG syntax.* The BNF provides a means for constructing a syntactically correct parser to interpret the compressed data stream.[36] This description represents our interpretation of the specification; it has not been sanctioned by JPEG.

The JPEG compressed data stream syntax may be thought of as a "language" for describing images. Any valid data stream is a "sentence" in the language. The set of syntactically valid sentences may be described by a grammar. The grammar can then be used to determine whether any given data stream is valid. The definition of a grammar serves two purposes. First, it provides proof that a language is viable. If ambiguities exist in the language, the process of describing its grammar will detect these ambiguities. Second, it provides the means for implementation of the grammar in a parser.

In order to formalize the concept of a grammar, it is necessary to define the production rules that describe the language and its syntax. This formal definition will be stated in Backus-Naur Form.[37] This language description form was first developed by Backus for describing the Algol language. The basis of the language description is a set of productions, or rules for forming complex objects from simpler ones. The simplest objects in the language are called *tokens* or *terminal symbols*, and objects built using the productions are called *nonterminal symbols*. A production is a rule of the form

$$x \ \longleftarrow \ y$$

where x is a nonterminal symbol and y is a sequence of one or more terminal and/or nonterminal symbols.

We can state a grammar $G(X)$ as a non-empty set of productions. X is a nonterminal symbol that is taken as the head of the grammar. Note that symbols that appear as the left part of any production are nonterminals, whereas those that do not are terminals. In order to make the BNF easier to interpret, nonterminal symbols are represented in lower case and are enclosed in brackets < ... >, whereas terminal symbols are represented in upper case. A special symbol, ϵ, denotes the empty string. The symbol | may be used as an "or" on the right side of a production, so that two productions, $x \longleftarrow a$ and $x \longleftarrow b$, can be combined into a single statement, $x \longleftarrow a \,|\, b$. In this case the right side of the production may be taken to be either a or b.

The process by which a sentence is recognized in the grammar is one of substitution. A sentence is built by starting with the head of the grammar and repeatedly substituting the right part of a production for a corresponding nonterminal symbol until only terminal symbols remain.

As an example, consider the grammar G(number):

---

* The authors are indebted to D. S. Thornton and K. L. Anderson for permission to include their work (and some of their words) and for their assistance in extending the BNF to include hierarchical compressed image data.

```
<number>    ←    <no>
<no>        ←    <no> <digit> | <digit>
<digit>     ←    0 | 1 | 2 | 3 | 4 | 5 | 6 | 7 | 8 | 9
```

This says that a <number> is a <no>, which in turn is either a single <digit> or a <no> followed by a <digit>. The terminal symbols are the decimal digits. The language thus consists of all strings of one or more decimal digits. The number 835 can be derived by the following reductions:

```
<number>    ←    <no>
            ←    <no> <digit>
            ←    <no> <digit> <digit>
            ←    <digit> <digit> <digit>
            ←    8 <digit> <digit>
            ←    83 <digit>
            ←    835
```

Given the grammar above we can establish a sentence 835 by showing that 835 reduces to <number>. We say that $v \leftarrow + w$ (i.e., v produces w), if, and only if, there exists a sequence of derivations such that $v \leftarrow u0 \leftarrow u1 \leftarrow u2 \leftarrow \ldots uN \leftarrow w$.

The last element that should be stated is the place for recursion in a grammar. A language can be described using either left or right recursion (or both). A derivation of the form $u \leftarrow + \cdots u$ is said to be right-recursive; a derivation of the form $u \leftarrow + u \ldots$ is left-recursive. A grammar without recursion is finite. A grammar with recursion is infinite and can be expressed clearly.

The following section presents a basic BNF for the JPEG syntax. This grammar is infinite and left-recursive. Its head is <jpeg_data>. The terminal symbols represent one-byte hexadecimal values, represented as BY*nn*, where *nn* is the hexadecimal value of the byte. The notation BY*xx* | ... | BY*yy* indicates any value between X'*xx*' and X'*yy*' (inclusive). Thus, in Figure 7-16, <anyb> represents any byte value, <non00b> indicates any byte value except 0, etc. References in parentheses are provided to the corresponding subclauses in Annex B of the JPEG DIS Part 1 in many cases.

In addition to terminal and nonterminal symbols, this JPEG BNF includes a number of functions denoted by "function()." These functions remove a specified number of bytes from the input stream for processing. The referenced subclauses of the DIS describe the data to be removed.

Some restrictions on the values of elements of the data stream cannot conveniently be reflected in the BNF. For example, many markers are followed by a <len> field indicating the length of the associated data in bytes; this length must be checked against the number of bytes actually supplied. As another example, the number of valid values for <ul> is finite but rather large; it is more efficient to have a parser accept whatever value is encountered and test its validity later, since an invalid value indicates

<anyb>	←	BY00 I ... I BYFF
<non00b>	←	BY01 I ... I BYFF
<nonFFb>	←	BY00 I ... I BYFE
<indexb>	←	BY00 I ... I BY3F
<kxindexb>	←	BY01 I ... I BY3F
<spechi0b>	←	BY00 I BY01 I BY02 I BY03
<spechi1b>	←	BY10 I BY11 I BY12 I BY13
<nsofb>	←	BYC0 I BYC1 I BYC2 I BYC3 I BYC9 I BYCA I BYCB
<dsofb>	←	BYC5 I BYC6 I BYC7 I BYCD I BYCE I BYCF
<rstmb>	←	BYD0 I ... I BYD7 (B.2.1)
<appnb>	←	BYE0 I ... I BYEF
<jpgnb>	←	BYF0 I ... I BYFD
<resb>	←	BY02 I ... I BYBF

Figure   7-16. BNF for byte groupings

an invalid data stream rather than an incorrect parsing of a valid data stream. In cases where such additional testing is appropriate, references to the relevant DIS subclauses are provided.

7.12 JPEG BNF ●

Figure 7-16 shows the terminal bytes for byte groups that are convenient for expressing the BNF. Every nonterminal symbol that ends in a "b" appears in this figure. In addition to the any-byte group (<anyb>) and the non-zero byte group (<non00b>) already described, this figure gives names to the non-hexadecimal-FF byte group (<nonFFb>), the 64 bytes that can be used to index into the zigzag sequence of DCT coefficients (<indexb>), the 63 bytes that are allowed values for Kx (<kxindexb>), the four byte groups used in specifications whose high order 4 bits are B'0000' (binary 0000), (<spechi0b>), or B'0001' (<spechi1b>), and the terminating byte for the nondifferential SOF_n markers (<nsofb>), the differential SOF_n markers (<dsofb>), the RST_m markers (<rstmb>), the APP_n markers (<appnb>), the JPG_n markers (<jpgnb>), and the reserved markers (<resb>) that are not yet defined.

Figure 7-17 defines the BNF for JPEG markers. The markers always have at least one X'FF' byte <prefix> followed by the appropriate terminating byte. <fill> is defined as an arbitrary string of X'FF' bytes, including the empty string (ϵ) of no bytes. It can precede the required prefix X'FF' byte of any marker. Most of the BNF for JPEG markers can be easily understood by comparing Figure 7-17 with Tables 7-4 and 7-5. The SOF_n markers have been grouped into nondifferential (<nsof>) and

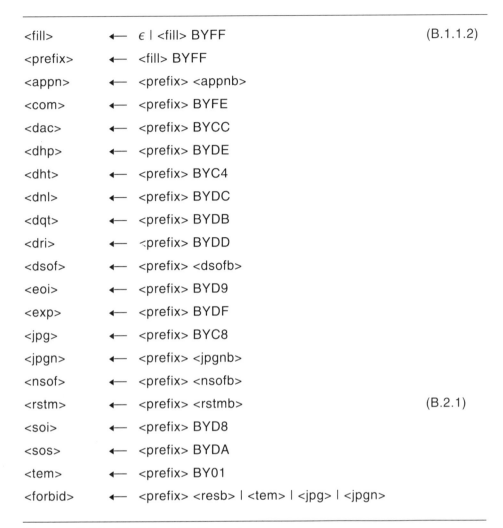

<fill>	←	ϵ | <fill> BYFF	(B.1.1.2)
<prefix>	←	<fill> BYFF	
<appn>	←	<prefix> <appnb>	
<com>	←	<prefix> BYFE	
<dac>	←	<prefix> BYCC	
<dhp>	←	<prefix> BYDE	
<dht>	←	<prefix> BYC4	
<dnl>	←	<prefix> BYDC	
<dqt>	←	<prefix> BYDB	
<dri>	←	<prefix> BYDD	
<dsof>	←	<prefix> <dsofb>	
<eoi>	←	<prefix> BYD9	
<exp>	←	<prefix> BYDF	
<jpg>	←	<prefix> BYC8	
<jpgn>	←	<prefix> <jpgnb>	
<nsof>	←	<prefix> <nsofb>	
<rstm>	←	<prefix> <rstmb>	(B.2.1)
<soi>	←	<prefix> BYD8	
<sos>	←	<prefix> BYDA	
<tem>	←	<prefix> BY01	
<forbid>	←	<prefix> <resb> | <tem> | <jpg> | <jpgn>	

Figure   7-17. BNF for JPEG markers

differential (<dsof>) frame markers. The TEM marker and any fill bytes which accompany it (<tem>) are not allowed in the interchange format. Therefore, the interchange format forbidden markers (<forbid>) are either <tem>, <jpg>, <jpgn>, or strings of X'FF' bytes followed by any of the reserved bytes.

Figure 7-18 gives the BNF for JPEG tables and miscellaneous marker segments. An integer value (<intval>) is defined as any two bytes. The length parameter (<len>) that follows any marker that starts a marker segment is an integer value. Actually, as mentioned earlier, the length parameter must properly describe the marker segment length (given in the tables in Annex B of the JPEG DIS Part 1), but that amount of checking is beyond the level of detail given here. The quantization table precision specifies either byte values <qbyt_spec> or integer values <qint_spec> for the quantization table entries. The actual quantization table (<qt_tbl>) is

<intval>	←	<anyb> <anyb>	
<len>	←	<intval>	(B.1.1.4)
<qbyt_spec>	←	<spechi0b>	
<qint_spec>	←	<spechi1b>	
<qt_tbl>	←	<qbyt_spec> s64chr() I <qint_spec> s64int()	(B.2.4.1)
<dqt_data>	←	<qt_tbl> I <dqt_data> <qt_tbl>	
<dc_spec>	←	<spechi0b>	
<ac_spec>	←	<spechi1b>	
<tabspec>	←	<dc_spec> I <ac_spec>	
<huff_tbl>	←	<tabspec> huffspec()	(B.2.4.2)
<dht_data>	←	<huff_tbl> I <dht_data> <huff_tbl>	
<ul>	←	<anyb>	(B.2.4.3)
<kx>	←	<kxindexb>	(B.2.4.3)
<dac_cond>	←	<dc_spec> <ul> I <ac_spec> <kx>	
<dac_data>	←	<dac_cond> I <dac_data> <dac_cond>	
<rstintv>	←	<intval>	(B.2.4.4)
<tblsmisc>	←	ε	(B.2.4)

<tblsmisc> ← ε   (B.2.4)
            I <tblsmisc> <dqt> <len> <dqt_data>
            I <tblsmisc> <dht> <len> <dht_data>
            I <tblsmisc> <dac> <len> <dac_data>
            I <tblsmisc> <dri> <len> <rstintv>
            I <tblsmisc> <com> <len> comspec()   (B.2.4.5)
            I <tblsmisc> <appn> <len> appspec()   (B.2.4.6)

Figure   7-18. BNF for JPEG tables and miscellaneous marker segments

made up of either the byte specification followed by 64 characters (s64chr()) or the integer specification followed by 64 integers (s64int()). The DQT marker's data (<dqt_data>) is one or more quantization tables. A Huffman table (<huff_tbl>) is specified by having the DC difference table specification (<dc_spec>) or the AC coefficient table specification (<ac_spec>) precede the actual table entries (huffspec()). The DHT marker's data (<dht_data>) is one or more Huffman tables.

The DAC marker's data (<dac_data>) is one or more arithmetic-coding conditioning tables. Each table is either a DC table specification byte followed by the upper U and lower L bounds (<ul>) in a byte or the AC table specification followed by the Kx parameter. The restart interval (<rstintv>) is an integer. The tables and miscellaneous marker segments (<tblsmisc>) consists of the empty string, or any combination of the DQT marker followed by a length and its data, the DHT marker followed by a

\<prec\>	← \<anyb\>	(B.2.2)
\<nlines\>	← \<intval\>	(B.2.2)
\<linlen\>	← \<intval\>	(B.2.2)
\<nfc\>	← \<non00b\>	(B.2.2)
\<fparms\>	← \<len\> \<prec\> \<nlines\> \<linlen\> \<nfc\> fspec()	(B.2.2)
\<nsc\>	← BY01 I BY02 I BY03 I BY04	(B.2.3)
\<ssstart\>	← \<indexb\>	(B.2.3)
\<ssend\>	← \<indexb\>	(B.2.3)
\<sapprox\>	← \<anyb\>	(B.2.3)
\<prog_spec\>	← \<ssstart\> \<ssend\> \<sapprox\>	
\<sparms\>	← \<len\> \<nsc\> sspec() \<prog_spec\>	(B.2.3)
\<ecdata\>	← ϵ I \<ecdata\> \<nonFFb\> I \<ecdata\> \<rstm\> I \<ecdata\> BYFF BY00	(B.1.1.5)
\<scan\>	← \<tblsmisc\> \<sos\> \<sparms\> \<ecdata\>	
\<ecdataf\>	← \<ecdata\> I \<ecdata\> \<dnl\> \<len\> \<nlines\>	(B.2.5)
\<scanf\>	← \<tblsmisc\> \<sos\> \<sparms\> \<ecdataf\>	
\<scanset\>	← \<scanf\> I \<scanset\> \<scan\>	
\<nframe\>	← \<tblsmisc\> \<nsof\> \<fparms\> \<scanset\>	
\<exp_data\>	← BY01 I BY10 I BY11	(B.3.3)
\<dtblsmisc\>	← \<tblsmisc\> I \<tblsmisc\> \<exp\> BY00 BY03 \<exp_data\> \<tblsmisc\>	
\<dframe\>	← \<dtblsmisc\> \<dsof\> \<fparms\> \<scanset\>	

Figure   7-19. BNF for JPEG frames, scans, and entropy-coded data

length and its data, the DAC marker followed by a length and its data, the DRI marker followed by a length and the restart interval value, the COM marker followed by a length and comment bytes (comspec()), and the APP$_n$ marker followed by a length and application bytes (appspec()).

Figure 7-19 gives the BNF for the frames, scans, and entropy-coded data. The frame parameters (\<fparms\>) consist of a length, sample precision (\<prec\>), number of lines Y (\<nlines\>), number of samples per line X (\<linlen\>), the number of frame components (\<nfc\>), and the frame component specification (fspec()).

The number of scan components (\<nsc\>) may only be 1, 2, 3, or 4. The progressive specification (\<prog_spec\>) has a spectral-selection-start index byte (\<ssstart\>), a spectral-selection-end index byte (\<ssend\>), and a byte of successive-approximation parameters (\<sapprox\>) (which doubles

\<xframes\>	⟵	ϵ I \<xframes\> \<dframe\> I \<xframes\> \<nframe\>	
\<fset\>	⟵	\<nframe\> \<xframes\>	
\<jpeg_data\>	⟵	\<soi\> \<nframe\> \<eoi\>	(B.2.1)
		I \<soi\> \<tblsmisc\> \<eoi\>	(B.5)
		I \<soi\> \<tblsmisc\> \<dhp\> \<fparms\> \<fset\> \<eoi\>	(B.3.1)
		I \<soi\> \<forbid\> \<eoi\>	(Undefined)

Figure   7-20. BNF for JPEG compressed data streams

as the predictor specification for lossless coding). The scan parameters (\<sparms\>) consist of a length, \<nsc\>, the scan component specification (sspec()), and progressive specification.

The entropy-coded data (\<ecdata\>) can be mixtures of non-X'FF' bytes, or X'FF' bytes followed by X'00' bytes, or restart markers, or even the empty string. A scan (\<scan\>) consists of the tables and miscellaneous markers, the SOS marker, scan parameters, and the entropy-coded data. The entropy-coded data for the first scan (\<ecdataf\>) may end with the DNL marker, its length, and the new number of lines in the image (\<nlines\>). The first scan (\<scanf\>) looks like the rest of the scans except that the entropy-coded data is allowed to be terminated by the DNL marker segment. A set of scans (\<scanset\>) is one or more scans in which only the first is allowed to terminate with the DNL marker segment. A nondifferential frame (\<nframe\>) may have tables and miscellaneous markers preceding the nondifferential SOF$_n$ marker which is followed by the frame parameters. The set of scans (\<scanset\>) completes a frame.

The tables and miscellaneous marker segments that can preceded a differential frame (\<dtblsmisc\>) in hierarchical mode include the EXP marker (\<exp\>) with its length and data (\<exp_data\>), as well as the marker segments allowed for nondifferential frames. The differential frame (\<dframe\>) has the same structure as a nondifferential frame.

Figure 7-20 shows the BNF for JPEG compressed data streams. Extra frames (\<xframes\>) consists of zero or more differential and nondifferential frames. A frame set (\<fset\>) always starts with at least one nondifferential frame. That is the only frame in nonhierarchical compressed image data.

JPEG compressed data (\<jpeg_data\>) consists of the SOI (\<soi\>) and EOI (\<eoi\>) markers surrounding one of three things: nonhierarchical compressed image data with a single nondifferential frame, abbreviated format compressed data with just tables and miscellaneous markers segments, or hierarchical compressed image data with multiple frames. The hierarchical data will always have a DHP marker (\<dhp\>) followed by the same parameters as would be used to describe the final image coded sequentially. This DHP marker segment precedes the first frame. The forbidden markers are undefined.

8

ENTROPY CODING
CONCEPTS

8.1 Entropy and information ○

The concept of entropy is derived from classical 19th century thermodynamics, in which it was found that this particular variable, entropy, always increased as an isolated system evolved toward equilibrium. As statistical mechanics was developed to explain the behavior of molecular systems, it was natural to link the results to earlier thermodynamic measures. Entropy turned out to be a particularly interesting thermodynamic parameter, for it was found to be a measure of the degree of "disorder" of the molecular system.

There are close parallels between entropy as used by physicists to describe disorder, and entropy as a measure of information. Both effectively give an idea of the degree of surprise—the degree to which things are unpredictable and unexpected. If a symbol occurs that is very improbable, we are surprised—and we would therefore expect that the information transferred in coding this symbol would be large. Conversely, if the symbol is very probable, we are not at all surprised and have learned very little: we already expected that symbol. Note that our degree of surprise has nothing to do with the meaning or significance of the information, only with its probability.

Table 8-1. Example of fixed code for four symbols

Weather	Binary code
Clear	00
Cloudy	01
Rainy	10
Snowy	11

This qualitative concept—the degree of surprise—is formally expressed by the following relationship. The amount of information I transferred in coding a symbol of probability p is given by:[38] *

$$I = \log_2(1/p) \tag{8-1}$$

Because $\log_2$ is the base-2 logarithm, we express the information I in bits. This measure of information has the right properties. Probabilities are numbers between 0 and 1. If the probability is one, the logarithm of $1/p$ is zero. If the probability becomes very small, $\log_2(1/p)$ becomes large. Therefore, symbols that are very probable convey a small amount of information, whereas symbols that are improbable convey a large amount of information.

This concept of information is extremely important in data compression. The number of bits of information for a symbol is equal to the ideal code length—the number of bits in the optimum code for the symbol. Consequently, the information for the symbols in our alphabet tells us how to design our code book.

The entropy H is the average information per symbol—that is, the average code length per symbol. It is the sum of the information for each symbol, s, weighted by the probability $p(s)$ of that symbol:

$$H = \sum_s p(s) \log_2(1/p(s)) \tag{8-2}$$

---

* When we calculate the information for a symbol we need to know the probability of that symbol. Generally, probabilities are measured by counting symbols, but the rules for doing this are not very rigid. For example, the probabilities might be measured for a huge group of symbols in order to derive typical values suitable for that whole group. Conversely, the probabilities might be derived from small groups of symbols in order to allow the probabilities to adapt to changing conditions. In both cases the probabilities or the code tables derived from the probabilities must be known to both the encoder and decoder if we are to use them for data compression.

The entropy is also extremely useful for another purpose. In "stationary" systems—systems where the probabilities are fixed—it provides a fundamental lower bound for the compression that can be achieved with a given alphabet of symbols. The entropy is therefore a very convenient measure of the performance of a coding system.

8.2 An example to illustrate entropy coding ○

Without trying to formally prove the mathematical relationships describing information and entropy, we shall explore their meaning using the following example.

Suppose we need to transmit weather information from remote sensing stations. For simplicity, we assume that there are four possible weather conditions: clear, cloudy, rainy and snowy. Given four possible conditions—four symbols in our alphabet—we need at most a two-bit number to express the current weather. If we simply use these two-bit numbers to represent our symbols, we have (by definition) a "code table" or "code book" with codes of "length" 2 as shown in Table 8-1. This is an example of fixed-length codes. Note that when four symbols are equally likely, their probabilities are each 0.25 and the information is 2 bits/symbol. Therefore, adopting a fixed-length code is equivalent to assuming a uniform probability for all symbols.

8.3 Variable-length code words ○

The objective is to transmit these weather samples at periodic intervals as efficiently as possible. Once we have defined our alphabet of symbols, we can improve the efficiency of transmission by using shorter code words for the more probable symbols and longer code words for the less probable symbols—that is, we can use variable-length code words. JPEG defines two entropy-coding techniques that do this, Huffman coding and arithmetic coding. Huffman coding is an optimal way of coding with integer-length code words; arithmetic coding is an optimal coding procedure that is not constrained to integer-length codes.

Suppose that over a long period of time observers find that, on the average, the remote weather stations in our example report clear weather 1/2 of the time, cloudy weather 1/4 of the time, rainy weather 1/8 of the time, and snowy weather the remaining 1/8 of the time. Table 8-2 shows the symbols, probabilities, and associated information gained when weather conditions are transmitted: if codes are assigned with lengths exactly matching the information for the symbol, the weather can be transmitted at an average of 1.75 bits/symbol. That is, the entropy is:

$$H = (1/2) \times 1 + (1/4) \times 2 + (1/8) \times 3 + (1/8) \times 3 = 1.75 \text{ bits/symbol}$$

The table column labelled "Integer code" gives a possible set of code assignments that is actually a Huffman code table. (In the next section we shall discuss how this table is generated.)

These codes have interesting properties. First, the code lengths are variable and matched to the information for the symbol. Second, the codes

Table   8-2. Weather symbols with variable-length (Huffman) codes

Weather	Probability	Information (ideal length)	Integer code
Clear	1/2	1 bit	0
Cloudy	1/4	2 bits	10
Rainy	1/8	3 bits	110
Snowy	1/8	3 bits	111

are unique—even though the lengths are not all the same, there is no possible way, given a known starting point, of decoding the wrong symbol.

Note that the probabilities for all the symbols in the alphabet sum to one. Note also that unless the probabilities are integer powers of two (our example is selected to have this property) the ideal code length will not be an integer. This presents us with a problem: either we have to select an optimum set of integer-length codes (Huffman coding), or we have to use a coding technique that allows non-integer code lengths (arithmetic coding).

8.3.1 Huffman coding ○

If a set of integer-length codes is to be selected, the assignment of codes is usually done by a procedure called Huffman coding.[35]   This procedure produces a "compact" code. For a particular set of symbols and probabilities, no other integer code can be found that will give better coding performance than this compact code.

Consider the situation in Table 8-3. The entropy—the average ideal code length required to transmit the weather—is given by

$$H = (3/4) \times 0.415 + (1/8) \times 3 + (1/16) \times 4 + (1/16) \times 4 = 1.186 \text{ bits/symbol}$$

However, fractional-bit lengths are not allowed, so the lengths of the codes listed in the column to the right do not match the ideal information. Since an integer code always needs at least one bit, increasing the code for

Table   8-3. Weather symbols with ideal and integer code lengths

Weather	Probability	Information (ideal length)	Integer code
Clear	3/4	0.415 bits	0
Cloudy	1/8	3.000 bits	10
Rainy	1/16	4.000 bits	110
Snowy	1/16	4.000 bits	111

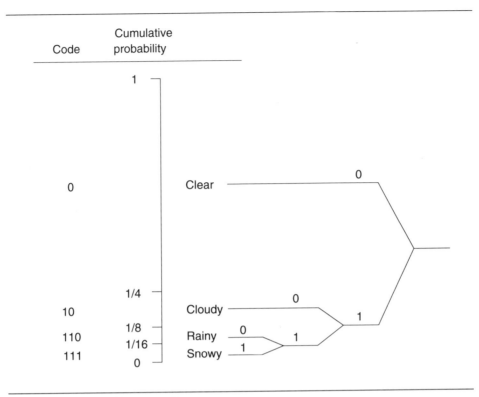

Figure   8-1. Huffman coding tree

"clear" to one bit seems logical. Why then are the other codes one bit shorter than the ideal length?

In general, when one code word must be lengthened, other code words can often be shortened. All that is necessary is that the codes be unique and compact; in this case the most compact code table consists of a one-bit code, a two-bit code and two three-bit codes. Huffman coding provides us (and JPEG) with a procedure for deriving these compact integer codes.

The Huffman code assignment procedure is based on a coding "tree" structure. This tree is developed by a sequence of pairing operations in which the two least probable symbols are joined at a "node" to form two "branches" of the tree. As the tree is constructed, each node at which two branches meet is treated as a single symbol with a combined probability that is the sum of the probabilities for all symbols combined at that node.

Figure 8-1 shows a Huffman code pairing sequence for the four-symbol case in Table 8-3. In this figure the four symbols are placed on the number line from 0 to 1 in order of increasing probability. The cumulative sum of the symbol probabilities is shown at the left. The two smallest probability intervals are paired, leaving three probability intervals of size 1/8, 1/8, and 3/4. We establish the next branch in the tree by again pairing the two smallest probability intervals, 1/8 and 1/8, leaving two probability intervals, 1/4 and 3/4. Finally, we complete the tree by pairing the 1/4 and 3/4 intervals.

To create the code word for each symbol, we assign a 0 and 1, respectively (the order is arbitrary), to each branch of the tree. We then concatenate the bits assigned to these branches, starting at the "root" (at the right of the tree) and following the branches back to the "leaf" for each symbol (at the far left). Note that each node in this tree requires a binary decision—a choice between the two possibilities—and therefore appends one bit to the code word.

One of the problems with Huffman coding is that symbols with probabilities greater than 0.5 still require a code word of length one. This leads to less efficient coding, as can be seen for the codes in Table 8-3. The coding rate R achieved with Huffman codes in this case is as follows:

$$R = (3/4) \times 1 + (1/8) \times 2 + (1/16) \times 3 + (1/16) \times 3 = 1.375 \text{ bits/symbol}$$

This rate, when compared to the entropy bound of 1.186 bits/pixel, represents an efficiency of 86%. If the inefficiency gets too large, the statistical model can be modified so that more efficient codes become possible. We shall address this point in section 8.4.1.3.

8.3.2 Arithmetic coding ○

Arithmetic coding is another method of coding that approaches the entropy limit. It is conceptually different from Huffman coding and does not require integer-length codes.

8.3.2.1 Basic principles of arithmetic coding ○

In arithmetic coding the symbols are ordered on the number line in the probability interval from 0 to 1 in a sequence that is known to both encoder and decoder. Each symbol is assigned a subinterval equal to its probability. Note that since the symbol probabilities sum to one, the subintervals precisely fill the interval from 0 to 1. Figure 8-2 illustrates a possible ordering for the symbol probabilities in Table 8-3.

The objective in arithmetic coding is to create a code stream that is a binary fraction pointing to the interval for the symbol being coded. Thus, if the symbol is "clear," the code stream is a binary fraction greater than or equal to binary 0.01 (decimal 0.25), but less than binary 1.0. If the symbol is "cloudy," the code stream is greater than or equal to binary 0.001, but less than binary 0.010. If the symbol is "rainy," the code stream is greater than or equal to binary 0.0001, but less than binary 0.0010. Finally, if the symbol is "snowy," the code stream is greater than or equal to binary 0.0000, but less than binary 0.0001. If the code stream follows these rules, a decoder can see which subinterval is pointed to by the code stream and decode the appropriate symbol.

Coding additional symbols is a matter of subdividing the probability interval into smaller and smaller subintervals, always in proportion to the probability of the particular symbol sequence. As long as we follow the rules and never allow the code stream to point outside the subinterval assigned to the sequence of symbols, the decoder will decode that sequence.

Note that the boundary between two intervals is always assigned to one of the intervals. Here we have chosen to assign it to the upper symbol.

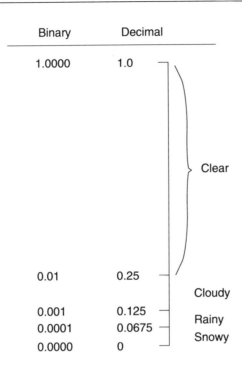

Figure   8-2. Partitioning the number line into subintervals

However, the opposite convention is also possible and leads to a very useful alternative implementation of arithmetic coding.

After we have coded many symbols, the interval P is the product of the probabilities of all the symbols coded, and the interval precision, the number of bits required to express an interval of that size, is given approximately by $-\log_2(P)$. Therefore, since

$$P = p_1 \times p_2 \times p_3 \times \cdots \times p_N \qquad [8-3]$$

the number of bits of precision is approximately:

$$-\log_2(P) = -(\log_2(p_1) + \log_2(p_2) + \log_2(p_3) + \cdots + \log_2(p_N)) \qquad [8-4]$$

Therefore, we conclude that the code stream length will be very nearly equal to the information for the individual symbol probabilities, and the average number of bits/symbol will be very close to the bound computed from the entropy. Although the code stream precision must be a few bits greater than $-\log_2(P)$ in order to guarantee that the final interval is correctly decoded, this slight inefficiency is negligible compared to the total length of the code stream.*

If we had to compute all possible interval subdivisions for a long sequence of symbols, the computational task would grow exponentially. Fortunately, however, the subdivision can be done recursively such that only one interval is further subdivided for each symbol coded. This recursive form is known as Elias coding.[39]

8.3.2.2 *Binary arithmetic coding* ○

The form of arithmetic coding adopted by JPEG (and JBIG) is restricted to binary coding decisions. Binary arithmetic coding works exactly as in the preceding section, except that the alphabet contains only two symbols. Note that a binary coder would not be practical if it could not efficiently code symbols with probabilities greater than 0.5.

Coding models usually have alphabets that have more than two symbols. Restricting the alphabet to only two symbols does not affect the coding efficiency, but it does require a translation from the multi-symbol alphabet to a sequence of binary decisions. Such translations take the form of a binary decision tree that is exemplified by the Huffman coding tree shown in Figure 8-1. Of course, pairing of symbols according to their probabilities is no longer needed. With an appropriate set of probabilities any properly constructed binary decision tree will give essentially the same coding efficiency.

Binary arithmetic coding was chosen for two reasons. First, the binary alphabet allows one to make some simple approximations in the interval scaling that eliminate the need for multiplication. Second, a very simple probability estimation technique has been developed for binary arithmetic coders that allows the coders to adapt to changing probabilities.

8.3.2.3 *Additional details about arithmetic coding* ○

In the later chapters of this book, we shall go into much more detail about the JPEG binary arithmetic coder. Several other features are added to this arithmetic coder that make it practical and enhance its effectiveness:

1. Conventions are chosen for symbol ordering and for how the code stream is positioned in the interval.

2. An approximation is developed for the multiplications needed for the interval scaling.

3. The probability interval and code stream are renormalized periodically by doubling them. This is done to keep the interval close to one, thereby making it possible to use fixed-precision integer arithmetic in the coding process. Renormalization is the mechanism by which bits are produced by the arithmetic coder.

---

* With the right terminating conventions for the code stream one can guarantee that only two additional bits of precision are needed relative to the precision of the final interval. In practice, terminating conventions may take a few more bits.

4. A mechanism is provided for resolving carry-over as the code stream is modified to keep it pointing at the correct interval.

5. A probability estimation procedure is incorporated that makes the coder adaptive. This estimation procedure uses the renormalization in the arithmetic coder to approximately count the symbol occurrences and thereby dynamically adjust the probability estimates used in the coding.

6. Coding conventions are defined that reserve code space for unique "markers" (this is true for Huffman coding as well). These markers can be located in the compressed data without decoding the data, and are useful for a variety of data management tasks.

8.3.3 Expansion ○

One of the dangers of variable-length coding is that it can lead to expansion rather than compression. Consider the variable-length-coding example of Table 8-2 again. If it rains for a week at a particular site, the codes sent for that week would all be three bits; during that week the variable-length coding would generate more bits than fixed codes would require. In this case the expansion is temporary, and presumably would be compensated by even better coding efficiency during dry spells. Data compression systems must be able to tolerate this kind of local expansion. It is an intrinsic property of systems that use variable-length codes.

Suppose that the remote sensing site was located in a rain forest. Then, three-bit codes would be used almost all of the time. In this second case the expansion is caused by a mismatch between the actual probabilities experienced by the particular site and the probabilities assumed for the code assignments, and the only cure is to use a different code book. This illustrates the value of adaptive coding.

8.4 Statistical modeling ○

Entropy coders use a "statistical model" to convert descriptors into symbols and assign probabilities (arithmetic coding) or code words (Huffman coding) to the symbols. The statistical model is an important part of the coding system, as it determines the symbol alphabet, and that in turn determines the entropy bound for the coding rate.

If symbol probabilities are stationary (i.e., they do not change), the entropy for a given set of symbols provides a theoretical lower bound for the achievable compression with that alphabet. However, a number of factors can be exploited to improve either the coding efficiency relative to that bound or the entropy bound itself.

8.4.1.1 Improving the entropy by changing the alphabet ○

In the earlier sections of this chapter we considered several variations of a coding problem in which one of four weather conditions—clear, cloudy, rainy, or snowy—was transmitted periodically from a remote station.

Suppose that our diligent observers notice a correlation between successive weather reports. They find that given a particular weather condi-

Table   8-4. New weather symbols with ideal and integer code lengths

Weather	Probability	Information (ideal length)	Integer code
Same	3/4	0.415 bits	0
Change to clear	1/16	4.000 bits	100
Change to cloudy	1/16	4.000 bits	101
Change to rainy	1/16	4.000 bits	110
Change to snowy	1/16	4.000 bits	111

tion, the weather has a 75% chance of being the same at the next measurement. They also find that when change occurs, the alternative weather conditions are equally likely. We then might adopt a different set of symbols—symbols that took advantage of this correlation as shown in Table 8-4. The entropy is given by

$$H = (3/4) \times 0.415 + (1/16) \times 4 + (1/16) \times 4 + (1/16) \times 4 + (1/16) \times 4$$
$$= 1.31 \text{ bits/symbol}$$

Note that the ideal code length for "same" is not an integer. In entropy coders that require integer code words the 0.415-bit length must be increased to one bit. As we have already seen, when one code word must be lengthened, other code words can often be shortened. In this case the other four choices can be coded with three-bit instead of four-bit codes. When we calculate the actual bits per symbol we get:

$$(3/4) \times 1 + (1/16) \times 3 + (1/16) \times 3 + (1/16) \times 3 + (1/16) \times 3 = 1.5 \text{ bits/symbol}$$

8.4.1.2 Improving the entropy by removing redundancy ○

The reader may have already noticed one problem with the previous example. For each current weather condition there is one "change to" code that will never be used. Clearly, if it is rainy, then the weather cannot change to rainy. Therefore, we should not have a code word for that situation. Rather, we should have three "change x" code words that signal a change to a new weather condition, but that have different meanings depending on the weather in effect at the last measurement. The system is more complicated, but the principles above still apply. Assuming that the "change x" symbols are all equally likely when a change occurs, our coding system is shown in Table 8-5. The entropy is then given by

$$H = (3/4) \times 0.415 + (1/12) \times 3.585 + (1/12) \times 3.585 + (1/12) \times 3.585$$
$$= 1.208 \text{ bits/symbol}$$

Table   8-5. Weather symbols with non-integer ideal and integer code lengths

Weather	Probability	Information (ideal length)	Integer code
Same	3/4	0.415 bits	0
Change 1	1/12	3.585 bits	10
Change 2	1/12	3.585 bits	110
Change 3	1/12	3.585 bits	111

Again, the code length for "same" has to be increased to one bit, but the three remaining symbols can be uniquely coded with one two-bit code and two three-bit codes. When we calculate the actual bits per symbol we get:

$$(3/4) \times 1 + (1/12) \times 2 + (1/12) \times 3 + (1/12) \times 3 = 1.417 \text{ bits/symbol}$$

Both the entropy and our actual coding rate improve when we remove the unnecessary symbol.

8.4.1.3 Improving the coding efficiency by grouping symbols ○

In the examples we have considered so far, whenever a symbol probability is greater than 0.5, we have increased the code length to one bit. A one-bit code word is, of course, the shortest code word we can assign. There is, however, a better way to handle this problem.

If we collect symbols with probabilities greater than 0.5 into combinations that better fit the constraint of integer lengths, we can significantly improve the coding efficiency and get closer to the ideal coding rate. The model we use to illustrate this uses a code for two measurements of the "same" weather in succession. A complete alphabet for all possible combinations can then be constructed by adding three codes for "same" followed by a change to each of the three other types of weather and three codes for an immediate change to the other three types of weather. Note that some of the symbols in this alphabet code more than one weather symbol as shown in Table 8-6.

Assuming independent probabilities, the probability of having two measurements of the same weather in succession is the product of the individual probabilities for two "same" measurements. The entropy would then be given by

$$\begin{aligned} H &= (9/16) \times 0.830 + (1/16) \times 4 + (1/16) \times 4 + (1/16) \times 4 \\ &\quad + (1/12) \times 3.585 + (1/12) \times 3.585 + (1/12) \times 3.585 \\ &= 2.113 \text{ bits/composite symbol} \end{aligned}$$

This seems to be a step backward, in that our entropy has increased! However, we now are coding more than one weather symbol with each of these new symbols.

Table   8-6. Conditional coding of weather symbols

Composite symbol	Probability	Information (ideal length)	Symbols	Integer code
Same-same	(3/4)×(3/4) = 9/16	0.830 bits	2	1
Same-change 1	(3/4)×(1/12) = 1/16	4.000 bits	2	0111
Same-change 2	(3/4)×(1/12) = 1/16	4.000 bits	2	0110
Same-change 3	(3/4)×(1/12) = 1/16	4.000 bits	2	0101
Change 1	(1/12) = 1/12	3.585 bits	1	0100
Change 2	(1/12) = 1/12	3.585 bits	1	001
Change 3	(1/12) = 1/12	3.585 bits	1	000

To compare with our previous codes we need to know how many weather symbols are coded on the average for each composite symbol. We calculate this from the product of the number of weather symbols in each composite symbol and the probability that that composite symbol will be coded. This gives us 1.208 bits/weather symbol, which is the same as in section 8.4.1.2.

If the entropy is unchanged by the use of these composite symbols, why then should we bother with this more complex model? The answer becomes obvious when we assign integer code words, as shown in the table. When we calculate the actual bits per symbol we get:

$$(9/16) \times 1 + (1/16) \times 4 + (1/16) \times 4 + (1/16) \times 4 + (1/12) \times 4$$
$$+ (1/12) \times 3 + (1/12) \times 3 = 2.145 \text{ bits/composite symbol}$$

Our coding efficiency improves dramatically with this scheme. Our new code is 98.5% efficient and gives a coding rate of 1.226 bits/weather symbol. By grouping some of the symbols into pairs we are able to improve our compression performance dramatically. In Huffman coding this principle is often used as part of "run-length" coding, in which runs of symbols are coded with a single symbol.

The entropy is the best coding rate we can achieve for a given alphabet of symbols and probabilities. If, as we hypothesized here, grouping of symbols to create a new alphabet does not change the entropy, our future improvement is limited to another 1.5%. We need to look for ways to change the probabilities or find a better alphabet of symbols to describe the weather.

8.4.1.4 Improving the entropy with conditional probabilities ○

So far, all of our examples have used alphabets of independent symbols and as such, are examples of "memoryless sources." In this section we shall consider conditional probabilities—probabilities that are a function of the state of symbols already coded. These conditional systems are sometimes called Markov sources.

In different parts of the world and at different seasons of the year, the most frequent types of weather change. Instead of collecting counts of independent symbols, we can collect symbol counts that are dependent on conditions known to both encoder and decoder. From these counts we can estimate "conditional probabilities." For example, the probability of rain may be very dependent on whether the current season is the rainy or the dry season. If we knew how the probabilities varied as a function of the season, we could adjust our codes accordingly.

As a more subtle example, the probability of "same" might be higher if the weather sent in the previous transmission was also the "same." If so, grouping of symbols into pairs and perhaps longer runs would decrease the entropy.

In general, when conditional probabilities are incorporated into the statistical model, the entropy of the system is decreased. Some Huffman coding systems switch code words or even entire code tables, based on past history. The JPEG arithmetic coder explicitly uses conditional probabilities for most of the important binary coding decisions, and this contributes significantly to its compression performance.

In the JPEG arithmetic coder the past events and coding decisions that influence the probability are termed the "context" of the decision, a term that is used rigorously in arithmetic coders to define an index for selection of the probability estimates. The "context index" selects a particular estimate from a set of estimates maintained in a set of storage locations that we call a "statistics area." We call the individual locations in this area "statistics bins."

8.5 Adaptive coding ○

We can improve our compression still more by adapting our probabilities to different regions, seasonal variations, and even local cyclical variations in the weather. As probabilities change, our entropy bound changes and we would like our coder to adapt to these changes. This ability is termed adaptive coding.

Although adaptive coding is not normally regarded as a part of the statistical model, it can be a significant factor in improving compression performance. Both of the JPEG entropy coders allow for adaptive coding. The Huffman coder can use custom Huffman tables for each image; the arithmetic coder dynamically estimates probabilities as it codes and therefore is intrinsically adaptive, even within an image.

9

JPEG BINARY
ARITHMETIC CODING

This chapter describes the binary arithmetic-coding procedures that are used by both JPEG and JBIG. Our treatment here supplements the sections in the JPEG DIS (Annex D), and readers may find it instructive to refer occasionally to the DIS as they read this chapter. We assume the reader has read Chapter 8, and is already familiar with the basic concept of recursive interval subdivision, which is at the heart of arithmetic coding.

The JPEG/JBIG arithmetic-coding algorithm is also known as the QM-coder, a name that was coined to describe the coder's technical ancestry. The QM-coder is a lineal descendent of the Q-coder,[40-44] but is significantly enhanced by improvements in the interval subdivision[45] and probability estimation.[46]

The QM-coder also derives many technical elements from arithmetic coders such as the skew coder,[47] other early work by Langdon and Rissanen,[48] a paper by Jones,[49] and the first papers on arithmetic coding—by Rissanen,[50] Pasco,[51] Rubin,[52] and Rissanen and Langdon.[53] A discussion of this early work can be found in Langdon's tutorial on arithmetic coding.[54]

The QM-coder is the result of an effort by JPEG and JBIG to combine the best features of these various arithmetic coders. It uses the basic coding conventions and interval subdivision of the skew coder and Q-coder and the renormalization-driven probability estimation of the Q-coder. However, the approximation to the multiplication used for interval subdivision in the Q-coder is enhanced by the incorporation of conditional exchange,[45] and the state machine for the probability estimation is improved by the incorporation of rapid initial learning based on Bayesian estimation principles.[46]*

The QM-coder resolves the carry-over in the encoder in a manner similar to the technique of Jones[49] (see also Langdon[58]), thereby avoiding the need for the more complex bit-stuffing technique used in the Q-coder and skew coder.† The coding conventions and symbol ordering in the QM-coder are defined for optimal software performance, but, as with the Q-coder,[41] alternative coding conventions and inverted symbol ordering are easily implemented.

The QM-coder is a binary arithmetic coder, which means that there are only two symbols, 1 and 0, to code. When a descriptor can have many values, it is decomposed into a tree of binary decisions and each binary decision is assigned a particular context-index S, as described in Chapter 12. In that chapter we also list a number of changes that must be made to the idealized version of arithmetic coding discussed there to make it a practical coding procedure. These are:

1. Conventions for symbol ordering and positioning of the code stream in the interval must be defined.

2. An approximation is defined for the multiplication that is needed to scale the probability interval. To mitigate some of the impact on coding efficiency caused by this approximation, a symbol assignment interchange known as conditional exchange is defined.

3. A renormalization of the probability interval is defined so that coding can be done with fixed-precision integer arithmetic.

4. An adaptive probability estimation technique is integrated into the coding procedure.

5. A method for resolving carry-over in the encoder is defined that limits carry propagation when the code stream position in the interval is modified.

6. Zero bytes are stuffed after each X'FF' in the code stream in order to avoid accidental creation of markers.

---

* The concept of conditional exchange is related to the "over-half processing" of MELCODE,[55] a coding system that was originally a form of Golomb coding,[56] and that was extended later to a form of arithmetic coding. Bayesian estimation in arithmetic coders was first used by Chamzas and Duttweiler in their "Minimax" arithmetic coder, a variant of the Cleary, Witten, and Neal arithmetic coder.[57]

† The use of this method for resolving carry-over in the QM-coder was suggested by Chamzas.[59]

7. A termination procedure is defined that guarantees that the marker at the end of the entropy-coded segment is intercepted and interpreted before the decoding of the segment is complete.[60]

These topics will be discussed in detail in this chapter as we develop the flowcharts of the arithmetic-coding procedures. Although we cover the procedures for probability estimation here, we leave a thorough discussion of the principles of this estimation technique for Chapter 14. We also defer some of the practical details about arithmetic-coding implementations to Chapter 13.

9.1 The QM-encoder ◖

As described in Chapter 12, the QM-encoder uses four procedures, Initenc, Code_0(S), Code_1(S) and Flush. These four procedures initialize the encoder, code a 0 decision, code a 1 decision and empty the encoder register at the end of the entropy-coded segment. In this section we shall first discuss the principles and coding conventions for the Code_0 and Code_1 procedures; then we shall discuss the initialization and termination procedures.

9.1.1 Symbol ordering and code stream conventions ◖

When the QM-coder codes a 1 or 0 decision, it does not directly assign intervals to these symbols. Rather, it assigns intervals to the more probable symbol (MPS) and less probable symbol (LPS), and orders the symbols on the number line such that the LPS subinterval is above the MPS subinterval. If the interval is A and the LPS probability estimate is Qe, the MPS probability estimate should ideally be $(1 - Qe)$. The respective subintervals are then $A \times Qe$ and $A(1 - Qe)$. This ideal subdivision and symbol ordering are shown in Figure 9-1.

Ideally, the code stream C needs only to point somewhere within the current interval, but for simplicity the QM-coder uses the convention that the code stream points to the bottom of the current interval. Then, we need only to add to the code stream when our coding decision requires us to select the upper (LPS) subinterval. Note that if the bottom of the interval is included within the interval, the top must be excluded.

If we follow this ideal scheme, coding a symbol changes the interval and code stream as follows:

After MPS:

C is unchanged
$$A = A(1 - Qe) = A - A \times Qe$$

After LPS:

$$C = C + A(1 - Qe) = C + A - A \times Qe$$
$$A = A \times Qe$$

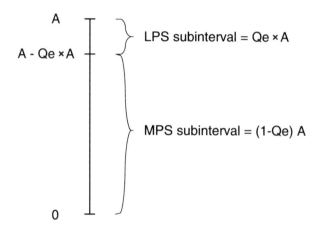

Figure   9-1. Illustration of symbol ordering and ideal interval subdivision. The LPS probability is Qe, and therefore the MPS probability is $(1 - Qe)$.

Note that the procedures for subdividing the interval and moving the code stream waste no code space (the MPS and LPS probabilities sum to 1); note also that the definition of the code stream convention allows no ambiguity about which interval the code stream points to.

9.1.2 Renormalization ◑

There are two problems with this ideal procedure as sketched above. First, we need a potentially unbounded precision for A, and second, we need a multiplication, $A \times Qe$, in the interval subdivision—a potentially costly operation in either hardware or software. These two problems are resolved by periodic renormalization of A, and a matching renormalization of C. The renormalization is done by doubling A each time it drops below some convenient minimum value, which we shall shortly require to be 0.75.[61] Each time A is doubled, C is also doubled in order to maintain its identity as a pointer to the subinterval.

The logic behind this renormalization is illustrated by the following example. Suppose we start our interval at 1.0 (the interval assigned to all possible sequences of symbols) and we code a sequence of LPS of probability 0.5. Then, we see the following behavior for A and C.

No. symbols coded	Decimal notation ($Qe = 0.5$)		Binary notation ($Qe = 0.1$)	
	A	C	A	C
0	1.0	0	1.0	0
1	0.5→1	0.5→1	0.1→1	0.1→1
2	0.5→1	1.5→3	0.1→1	1.1→11
3	0.5→1	3.5→7	0.1→1	11.1→111
4	0.5→1	7.5→15	0.1→1	111.1→1111

With each renormalization (indicated by →) we add a bit to the integer part of the code stream, leaving the fractional part for the coding of future symbols. Note that if we were to code an MPS, we would not add 0.5 to C, and would therefore get a zero bit after renormalization.

9.1.3 Approximating the multiplication ◑

The minimum value of 0.75 for the interval is motivated by the need to replace the multiplication by a simple approximation that requires A to be of order 1, i.e., in the range $1.5 > A \geq 0.75$. Then, $A \times Qe \approx Qe$ and we can use:

After MPS:

C is unchanged
$A = A(1 - Qe) \approx A - Qe$

After LPS:

$C = C + A(1 - Qe) \approx C + A - Qe$
$A \approx Qe$

With this approximation the multiplications are replaced by simple addition and subtraction.

9.1.4 Integer representation ◑

Renormalization allows the QM-coder to use fixed-precision integer arithmetic in the coding operations. For the representation chosen for the QM-coder, X'10000' is defined as the exclusive upper bound for the interval A (1.5), and X'8000' is defined as the inclusive lower bound for A (0.75). The encoder therefore is as follows:

After MPS:

C is unchanged   /* C points to base of MPS subinterval */
$A = A - Qe$ /* Calculate MPS subinterval   */
if $A <$ X'8000'   /* If renormalization needed   */
    renormalize A and C
end

After LPS:

$C = C + A - Qe$ /* Point C at base of LPS subinterval */
$A = Qe$ /* Set interval to LPS subinterval   */
renormalize A and C /* Renormalization always needed   */

The relationship between integer and decimal values we have been using is dictated by the approximation to the multiplication, and would suggest that a value of 1.0 be represented by (4/3)(X'8000') or X'AAAA'. However, this decimal equivalency is used only to provide an interpretation of the integer values. What counts is the accuracy of the interval subdivision, and this is determined by the relative values of Qe and A. Assuming Qe is matched to the statistics of the decision sequence, the average value of A defines the integer equivalent of decimal 1.0. Experimentally, we find that the average value of A is X'B55A' for JPEG compression, and this is only slightly higher than the optimal value of X'AAAA'.* As a convenience, we shall continue through the next section to use 0.75 as the decimal equivalent of X'8000'. However, later in this chapter we shall use the measured average value in computing the decimal equivalents of the integer Qe values in the state machine.

9.1.5 Conditional exchange ◑

One of the problems with the approximation to the multiplication is that when Qe is of order 0.5 (X'5555'), the size of the subinterval allocated to the MPS can be as small as 0.25 (X'2AAB'). To avoid this interval size inversion, the assignment of LPS and MPS to the two intervals is interchanged whenever the LPS subinterval becomes larger than the MPS subinterval. This is known as "conditional exchange," and the term "conditional" comes from the fact that the interval reassignment is only carried out when the LPS probability occupies more than half of the total interval available. A sketch of two possible cases, one without, the other with conditional exchange, is shown in Figure 9-2. Note that when conditional exchange occurs, $0.5 \geq Qe > (A - Qe)$. Both subintervals are clearly less than 0.75 (X'8000'), and renormalization must occur. Consequently, the test for conditional exchange is performed only after the encoder has determined that a renormalization is needed. The encoder is therefore:

---

* A slightly different average, X'B893', is given in the JBIG DIS.

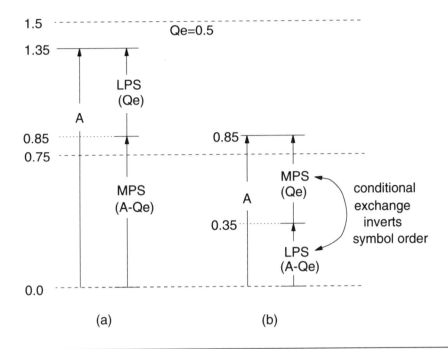

Figure   9-2. Illustration of interval subdivision. (a) without and (b) with
             conditional exchange.

After MPS:

```
C is unchanged
A = A − Qe          /* Calculate MPS subinterval          */
if A < X'8000'      /* If renormalization needed          */
  if A < Qe         /* If interval sizes inverted         */
    C = C + A       /* Point to LPS subinterval base      */
    A = Qe          /* Set interval to LPS subinterval    */
  end
  renormalize A and C
end
```

After LPS:

```
A = A − Qe          /* Calculate MPS subinterval          */
if A ≥ Qe           /* If interval sizes not inverted     */
  C = C + A         /* Point to LPS subinterval base      */
  A = Qe            /* Set interval to LPS subinterval    */
end
renormalize A and C
```

9.1.6 Probability estimation ◑

Adaptive probability estimation has been used in a number of arithmetic coders.[47, 48, 62] The probability estimation used in the QM-coder is based on the renormalization-driven estimation developed for the Q-coder.[42] However, it has been enhanced by a better estimator state machine structure based on the Bayesian estimation principles incorporated in the Minimax coder.[46]*

The estimation process is based on a form of approximate counting[64] in which the interval register renormalization is used to estimate the MPS and LPS symbol counts. Whenever either MPS or LPS renormalization occurs, the count of MPS is cleared (effectively), and a new estimate is obtained from a table that provides a bigger Qe value when the LPS renormalization occurs and a smaller Qe value when the MPS renormalization occurs.

Although the system is stochastic, the estimation state machine naturally tends to move toward the correct estimate. If the Qe value is too large, MPS renormalization is more probable than LPS renormalization and the Qe value is likely to decrease. Conversely, if the Qe value is too small, MPS renormalization is less probable than LPS renormalization and the Qe value is likely to increase. If the sense of the MPS is wrong, Qe will increase until it reaches approximately 0.5, at which point the sense of MPS and LPS are interchanged. Note that the estimation is done for whichever context is being coded at the time that the renormalization occurs.

The detailed analysis of the estimation technique will be described in Chapter 14. What is important here is to understand the implementation of the estimation procedure. Table D.2 of the JPEG DIS contains the complete estimator state machine. For each index in the table there are four columns: Qe, $Next\_Index\_LPS$, $Next\_Index\_MPS$, and $Switch\_MPS$. Qe is the LPS probability estimate. $Next\_Index\_LPS$ is the index to the new probability estimate following an LPS renormalization, and $Next\_Index\_MPS$ is the index to the new probability estimate following an MPS renormalization. $Switch\_MPS$ controls the changing of the sense of the MPS decision; if $Switch\_MPS$ is not zero and the LPS renormalization is required, the sense of the MPS must be changed before moving to $Next\_Index\_LPS$. The steps needed to reach a new probability estimate after the MPS renormalization (see Figure D.5 of the JPEG DIS) are:

```
I = Index(S)             /* Current index for context S          */
I = Next_Index_MPS(I)    /* New index for context S              */
Index(S) = I             /* Save this index at context S         */
Qe(S) = Qe_Value(I)      /* New probability estimate for context S */
```

---

* This merging of renormalization-driven estimation with the Bayesian estimation combines the coding performance of the Minimax coder with the computational simplicity of the Q-coder.[63]

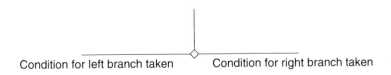

Figure   9-3. Decision tree conventions

For the LPS renormalization path (see Figure D.6 of the JPEG DIS):

$I = Index(S)$ /* Current index for context S */
if $Switch\_MPS(I) = 1$ /* If changing MPS sense for context S */
    $MPS(S) = 1 - MPS(S)$ /* Exchange MPS sense (1 to 0, 0 to 1)   */
end
$I = Next\_Index\_LPS(I)$ /* New index for context S */
$Index(S) = I$ /* Save this index at context S */
$Qe(S) = Qe\_Value(I)$ /* New probability estimate for context S */

On the MPS path the state machine index for the current context is used
to look up a new index, $Next\_Index\_MPS$, which is saved in the context
store. Optionally, the Qe value for that index can also be saved in the
context store, eliminating the need for an indirect storage access to get the
value of Qe. The LPS path procedure is similar, except that $Switch\_MPS$
must be checked to see whether the sense of the MPS must be changed
before getting the new estimate. A different organization of the estimation
state machine that avoids explicit testing of the $Switch\_MPS$ bit will be
described in Chapter 13.

9.1.7 Compact decision tree notation ❶

The compact decision tree notation illustrated in Figure 9-3 is used for the
flowcharts in this chapter. An extended version of this notation is used in
Chapter 12 and Chapter 13.

9.1.8 Decision tree for encoding ❶

The decision tree for the Code_$D(S)$ coding procedure is shown in Figure
9-4. If $D = 0$, this is the Code_0(S) procedure of the JPEG DIS, whereas
if $D = 1$, it is the Code_1(S) procedure. Note that very few operations are
required on the most likely path in which the MPS is coded without
renormalization.

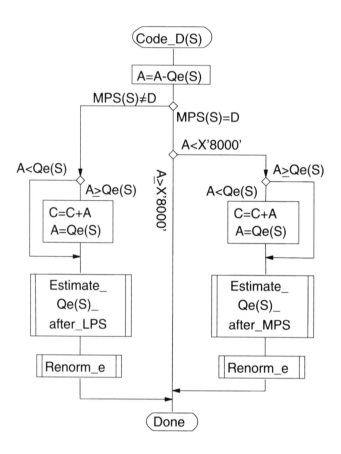

Figure   9-4. Encoder decision tree for coding a decision *D*

9.1.9 Output procedure for the encoder ◗

Output bits are created by the renormalization procedure in Figure 9-5 (see Figure D.7 of the JPEG DIS). A counter, *CT*, counts the number of times *A* and *C* are doubled by the shift-left-logical (SLL) operation, and thus counts the number of bits produced. When *CT* is zero, Byte_out is called to transfer a byte of compressed data from the code register to the code buffer. Byte_out, shown in Figure 9-6, contains the logic to resolve carry-overs so that they cannot propagate through more than one byte of data in the output buffer.

The carry-over control in Byte_out works on the principle that only X'FF' bytes can propagate a carry and therefore, if they are produced, they are stacked—i.e. counted—until the carry can be resolved. The carry is resolved either when a carry-over explicitly occurs or when the next output byte is less than X'FF'. At that point stacked data can be transferred to the compressed data buffer.

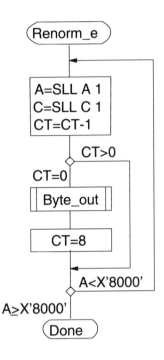

Figure   9-5. Renormalization in the encoder

The code register design determines how the output byte is removed. For the flowcharts given in the DIS, the register has the following bit assignment:

C register: 0000 cbbb bbbb bsss xxxx xxxx xxxx xxxx

A register: 0 aaaa aaaa aaaa aaaa

The "x" bits are the fractional code register bits, "s" denotes three spacer bits, "b" bits are the bits of the byte that will be removed from the register after each 8 renormalizations, "c" is a carry bit that occasionally gets set because of carry-over in the register arithmetic, and "a" is an interval register bit. Normally bit 16 of the A register is zero, as shown—however, see the section on initialization below.

Spacer bits delay the output, thereby giving more chance for carries to be resolved before the byte of data is removed from the register. Note, however, that there is no way for carry-over to exceed the capacity of the register as defined here. The bounds on carry-over are dictated by the maximum possible sum of C and A. Immediately after a byte has been removed from the code register, C is less than X'80000' and A is less than X'FFFF'. The sum is therefore less than X'8FFFF', and after eight renormalizations it is less than X'8FFFF00'. Therefore, the upper four bits of the C register can never be other than zero.

Referring to Figure 9-6, the branch to the left at the first test is taken if the carry bit is set. The carry bit must be added to the data written in earlier calls to Byte_out, including any stacked X'FF' bytes. The addition of the carry converts the stacked X'FF's to zeros, and the carry then propagates to the last byte B actually written to the code buffer. The carry bit is therefore added to the byte in the code buffer (which is always less than X'FF'), and if that addition produces X'FF', a zero byte is inserted (stuffed) immediately following. Any stacked bytes (now zero) are then placed in the compressed data buffer, followed by the new byte of compressed data. Note that when the carry occurs, the byte defined by 'bbbbbbbb' in the code register is X'1F' or less, and therefore can be written directly to the code buffer without testing for the X'FF' value.[*]

Again referring to Figure 9-6, the branch down from the first test is taken if the carry bit is not set. If the new byte is X'FF', the second branch down must be taken; the byte must be stacked until future carries can no longer propagate through it. The stack count SC is incremented, but no further action is taken except to clear the high order bits of the C-register.

If the new byte is less than X'FF', the branch to the right in Figure 9-6 is taken. At this point we can guarantee than no carry will propagate through the stacked X'FF' bytes, and they are therefore placed in the compressed data buffer with stuffed zero bytes appended. The new byte of data is then written immediately following.

If the compressed data is purely random, the probability of the stack count exceeding 3 or 4 is quite small. However, there are special conditions in which repeating patterns of coding decisions and contexts can cause SC to reach much larger values. The only rigorous upper bound on SC is the size of the compressed data set itself, and it would be dangerous to assume that SC can never approach this bound. We have observed stack counts of more than 80 in JBIG sequential compression of binary halftone images, and there is some reason to suspect that a worst-case stack count exceeding a 16-bit counter capacity is possible. However, a 32-bit counter represents such a vast amount of compressed data that it should be able to handle any situation that might reasonably be encountered in either JPEG or JBIG.

9.1.10 Initialization of the QM-encoder ❶

The QM-encoder is initialized with A = X'10000' and $C = 0$. The starting value for A is the exclusive upper bound of the interval, and code values from X'0000000....00' to X'FFFFFFF...FF' are possible (ignoring stuffed

---

[*] This would not be true if there were no spacer bits. Without spacer bits the maximum value for both C and A is X'FFFF' when starting a new byte. Therefore, the sum is X'1FFFE', which after shifting by eight bits becomes X'1FFFE00'. The carry bit is set, and the next eight bits can be X'FF'.

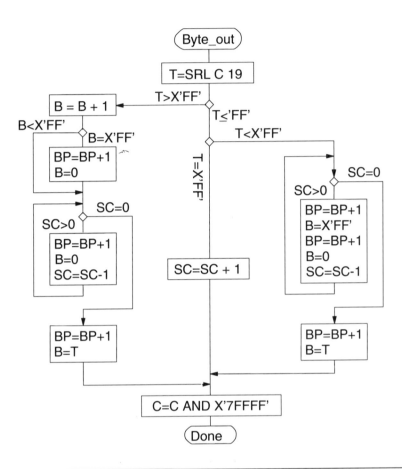

Figure   9-6. Outputting a byte in the encoder

zero bytes). Note that in 16-bit implementations A can be set to 0, and if normal underflow in the arithmetic can occur, it will have the correct value after the first symbol is coded. The Byte_out renormalization shift count, CT, is set to 11, reflecting the fact that the first byte of data will be complete when both the output byte and three spacer bits have valid data.

The pointer to the compressed data is initialized to point to a position one byte before the first byte of compressed data. Byte_out will increment this pointer before writing the first byte to the buffer.

The statistics areas are initialized to an estimation table index of zero. This starts each context in the correct rapid initial learning or "fast attack" state. The $Qe(S)$ values are initialized to X'5A1D', the probability value for an index of zero (see Table D.2 of the JPEG DIS). $MPS(S)$ values are set to zero.

There is an interesting question concerning the interpretation that should be given to the starting initial interval. If we use the relationship that decimal 1.0 is equivalent to X'B55A', we find that X'10000' is equivalent to decimal 1.41. Yet clearly, we cannot have a probability interval of 1.41. The difficulty here lies in our attempting to assign rigorously a decimal equivalent to each integer value. These assignments have meaning on the average, but the initial startup condition is a special situation in which we can assign meaning relative only to the starting Qe. Since that first Qe value is X'5A1D', our choosing A=X'10000' simply means that in using a starting value for A that permits us the full range of compressed data output, we must use a less than optimal first interval subdivision. However, because we do not know what that first Qe should be anyhow, this less than optimal subdivision is of no practical consequence.

9.1.11 Termination of the entropy-coded segment ◗

After the last symbol has been coded, the information still in the code register must be transferred to the compressed data buffer. This is done by the Flush routine shown in Figure 9-7.

In order to understand the operation of Flush, we must first understand a convention adopted for the decoder. If the decoder encounters a marker, it must continue to supply zero bytes to the decoding procedures until the correct number of MCU have been decoded and decoding is complete. The strategy adopted in Flush, therefore, is to create as many trailing zero bits in the code register as possible and to write only two bytes to the compressed data buffer. The rest of the bits are guaranteed to be zero, and therefore will be regenerated in the decoder when it encounters the terminating marker. Furthermore, because the final zero bytes that were discarded are needed in order to have enough precision to decode the last few symbols, the decoder will try to read them and will therefore encounter the terminating marker before it completes decoding. This terminating marker can therefore provide information needed to terminate the decoding process.

Then, optionally, the encoder may discard as many more trailing zero bytes as possible from the code buffer, nibbling away until it encounters non-zero data, a stuffed zero, or the start of the buffer.* These trailing zeros can be discarded because the decoder will regenerate them when it encounters the marker at the end of the segment. This can be done in a number of ways, and the reader is referred to the JPEG DIS Figure D.15 for one possible procedure.

9.2 The QM-decoder ◗

An arithmetic decoder decodes a binary decision by determining which subinterval is pointed to by the code stream. When fixed-precision integer arithmetic is used, however, the A register is not large enough to contain more than the current subinterval available for decoding new symbols. The code stream must therefore be kept relative to this subinterval by subtracting from it each subinterval that was added by the encoder. Then,

---

* This nibbling away of trailing zeros is nicknamed "Pacman" because of the similarity to the popular computer game of that name.

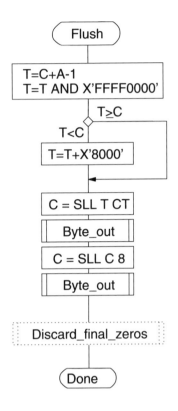

Figure   9-7. Termination of the entropy-coded segment

the code stream becomes a pointer into the current interval relative to the base of the current interval and is guaranteed to have a value within the current interval.

9.2.1 Register bit assignment in the decoder ◑

The decoder makes use of two registers, A and C. A is the interval register and serves exactly the same function as in the encoder. It also must always have exactly the same value before decoding a given symbol that the encoder had before encoding that symbol, and therefore must undergo the same arithmetic and renormalization procedures. C is segmented into two parts, Cx and $C\text{-}low$. Cx is the portion of the code register containing the offset or pointer to the subinterval and $C\text{-}low$ contains up to eight bits of new data. Each time A is renormalized, Cx and $C\text{-}low$ are shifted as a pair such that the MSB of $C\text{-}low$ becomes the LSB of Cx. For 16-bit architectures where Cx and $C\text{-}low$ are separate registers, the bit assignments for A, Cx and $C\text{-}low$ are as follows:

```
        MSB                    LSB
        |                      |
A       aaaa aaaa aaaa aaaa
Cx      cccc cccc cccc cccc
C-low   bbbb bbbb 0000 0000
```

In 32-bit architectures Cx and $C\text{-}low$ can be implemented as one 32-bit register:

```
                    Cx                  C-low
Cx.C-low    cccc cccc cccc cccc bbbb bbbb 0000 0000
       A    aaaa aaaa aaaa aaaa 0000 0000 0000 0000
```

If C is a 32-bit register, A should also be a 32-bit register, in order to facilitate the comparisons between Cx and A. Note that the bits in $C\text{-}low$ cannot affect the comparison of A and Cx, because there are no tests for equality.

9.2.2 Decision tree for decoding ◑

The decision tree for the Decode(S) procedure is shown in Figure 9-8. The code register is, by definition, a pointer to the subinterval relative to the base of the current interval. Therefore, by comparing Cx to the size of the lower subinterval, $A - Qe$, the first decision determines whether the upper or the lower subinterval has been decoded. If the upper subinterval is decoded, the encoder must have added the lower subinterval to the code stream; the decoder must therefore subtract that amount. The value left in the code register, Cx—the offset into the interval—thus remains a pointer to the subinterval for the symbols that have not yet been decoded.

Decoding the MPS or LPS subinterval does not yet completely determine which symbol was decoded. If the MPS subinterval is decoded and no renormalization is needed, the result is known. However, whenever renormalization is needed, the interval assignments might have undergone conditional exchange. The decoder must then test for conditional exchange and interchange interval assignments if needed. It must then do the estimation in a manner appropriate for the symbol decoded (i.e., exactly as the encoder did it) and renormalize. As in the encoder, very few operations are required on the path in which the MPS is decoded without renormalization, and this is the most likely path.

If renormalization is needed, both the code register and the interval register are shifted left until the most significant bit of A is again set. This is shown in Figure 9-9. The counter CT determines when $C\text{-}low$ is empty and Byte_in must be called to insert a new byte of data in $C\text{-}low$.

9.2.3 Input procedure for the decoder ◑

The input procedure for the decoder is greatly simplified relative to the output procedure (Byte_out) of the encoder by the carry resolution in the

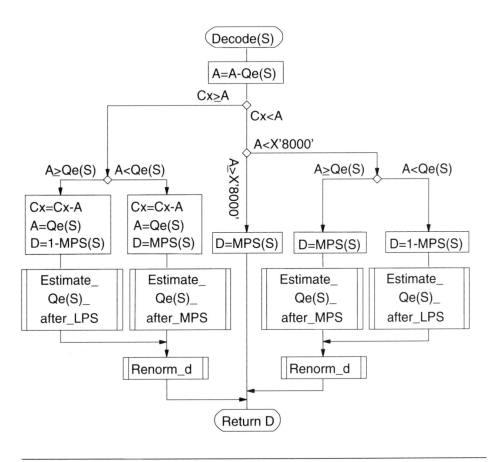

Figure   9-8. Decoder decision tree for decoding a decision *D*

encoder. As shown in the flowchart of Figure 9-10, the only complication is caused by the need to check for X'FF' bytes and to "unstuff" (i.e., discard) the zero byte that often follows that byte. As part of doing that, however, the decoder must check to make sure the byte following the X'FF' is actually zero. If the byte following an X'FF' is not zero, the marker code that terminates the entropy-coded segment has been encountered. The decoder must then, by definition, be provided with zero bytes until decoding is complete. Note that several zero bytes may be needed because of the "Pacman" termination allowed in the encoder.

Because the decoder may request as many zero bytes as it needs, decoding is not self-terminating. The decoder must determine by explicit testing when it has decoded the correct amount of data. Note that the decoder will always need at least one zero byte after encountering the marker. It therefore is guaranteed to intercept and interpret the JPEG DNL marker (which defines the number of lines in the image) in time to terminate decoding correctly.

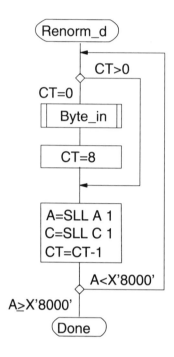

Figure   9-9. Decoder renormalization

9.2.4 Initialization of the decoder ◑

Initialization of the interval register is the same for encoder and decoder. As in the encoder, a 16-bit precision A register should be initialized to zero; the first decoding operation will cause it to underflow to the correct 16-bit value. The high-order part of the code register, Cx, is simply loaded with the first two bytes of compressed data and the count CT is cleared so the first renormalization will cause a new byte of data to be loaded into C-low.

9.3 More about the QM-coder ◑

In this chapter we have presented the basic structure of the QM-coder. Further details for the QM-coder that allow for efficient software and hardware are given in Chapter 13; Chapter 14 contains a discussion of renormalization-driven probability estimation.

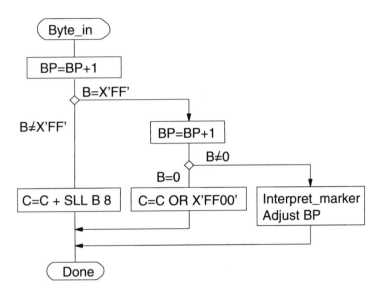

Figure   9-10. Decoder input procedure

10

JPEG CODING MODELS

In this chapter we describe the JPEG encoder and decoder coding models. The reader is reminded that the segmentation of the system into a model and the JPEG entropy coder is at a point that keeps the model independent of the choice of entropy coder. Consequently, all the topics discussed in this chapter apply equally to either Huffman coding or arithmetic coding. Because it is dependent on the entropy coder, the statistical modeling is imbedded in the JPEG entropy coder part of the system; this is discussed in Chapter 5.

The models discussed here fall into two general classes: those relating to predictive coding (DPCM) and those relating to transform coding. There are many similarities and shared elements in these models, and that sharing of concepts and coding structures makes the total system much less complex than it might at first seem to be, given the number of different coding processes that JPEG allows.

The sequential DCT-based encoder and decoder models are at this point quite well-known because of the popularity of the JPEG baseline system. What may not be so well-known is that the coding models used in the other JPEG modes of operation are extensions of the baseline models and retain many of the baseline features.

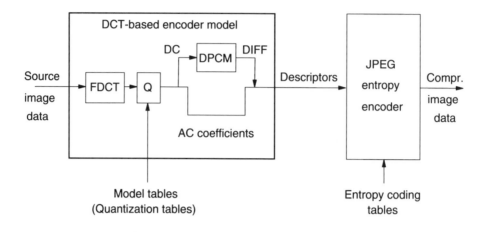

Figure 10-1. DCT-based encoder model

10.1 JPEG sequential DCT-based coding models ◑

10.1.1 Encoder structure ◑

The encoder model converts source image data into descriptors (the conversion of descriptors to the more classic "symbols" is done by the statistical model inside the JPEG entropy encoder). Figure 10-1 illustrates the separation of the encoder model for DCT-based systems into a FDCT and quantizer (Q), and a DPCM model for the quantized DC coefficients. The model tables supply the quantization values for the DCT quantization. In

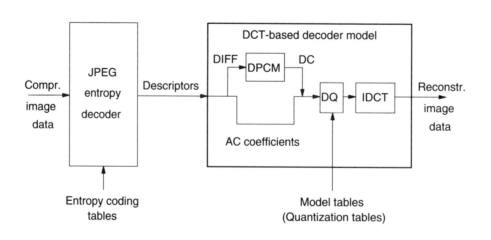

Figure 10-2. DCT-based decoder model

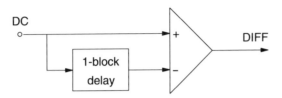

Figure 10-3. Diagram of the DPCM model for DC coefficient encoding

the sequential DCT-based mode, the descriptors for a block are the DC
DIFF followed by the quantized AC coefficients in zigzag order.

10.1.2 Decoder structure ◑

The JPEG entropy decoder and decoder model are inverses of the JPEG
entropy encoder and encoder model, respectively. The JPEG entropy
decoder decodes the descriptors, and the decoder model converts the
descriptors into reconstructed image data. Figure 10-2 divides the decoder
model for DCT systems into a DPCM model to reconstruct the quantized
DC coefficients, a dequantizer (DQ), and an IDCT.

10.1.3 Coding model for DC coefficients ◑

The DC coefficients are DPCM coded, using a 1-D predictor that is the
DC value of the previous 8×8 block coded from the same component. This
choice of predictor holds for both interleaved and non-interleaved data.
The DPCM block is illustrated in Figure 10-3, which reveals the simplicity
provided by this predictor definition.

JPEG explored other variations in predictors, but the best of them
gave less than 1% better compression overall (less than 10% relative to the
DC compression alone). Note that when vertical sampling factors are
greater than one, the order of coding of blocks, and therefore the predictor
for a given block, may be dependent on which components are present in
a scan. This is discussed at length in Chapter 7.

At the beginning of the scan and at the beginning of each restart
interval, the prediction for the DC coefficient, *PRED*, is initialized to 0 for
each component. This seems like an odd choice for an initial value, until
it is recognized that 0 is actually a neutral gray. In computing the FDCT
the input data are level-shifted to a signed representation by subtracting
$2^{(P-1)}$, in which P is the precision specified for the input data. Thus, for
the baseline system, the input data are level-shifted by subtracting 128.
In principle this level shift affects only the DC coefficient, shifting a neutral gray intensity to zero. Difference values between prediction and DC
value are also unaffected, so the only effect of the level shift is to change
the most logical initial starting value for each component to zero (rather
than $2^{(P-1)}$ as in the lossless mode).

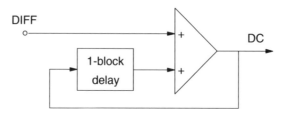

Figure 10-4. Diagram of the DPCM model for DC coefficient decoding

The DPCM decoder does exactly the inverse of the encoder. As illustrated in Figure 10-4, the DPCM difference, *DIFF*, is decoded and added to the prediction *PRED*, where *PRED* is the DC coefficient from the previous 8×8 block decoded for the same component.

After computation of the IDCT the signed *P*-bit precision output data are level-shifted by adding $2^{(P-1)}$, converting the output to an unsigned representation. For the baseline system, the output data are level-shifted by adding 128.

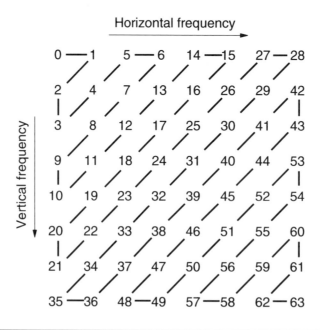

Figure 10-5. Zigzag ordering of DCT coefficients

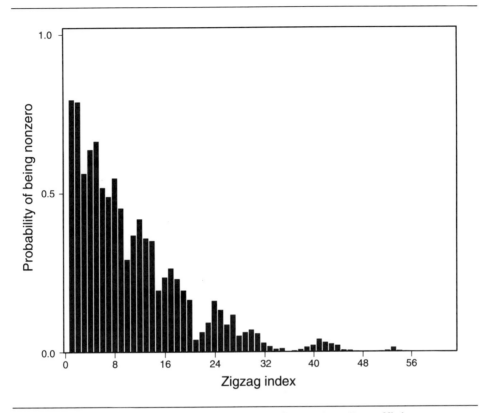

Figure 10-6. Probability of being nonzero for each AC coefficient

10.1.4 Coding model for AC coefficients ◑

The 2-D array of DCT coefficients is rearranged into a 1-D linear array
or vector, $ZZ(0, \ldots, 63)$, in which $ZZ(0)$ is the DC coefficient. The zigzag
ordering in ZZ relative to the normal 2-D coefficient array is shown in
Figure 10-5. This zigzag index sequence, which is described in Tescher's
1978 report,[65] creates a 1-D vector of coefficients, where the lower DCT
frequencies tend to be at lower indices. This zigzag sequence is an
important part of the coding model, as it affects the statistics of the symbols used in the statistical model. When the coefficients are ordered in this
fashion the probability of coefficients being zero is an approximately
monotonic increasing function of the index, as can be seen in Figure
10-6. The authors experimented with some alternative orderings, but
found that the zigzag order was always essentially equivalent or better from
a coding efficiency standpoint. With the arithmetic entropy coder the
compression is very insensitive to the order of coefficient coding.

10.2 Models for progressive DCT-based coding ◑

The extensions to the DCT-based sequential system defined in this section
provide a technique for progressive coding of the DCT coefficients. JPEG

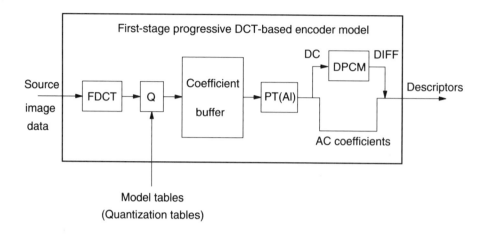

Figure 10-7. First-stage progressive DCT-based encoder model

defines two complementary progressive coding techniques: successive approximation and spectral selection.

In the spectral-selection technique contiguous coefficients in the zigzag sequence are grouped into bands, and each band is sent in a separate scan. In the early scans only the DC and perhaps a few low-frequency AC coefficients provide a blurred but recognizable rendition of the image.

In the successive-approximation technique the coefficients are first coded with reduced precision, such that the important (large magnitude) coefficients are represented approximately. The reduced precision of the coefficients introduces noise and artifacts, but all major features of the image can be recognized.

In the progressive DCT-based mode the DC coefficients are always sent in a separate scan. The DC coefficients from several components may be interleaved, whereas AC coefficients from different components may not be interleaved.

10.2.1 Spectral-selection models ◑

In the spectral-selection mode the zigzag sequence of DCT coefficients is segmented into contiguous bands, and each band is sent in a separate scan. The DC coefficients are always sent in a separate band, and only the DC coefficient scans may contain interleaved data from more than one component. The DC coefficients for a given component must always be sent first, but the bands that follow do not have to occur in any set order.

The band is specified in the scan header by the parameters Ss and Se. Ss is the starting index in $ZZ(0, \ldots, 63)$ and Se is the ending index. These indices are inclusive, so if $Ss = 2$ and $Se = 5$, coefficients 2, 3, 4, and 5 are in the band.

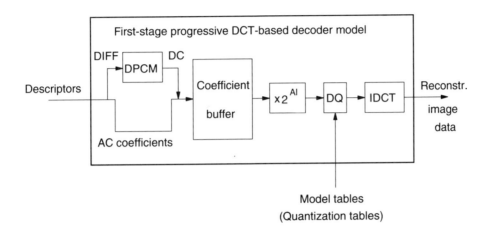

Figure 10-8. First-stage progressive DCT-based decoder model

10.2.2 First-stage successive-approximation models ◑

In a first stage* of successive approximation the DCT coefficients are reduced in precision by a point transform. Note that several spectral-selection scans can be used within each stage of successive approximation.†

Except for the segmentation of the coefficients into bands, the point transform to reduce the coefficient precision, and the prohibition on interleaving data when coding AC coefficients in a progressive scan, the first stage DC and AC coding models are the same as in the sequential mode.

The first-stage progressive DCT-based encoder model adds two new functional blocks to the sequential DCT-based encoder model: namely, a coefficient buffer and a point transform. The encoder diagram is shown in Figure 10-7, and the corresponding decoder is shown in Figure 10-8. The decoder also needs a coefficient buffer and an inverse point transform. For convenience in later stages of successive approximation, the coeffi-

---

* We use the term "first stage" to indicate that these models are used for the first scan of any band of coefficients. We shall apply the term "later stage" to models that code bands already coded at lower precision in a preceding scan.

† Our view that spectral selection is applied within a stage of successive approximation is really a reflection of the fact that the coding models for the first stage of successive approximation are different from the models for later stages, whereas spectral selection is primarily a question of setting starting and stopping indices. However, as the French members of JPEG pointed out with impeccable logic, the converse is also true. One can also regard successive approximation as being applied within a stage of spectral selection.

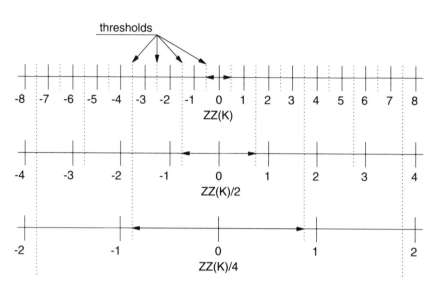

Figure 10-9. Illustration of "fat zero" produced by point transform of AC
        coefficients

cients are shown stored at reduced precision in the decoder, but other
schemes are possible.

    In hardware implementations the encoder coefficient buffer can be
eliminated if the DCT can be recomputed from data in an image buffer.
That would be an ill-advised strategy in a decoder, as the recomputed DCT
would not always match the encoder's DCT.

    For DC coefficients the point transform is defined to be an
arithmetic-shift-right by Al bits. The arithmetic-shift-right uniformly
truncates the low order bits, without regard to the sign of the level-shifted
DC coefficient. For AC coefficients the point transform is defined to be
a signed division by 2^{Al}, in which Al is the integer specified by "successive
approximation bit position low" in the scan header. This division by 2^{Al}
produces a "fat zero" in the quantization, as is illustrated by Figure 10-9.

    In this figure the threshold levels, indicated by dotted lines, are
spaced uniformly. As the point-transform parameter Al is increased, the
intervals between threshold levels for the reduced precision coefficients
grow larger. However, because the point transform truncates the least
significant bits of the magnitude, the interval around zero grows larger than
the other intervals, becoming twice the other intervals in the limit of large
Al—thus, the term "fat zero."

    The fat zero would be a problem for the DC coefficients, because it
would produce a discontinuity in the quantization at level-shifted DC
values of 0. This is the reason that the point transform of the DC is

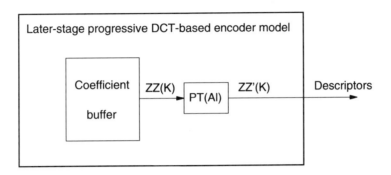

Figure 10-10. Later-stage progressive encoder model

defined as a shift-right-arithmetic operation. The inverse point transform in the decoder model is always a multiplication by 2^{Al}, even for the DC coefficient.

10.2.3 Successive-approximation models for later stages ◑

In the stages of successive approximation that follow the first stage for a given component, the precision of the coefficient values is progressively improved by decreasing Al by 1 for each stage. When the scans for $Al = 0$ are complete, the DCT coefficient values are at full precision. Note that the "successive approximation bit position high" parameter, Ah, in the header is always set to the value of Al used in the preceding successive-approximation scan. If there is no preceding stage, it is set to zero. Although in principle the encoder could recompute the DCT for each stage, we assume a coefficient buffer for this model, as shown in Figure 10-10.

The decoder for later stages of successive approximation also requires the coefficient buffer, as shown in Figure 10-11. In addition, the decoder requires feedback of AC coefficient values from the coefficient buffer to the entropy decoder.

10.2.3.1 Later stages of successive approximation for DC coefficients ◑

If the point transform is used to reduce the precision of the DC coefficients in the first stage, the least significant bits of the DC coefficients are sent in later scans of the progression, one bit plane at a time. Figure 10-10 illustrates how the DC values are extracted from the coefficient buffer and point-transformed into the descriptors. Because the low-order bits of the DC coefficient are almost completely random, the bit values are coded directly, without any attempt to predict them. Figure 10-12 illustrates the coding of DC values in successive-approximation stages. Note that DC coefficients may be interleaved in these later stages.

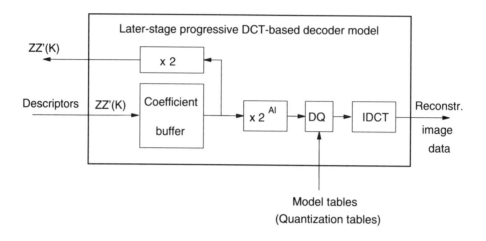

Figure 10-11. Later-stage progressive decoder model

10.2.4 Later stages of successive approximation for AC coefficients ◑

Figure 10-11 illustrates the decoder structure for later stages of successive approximation. As in the sequential mode, the descriptors for AC coding are the AC coefficients in zigzag order. The feedback of buffered decoded coefficient values from the decoder model to the decoder statistical model is essential. This supplies to the decoder statistical model the coefficient values that were decoded in prior scans. Information from prior scans is needed in order to discriminate between coefficients that are known to be non-zero and those that may still be zero.

The encoder must also distinguish between coefficients coded as non-zero and coefficients coded as zero in previous scans. For coefficients known to be non-zero from previous scans only the least significant bit of the point transformed coefficient is coded. The rest of the coefficients in the band can have one of three values, 0, −1, and +1.

10.2.5 IDCT implementations for progressive decoders ◑

One of the problems with progressive decoding is the repeated computation of the IDCT. This may be of little consequence in hardware implementations, but can be a major problem for software. One possibility is to use "fast-path" structures in which sparse DCT arrays are computed by special implementations with a knowledge that certain coefficients are zero. Another alternative is the Goertzel IDCT,[66] in which each coefficient is transformed separately. The spatial representation must include a significant number of undisplayed fractional bits, however, or the round-off

Decoded DC' bit values

Bit position:	76543210	Comment
Original value:	xxxxxxyz	Binary representation
1st stage, Al = 2:	00xxxxxx	xxxxx coded differentially
2nd stage, Al = 1:	y	Bit y coded
3nd stage, Al = 0:	z	Bit z coded

Figure 10-12. Successive-approximation coding of DC values

errors in this form of IDCT will grow large enough to have a serious impact on visual quality.[*]

10.2.6 Visibility of distortion in low bit-rate progressions ◐

Figure 10-13 gives a plot of the average number of bytes of compressed data allocated to the AC coefficients as a function of the progression parameters. The values listed are the averages for the luminance components of the nine JPEG test images, and the suggested quantization table in the JPEG DIS was used in obtaining the data. The compressed bytes for the DC coefficients would add a constant offset of 4,041 bytes to both curves. These curves provide us with a means to compare the compression for spectral selection to that for successive approximation, as a function of their respective controlling parameters.

A qualitative understanding of the greater effectiveness of successive approximation can be gained from the contrast sensitivity function (CSF) for luminance. Assuming that the full-precision DCT is quantized at the threshold for perception, suprathreshold contrast response measurements[11] would suggest that large distortions should be somewhat less dependent on spatial frequency than the CSF (which is at the threshold of visibility). Therefore, we should expect that curves for a given degree of perceived distortion as a function of spatial frequency should be qualitatively as shown in Figure 10-14. As we depart further from the threshold CSF, the distortion in the image should increase monotonically, thereby decreasing the perceived image quality.[†] Because the curves are separated approximately by factors of two, they should correspond roughly to scans of successive approximation.

---

[*] Although the details were never made public, the authors believe that some form of IDCT similar to the Goertzel IDCT was used in J. Vaaben's impressive Intel 80386-based software demonstration of hierarchical DCT progression at the January 1988 JPEG meeting.

[†] Subjective measurements of JPEG image quality have been made for the hierarchical mode.[67] However, the relationship between the curves we postulate to be at constant distortion and the measures of perceived distortion and image quality is unknown, and even the shape of the suprathreshold distortion curves is open to question.[11]

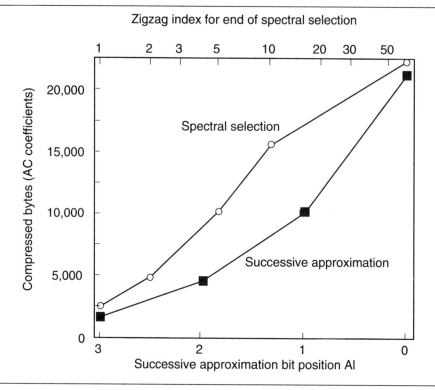

Figure 10-13. Compressed bytes for the AC coefficients as a function of the progression parameters. The upper curve, for spectral selection, uses the abscissa above the plot. The lower curve, for successive approximation, uses the abscissa below the plot. Note that the compressed data for the DC coefficient are not included in the compressed byte values.

The shaded areas in Figure 10-14 delineate the conditions under which distortion should be visible. The two cases illustrated are for comparable bit rates for the two progressive techniques. Successive approximation produces relatively uniform distortion across the full AC coefficient spatial-frequency spectrum of the DCT block. Spectral selection gives a sharp cutoff in the pass-band. Although no visible distortion is introduced below the cutoff, the distortion is large above the cutoff. A clear example of this can be seen in Figure 10-15a.

Because the distortion is large above the cutoff frequence and so many bits are needed to code the lowest frequency terms to full precision, pure spectral selection is far from optimal for good-quality low bit-rate progressive coding. However, spectral selection is useful if one wants reduced-resolution output and, as will be seen in section 10.2.7, the two modes together give a very powerful and graceful progressive coding scheme.

This progressive DCT-based system is remarkable from one standpoint: The coding efficiency is not degraded significantly, relative to the

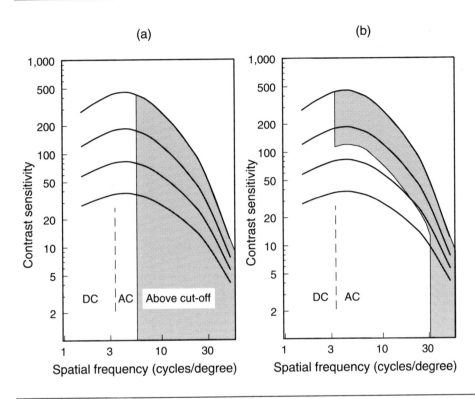

Figure 10-14. Curves of constant distortion as function of spatial frequency. (a) Relationship to spectral-selection distortion. (b) Relationship to successive-approximation distortion. The uppermost curve in both plots is the luminance contrast sensitivity function.

sequential system. Most progressive-coding schemes have to trade off overall compression performance to get the best compression performance in the early stages. For example, subsampling is a classic method for achieving progression within spatial predictive algorithms. However, simple subsampling patterns that do not require recoding of pixels invariably lead to major problems with aliasing. If the aliasing is suppressed by low-pass filtering, pixels have to be coded more than once and coding inefficiency results. The authors are not aware of any predictive progressive algorithms that can match compression performance of the sequential equivalent without producing aliasing.

In the DCT frequency domain, however, a subsampling strategy is more successful. Sending only a fraction of the coefficients gives blurred images lacking detail, but without significant aliasing. Sending the coefficients at reduced precision tends to produce blocking artifacts and visible "ringing" at edges, but again without significant aliasing.* Given that the DCT is effectively an approximation to a set of pass-band filters, perhaps this result is not too surprising. In any event, the DCT-based progressive mode is quite effective, and can sometimes even achieve slightly better compression than that obtained with the sequential mode. Note that when the progression is complete, the DCT is identical to the DCT transmitted in the sequential mode.

10.2.7 Mixing successive approximation and spectral selection ◑

The illustrations given in Chapter 6 show that successive approximation is considerably more effective than spectral selection at very low bit rates. However, the two modes used in combination give even better results. For the "Barbara 2" image from the JPEG test set, the spectral-selection mode requires about 0.36 bits/pixel to code the DC and the first five AC coefficients in the zigzag sequence. The result is shown in Figure 10-15a. Successive approximation with the DC coefficients at full precision and the AC coefficients divided by 4 requires 0.24 bits/pixel, and the reconstructed image quality (not shown) is in most respects already superior to that of Figure 10-15a. However, if another scan is used to improve the precision of the first 9 AC coefficients by one bit, the result is the 0.36 bits/pixel reconstruction in Figure 10-15b. Improving the precision of the low frequency terms helps to suppress some of the blocking artifacts, thereby producing results superior to either pure spectral selection or pure successive approximation. These results were obtained with the arithmetic entropy coder.

10.3 Coding model for lossless coding ◑

The coding model developed for coding the DC coefficients of the DCT is extended to allow a number of 1-D and 2-D predictors for the lossless coding function. Each component in the scan uses an independent predictor; the predictor selection value for the scan is set in the scan header.

10.3.1 Prediction ◑

Figure 10-16 shows the relationship between the neighboring values used for prediction and the sample being coded. x is the sample to be predicted and a, b, and c are the samples immediately to the left, immediately above, and diagonally above and to the left of the current sample in the 2-D sample array. Note that only samples previously coded and therefore available to both encoder and decoder may be used for prediction. The allowed predictors, one of which is selected in the scan header, are listed in Table 10-1.

Selection-value 0 is reserved for differential coding in the hierarchical progressive mode. Selections 1, 2, and 3 are 1-D predictors and selections

---

* Postprocessing can be used to suppress some of the blocking and ringing artifacts. These are discussed in Chapter 16.

(a)

(b)

Figure 10-15. Illustrations of "Barbara 2" test image at 0.36 bits/pixel. (a)
Pure spectral selection, AC coefficients 1–5. (b) Successive
approximation mixed with spectral selection.

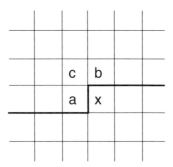

Figure 10-16. Relationship between sample and prediction samples

4, 5, 6, and 7 are 2-D predictors. The division by 2 in the prediction equations is done by an arithmetic- or a logical-right-shift of the integer values, depending on whether the quantity is a signed difference or an unsigned sum.

10.3.2 Prediction at the array edges ◑

The predictions used at the edges of the array are as follows: The 1-D horizontal predictor (prediction sample a) is used for the first raster line of samples at the start of the scan and at the beginning of each restart interval. The prediction for the first sample of this line is $2^{(P - Pt - 1)}$, in which P is the precision and Pt is the point-transform shift. The selected predictor is used for all subsequent raster lines, and the prediction for the first sample of these subsequent lines is prediction sample b from the line above. Note that in lossless coding the point transform is an unsigned division.

Table 10-1. Predictors for lossless coding

Selection-value	Prediction
0	no prediction (differential coding)
1	a
2	b
3	c
4	$a + b - c$
5	$a + (b - c)/2$
6	$b + (a - c)/2$
7	$(a + b)/2$

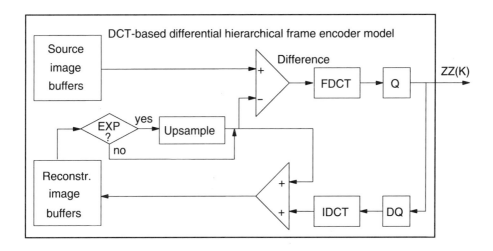

Figure 10-17. Hierarchical mode coding model for DCT-based differential frames

Each prediction is calculated with full integer arithmetic precision, and without clamping of either underflow or overflow beyond the input precision bounds. For example, if a and b are both 16-bit integers, the sum $a + b$ is a 17-bit integer. After dividing the sum by 2 (predictor 7), the prediction is a 16-bit integer.

10.3.3 Modulo calculation of 16-bit DPCM differences ◑

The difference between the prediction and the input is calculated modulo 65,536. Therefore, the prediction can also be treated as a modulo-65,536 value. In the decoder the difference is decoded and added, modulo 65,536, to the prediction.

10.4 Models for hierarchical coding ◑

In the hierarchical mode of operation the source image may be downsampled before the coding of the first stage. Once the first frame for a component has established a reference image, subsequent frames code the differences between the source and reconstructed samples. This means that the encoder must create reconstructions whenever a following frame requires a reference component. Because the differences are already based on a prediction—the referenced component—the coding models are modified to remove the prediction from neighboring samples and DC coefficient values. For nondifferential frames the coding models are as already specified for each of the JPEG modes.

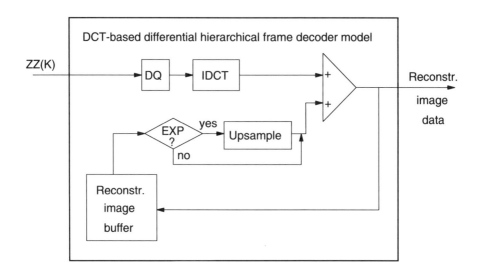

Figure 10-18. Hierarchical mode decoding model for DCT-based differential frames

10.4.1 Model for hierarchical DCT-based mode differential frames ◑

In differential DCT-based frames the DCT is of the differences between the current input arrays and reference arrays, where these reference arrays are outputs (possibly upsampled) of a previous hierarchical frame. Since the data are already differential, the input is not level-shifted before calculation of the FDCT. In addition, the DC values are coded directly rather than differentially. These modifications to the coding model hold for both sequential and progressive DCT-based differential frames.

Figure 10-17 illustrates the system structure for a sequential DCT-based differential frame. Figure 10-17 assumes the source image is at the correct resolution to match the upsampled reconstructed image. If not, downsampling would be needed on the source image input path.

The sequential and progressive DCT statistical models (see Chapter 11 and Chapter 12) are used without modification. The DC coefficient is compressed using the DC statistical models, and the AC coefficients are compressed using the AC statistical models.

Figure 10-18 shows the hierarchical decoder for differential DCT-based frames. As with the other DCT-based modes, precision of source images is limited to either eight-bit or 12-bit samples. The precision of the differential image is always one bit greater, requiring difference precisions of nine bits or 13 bits. The precision of the DCT descriptors is correspondingly increased to a maximum of 12 bits or 16 bits. Note that because the DC term is coded directly rather than differentially, its precision is the same as in the nondifferential frames.

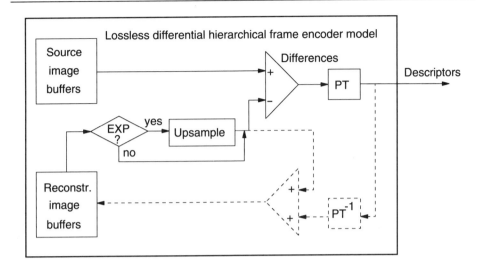

Figure 10-19. Hierarchical mode coding model for spatial differential
            frames. If more than one differential frame is used, the parts
            sketched with dashed lines are needed. The point transform
            is performed in the box labelled PT, whereas the inverse
            point transform is performed in the box labelled PT^{-1}.

10.4.2 Model for hierarchical-mode lossless corrections ◑

The encoder model for hierarchical-mode lossless corrections is shown in
Figure 10-19. The coding model for hierarchical-mode lossless corrections
is similar to the DPCM model, except that the predictions are derived from
the reference image. The prediction selection in the scan header is therefore always zero in a lossless differential frame. The differences are point
transformed by a signed divide by 2^{Al} and fed directly to either the
Huffman or arithmetic JPEG entropy coder.

The corresponding lossless differential hierarchical frame decoder
model is shown in Figure 10-20. The differences are decoded by the
appropriate JPEG entropy decoder, inverse point transformed, and added
to the samples from the reference image.

Unless the lossless differential frame is the last frame of a DCT-based
hierarchical progression, the precision of the source image samples may
be from two to 16 bits/sample. The differential input to the entropy coder
has one bit more precision, except for 16-bit precision samples. In that
case the differences are calculated modulo 65,536, keeping them at 16-bit
precision. Decoder outputs for 16-bit precision are also calculated modulo
65,536.

10.4.3 Upsampling and downsampling ◑

The bilinear interpolation upsampling filter for the hierarchical mode is
precisely specified in the DIS in Annex J, and the reader is referred to that

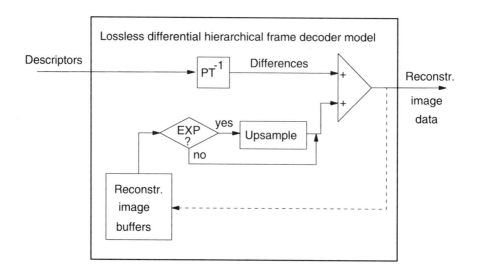

Figure 10-20. Hierarchical mode decoding model for DCT-based differential frames

section for details. The downsampling is not specified, but a cautionary note states that the filter should "be consistent with the upsampling filter." What this means is that the upsampling produces interpolated values on the grid points between the lower-resolution samples, and downsampling filters should produce an output aligned such that subsequent upsampling will minimize the differences to be coded. Otherwise the output will shift in a disconcerting manner when later stages are overlaid onto earlier ones during decoding.

Subclause K.5 of the DIS describes one simple downsampling filter that has the correct alignment. However, the authors' experience with hierarchical progression suggests that using filters with a sharper cut-off than the simple filter in subclause K.5 of the DIS will significantly improve the prediction of later stages, and therefore the overall bit rate.

11

JPEG HUFFMAN
ENTROPY CODING

This chapter contains a discussion of the JPEG Huffman entropy coder
implementations for all of the JPEG modes of operation. As shown in
Figure 5-2 in Chapter 5, this functional block of the system includes both
the conversion of descriptors to symbols in the statistical models and the
conversion of symbols to compressed image data by the Huffman entropy
encoders. The generation of descriptors is covered in Chapter 10. The
material in this chapter is drawn from a number of Annexes of the DIS,
but it is presented in a different order, that emphasizes the similarities in
the various statistical models.

Chapter 12 presents the same material for the JPEG arithmetic
entropy coder. The reader should be aware that the statistical models for
the two JPEG entropy coding systems have much in common. JPEG made
a concerted effort, in fact, to keep the structure of the statistical models
as common as possible, in order to facilitate transcoding between the two
entropy-coded data streams.

The statistical models for both the Huffman and arithmetic coding
of DCT coefficients depend on two basic attributes of the model that creates the descriptors. First, the DCT basis functions are almost completely
decorrelated, such that they can be compressed independently without

Table 11-1. Huffman coding of DPCM differences

SSSS	DPCM difference	Precision DCT–based		Lossless
0	0	8	12	2–16
1	–1,1	8	12	2–16
2	–3,–2,2,3	8	12	2–16
3	–7,...,–4,4,...,7	8	12	3–16
4	–15,...,–8,8,...,15	8	12	4–16
5	–31,...,–16,16,...,31	8	12	5–16
6	–63,...,–32,32,...,63	8	12	6–16
7	–127,...,–64,64,...,127	8	12	7–16
8	–255,...,–128,128,...,255	8	12	8–16
9	–511,...,–256,256,...,511	8	12	9–16
10	–1023,...,–512,512,...,1023	8	12	10–16
11	–2047,...,–1024,1024,...,2047	8	12	11–16
12	–4095,...,–2048,2048,...,4095		12	12–16
13	–8191,...,–4096,4096,...,8191		12	13–16
14	–16383,...,–8192,8192,...,16383		12	14–16
15	–32767,...,–16384,16384,...,32767		12	15–16
16	32768			16

concern about correlation between coefficients. Second, the zigzag sequence, $ZZ(1, ..., 63)$, delivers the coefficients to the entropy coder approximately ordered according to decreasing probability of occurrence. The special symbol for the long run of zeros that usually occurs at the end of the zigzag sequence is an important aspect of the statistical models.

11.1 Statistical models for the Huffman DCT-based sequential mode ◑

Two statistical models are used in the Huffman DCT-based sequential mode. One applies to the coding of DC differences generated by the DPCM model and one applies to the coding of AC coefficients.

The DC coefficient of the 8×8 block is always coded first. The reader should recall (see Chapter 10) that the DC is coded using a DPCM model in which the prediction is the DC coefficient of the most recently coded block from the same component; the difference between the DC value and the prediction is the descriptor fed to the entropy coder. The AC coefficients (the rest of the descriptors for the block) are then coded in zigzag sequence order.

11.1.1 Statistical model for DPCM difference coding ◑

The Huffman statistical model for coding of DPCM differences segments the difference values into a set of (approximately) logarithmically increasing magnitude categories. Because this model is used in all of the JPEG modes of operation that use DPCM coding, the difference magnitude cat-

Table 11-2. Additional bits for sign and magnitude

SSSS	DPCM difference	Additional bits (binary)
0	0	–
1	–1,1	0,1
2	–3,–2,2,3	00,01,10,11
3	–7,...,–4,4,...,7	000,...,011,100,...,111
4	–15,...,–8,8,...,15	0000,...,0111,1000,...,1111
⋮	⋮	⋮
16	32768	–

egories in Table 11-1 are listed for the full range of precisions that might be encountered in JPEG implementations. Each of the difference categories is a symbol, and is therefore assigned a Huffman code. The model, although not quite optimal in terms of the entropy, has a very small alphabet of symbols; the overhead to specify the Huffman table with a DHT marker segment is therefore small.

Except for zero differences, the difference category codes do not fully describe the difference. Therefore, immediately following the Huffman code for non-zero difference categories, $SSSS$ additional bits are appended to the code stream to identify the sign and fully specify the magnitude of the difference. There is one exception to this: $SSSS$ = 16 is not followed by any additional bits. This special case occurs because of the modulo-65,536 arithmetic in the calculation of the differences.

The rule for these additional bits is as follows. If the DPCM difference is positive, append the $SSSS$ low order bits of the difference; if the DPCM difference is negative, subtract one from the difference and append the $SSSS$ low-order bits of this result. The additional bits are appended most significant bit first. Table 11-2 lists additional bit sequences for a few of the magnitude categories. Note that the leading bit of the additional bits is 1 if the difference is positive, and 0 if the difference is negative. As shown, Table 11-1 has enough entries to code any DPCM difference precision that might be required in any JPEG mode of operation. Baseline systems allow only for eight-bit precision samples and therefore do not need to support $SSSS$ values greater than 11. Table 11-3 provides a few

Table 11-3. Example of Huffman coding for DPCM differences

Quantized DC value	+8	+9	+8	–6	–8	–3	+3	+3
DPCM difference	0	+1	–1	–14	–2	+5	+6	0
SSSS	0	1	1	4	2	3	3	0
Additional bits	–	1	0	0001	00	101	110	–

		0	1	2	3	4	5	6	7	8	9	10	11*	12*	13*	14*	15*
	0	EOB	01	02	03	04	05	06	07	08	09	0A	0B	0C	0D	0E	0F
	1	N/A	11	12	13	14	15	16	17	18	19	1A	1B	1C	1D	1E	1F
	2	N/A	21	22	23	24	25	26	27	28	29	2A	2B	2C	2D	2E	2F
	3	N/A	31	32	33	34	35	36	37	38	39	3A	3B	3C	3D	3E	3F
	4	N/A	41	42	43	44	45	46	47	48	49	4A	4B	4C	4D	4E	4F
	5	N/A	51	52	53	54	55	56	57	58	59	5A	5B	5C	5D	5E	5F
	6	N/A	61	62	63	64	65	66	67	68	69	6A	6B	6C	6D	6E	6F
	7	N/A	71	72	73	74	75	76	77	78	79	7A	7B	7C	7D	7E	7F
RRRR	8	N/A	81	82	83	84	85	86	87	88	89	8A	8B	8C	8D	8E	8F
	9	N/A	91	92	93	94	95	96	97	98	99	9A	9B	9C	9D	9E	9F
	10	N/A	A1	A2	A3	A4	A5	A6	A7	A8	A9	AA	AB	AC	AD	AE	AF
	11	N/A	B1	B2	B3	B4	B5	B6	B7	B8	B9	BA	BB	BC	BD	BE	BF
	12	N/A	C1	C2	C3	C4	C5	C6	C7	C8	C9	CA	CB	CC	CD	CE	CF
	13	N/A	D1	D2	D3	D4	D5	D6	D7	D8	D9	DA	DB	DC	DD	DE	DF
	14	N/A	E1	E2	E3	E4	E5	E6	E7	E8	E9	EA	EB	EC	ED	EE	EF
	15	ZRL	F1	F2	F3	F4	F5	F6	F7	F8	F9	FA	FB	FC	FD	FE	FF

(SSSS labels the columns 0–15; RRRR labels the rows 0–15)

*   Not used in sequential mode including baseline with 8 bit input
N/A   Not applicable for sequential mode

Figure 11-1. Huffman AC statistical model run-length/amplitude combinations

examples of the conversion of the DPCM descriptors to symbols for JPEG Huffman entropy coding.

11.1.2 Statistical model for coding of AC coefficients ◗

The coefficients in ZZ are ordered such that the lower DCT frequencies tend to occur first. Because high frequencies are highly likely to be zero (especially after quantization), the higher-index elements of ZZ are usually zero. This is directly incorporated into the AC statistical model by means of a special symbol or condition called the end-of-block (EOB). EOB means that the rest of the coefficients in the block are zero.

In order to get coding efficiencies approaching the entropy, the Huffman coder aggregates zero coefficients into runs of zeros. However, even this strategy is not quite good enough, so the Huffman coder statistical model uses symbols that combine the run of zeros with magnitude categories for the non-zero coefficients that terminate the runs. These magnitude categories increase logarithmically in exactly the same way as the DPCM difference categories, and the coding strategy is therefore very similar. Consequently, except for very long runs of zeros, Huffman code words are assigned to all possible combinations of runs of zeros and magnitude categories. These code words are followed by appended bits that fully specify the sign and magnitude of the non-zero coefficient.

Table 11-4. Huffman coding of AC coefficients

SSSS	AC coefficients	Precision	
1	−1,1	8	12
2	−3,−2,2,3	8	12
3	−7,...,−4,4,...,7	8	12
4	−15,...,−8,8,...,15	8	12
5	−31,...,−16,16,...,31	8	12
6	−63,...,−32,32,...,63	8	12
7	−127,...,−64,64,...,127	8	12
8	−255,...,−128,128,...,255	8	12
9	−511,...,−256,256,...,511	8	12
10	−1023,...,−512,512,...,1023	8	12
11	−2047,...,−1024,1024,...,2047	8	12
12	−4095,...,−2048,2048,...,4095	8*	12
13	−8191,...,−4096,4096,...,8191		12
14	−16383,...,−8192,8192,...,16383		12
15	−32767,...,−16384,16384,...,32767		12*

* Defined only for the differential frames in hierarchical mode

The 2-D code table structure is shown in Figure 11-1. Columns marked with an asterisk are not required in the baseline system. *RRRR* is the run length of zero coefficients before the non-zero coefficient. *SSSS* has the same function as in coding of DPCM differences, and is the number of additional bits required to specify the sign and amplitude. The scheme for determining these bits is exactly the same scheme as for DPCM difference coding, and the pattern for these bits is found in Table 11-2. They are appended to the code stream following the Huffman code for the *RUN-SIZE* symbol. Entries marked N/A in Figure 11-1 are not applicable for the sequential DCT-based mode. These entries are used for EOB run length codes in the progressive DCT-based modes of the extended system.

The composite *RUN-SIZE* value is $(16 \times RRRR) + SSSS$, and each *RUN-SIZE* value is one symbol in the alphabet. Table 11-4 gives the values assigned to each amplitude range specified by *SSSS*.

The *SSSS* = 0 column is reserved for two special symbols, EOB and ZRL. EOB codes the end-of-block condition, and ZRL (zero run length) is used for the rare case in which the run of zeros is larger than 15. ZRL codes a run of 16 zeros (it can be interpreted as a run of 15 zeros followed by a coefficient of zero amplitude). In the rare case in which the 63rd coefficient is not zero, the EOB is not coded. The flowchart in Figure F.2 of the DIS illustrates how the symbols are identified and coded.

Note that Table 11-1 and Table 11-4, which identify the *SSSS* categories for the DC Huffman differences and the *SSSS* categories of the AC coefficients, respectively, are identical except for the number of categories required for each precision. This commonality is not accidental. JPEG made a concerted effort to retain as many common elements as possible between the different parts of the system.

Table 11-5. Example of Huffman symbol assignments to AC descriptors

Zigzag index	1	2	3	4	5	6	7	8	9	10	11	...	63
AC descriptor	0	0	0	0	–14	0	0	+1	0	0	0	...	0
RRRR		←— 4 —→					← 2 →			←—— EOB ——→		
SSSS					4			1			0		
RUN-SIZE					68			33			0		
Additional bits					0001			1			–		

The example in Table 11-5 may help to illustrate this 2-D model. Suppose we have the set of AC coefficients shown there in zigzag order. For the first symbol we select the code for RUN-$SIZE = 68$, corresponding to $RRRR = 4$ and $SSSS = 4$. This is followed by four extra bits completing the coding of the sign and magnitude (see Table 11-2). For the second symbol we select the code for RUN-$SIZE = 33$, corresponding to $RRRR = 2$ and $SSSS = 1$. This is followed by one extra bit to specify the sign. The block is then completed by coding the EOB symbol.

11.2 Statistical models for progressive DCT-based coding ◑

The statistical models for spectral selection and for the first stage of successive approximation are identical. Much of the statistical modeling for the sequential DCT mode carries over to the progressive mode. However, the reader should recall (see Chapter 10) that in spectral selection the coefficients are sent in bands. Consequently, the EOB symbol changes meaning subtly, becoming end-of-band rather than end-of-block.

Coding bands of coefficients and coding reduced-precision coefficients make the EOB so much more probable that significant coding inefficiency may be encountered when only a single EOB code is used. For this reason a set of EOB run-length symbols is incorporated into the progressive DCT statistical models.

The statistical models for later stages of successive approximation are similar to those for the first stage, but are modified to a form discussed in the next sections that allows information about incremental increases in precision to be added to the existing DCT coefficient data already sent.

11.2.1 Statistical model for the first stage of DC progressive coding ◑

In the progressive DCT-based mode of operation the DC coefficients are always sent in a separate scan. The statistical model for the first stage of successive approximation is the same as the statistical model used for the DC coefficients in sequential coding.

Table 11-6 uses the example of DPCM descriptor to symbol conversion for Huffman coding given in Table 11-3 to illustrate a first stage of DC progressive coding that does not code the final bit.

		SSSS														
													*	*	*	*	*
		0	1	2	3	4	5	6	7	8	9	10	11	12	13	14	15
	0	EOB0	01	02	03	04	05	06	07	08	09	0A	0B	0C	0D	0E	0F
	1	EOB1	11	12	13	14	15	16	17	18	19	1A	1B	1C	1D	1E	1F
	2	EOB2	21	22	23	24	25	26	27	28	29	2A	2B	2C	2D	2E	2F
	3	EOB3	31	32	33	34	35	36	37	38	39	3A	3B	3C	3D	3E	3F
	4	EOB4	41	42	43	44	45	46	47	48	49	4A	4B	4C	4D	4E	4F
	5	EOB5	51	52	53	54	55	56	57	58	59	5A	5B	5C	5D	5E	5F
	6	EOB6	61	62	63	64	65	66	67	68	69	6A	6B	6C	6D	6E	6F
	7	EOB7	71	72	73	74	75	76	77	78	79	7A	7B	7C	7D	7E	7F
RRRR	8	EOB8	81	82	83	84	85	86	87	88	89	8A	8B	8C	8D	8E	8F
	9	EOB9	91	92	93	94	95	96	97	98	99	9A	9B	9C	9D	9E	9F
	10	EOB10	A1	A2	A3	A4	A5	A6	A7	A8	A9	AA	AB	AC	AD	AE	AF
	11	EOB11	B1	B2	B3	B4	B5	B6	B7	B8	B9	BA	BB	BC	BD	BE	BF
	12	EOB12	C1	C2	C3	C4	C5	C6	C7	C8	C9	CA	CB	CC	CD	CE	CF
	13	EOB13	D1	D2	D3	D4	D5	D6	D7	D8	D9	DA	DB	DC	DD	DE	DF
	14	EOB14	E1	E2	E3	E4	E5	E6	E7	E8	E9	EA	EB	EC	ED	EE	EF
	15	ZRL	F1	F2	F3	F4	F5	F6	F7	F8	F9	FA	FB	FC	FD	FE	FF

* Not used with 8 bit input

Figure 11-2. Huffman statistical model run-length/amplitude combinations for the progressive DCT-based mode

11.2.2 Statistical model for the first stage of AC progressive coding ◑

For pure spectral selection and for the first stage of successive approximation, the sequential DCT statistical model is extended to include run-length codes for EOB. The symbol table with these additions is shown in Figure 11-2. Each EOB*n* code codes a logarithmic run-length category. After each EOB run-length code except EOB0, additional bits are appended to specify fully the exact run-length. Table 11-7 shows the values in each category. EOB runs greater than 32,767 are coded by a sequence of EOB runs of length 32,767, followed, if necessary, by a final EOB run of the correct size to complete the run. Note that the ZRL code may be used only to specify a run of zeros within a single block.

Table 11-6. Example of Huffman DPCM progressive coding (*Ah*=0, *Al*=1)

Quantized DC value	+8	+9	+8	−6	−8	−3	+3	+3
Point transformed DC	+4	+4	+4	−3	−4	−2	+1	+1
DPCM difference	0	0	0	−7	−1	+2	+3	0
SSSS	0	0	0	3	1	2	2	0
Additional bits	–	–	–	000	0	10	11	–

Table 11-7. EOBn run-length codes and additional bits

EOBn code	Additional bits	EOB run length
EOB0	0	1
EOB1	1	2,3
EOB2	2	4,...,7
EOB3	3	8,...,15
EOB4	4	16,...,31
EOB5	5	32,...,63
EOB6	6	64,...,127
EOB7	7	128,...,255
EOB8	8	256,...,511
EOB9	9	512,...,1023
EOB10	10	1024,...,2047
EOB11	11	2048,...,4095
EOB12	12	4096,...,8191
EOB13	13	8192,...,16383
EOB14	14	16384,...,32767

Table 11-8 gives an example of Huffman code symbol generation for a point transform parameter Al of 1. In principle, the Huffman tables provided in Annex K of the JPEG DIS could be used to encode all the progressive scans. However, these tables would give very poor coding efficiency, because they have no EOB run codes. In addition, the encoding procedure would not follow the flow charts for the progressive mode if EOB run codes were not allowed. The statistics for the progressive modes are very different, and custom Huffman tables are needed if competitive efficiencies are to be achieved.

11.2.3 Statistical model for later stages of DC successive approximation ◑

The statistical model for later stages of successive approximation of the DC coefficient is very simple. The DC precision is increased by one bit,

Table 11-8. Example of first-stage Huffman AC successive approximation

Zigzag index	1	2	3	4	5	6	7	8	9	10	11	...	63
AC descriptor	0	0	0	0	–14	0	0	+1	0	0	0	...	0
Without final bit	0	0	0	0	–7	0	0	0	0	0	0	...	0
RRRR		←——— 4 ———→			← ———	EOB0	——→			
SSSS					3				0				
RUN-SIZE					67				0				
Additional bits					000				–				

Table 11-9. Example of Huffman coding for final DC bit ($Ah=1$, $Al=0$)

Quantized DC value	+8	+9	+8	−6	−8	−3	+3	+3
Already coded × 2	+8	+8	+8	−6	−8	−4	+2	+2
Final bit coded	0	1	0	0	0	1	1	1

and the low-order bit is coded. The decoder adds this bit to the existing decoded DC value at the correct bit position. In Huffman coding of later stages of DC successive approximation the Huffman code words are the bit values. Table 11-9 illustrates a stage of progressive coding in which corrections are coded that, when added to the point-transformed reconstructed DC values in Table 11-6, reconstruct the original DC values coded in Table 11-3.

11.2.4 Statistical model for later stages of AC successive approximation ❶

The later stages of successive approximation increase the magnitude precision by one bit in each stage. Two categories of coding operations are needed for this: corrections to coefficients already known to be non-zero, and the sign and position of newly non-zero coefficients.

If the coefficient is already known to be non-zero, the next least significant bit of the point-transformed coefficient is coded. The Huffman code for this is the bit value itself.

If a coefficient is newly non-zero because of the increase in precision, the position and sign are coded using the run-amplitude structure shown in Figure 11-2. However, only the *SSSS* amplitude categories of 0 and 1 are relevant; the rest cannot occur. The additional bit to specify the sign is appended to the code stream immediately following the Huffman code for the *RUN-SIZE* symbol. Note that the run length of zero coefficients does not include the coefficients that are already known to be non-zero.

Conceptually, newly non-zero coefficients are coded completely separately from corrections to coefficients already known to be non-zero. However, the data streams for the two coding operations are interleaved by appending all correction bits that occur within the run of zero coefficients immediately after the *RUN-SIZE* code and the additional bit specifying the sign. An example is given in Table 11-10 for a final correction stage ($Ah = 1$, $Al = 0$). The EOB run count is assumed to be non-zero at the start of this block; therefore, the first non-zero coefficient causes an EOB code to be generated. The rule for appending correction bits is therefore as follows. All correction bits accumulated in the process of coding a *RUN-SIZE/SSSS* symbol are appended immediately following the code for that symbol (following the *SSSS* bits if any). At that point the number of correction bits within the zero run is known and a decoder is able to remove the appropriate number of correction bits from the data stream. Each ZRL code and EOB*n* code is also followed by appended bits for cor-

Table 11-10. Example of last-stage Huffman AC successive approximation

	–	1	2	3	4	5	6	7	8	9	10	...	63	
Zigzag index	–	1	2	3	4	5	6	7	8	9	10	...	63	
AC descriptor	–	-1	0	-3	0	+8	0	+1	0	+3	0	...	0	
Already coded × 2	–	0	0	-2	0	+8	0	0	0	+2	0	...	0	
Correction bits	–	–	–	1	–	0	–	–	–	0	–	...	–	
Zero run count	0	0	1	1	2	2	3	3	1	1	2	...	55	
EOB run count	N	*	0	0	0	0	0	0	0	0	0	...	0	1
Pending cor. bits				1		0				0				
SSSS		1						1						
RUN-SIZE		1						49						
Additional bits		0						1						
Appended bits								10						

* Code EOB for run length
N followed by any pending correction bits
    before coding the index 1 coefficient.

rections made to non-zero coefficients within the run of zeros covered by that code. This order leads to a relatively complex encoder (see Figure G.7 in Annex G of the JPEG DIS), but provides the data at the right time for the much simpler decoder.

As in the spectral-selection case, the EOB run count is incremented only when the EOB occurs before the last coefficient in the band. Note that the EOB run count is not incremented if the final coefficient is newly non-zero (zero in earlier stages). It is incremented, however, if the EOB occurred at a lower zigzag index and correction bits must be coded for the final coefficient. For example, if all the AC coefficients are non-zero and need only correction bits, the EOB would occur before ZZ(1) and the EOB run count would be incremented.

At the end of the scan or restart interval, any pending EOB run is coded, followed by any pending correction bits.

The reader is referred to subclause G.1.2.3 of the DIS for detailed flow charts for the Huffman successive-approximation progressive mode.

11.3 Statistical models for lossless coding and hierarchical mode spatial corrections ◑

The Huffman coding statistical model for the DC difference coding (section 11.1.1) is used without modification in both the JPEG DPCM-based lossless mode and hierarchical mode spatial corrections.

11.4 Generation of Huffman tables ◑

Figure 11-3 gives an overview of the procedures defined by JPEG for all aspects of Huffman table generation, and references the annex in the DIS where they are found.

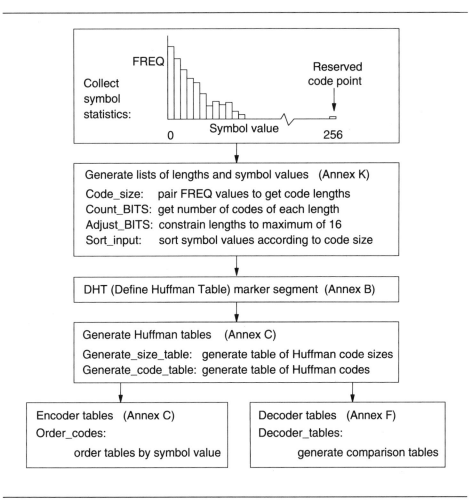

Figure 11-3. JPEG procedures for Huffman table generation

At the top of Figure 11-3 is a box with the legend "Collect symbol statistics," that provides the frequency of occurrence (*FREQ*) for each symbol value. As discussed in Chapter 8, the probabilities of all of the symbols in the alphabet must be known before a Huffman table can be constructed. This is done by counting symbol occurrences, usually for a large group of images that are considered to be "typical" for the application. Provided that a code word is defined for every possible symbol, the table can then be reused from one image to the next.* When maximum coding efficiency is desired, however, the symbol counts can be gathered

---

* The discussion of Bayesian estimation in Chapter 14 applies to Huffman coding as well as arithmetic coding. Reserving a code word for all symbols is usually done by initializing all symbol counts with 1 instead of 0. This is a form of Bayesian estimation.

for each image separately, and a custom Huffman table can be generated just for that image or even for one scan. In either case, the control system must make certain that the decoder has the correct Huffman tables for each image.

11.4.1 Creating a Huffman table specification ◑

The Huffman tables are defined by the DHT segment, and JPEG provides only procedures to create the actual encoder and decoder tables from this segment. Some guidance is provided, however, in an informative annex of the DIS (Annex K, subclause K.2) on how the DHT segment itself is created.

The syntax for the DHT segment is described in section 7.8.1. We see there that the Huffman table is specified by the count of the number of codes of each code length, and the list of values for each code of each length. We note that only lengths from 1 to 16 are allowed by the DHT syntax, and, therefore, no Huffman code greater than 16 bits in length can be specified. We also note that the symbols generated by the Huffman statistical models are assigned values that always fall in the range from 0 to 255 (for example, see Figure 11-1).

Starting from the vector, $FREQ(V)$, that contains the number of occurrences of each symbol value V, the procedure Code_size (see Figure K.2 of the DIS) pairs probabilities to create the Huffman code tree. The Code_size procedure uses a vector, $CODESIZE(V)$, to accumulate the code lengths as the pairing of symbols proceeds, and a vector $OTHERS(V)$ to record the chains of paired symbol values. A careful study of Figure K.2 will reveal that the procedure Code_size implements the classical Huffman symbol pairing algorithm. Note that one code point is reserved by forcing $FREQ(256) = 1$. This is seen in Figure 11-3. Reserving this code point for an illegal symbol value eliminates from the Huffman code table the code word comprised entirely of one-bits. Note also that any $FREQ(V)$ element that is zero is bypassed in the pairing search, and code words for that V will not be generated.

The procedure Count_BITS in Figure K.2 of the DIS creates the list, BITS, that contains the number of codes of each length. As noted in the DIS, BITS may have more than the 16 entries allowed in the DHT segment, and the procedure Adjust_BITS is invoked to shorten any codes that are too long. The reader is referred to the DIS for this procedure, as well as for the Sort_input procedure that sorts the values into code length categories. The length-1 to length-16 entries in BITS and the sorted value sequence make up the Huffman code table specification in the DHT segment.

Note that although the procedures given in Annex K of the DIS happen to produce the symbol values for each code length in monotonically increasing order, the DIS explicitly states that other orderings may be used.

11.4.2 Generating Huffman tables from the Huffman table specification ◑

Referring to Figure 11-3, the Huffman tables are generated using two procedures documented in Annex C of the DIS. The first procedure, Generate_size_table, creates the table of Huffman code sizes, *HUFFSIZE*, for each symbol value, following the symbol order of the DHT segment. The second procedure, Generate_code_table, generates the table of Huffman codes, *HUFFCODE*, again following the symbol order of the DHT marker segment. The Huffman codes are generated using a counting procedure that produces a monotonically increasing sequence of codes for each length, the final increment for a given length creating the unique prefix for the first code in the next length category.

11.4.3 Huffman tables for the encoder ◑

A third procedure in Annex C of the DIS, Order_codes, creates two tables, *EHUFCO* and *EHUFSI*, in which the codes and sizes are sorted in order of symbol value. A symbol fed to the entropy coder can be used to look up the code and its length in these tables.

11.4.4 Huffman tables for the decoder ◑

The implementation of the decoder provided by JPEG is an example, but is normative only in a functional sense. This particular decoder makes use of the fact that the JPEG Huffman codes of each length in *HUFFCODE* are ordered by numeric value of the code and the maximum code value for any given length is a prefix to codes of longer length. Two tables, *MAXCODE* and *MINCODE*, indexed by code length, are generated by the procedure Decoder_tables in Figure F.15 of the DIS. As can be seen in Figure F.16, *MAXCODE* is used to determine when the correct number of bits has been received, and *MINCODE* is used to compute the index to the list of values relative to the first value with a code of that length.

Other decoder implementations that may be computationally more efficient are possible.

12

ARITHMETIC CODING
STATISTICAL MODELS

The arithmetic-coding version of the JPEG entropy coder includes both the statistical model and the binary arithmetic coding procedures (Figure 5-2). The subject of this chapter is the statistical models for arithmetic coding. These models are closely related to the statistical model Langdon developed for lossless coding of grayscale images.[68]

The task of the statistical model is to convert the descriptors delivered by the model into symbols—in this case, binary symbols. The binary arithmetic coding procedures compress these symbols to form the compressed data. The binary arithmetic coding procedures have already been described in Chapter 9, and will be considered to be known procedures in this chapter. We provide a brief summary of them below.

12.1 Overview of JPEG binary arithmetic-coding procedures ◑

In Chapter 9 six procedures are developed that are invoked either in the external control code or by the statistical model. These six procedures are as follows:

Procedure	Function
Encoder:	
Initenc	Initialize the encoder
Code_0(S)	Code a 0-decision for context index S
Code_1(S)	Code a 1-decision for context index S
Flush	Terminate encoder's entropy-coded segment
Decoder:	
Initdec	Initialize the decoder
Decode(S)	Decode a binary decision for context index S

In the encoder "Initenc" initializes the entropy encoder, "Code_0(S)" and "Code_1(S)" code 0-decisions and 1-decisions, respectively, for context-index S, and "Flush" empties the encoder code register at the end of the restart interval or scan. Note that Code_0(S) and Code_1(S) always occur in pairs in the encoder.

In the decoder "Initdec" initializes the entropy decoder; "Decode(S)" decodes and returns a 0 or 1 binary decision for context-index S.

The probability estimation procedures that provide adaptive estimates of the probability for each context are imbedded in "Code_0(S)," "Code_1(S)," and "Decode(S)." These estimation procedures are identical for the encoder and the decoder. At the start of a scan and at each restart, all statistics bins are reinitialized to the standard default value by either Initenc or Initdec.

12.2 Decision trees and notation ◑

The decision tree of a statistical model describes the relationship between the descriptors and the equivalent sets of binary decisions. In the encoder the descriptor is translated to binary decisions; in the decoder the binary decisions are translated to descriptors.

The binary decision trees use the notation shown in Figure 12-1. (This notation was introduced in a limited form in Figure 9-3.) For encoders the solid diamond represents a test with a true/false or 1/0 result, in which true results in a Code_1(S) on the downward branch and false results in a Code_0(S) on the branch to the right. For decoders the solid diamond represents Decode(S), with the downward branch taken if a 1 is decoded, and the right branch taken if a 0 is decoded. Each binary coding decision in the tree is labelled with its context index, while to the left of the diamond the decision that determined true/false is listed. Input may be from the top or left.

Some decisions are made based on information known to both encoder and decoder and therefore do not require calls to the encoding/decoding procedures. A decision made without coding is represented by an open diamond; the criterion for taking a particular branch is given by text along that branch.

With some decisions coded and some not coded, structures using this notation are really hybrids of a decision tree and a flowchart; we devised

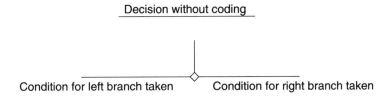

Decision without coding

Condition for left branch taken   Condition for right branch taken

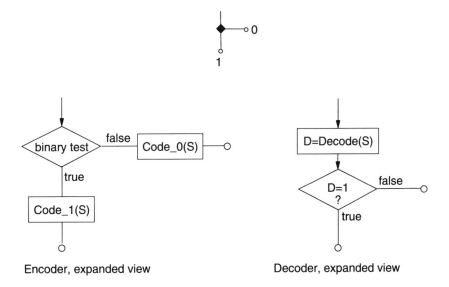

Arithmetic coding decision (encoder or decoder)

0

1

binary test   false   Code_0(S)

true

Code_1(S)

Encoder, expanded view

D=Decode(S)

D=1
?   false

true

Decoder, expanded view

Figure 12-1. Notation for binary decisions. An expanded view of the
decision notation is shown in the lower half of the figure.
Encoders test the binary decision and branch to the appropriate coding procedure. Decoders decode the binary decision and branch appropriately.

this hybrid in an effort to illustrate the decision sequences as compactly
as possible.

A benefit of this compact notation is that these decision trees apply
to both encoders and decoders. In the encoder the value or condition
needed for the test is provided by the encoder model, whereas in the
decoder the decision decoded is used to reconstruct the value or condition.

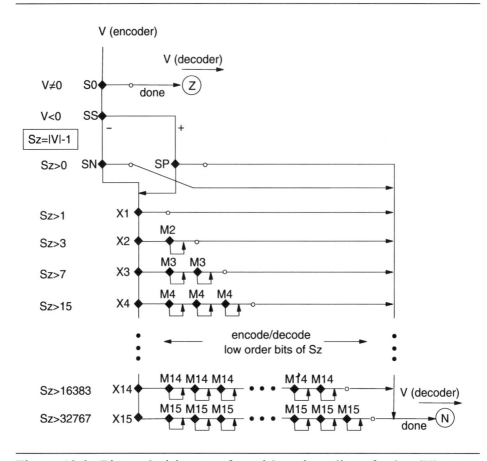

Figure 12-2. Binary decision tree for arithmetic coding of value (V)

12.3 Statistical models for the DCT-based sequential mode with arithmetic coding ◑

12.3.1 Statistical model for DPCM differences ◑

The statistical model for DPCM differences consists of two parts, the binary decision tree for converting the descriptors into binary symbols (binary decisions) and the conditioning of each binary decision in that tree.

12.3.1.1 *Binary decision tree for coding DPCM differences* ◑

The tree of binary decisions and associated context labels for coding DPCM differences is shown in Figure 12-2. This tree provides a somewhat different view of the encoding procedures (Figures F.4, F.6, F.7, and F.8) and the decoding procedures (Figures F.19, F.21, F.22, and F.23) in the DIS.

In encoding the DPCM difference, *DIFF* (see Figure F.4) is assigned to the variable V. The first binary decision, labelled $S0$, determines whether V is zero or non-zero. If V is zero, the coding is complete for that

Table 12-1. Binary decision sequences for arithmetic coding of DPCM differences

DIFF	Binary decision	Precision DCT Lossl.		
0	0	8	12	2–16
Sz				
0	1S0	8	12	2–16
1	1S10	8	12	2–16
2,3	1S110M	8	12	3–16
4–7	1S1110MM	8	12	4–16
8–15	1S11110MMM	8	12	5–16
16–31	1S111110MMMM	8	12	6–16
32–63	1S1111110MMMMM	8	12	7–16
64–127	1S11111110MMMMMM	8	12	8–16
128–255	1S111111110MMMMMMM	8	12	9–16
256–511	1S1111111110MMMMMMMM	8	12	10–16
512–1023	1S11111111110MMMMMMMMM	8	12	11–16
1024–2047	1S111111111110MMMMMMMMMM	8	12	12–16
2048–4095	1S1111111111110MMMMMMMMMMM		12	13–16
4096–8191	1S11111111111110MMMMMMMMMMMM		12	14–16
8192–16383	1S111111111111110MMMMMMMMMMMMM		12	15–16
16384–32767	1S1111111111111110MMMMMMMMMMMMMM		12	16
	1S1111111111111111		Forbidden	

S = 0 if positive and 1 if negative.

difference and the exit marked Z is taken. All other paths eventually reach exit N. As it turns out, Z and N are identical for DPCM difference coding, but the same tree can also be used for AC coefficient coding if Z and N are kept distinct.

If V is non-zero, the sign is then coded with context SS. At this point V is converted to Sz, $(Sz = |V| - 1)$. The next decision (Is $Sz > 0$?) uses either context SN or context SP, depending on the sign. In DPCM coding contexts $S0$, SS, SN and SP are all conditioned on DPCM differences already coded.

If Sz is not zero, the Xn binary decisions (is $Sz > (2^n - 1)$?) are then coded to identify the logarithmic magnitude category of Sz. This is followed by Mn decisions giving, in MSB to LSB order, the additional low-order bits needed to complete the magnitude specification. The bit sequences of the binary decisions for each magnitude category are shown in Table 12-1.

In this table S is 1 if the difference is negative and 0 if positive. $X1$, $X2$, ..., Xn (1, 1, ..., 0) is the magnitude category coding decision sequence (i.e., a string of 1-decisions followed by a 0-decision) and Mn, Mn, ... is the decision sequence for coding the least significant bits of Sz. The columns to the right show which categories are needed for a given precision. Note

Table 12-2. Lower and upper bounds for |DIFF| for "small" conditioning category

L	Lower bound	U	Upper bound
0	>0	0	≤1
1	>1	1	≤2
2	>2	2	≤4
3	>4	3	≤8
4	>8	4	≤16
5	>16	5	≤32
⋮	⋮	⋮	⋮
15	>16384	15	≤32768

that the number of Xn decisions is bounded by the precision. Note also that the decision sequence itself is already a relatively compact code.

12.3.1.2 Conditioning of DPCM binary decisions ◑

One key difference between the Huffman coding and arithmetic-coding statistical models is the use of conditional probabilities in the arithmetic-coding version. Conditional probabilities are used in a number of places in the binary decision sequences shown in Table 12-1. The conditioning is on parameters such as the sign of the difference coded, the position in the decision tree, and prior differences coded (in all cases, the conditioning can use only information known to both encoder and decoder).

The probabilities used in coding the $S0$, SS, SP and SN decisions are conditioned on the previous difference, Da, coded for the same component. This difference is classified into five categories: large negative, small negative, zero, small positive and large positive. As shown in Figure F.10 of the JPEG DIS, the thresholds between these categories are established by two parameters, L and U, which are set before the start of coding. L and U default to 0 and 1, respectively, but can be set in a DAC marker segment to any integer value from 0 to 15. Note that L must always be less than or equal to U.

The exclusive lower bound for a |DIFF| value to be considered "small" (rather than "zero") is the smallest integer value of $2^{(L-1)}$. The inclusive upper bound for a |DIFF| value to be considered "small" (rather than "large") is 2^U. The bounds on the "small" conditioning category are shown in Table 12-2. The default values, $L=0$ and $U=1$, reflect the relatively tight distribution of DC differences around zero.

Note that the segmentation by L and U produces difference categories exactly aligned with the logarithmic segmentation in the decision tree. Consequently, the conditioning category for a given value can be determined solely from the position in the tree after the final decision is encoded or decoded. This fact can be used to minimize the calculations needed to establish conditioning categories.

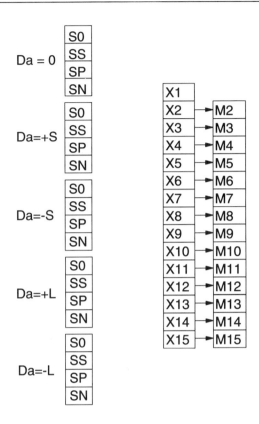

Figure 12-3. Structure of statistics area for DC DPCM difference coding

Figure 12-3 shows the structure of the statistics area for DC difference coding. Depending on the conditioning category for the previous difference, Da, one of five sets of statistics bins is selected for coding the $S0$, SS and SP/SN decisions. Since SP is used only on the positive path and SN only on the negative path, the first magnitude decision, $Sz > 0$, is coded with one of 10 probability estimates.

For magnitudes larger than one, additional decisions $X1$, $X2$, ..., Xn must be coded to establish the magnitude category. A different statistics bin is used for each Xn decision, and a corresponding Mn statistics bin is used to code all Mn decisions following the terminating (zero) Xn decision. The Mn decisions are coded in MSB to LSB bit order, starting with the bit after the leading 1-bit (which is already known from the $X1$, ..., Xn decisions) and ending with the least significant bit. For DC coefficient coding the Xn decisions are conditioned only on the position they occupy in the decision tree. The Mn decisions are conditioned only on the position of the terminating Xn decision. Note that each active DC conditioning table requires a statistics area with this structure.

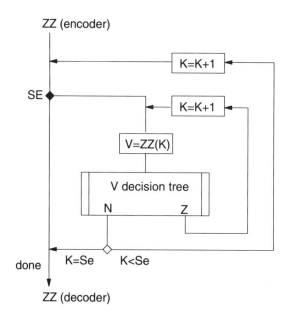

Figure 12-4. Arithmetic coding of AC coefficients

At the start of a scan and at the beginning of each restart interval, the difference for the previous DC value is defined to be zero in determining the conditioning state.

12.3.2 Statistical model for coding of AC coefficients ◑

The arithmetic coder has no problem efficiently coding symbols with probabilities greater than 0.5, and therefore codes each coefficient independently in zigzag sequence order. The EOB concept is an important aspect of the statistical model, however, and is carried over from the Huffman coding model in modified form.

The decision tree in Figure 12-2 also applies to the coding of an AC coefficient. However, the basic data unit for the DCT-based modes is the 8×8 block of coefficients, and additional decisions external to the tree in Figure 12-2 are required to code all of the coefficients in the 8×8 block. Figure 12-4 illustrates the full decision sequence for the 8×8 block; note that the decision tree in Figure 12-2 is imbedded in the box labelled "V decision tree."

As the AC coefficients (ZZ) are fed into the decision sequence, the first decision is the EOB decision—has the EOB been reached? If so, a 1-decision is coded and the coding of the data unit is complete. If not, a 0-decision is coded and the V decision tree procedure is entered.

Here is where the two exit paths in the V decision tree are used. Since we know that a non-zero coefficient remains in the block, we need only to code the EOB decision after each non-zero coefficient. Therefore, the Z

Table 12-3. AC binary decision sequences for arithmetic coding

ZZ(K)	Binary decision	Precision	
0	0	8	12
Sz			
0	1S0E	8	12
1	1S10E	8	12
2,3	1S110ME	8	12
4–7	1S1110MME	8	12
8–15	1S11110MMME	8	12
16–31	1S111110MMMME	8	12
32–63	1S1111110MMMMME	8	12
64–127	1S11111110MMMMMME	8	12
128–256	1S111111110MMMMMMME	8	12
256–511	1S1111111110MMMMMMMME	8	12
512–1023	1S11111111110MMMMMMMMME	8	12
1024–2047	1S111111111110MMMMMMMMMME	8*	12
2048–4095	1S1111111111110MMMMMMMMMMME		12
4096–8191	1S11111111111110MMMMMMMMMMMME		12
8192–16383	1S111111111111110MMMMMMMMMMMMME		12
16384–32767	1S1111111111111110MMMMMMMMMMMMMME		12*
	1S1111111111111111	Forbidden	

S = 0 if positive and 1 if negative.

* Defined only for the differential frames in hierarchical mode.

exit path (for $V=0$) bypasses the EOB decision, allowing coding of runs of zeros with only the zero/non-zero decision (decision S0 in Figure 12-2).

When a non-zero coefficient is encountered, the V decision tree is followed to exit N. At this exit the index is checked to see whether it is the last index in the block; if not, the decision tree returns to the EOB decision.

The sequence of binary decisions for coding the AC coefficients is given in Table 12-3. Except for bit E, the EOB decision coded after each non-zero coefficient, this table is identical to Table 12-1. Consequently, the decision tree structure defined for the sequence of binary decisions in section 12.3.1 also applies to Table 12-3.

Following the last magnitude bit, $E = 1$ is coded if the EOB has been reached, and $E = 0$ is coded if there are more non-zero coefficients remaining in the block. The precision shown in the table covers the full range required for coding the AC coefficients for all of the DCT-based modes.

12.3.2.1 Conditioning of AC binary decisions ◖

The SE, $S0$, SP/SN and $X1$ decisions are conditioned only on position in the zigzag sequence (i.e., on the current index value K). The nearly ideal decorrelation achieved by the DCT means that conditioning on other coefficient values in the block cannot significantly improve coding per-

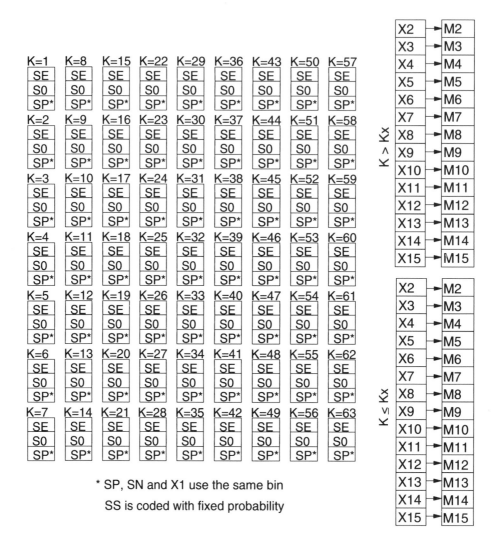

Figure 12-5. Structure of statistics area for AC sequential coding

formance. Experiments by the authors showed that there is some minor correlation with low-frequency DCT coefficients in adjacent blocks. However, the complexity of the conditioning computations was too great to justify a statistical model that included these effects, given the small (1%–2%) gains in compression.

The sign of the non-zero coefficients is random and evenly divided between positive and negative values. Therefore, the sign decision SS is coded with a fixed probability estimate of 0.5 (Qe= X'5A1D', MPS = 0). Since there is no statistical difference between positive and negative coef-

Table 12-4. Example of arithmetic coding binary symbol assignments to AC descriptors

Zigzag index	1	2	3	4	5	6	7	8	9	10	11	...	63
AC descriptor	0	0	0	0	–14	0	0	+1	0	0	0	...	0
SE	0					0			1		–		
$S0$	0	0	0	0	1	0	0	1		←—— EOB ——→			
SS					1			0			–		
Sz					13			0			–		
SN/SP					1			0			–		
$X1$					1			–			–		
$X2,X3,X4$					110			–			–		
$M4,M4,M4$					101			–			–		

ficients, it is reasonable to code the first magnitude decision, SP or SN, in the same statistics bin. What is not so obvious is that $X1$ should also be coded in that same bin. However, experiments by the authors showed that the statistical behavior of these three decisions was so similar that they could be assigned to the same bin.

The rest of the decisions, $X2, ..., Xn, Mn, ..., Mn$, are coded following the assignments given in Figure 12-2. The Xn estimates are conditioned, however, on whether or not the current index K is above a value (Kx) set before the start of coding. Kx should be set so as to separate approximately the DCT coefficients into "low frequency" coefficients that often have large magnitudes, and "high frequency" coefficients that usually have smaller magnitudes. Kx defaults to 5.

Other values of Kx may be signaled using the DAC marker code (section 7.8.2). Experiments done by the authors showed that good results for eight-bit precision source images are obtained when the conditioning of spectral selection bands is set to

$$Kx = Kmin + \text{SRL} \ (8 + Se - Kmin) \ 4$$

where SRL is the shift-right-logical operator and $Kmin$ is the index of the first AC coefficient in the band. Therefore, $Kmin$ is 1 for sequential modes and Ss for progressive modes. This expression for Kx produces the default value of 5 when the band is from index 1 to index 63.

Figure 12-5 shows the structure of the statistics bins for the AC sequential mode coding. Remember that the SP, SN, and $X1$ decisions all use the same bin.

12.3.2.2 Illustration of sequential DCT-based descriptor-to-symbol conversion ❶

We illustrate AC descriptor conversion to symbols in Table 12-4 with the same example used to illustrate Huffman symbols in Table 11-5. For the first four descriptors we code 0-decisions. For the next descriptor we code a 1-decision, followed by a 1-decision for the negative sign. We subtract 1 from the magnitude to get $Sz = 13$, and then code binary '11110' to identify the magnitude category. We code the three least significant bits of 13, binary '101' to complete the magnitude. We follow this with a 0-decision indicating that the EOB has not yet been reached. Note that the $X2$, $X3$, $X4$ and $M4$ bins are from the set defined for $K \le Kx$, assuming Kx is defaulted to 5. The remaining two zero coefficients, the non-zero coefficient just before the EOB, and the terminating EOB decision are then coded.

12.4 Statistical models for progressive DCT-based coding ◑

The primary difference between the sequential mode and the first pass of successive approximation with spectral selection is the grouping of the coefficients into bands. If successive approximation is used, the coefficients are also point-transformed to a reduced precision. The probability of zero coefficients is increased when the successive-approximation point transform is used, and the probability of encountering the EOB is increased when more coefficients are zero or the band is very small. However, because of the adaptive nature of the JPEG binary arithmetic coder and its ability to code fractional-bit code lengths, the statistical model developed for the sequential DCT-based mode can be used without modification for both pure spectral selection and the first stage of successive approximation. With a change in interpretation of EOB to mean end-of-band rather than end-of-block, the statistical model for the sequential mode described in Figures 12-2, 12-4 and 12-5 also applies to the first stage of the progressive mode.

Subsequent stages of successive approximation do require a modification of the statistical model. The coefficients are now divided into two categories: those already known to be non-zero from the previous stage, and those that may be zero or non-zero at the higher precision. The statistical model must treat these categories differently.

12.4.1 Extensions to the statistical models for DC coding ◑

The statistical model for the sequential mode DC coding is used without modification in the first stage of DC coding in the progressive DCT-based mode. Note that the point transform is on the input DC values, not on the $DIFF$ value. Note also that the point transform is an arithmetic shift, not a division, and therefore simply truncates the low-order bit. Consequently, later stages of successive approximation need only to code the low-order bit of the point transformed original DC value, $ZZ(0)$, sending a 1-decision if $ZZ(0)$ is odd and a 0-decision if $ZZ(0)$ is even. These decisions are coded using a fixed probability estimate of 0.5 ($Qe = X'5A1D'$, $MPS = 0$).

12.4.2 Statistical model for later stages of AC successive-approximation coding ◑

The point transform for AC coefficients is a division by a power of 2 that truncates the low-order magnitude bits. With each later stage of successive

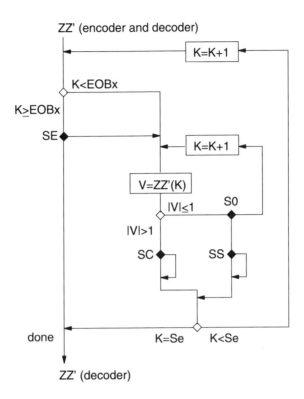

Figure 12-6. Decision tree for a later stage of successive approximation

approximation the precision is increased by one bit. Consequently, for coefficients coded as non-zero in earlier stages, later stages of successive approximation must send the low-order bit as a correction bit (is the coefficient odd or even?). For coefficients that have remained zero up to the current scan, a zero/non-zero decision (and a sign decision if non-zero) are needed. Figure 12-6 shows the decision sequence for a later stage of successive approximation.

The decision tree in Figure 12-6 is somewhat similar to the tree for the first-stage coding of AC coefficients (see Figure 12-4), but there are two major differences: First, a test is added that bypasses the coding of the EOB decision whenever the zigzag index is less than the EOB position of the previous successive-approximation stage. Second, the coding of the non-zero coefficients is divided into two categories: one where only a correction bit is needed (decision *SC*) and the other where a zero/non-zero decision (*S0*) is made, followed by a sign decision (*SS*) for all newly non-zero coefficients. Note that the latter category is equivalent to a truncated version of the *V* decision tree of Figure 12-2.

The conditioning used for the decisions in Figure 12-6 is similar to the first-stage AC coefficient coding. The *SE*, *S0*, and *SC* decisions are conditioned on zigzag index *K*, and *SS* is coded with a fixed probability

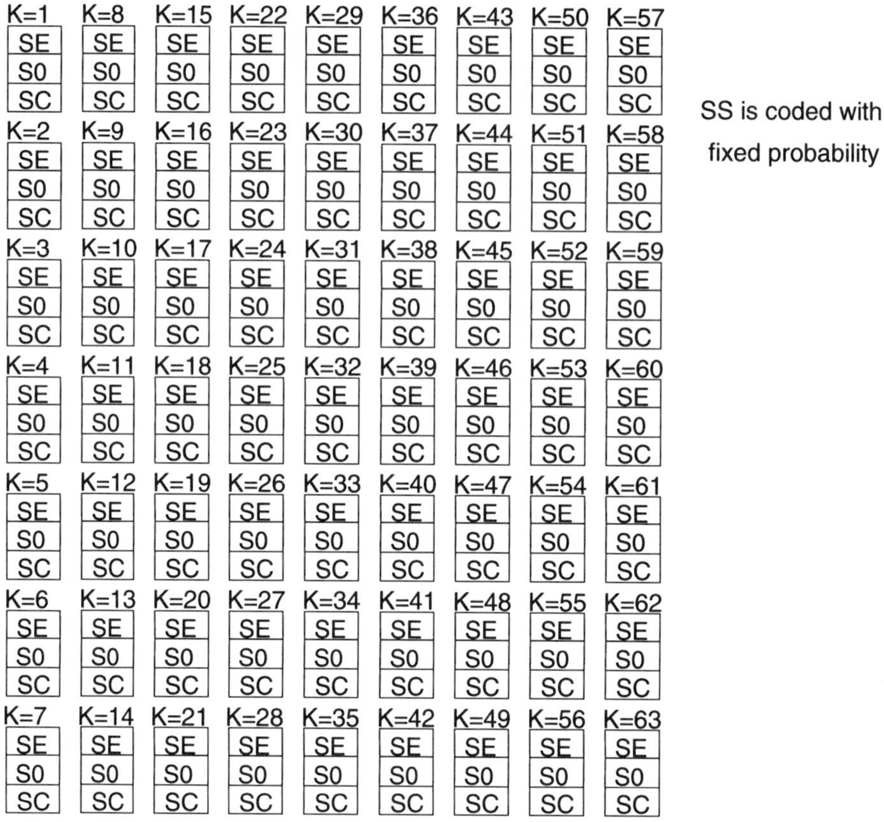

K=1	K=8	K=15	K=22	K=29	K=36	K=43	K=50	K=57
SE	SE	SE	SE	SE	SE	SE	SE	SE
S0	S0	S0	S0	S0	S0	S0	S0	S0
SC	SC	SC	SC	SC	SC	SC	SC	SC
K=2	K=9	K=16	K=23	K=30	K=37	K=44	K=51	K=58
SE	SE	SE	SE	SE	SE	SE	SE	SE
S0	S0	S0	S0	S0	S0	S0	S0	S0
SC	SC	SC	SC	SC	SC	SC	SC	SC
K=3	K=10	K=17	K=24	K=31	K=38	K=45	K=52	K=59
SE	SE	SE	SE	SE	SE	SE	SE	SE
S0	S0	S0	S0	S0	S0	S0	S0	S0
SC	SC	SC	SC	SC	SC	SC	SC	SC
K=4	K=11	K=18	K=25	K=32	K=39	K=46	K=53	K=60
SE	SE	SE	SE	SE	SE	SE	SE	SE
S0	S0	S0	S0	S0	S0	S0	S0	S0
SC	SC	SC	SC	SC	SC	SC	SC	SC
K=5	K=12	K=19	K=26	K=33	K=40	K=47	K=54	K=61
SE	SE	SE	SE	SE	SE	SE	SE	SE
S0	S0	S0	S0	S0	S0	S0	S0	S0
SC	SC	SC	SC	SC	SC	SC	SC	SC
K=6	K=13	K=20	K=27	K=34	K=41	K=48	K=55	K=62
SE	SE	SE	SE	SE	SE	SE	SE	SE
S0	S0	S0	S0	S0	S0	S0	S0	S0
SC	SC	SC	SC	SC	SC	SC	SC	SC
K=7	K=14	K=21	K=28	K=35	K=42	K=49	K=56	K=63
SE	SE	SE	SE	SE	SE	SE	SE	SE
S0	S0	S0	S0	S0	S0	S0	S0	S0
SC	SC	SC	SC	SC	SC	SC	SC	SC

SS is coded with fixed probability

Figure 12-7. Structure of the statistics area for later stages of successive approximation

of 0.5 (Qe = X'5A1D', $MPS = 0$). The statistics area for later stages of successive approximation is illustrated in Figure 12-7.

12.5 Statistical models for lossless coding and hierarchical-mode spatial corrections ◑

The arithmetic-coding statistical model used in the lossless mode and for hierarchical mode spatial corrections closely parallels the arithmetic-coding statistical model for the DCT DC difference coding. The decision tree in Figure 12-2 is adopted without modification. However, since some of the predictors are 2-D in the JPEG lossless mode of operation, the arithmetic coding conditioning is also extended to a 2-D form.

12.5.1 Two-dimensional conditioning statistical model ◑

The statistical model for the DC coefficient DPCM difference coding (section 12.3.1) is extended to a 2-D form by conditioning the S0, SS and

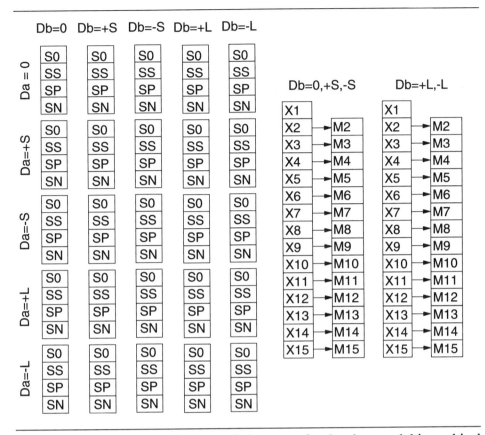

Figure 12-8. Structure of the statistics area for lossless and hierarchical
mode spatial corrections

SP/SN coding decisions on both the difference for the sample to the left
(*Da*) and the difference for the sample immediately above (*Db*). The *Da*
and *Db* differences are each segmented into five categories, following the
same rules that were defined segmenting *Da* in the 1-D DPCM coding
model for the DC coefficients (section 12.3.1.2). When combined, these
two independent conditioning parameters provide 25 conditioning states.

Figure 12-8 shows the statistics area needed for this 2-D model. Note
that the conditioning of the $X1, ..., Xn, Mn, ..., Mn$ decision sequence is
also enriched in that these decisions are now conditioned on the magnitude
category derived for *Db*.*

When conditioning is derived from neighboring values, conditioning
at the edges of the sample array must be defined. At the beginning of each
scan and restart interval the conditioning derived from the line above is

---

* Conditioning these decisions on *Da* gave no significant improvement in coding
efficiency.

set to zero. The value of *Da* used for conditioning at the left edge of each row of samples is also set to zero.

12.6 Arithmetic coding conditioning tables ◐

The DAC segment can be used to set new values for the conditioning parameters *L*, *U* and *Kx*. The syntax for doing this is given in Chapter 7. The SOI marker always resets these parameters to their default values of 0, 1, and 5, respectively. Unlike Huffman tables, they cannot be inherited; unlike Huffman tables, there is a default if nothing is explicitly transmitted.

13

MORE ON
ARITHMETIC CODING

In this chapter we discuss the characteristics that make the QM-coder efficient in both hardware and software. Most of these concepts go back to the Q-coder,[40-44] because the QM-coder is a "lineal descendent" of this arithmetic coder. The Q-coder was developed in direct response to the need for an arithmetic coder that: (1) required no multiplications, (2) could code efficiently across the range of probabilities typically encountered in image coding, (3) could estimate probabilities without significant additional computations, and (4) was well suited for both software and single-chip hardware implementations.

Many of the concepts in the Q-coder were developed as a result of earlier work. The approximation to the multiplication used in the Q-coder was taken directly from the skew coder (but with a subtle reinterpretation of the interval subdivision[61]). The concept of adaptive probability estimation was also well known at the time—for example, the Helman, et al., "Monte Carlo" estimator based on approximate counting,[62] and Mitchell and Goertzel's estimator based on periodically renormalized symbol counts.[69]

The Q-coder broke new ground in two places: the development of compatible but different coding procedures that allowed for separately optimized hardware and software implementations,[41, 70] and the integration of a very effective probability estimation technique into the arithmetic coding operations.[42, 71]

13.1 Optimal procedures for hardware and software ●

Compatible optimal coding procedures for hardware and software were developed because of a perceived conflict between the best hardware and the best software implementations. In general two symbol orderings are possible in binary arithmetic coding: LPS subinterval above MPS subinterval and MPS subinterval above LPS subinterval. If the convention is adopted that the code stream points to the bottom of the interval, the arithmetic-coding procedures for the two symbol orderings are as follows:

Convention (a):   [LPS above MPS]

```
    A = A − Qe              /* MPS subinterval                       */
    if MPS                  /* C is unchanged on MPS path            */
        renormalize if needed
    else                    /* LPS                                   */
        C = C + A           /* Point to base of upper (LPS) subinterval */
        A = Qe              /* New interval is LPS subinterval       */
        renormalize
    end
```

Convention (b):   [MPS above LPS]

```
    if LPS                  /* New interval is LPS subinterval       */
        A = Qe              /* C is unchanged on LPS path            */
        renormalize
    else                    /* MPS                                   */
        A = A − Qe          /* MPS subinterval                       */
        C = C + Qe          /* Point to base of upper (MPS) subinterval */
        renormalize if needed
    end
```

As in Chapter 9, A = interval, C = code stream, Qe = LPS probability estimate, and we approximate the multiplication as in the Q-coder. Conditional exchange is readily added to any of the coding conventions discussed in this chapter.

Convention (a) is better suited to software, because there is less computation on the more probable path. However, in hardware the LPS path has two operations that must be done serially, and this increases the maximum circuit delay.

Convention (b) is better suited to hardware, in that the two MPS path operations can be done in parallel. However, in software more computations are required on the more probable path, decreasing the software efficiency.

The symbol-ordering convention ([MPS above LPS] or [LPS above MPS] in the interval) and the code stream convention (where the code stream points into the interval) determine the meaning of the code stream. Including cases (a) and (b) above, there are actually four cases of interest for a Q-coder or QM-coder implementation:

Case	Symbol ordering	Code stream points to:
(a)	LPS above MPS	Bottom of interval
(b)	MPS above LPS	Bottom of interval
(c)	LPS above MPS	Top of interval
(d)	MPS above LPS	Top of interval

Figure 13-1 illustrates these four cases. Case (a) is the convention adopted by the QM-coder; case (b) is the convention used in the Q-coder. The two other procedures are:

Convention (c):   [LPS above MPS]

```
if LPS              /* C is unchanged on LPS path            */
   A = Qe           /* New interval is LPS subinterval       */
   renormalize
else                /* MPS                                   */
   A = A − Qe       /* MPS subinterval                       */
   C = C − Qe       /* Point to top of lower (MPS) subinterval */
   renormalize if needed
end
```

Convention (d):   [MPS above LPS]

```
A = A − Qe          /* MPS subinterval                       */
if MPS              /* C is unchanged on MPS path            */
   renormalize if needed
else                /* LPS                                   */
   C = C − A        /* Point to top of lower (LPS) subinterval */
   A = Qe           /* New interval is LPS subinterval       */
   renormalize
end
```

We shall first consider how case (c) can be made to generate the same code stream as case (a), and therefore how case (d) can be made to generate the same code stream as case (b). We shall then consider how the code stream for case (b) can be converted to the code stream for case (a), and vice versa.

13.1.1 Same code stream with different code-stream conventions ●

Let us first consider the two code stream conventions of Figure 13-1. We see that choices (a) and (c) converge in the limit of an infinite number of symbols to the same point on the number line. From an operational point

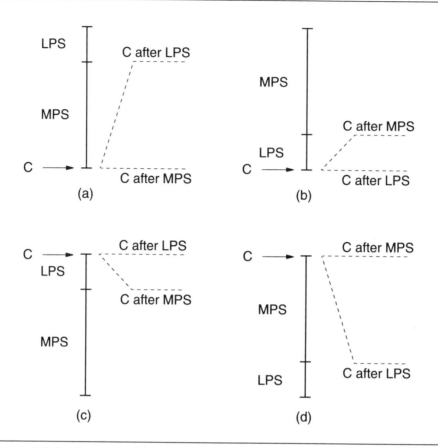

Figure 13-1. Four combinations of symbol orderings and code stream
          conventions

of view the two coding procedures are very different, yet the difference
between the two code streams is always equal to the remaining probability
interval A. Similarly, (b) and (d) converge to the same point. Therefore,
if at the end of coding the final interval is subtracted from the code stream,
the two procedures will give identical results for a given symbol ordering.

  Now let us take a closer look at these coding procedures. In procedure
(a) we add to the code register, and therefore need to guard against carry
propagation. For the "pure" output convention of the QM-coder, we need
to hold or stack X'FF' bytes until the carry has been resolved. The carry
is resolved either when it occurs and converts the stacked X'FF' bytes to
zeros or when the next byte is less than X'FF'. This guarantees that if a
carry occurs, the byte in the compressed buffer can absorb it and not
propagate it.

  In procedure (c) we subtract from the code register and therefore need
to guard against borrow propagation. This means that we need to hold or
stack X'00' bytes until the borrow has been resolved. The borrow is
resolved either when it occurs and converts the X'00' bytes to X'FF' or

C register:	0000 cbbb bbbb bsss xxxx xxxx xxxx xxxx
A register:	0 aaaa aaaa aaaa aaaa

Figure 13-2. Code register bit assignments

when the next byte is greater than X'00'. In direct analogy to blocking of carry propagation, if a borrow occurs, the byte in the buffer can supply it rather than propagating it.*

In section 9.1.9 of Chapter 9 we defined the bit assignments shown in Figure Figure 13-2 for the C and A registers. Bit "c" now takes on a broader interpretation, being a "carry bit" for convention (a) and a "borrow bit" bit for convention (c). The "c" bit is normally clear for convention (a); if set, a carry must be propagated to the byte in the buffer. By analogy, the "c" bit is normally set for convention (c); if clear, a borrow must be propagated to the byte in the buffer.

13.1.1.1 Initialization for case (c) ●

The initialization of the code register in procedure (c) must be done carefully. Ideally, we start both the interval register and the code register at 1.0000.... For the integer representation used in the QM-coder we have already noted that the A register can be set to X'10000', or, for a 16-bit register, to X'0000' (where the first subtraction automatically underflows to the correct value). We also must set the code register to this same value, but this introduces an interesting problem in the initialization. Where should the leading 1-bit be placed?

By convention, the borrow bit is set at the start of each new byte, and we therefore set the starting value of C and A to:

C register:	0000 0000 0000 0001 0000 0000 0000 0000
A register:	1 0000 0000 0000 0000

In addition, however, we must add 1 to the byte of data immediately preceding the start of the entropy-coded data. Note that for the register conventions that we have been using, 11 shifts are needed before the first byte is removed from the C register. This aligns the "c" bit with the position shown in Figure Figure 13-2.

---

* The description here is for the "pure" output conventions of the QM-coder. The concept of compatible, but different, coding conventions for hardware and software was first developed in the Q-coder,[41] and the Q-coder used bit-stuffing to block carry propagation instead of delaying the output until the carry was resolved. The various coding conventions discussed in this chapter can be used with either method for controlling carry propagation.

If the "c" bit is clear when the first byte is complete after 11 shifts (eight bits plus three spacer bits), it was borrowed during the coding of that byte and the borrow must be propagated to the byte in the code buffer. If the "c" bit is still set when the first byte is complete, nothing was borrowed, and all low-order bits, including the output byte bits (the "b" bits), must be zero. In this case the zero output byte must be stacked until the borrow is resolved. Eventually a borrow must occur (perhaps only at the end of coding), and for this reason the understood leading 1-bit is added to the byte just preceding the start of the entropy-coded data. Borrow resolution will always remove that bit.

13.1.1.2 Termination of the code stream ●

After the final symbol is coded, the two code streams are still exactly A apart. Therefore, if we subtract A from the code stream produced by convention (c), we have identical code streams. The final termination can then be done exactly the same way as in convention (a).

13.1.1.3 Byte stuffing ●

Byte stuffing after X'FF' is a postprocessing step, and is done exactly the same way in both procedures.

13.1.2 Converting between symbol orderings ●

Referring to Figure 13-1, we see by inspection that if we code a particular sequence of symbols following case (a), the resulting code stream is a binary fraction that is the sum of the LPS subintervals for all LPS coded. If we code with the opposite symbol ordering but with the code stream pointing to the top of the interval (case (d)), the cumulative offset from the top of the interval will be the same as the offset from the bottom of the interval in case (a). Therefore, regarding the code stream as a fraction pointing to a particular subinterval, the code stream for case (a), C_a, is related to the code stream for case (d), C_d, by:

$$C_a = 1 - C_d \qquad [13\text{-}1]$$

When fixed-precision arithmetic is used, we must convert this to a fractional form as follows:

$$C_a = \text{X'FFFF ... FF'} - C_d \qquad [13\text{-}2]$$

Each byte of C_d is subtracted from X'FF' to invert it (which can be done by exclusive ORing each C_d byte with X'FF' if that is more convenient). Stuffed bytes are removed from C_d before the conversion and inserted in C_a after the conversion.

For inverted code streams the "Pacman" termination must be done differently. For the standard conventions of LPS over MPS and code stream at the bottom of the interval, the code stream is adjusted to clear as many trailing zero bits as possible, and trailing zero bytes are then discarded. When the symbol ordering is inverted, the code stream is adjusted to set as many trailing 1-bits as possible and trailing X'FF' bytes are then discarded. In both cases the bits discarded are produced by coding of MPS. The standard convention decoder regenerates trailing 0-bits to complete decoding, whereas the inverted code stream decoder regenerates trailing 1-bits—the equivalent of 0-bits after inversion.

13.2 Fast software encoder implementations ●

The QM-coder is structured to permit fast software. Although many of the "tricks" one uses are very dependent on specific details of the instruction set, there are some general aspects that apply to virtually all computer architectures. There are also a number of things about the QM-coder that are designed for efficient software, and most of this section will be devoted to those aspects.

One of the obvious things one must do to get to fast software is to map the key variables, $Qe(S)$, A, and C, to machine registers and keep those variables in registers as much as possible. Branching is potentially a costly operation (although some modern architectures that incorporate speculative execution significantly reduce the penalty for branching). Note that it is usually quite advantageous to structure the branching such that the branch is not taken for the more probable case. However, the specifics of these aspects of optimization are so architecture-dependent that it is inappropriate for us to go into further detail here.

In most processors a special register, the condition code register, contains bits which are set appropriately for carry/borrow, sign, overflow, and other conditions following certain arithmetic and logical operations. Although the details are quite processor-dependent, these condition codes can usually improve software efficiency significantly when they are used to control branching. In this respect it is important to note that many of the branching decisions in the QM-coder flowchart depend on the sign of the result of an arithmetic operation that immediately precedes the branch. As one example, note that renormalization is required whenever the interval register is less than X'8000'. This means that the test of a single bit, typically the sign bit, determines when the renormalization is needed.[*]

Another "trick" can be used to improve the computational efficiency of the probability estimation. In Chapter 9 the state machine estimation is described as it is documented in the JPEG DIS. There is, however, a functional equivalent of these procedures which removes the need for a test of *Switch_MPS* on the LPS renormalization path. If the sense of the MPS and the Qe value are combined into one 16-bit entity in which the MPS is determined by the sign bit, the sense of the MPS can then be determined from the composite value. This avoids separate storage accesses for these two quantities.

---

[*] This single-bit test is derived from the skew coder,[47] where it was used to simplify hardware design.

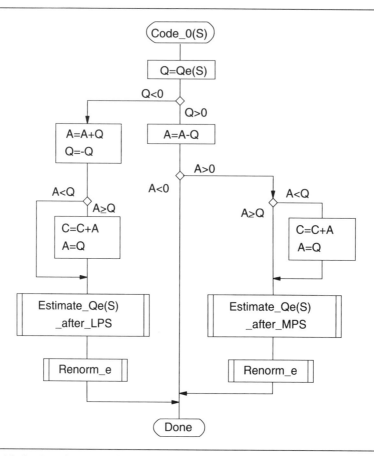

Figure 13-3. Implementation of Code_0 for modified estimator state machine

There are at least two ways of doing this. The simplest we have found is to make *Qe* positive if the MPS sense is zero, and negative if the MPS sense is 1.[72] Then, the state machine is doubled in size, as half is for positive MPS, and half is for negative MPS. Whenever the *Switch_MPS* bit is 1 on the LPS renormalization path, the transition is to the appropriate *Qe* in the other half of the state machine. With this doubled state machine, estimation on the LPS and MPS paths is essentially identical, the only difference being the direction of movement in the state machine.*

---

* Note that in doubling the state machine, one state can never be reached when the system is started at the mandated initial state. Just as the estimator can never return to the starting state in the fast-attack ladder, it can never get to the corresponding state for negative *Qe* values. Consequently, that state can be replaced by a fixed probability with no adaptation (where the *Next_index_LPS* and *Next_index_MPS* are the same as the state index). This state can be used for coding fixed probabilities such as the DCT AC coefficient sign decision.

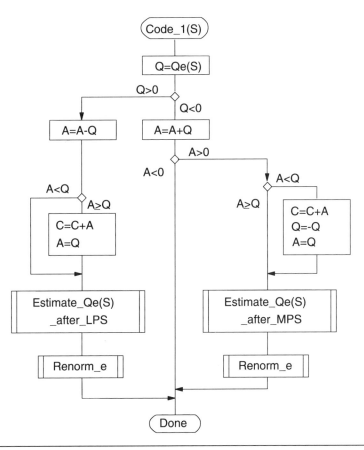

Figure 13-4. Implementation of Code_1 for modified estimator state machine

Figures 13-3 and 13-4 illustrate the implementation of Code_0(S) and Code_1(S) with this doubled state machine; these figures also illustrate a number of other things that lead to efficient software. They assume that A, C and the context index S are already in dedicated registers, and that another register, the Q register, is available to hold the Qe value.

Referring to Figures 13-3 and 13-4, the value of $Qe(S)$ is first loaded into the Q register and the sense of the MPS is determined by testing the sign of Q. Because the sense of the decision being coded is known at the entry to the procedure, the appropriate branch can then be taken. If the branch is taken for negative Q in either figure, the MPS subinterval is calculated by adding the negative Q to the A register. There is no need, therefore, to negate the Q value when following the MPS path for Code_1(S). Note that after A is decreased by subtracting Q (or adding a negative Q), the condition code for positive or negative result is usually set. There is no need, therefore, in most processors to test the sign of A explicitly.

Once the branch into a renormalization path is taken, Q is negated if it is negative and the Q register is then compared to the A register to see whether conditional exchange is needed. After the C register and A register have their final value, estimation and renormalization complete the coding on these paths. Note that the condition code can also be used in the renormalization loop to determine when renormalization is complete. If the shift-left-logical of A does not set the condition code, the same effect can be achieved by adding the A register to itself. In most processors that sets the condition code appropriately.

The counter used to count the renormalization shifts can also be eliminated. A flag bit can be placed in the encoder code register such that the register has the following structure:

	C-register						
	MSB						LSB
Starting new byte	0000	0000	f000	0sss	xxxx	xxxx	xxxx	xxxx
Byte complete	f000	cbbb	bbbb	bsss	xxxx	xxxx	xxxx	xxxx

Then, if after a shift left (or add of C to itself) the condition code indicates that the sign bit is set, a new byte of data is complete. The branch to Byte_out in Figure 9-5 is then taken. A 32-bit architecture is assumed here, but similar implementations are possible in 16-bit architectures with a pair of registers or with a register and a storage location.

13.3 Fast software decoder implementations ●

Although most aspects of fast software are common to both encoder and decoder, there is one not-so-obvious aspect of decoder software that deserves mentioning. In the encoder the symbol-ordering convention determines whether the encoder organization is appropriate for serial or parallel operation; encoding with serial operations is usually better for software, whereas encoding with parallel operations is usually better for hardware. The parallel-operation decoder, however, can be used in both hardware and software.

Decoding with serial operations and a standard code stream is done as follows for the $MPS = 0$ case:

```
A = A - Q            /* Calculate MPS subinterval        */
if C < A             /* If C within MPS subinterval       */
   decode MPS subinterval
else
   decode LPS subinterval
end
```

Decoding with parallel operations is carried out with the inverted code stream and symbol ordering:

```
    C = C − Q                /* Subtract LPS interval from C      */
    if C ≥ 0                 /* If C did not underflow            */
        A = A − Q            /* New interval is MPS subinterval   */
      decode MPS subinterval
    else
        C = C + Q            /* Restore C register                */
        A = Q                /* New interval is LPS subinterval   */
      decode LPS subinterval
    end
```

The parallel conventions seem to involve an inefficiency in that the subtraction of Qe from C must be undone on the LPS subinterval path. However, because condition codes can now be used in place of explicit tests in both the decoding and renormalization comparisons, there is actually a net saving.

13.4 Conditional exchange ●

The additional computations for conditional exchange are minimized by the fact that conditional exchange can only occur and therefore must only be tested for when renormalization is needed.

While the net effect of conditional exchange is to increase modestly the complexity of the QM-coder, there are some trade-offs relative to the simpler conventions of the Q-coder. First, by introducing conditional exchange we guarantee that the interval register for the MPS symbol is always greater than decimal 0.375. Consequently, renormalization after coding the MPS symbol never requires more than one renormalization shift and the test in the renormalization loop can be eliminated on the MPS path. Second, Arps[73] has pointed out that in hardware implementations the test for conditional exchange can be converted from

$$A − Q < Q$$

to

$$A < 2Q$$

and can therefore be done in parallel with other decisions in both encoder and decoder.

13.5 QM-coder versus Q-coder ●

Because the QM-coder is so close to the Q-coder in structure and operation, it may be of some interest to compare the two and understand the relative advantages of one versus the other. The three areas in which there are significant differences are in the interval subdivision, the output path, and the probability estimation tables.

The interval subdivision in the QM-coder is improved over the Q-coder by the addition of conditional exchange to the QM-coder. This increases the complexity modestly (<10% for software), but provides about 0.5% better compression.

The carry-over control in the QM-coder is accomplished by always resolving the carry-over in the encoder. Quite separately from this, the QM-coder stuffs zero bytes after each X'FF' in order to guarantee that markers can be located without decoding. The Q-coder uses bit-stuffing to provide a carry receiver, thereby blocking carry propagation without fully resolving the carry. The register conventions defined for the Q-coder guarantee that any value from X'90' to X'FF' can never be produced immediately following an X'FF' byte by a correctly functioning Q-coder.

The QM-coder carry resolution is significantly simpler, but does lead to a slight loss in coding efficiency. This is actually not due to the carry resolution (which requires no bit stuffing at all), but to the byte stuffing defined to keep the markers unique. The one significant problem with the QM-coder technique for handling carry-over is output latency. Under certain conditions strings of X'FF' bytes can occur, and the output must be stacked until the carry is resolved. The only true upper bound on the number of stacked bytes is the length of the compressed data, and it is not at all difficult to create conditions in which stack counts in excess of 80 occur.

The third area in which there are significant differences between the QM-coder and the Q-coder is the estimation tables. Both estimators use the Q-coder renormalization-driven estimation, but the QM-coder uses an initial (fast attack) sequence of estimator states to arrive quickly at approximately the right probability estimate. When initial learning over-head is a significant factor, this provides the QM-coder with significantly improved performance.

The Q-coder uses a simpler state machine that was optimized for sequential compression algorithms. Because there are many fewer states, this state machine is more responsive to unstable statistics; it can outperform the QM-coder when the coding model is poor and the statistics unstable. The effect is most observable in the JBIG sequential mode, where under some conditions the Q-coder performance on binary halftone images can be significantly better. This is especially true when fast-attack states are added to the state machine, as then the overhead for initial learning is greatly reduced. Data on a state machine with a granularity close to that of the Q-coder will be presented in Chapter 14.

One aspect of the Q-coder estimator state machine that was not carried over into the QM-coder was the constraint on bits set in the estimated probability values. This is unfortunate, in that this constraint, suggested by Langdon,[74] would have permitted significantly faster decoding of the state machine index in hardware implementations.

13.6 Resynchronization of decoders ●

The QM-coder marker structure and termination procedures were designed with two specific properties: first, the ability to identify uniquely any marker without decoding the entropy-coded segment, and, second, the ability to guarantee that the marker following the last byte in the entropy-coded segment would be encountered before the segment was completely decoded.

Assuming restart is enabled, the first property allows a decoder to periodically recover from bit errors in the entropy-coded segment. Once the decoder has determined that an error condition exists, it can scan the entropy-coded segment to locate the next restart marker and resume decoding there.

Two symptoms of error conditions are: completing decoding of the restart interval before encountering the restart marker, and decoding physically impossible data. Determining that an error condition exists is not always possible, because some compressed data bit errors do not create detectable output errors.

Bit errors can also create or destroy markers in the entropy-coded data. In the first case, the code is highly likely to be illegal, and error recovery then consists of scanning for the next restart marker. In the second case, the expected marker at the end of the entropy-coded segment will not be found, and error recovery again consists of scanning for the next restart marker. In this case the modulo-8 count imbedded in the restart marker permits identification of the MCU at which recovery is possible. The intervening data are corrupted, of course, just as they are in CCITT-G3 facsimile error recovery.

13.7 Speedup mode ●

The QM-coder normally requires one complete cycle per symbol encoded/decoded. In some applications such as the JBIG bi-level image compression system, however, long sequences of more probable symbols are encoded/decoded using the same context. It is possible then to use a "speedup mode"[75] to code these runs of symbols and contexts in larger groups.

Suppose that it is known that a sequence of N more probable symbols is to be coded for a particular context. This may be done in the conventional QM-coder implementation by reducing the interval A for each decision coded, comparing A with the lower bound on A after each reduction, and renormalizing whenever A becomes too small. However, if Qe (the amount by which A is reduced) is small, it may be possible to code a large number of symbols before renormalization is needed. In fact, if Qe is too small for conditional exchange to occur ($Qe \leq$ X'4000'), the number of symbols which can be coded before renormalizing is one more than the integer quotient $(A-$ X'8000'$)/Qe$. If this quotient is greater than or equal to the number N of symbols to be coded, A can be reduced by $N \times Qe$ before testing for renormalization.

If the quotient is less than N, then the number of symbols specified by the quotient can be coded by reducing A by the product of the quotient and Qe. Alternatively, the A value can be reset to the remainder plus X'8000'. The number of symbols coded is then subtracted from N and one additional coding operation is performed; this coding operation causes a renormalization. Any remaining symbols with the same context can then be coded by repeating the process. As long as $Qe \leq$ X'4000' or N is greater than one, no conditional exchange can occur. Note that because only the MPS is being coded, renormalizations will make Qe smaller.

A similar logic applies to the decoder. If long sequences of more probable symbols are expected to occur in a given application and if the decoding of such symbols does not cause the model to change the context within the sequence, the QM-decoder can be modified to decode more than one decision at a time. As long as $Qe \leq X'4000'$ and Qe can be subtracted from A without resulting in either of the conditions $C \geq A$ or $A < X'8000'$, more probable symbols will be decoded. Therefore, if some suitable multiple of Qe ($K \times Qe$ for some K greater than one) is chosen, a group of K more probable symbols can be decoded in one operation.

This "speedup mode" makes it computationally practical to use compression models that tend to produce runs of MPS symbols in a given context. Such runs can be encoded and decoded by the QM-coder without the necessity of coding each decision individually. The compressed data stream is the same as the one the conventional QM-coder would produce.

For sequential neighborhood template models such as those adopted by JBIG, speedup mode improves software throughput by a factor of about five. Similar concepts can be used to improve hardware throughput.

14

PROBABILITY ESTIMATION

This chapter is devoted to the probability estimation technique used in the JPEG/JBIG arithmetic coder (QM-coder). The probability estimation is directly imbedded in the arithmetic-coding procedures, and uses renormalization of the interval register to "count" the symbols. The basic concept of renormalization-driven estimation is derived from the Q-coder,[40] but there are conceptual links to other probability estimation techniques that are also based on approximate counting.[62, 64]

An important aspect of the estimation technique used in the Minimax arithmetic coder[46] is also incorporated in the QM-coder estimator. The Minimax arithmetic coder uses probabilities taken from Bayesian estimates generated from direct counts of symbols. This, in combination with appropriate assumptions about expected probabilities, is very effective in minimizing the impact on coding efficiency due to the coding of the initial decisions for a context. When coding the initial decisions, the coder "learns" about the probability, and Bayesian estimation provides an optimal strategy for this initial learning stage.

When the Bayesian estimation concept was incorporated into renormalization-driven estimation by adding states for initial learning,[63] it was found that approximate counting produced a coding performance

233

nearly as good as that achieved with complete symbol counting. Consequently, the combination of renormalization-driven estimation and Bayesian estimation, together with approximate subdivision of the interval, conditional exchange, and "pure" carry-over resolution, was adopted by both JPEG and JBIG for the arithmetic-coding technique. Further optimization of the estimation state machine by JBIG produced the probability estimation technique used in the QM-coder.

14.1 Bayesian estimation ◐

In Huffman coding the codes are developed by counting the occurrence of each symbol. As was done in Chapter 11, the symbol probability is assumed to be proportional to the symbol count. However, some symbols occur very infrequently. If a code word must be available for each symbol in the alphabet, the count for each symbol is typically started at 1. In an N-ary alphabet, therefore, the initial assumed probabilities are:

$$P(1) = P(2) = ... = P(N) = 1/N \qquad\qquad [14\text{-}1]$$

which would give a code book of equal-length codes of length $\log_2(N)$. Of course, once the counts $C(M)$ are accumulated for each of the N symbols this estimate would be revised to:

$$P(M) = \frac{C(M) + 1}{\displaystyle\sum_{K=1}^{N} (C(K) + 1)} \qquad\qquad [14\text{-}2]$$

This is a particular form of Bayesian estimation in which no prior knowledge is assumed about symbol probabilities.[76] In the limit of large symbol counts the effect of prior information about symbol probabilities becomes vanishingly small. This is usually the case in Huffman coding, in which the primary effect of an assumption of prior information is to reserve a code word for all symbols.

More generally, when there is prior information about symbol probabilities, the starting numerical factors can be adjusted to account for that information:[77]

$$P(M) = \frac{C(M) + B(M)}{\displaystyle\sum_{K=1}^{N} (C(K) + B(K))} \qquad\qquad [14\text{-}3]$$

where $B(M)$ represents prior information about the expected probability for symbol M.

The following derivation of the above relationship for a binary alphabet is adapted from a more general treatment found in Williams.[78] Given symbols 0 and 1 with counts $C(0)$ and $C(1)$, respectively, and a linear relationship between probabilities and symbol counts,

$$P(0) = \frac{aC(0) + b}{c(C(0) + C(1)) + d} \qquad\qquad [14\text{-}4a]$$

$$1 - P(0) = \frac{aC(1) + b}{c(C(0) + C(1)) + d} \qquad\qquad [14\text{-}4b]$$

where a, b, c and d are constants. Since $C(0)$ and $C(1)$ are arbitrary, these equations are consistent only if $a = c$ and $d = 2b$. If we then define $t = b/a$, we get:

$$P(0) = \frac{C(0) + t}{C(0) + C(1) + 2t} \qquad\qquad [14\text{-}5]$$

where t represents prior information about the probabilities. In general, the terms $B(M)$ (Equation 14-3) and t (Equation 14-5) are greater than zero. However, if we are dealing solely with *post priori* estimates, a symbol probability can be zero; such is the case with custom Huffman tables.

14.2 Renormalization-driven estimation ❶

The derivation of probability estimates from arithmetic coder renormalization is based on the ability to make an approximate estimate of the counts of LPS and MPS symbols after each renormalization. Each time the arithmetic encoder or decoder is renormalized—that is, after every LPS and after each MPS that requires renormalization—a new probability estimate is generated.

Figure 14-1 illustrates how the interval is revised after each coding. If an LPS occurs, the LPS count is 1 and the MPS count is still too small to produce an MPS renormalization. This causes a transition of the estimator state machine to a larger LPS probability estimate. Conversely, if the MPS renormalization occurs, the LPS count is still zero; this causes a transition to a smaller LPS probability estimate. Assuming a simple nearest-state transition for both MPS and LPS, the state machine balances at an LPS probability estimate where the MPS and LPS renormalization probabilities are equal on average. If the state machine is correctly formulated and in equilibrium, the distribution of probability estimates should be centered about the desired value.

The driving forces that tend to keep the distribution centered about the desired values arise as follows. If the probability estimate is too high, the number of MPS that can be coded without MPS renormalization is too small and the probability of the MPS renormalization is too high. The state machine is therefore likely to make a transition to a smaller estimate. Conversely, if the probability estimate is too small, the number of MPS that can be coded without MPS renormalization is too large and the state machine is then likely to make a transition to a larger estimate.

The degree of change of the probability estimate depends on the granularity (the density of states) of the state machine. On average, coarser granularities require a larger change in the probability estimate for each transition. Although the larger change in the probability estimate implies

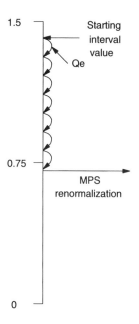

Figure 14-1. Subdivision of the interval for a sequence of MPS

a stronger restoring force toward the equilibrium value, the net effect of coarser granularity is a broadening of the distribution of probability estimates. Therefore, coding inefficiency increases with coarser granularity, even when the average probability estimate is still quite accurate. Conversely, a finer granularity leads to smaller coding inefficiencies.

The role of granularity can also be cast in terms of Bayesian estimation. If the granularity is coarse, less weight is placed on information from symbol counts prior to the current renormalization interval and the current counts therefore have a larger effect on the new estimate. A fine granularity has the opposite effect, increasing the weight given to symbols coded before the current renormalization interval.

14.3 Markov-chain modelling of the probability estimation ●

The state machine for the estimation process is a Markov chain containing one state for each probability estimate (note that the probability estimate includes the sense of the MPS). Except for the two special states at each end of the chain, each state can make transitions to two other states. The transition to a state with smaller LPS probability is made after each MPS renormalization, and the transition to a state with larger LPS probability (or a change in MPS sense) is made after each LPS renormalization. In turn, transitions into the state occur from one or more other states. By

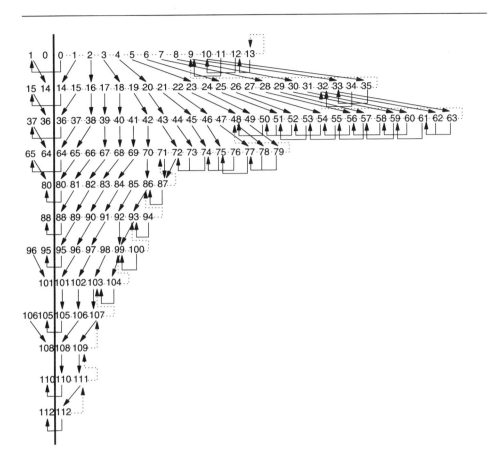

Figure 14-2. QM-coder estimation state machine

definition, the state machine has mirror symmetry about the change in the
sense of the MPS.

A sketch of the QM-coder estimation state machine is shown in Figure
14-2. The numbers marking each state are the index values in Table D.2
in the JPEG DIS. Many of the states in this estimator state machine are
transient states that are only populated during the learning stage of estimation. After a sufficient number of symbols have been coded, the occupation probability of these states becomes vanishingly small; only the
nontransient states that can be reentered from some other nontransient
state have a nonzero equilibrium population probability. These nontransient states, remapped to make the similarity to the original Q-coder state
machine clearer, are shown in Figure 14-3, and are also listed in Table
14-1. The *skew* referred to in that table is defined by $skew = \log_2(1/q)$.

Three models have been developed for the nontransient portion of this
renormalization-driven estimation procedure. The first, a simple approximate model, predicts the general character of the state machine transitions. The second, a precise modelling of the state machine behavior for

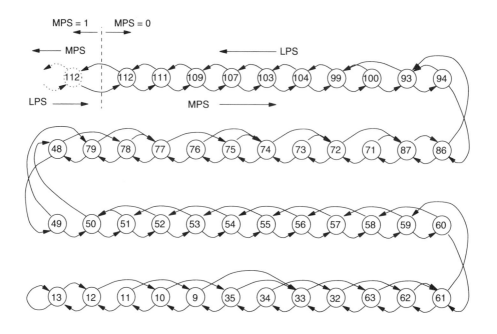

Figure 14-3. Nontransient probability states in the QM-coder state machine

a single context, provides accurate prediction of coding inefficiencies. The third, a more approximate model for mixed-context coding, allows semi-quantitative analysis of the more complex situation in which coding of many contexts is interleaved.

14.4 Approximate model ●

The following simple model for the state machine transitions is found in the Q-coder overview.[40]

The interval register is decremented by the current LPS estimate Qe following each MPS. After N_{MPS} symbols, the MPS renormalization occurs. This gives:

$$N_{MPS} = dA/Qe \qquad [14\text{-}6]$$

where dA is the change in the interval register between renormalizations. We adopt conventions for the finite state machine organization such that increasing index k corresponds to decreasing Qe. We also assume, as in Figure 14-3, that MPS transitions can occur only to the nearest neighbor. Thus, for each LPS transition,

$$N_{MPS} = [A - 0.75 + 0.75(dk - 1)]/Qe \qquad [14\text{-}7]$$

Table 14-1. Nontransient states of QM-coder estimation state machine

Index	*Qe* (hex)	*Qe* (decimal)	*Skew* (decimal)	dImps	dIlps	Xmps	DIS Index
0	59EB	0.49582	1.01211	1	0	1	112
1	5522	0.46944	1.09100	1	1	0	111
2	504F	0.44283	1.17516	1	1	0	109
3	4B85	0.41643	1.26387	1	1	0	107
4	4639	0.38722	1.36878	1	1	0	103
5	415E	0.36044	1.47215	1	1	0	104
6	3C3D	0.33216	1.59004	1	1	0	99
7	375E	0.30530	1.71169	1	1	0	100
8	32B4	0.27958	1.83864	1	2	0	93
9	2E17	0.25415	1.97627	1	1	0	94
10	299A	0.22940	2.12408	1	2	0	86
11	2516	0.20450	2.28985	1	1	0	87
12	1EDF	0.17023	2.55446	1	1	0	71
13	1AA9	0.14701	2.76603	1	2	0	72
14	174E	0.12851	2.96010	1	1	0	73
15	1424	0.11106	3.17061	1	2	0	74
16	119C	0.09710	3.36437	1	1	0	75
17	0F6B	0.08502	3.55610	1	2	0	76
18	0D51	0.07343	3.76751	1	2	0	77
19	0BB6	0.06458	3.95286	1	1	0	78
20	0A40	0.05652	4.14509	1	2	0	79
21	0861	0.04620	4.43588	1	2	0	48
22	0706	0.03873	4.69047	1	2	0	49
23	05CD	0.03199	4.96640	1	2	0	50
24	04DE	0.02684	5.21956	1	1	0	51
25	040F	0.02238	5.48167	1	2	0	52
26	0363	0.01867	5.74276	1	2	0	53
27	02D4	0.01559	6.00280	1	2	0	54
28	025C	0.01301	6.26424	1	2	0	55
29	01F8	0.01086	6.52537	1	2	0	56
30	01A4	0.00905	6.78840	1	2	0	57
31	0160	0.00758	7.04321	1	2	0	58
32	0125	0.00631	7.30789	1	2	0	59
33	00F6	0.00530	7.56013	1	2	0	60
34	00CB	0.00437	7.83731	1	2	0	61
35	00AB	0.00368	8.08479	1	1	0	62
36	008F	0.00308	8.34277	1	2	0	63
37	0068	0.00224	8.80221	1	2	0	32
38	004E	0.00168	9.21724	1	2	0	33
39	003B	0.00127	9.62000	1	2	0	34
40	002C	0.00095	10.04321	1	2	0	35
41	001A	0.00056	10.80221	1	3	0	9
42	000D	0.00028	11.80221	1	2	0	10
43	0006	0.00013	12.91768	1	2	0	11
44	0003	0.00006	13.91768	1	2	0	12
45	0001	0.00002	15.50265	0	1	0	13

where dk is the change in index after the LPS renormalization and A averages to X'B55A' (decimal 1.06). Thus, on average,

$$N_{\text{MPS}} = [0.31 + 0.75(dk - 1)]/Qe \qquad\qquad [14\text{-}8]$$

Because Qe does not change very much with each renormalization, the estimator will be balanced when the true LPS probability q is approximately equal to the estimate:

$$Qe \approx q = \frac{1}{N_{\text{MPS}} + 1} \qquad\qquad [14\text{-}9]$$

where prior symbol counts are also assumed to be consistent with q and can therefore be eliminated from the equation. Substituting for N_{MPS}, we get an approximate relationship that for q near 0.5 is best satisfied by $dk = 1$, whereas for small q the relationship is best satisfied by $dk = 2$. Note that two of the transitions in Figure 14-3 violate this rule. As will be seen, this causes problems.

14.5 Single- and mixed-context models ●

If the effects of conditional exchange are ignored (conditional exchange can occur only for probabilities from 0.5 to 0.375), the models developed for the Q-coder probability estimation[42] can be applied to the nontransient states of Figure 14-3. The formalism presented in this section applies to both the single-context and mixed-context models. Both of these models are Markov chain models[77] in which the occupation probabilities of each state can be calculated from a transition probability matrix.

In a Markov chain model the net transition probability from each state is zero when the system is in equilibrium. For all transitions relating to the ith state,

$$\sum_{j} Pij = 0 \qquad\qquad [14\text{-}10]$$

where $P_{i,j}$ is the transition probability from state i to state j. For each state i, if the transition is into the state, the transition probability is positive; if the transition is out of the state, the transition probability is negative.

The array of equations for $P_{i,j}$ is overdetermined as it stands, because only ratios of occupation probabilities can be determined. The full solution is obtained by replacing one equation in the array with the constraint that the total state occupation probability must be unity.

The problem is now to calculate the transition probabilities. For both models the transition probability from any state to the appropriate neighboring state after coding the LPS is given by the product of the occupation probability and the LPS symbol probability. The transition probability after coding the MPS is calculated differently, depending on whether the model is for a single context or mixed contexts.

14.6 Single-context model ●

For the single-context model each LPS transition "seeds" a starting interval value for an unbroken string of MPS events. The probability of a string of m MPS events of probability $(1 - q)$ is given by $(1 - q)^m$, which gives a decreasing probability for each MPS renormalization that follows. An example of such a chain of MPS renormalizations is shown in Figure 14-4. For each state in the chain there is a net MPS transition probability into the state given by the difference between the probability of reaching the state before an LPS occurs and leaving the state before an LPS occurs.

The problem is made slightly more complicated by the interchange of the sense of MPS at the middle of the state machine. We account for this by defining

$$X_j = q \qquad\qquad k \geq k_s \qquad\qquad\qquad\qquad [14\text{-}11a]$$

$$X_j = 1 - q \qquad\qquad k < k_s \qquad\qquad\qquad\qquad [14\text{-}11b]$$

where k_s is the index value at which the MPS sense is switched. Note that the state machine has mirror symmetry about the boundary between k_s and $k_s - 1$. Note also that interchange between the two halves of the machine can occur only by means of LPS transitions. Therefore, a decaying chain of MPS events can involve only one sense of the definition of X_j in Equations 14-11a and 14-11b.

The net MPS transition probability into the state is balanced by a probability of LPS transitions out of the state. Equation 14-10 therefore becomes:

$$\sum_j n_j X_j (1 - X_j)^{t_{j,k}} - n_j X_j (1 - X_j)^{r_{j,k}} = n_k X_k \qquad\qquad [14\text{-}12]$$

where n_j is the occupation probability of state j and $n_j X_j$ is therefore the probability of leaving state j by means of an LPS renormalization.

The terms in Equation 14-12 arise as follows. After each LPS event we assume that an ever-lengthening chain of MPS events occurs (with ever-decreasing probability). Since the starting LPS value is known, the LPS renormalization provides a known interval register value.* Therefore, we can calculate exactly how many MPS events must occur before we reach state k, starting from state j. As each symbol is coded, the probability of not undergoing an LPS transition is $(1 - X_j)$. Therefore, if $t_{j,k}$ MPS events are required to reach state k when starting from state j, $(1 - X_j)$ must be raised to the $t_{j,k}$ power to get the probability of reaching state k. Similarly,

---

* If conditional exchange is allowed, this is no longer true. The computation of the transition probabilities then becomes far more complex.

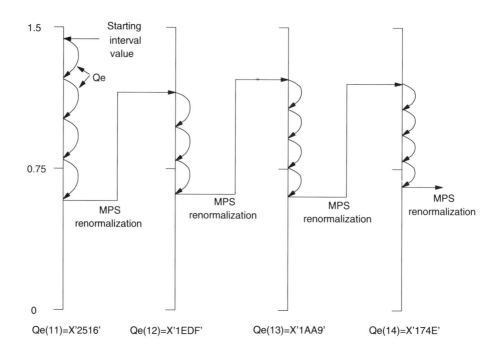

Figure 14-4. Example of a chain of MPS renormalizations. The Qe **values are for index values 11, 12, 13, and 14 of Table 14-1.**

if $r_{j,k}$ MPS events are required to leave state j, when starting from state j, $(1 - X_j)$ must be raised to the $r_{j,k}$ power to get the probability of leaving state k. Finally, to get a net transition probability of zero, we equate the net probability of entering state k because of MPS events to $n_k X_k$, the probability of leaving state k because of an LPS event.

For the single-context model the coding efficiency is calculated from a knowledge of the transition probabilities from each state due to renormalizations and the number of bits produced by that renormalization. Once the occupation probability n_k of state k is known, the bits per symbol for that state are given by

$$R_k = n_k X_k (B_{\text{LPS},k} + \sum_j B_{\text{MPS},k,j} (1 - X_j)^{r_{k,j}}) \qquad [14\text{-}13]$$

where $B_{\text{LPS},k}$ is the number of bits produced by an LPS renormalization out of state k, and $B_{\text{MPS},k,j}$ is the number of bits produced by MPS renormalization out of state j in the MPS chain that starts with an LPS from state k. For the Q-coder (no conditional exchange) some MPS renormalizations require two shifts. The conditional exchange in the QM-coder prevents this, but because the effect of conditional exchange is not included in this calculation, we must still allow for more than one shift.

The total coding rate in bits/symbol is the sum of R_k over all states.

$$R = \sum_k R_k \qquad\qquad [14\text{-}14]$$

The coding inefficiency, CI, is therefore given by:

$$CI = \frac{R - H}{H} \qquad\qquad [14\text{-}15]$$

where H is the entropy of the binary source with LPS probability q, i.e.,

$$H = - q \log_2(q) - (1 - q) \log_2(1 - q) \qquad\qquad [14\text{-}16]$$

The agreement between calculated and experimental coding inefficiencies is extremely good for this model, as should be expected. Except for truncation of the set of equations that we solve numerically to get the occupation probability of each state, there are no approximations.

14.7 Mixed-context model ●

For the mixed-context coding model the MPS renormalization probability is determined by the probability that the interval register, A, is small enough that the next MPS event will cause renormalization. Assume that the interval register values have a uniform distribution in the interval {0.75, 1.5}. Then, given that state k is occupied, the probability $Pl_{k,l}$ of undergoing an LPS renormalization to the appropriate neighboring state l is:

$$Pl_{k,l} = q \qquad\qquad [14\text{-}17]$$

and the probability $Pm_{k,m}$ of undergoing an MPS renormalization to the appropriate neighboring state m is:

$$Pm_{k,m} = (1 - q) \frac{Qe_k}{0.75} \qquad\qquad [14\text{-}18]$$

Equation 14-10 then becomes:

$$\sum_j n_j Pl_{j,k} + \sum_j n_j Pm_{j,k} = n_k Pm_{k,m} + n_k Pl_{k,l} \qquad\qquad [14\text{-}19]$$

where the two sums are only over states that can make LPS or MPS transitions, respectively, to state k.

The coding rate in bits per pixel is calculated from the partitioning of the interval. For a given estimate Qe_k and interval A, the code length for the LPS is $-\log_2(Qe_k/A)$ and the code length for the MPS is $-\log_2(1 - Qe_k/A)$. Assuming the A values are distributed uniformly, the coding rate for state k is given by:

$$R_k = \frac{1}{0.75} \int_{0.75}^{1.5} (-q \log_2(Qe_k/A) - (1 - q) \log_2(1 - Qe_k/A))dA \quad [14\text{-}20]$$

which can be integrated by standard forms to

$$R_k = qR_{\text{LPS},k} + (1 - q)R_{\text{MPS},k} \quad\quad\quad [14\text{-}21a]$$

$$R_{\text{LPS},k} = \log_2(X_0) + 2 - \log_2(e) \quad\quad\quad [14\text{-}21b]$$

$$R_{\text{MPS},k} = 2 \log_2\left(\frac{X_1}{X_1 - 1}\right) - \log_2\left(\frac{X_0}{X_0 - 1}\right) + \frac{1}{X_0} \log_2\left(\frac{X_1 - 1}{X_0 - 1}\right) \quad [14\text{-}21c]$$

where

$$X_0 = 0.75/Qe_k, \quad\quad\quad X_1 = 1.5/Qe_k \quad\quad\quad [14\text{-}21d]$$

The agreement between calculated and measured coding inefficiencies for this model is fairly reasonable at higher skews; it is not as good for probabilities closer to 0.5. In applying this model to the Q-coder[42] the assumption was made that the interval register values were uniformly distributed in the interval {0.75, 1.5}. For the QM-coder the measured distribution of interval register values favors lower values and is described in the JBIG DIS as having a $1/A$ dependency.

14.8 Application of the estimation models to the QM-coder ◑

Figure 14-5 shows the results of modelling and experiment for single-context coding. The agreement between the two is very close, except for probabilities near 0.5 and in the immediate vicinity of the rather large peak in coding inefficiency for skews above 10. There is some reason to suspect that the experimental data did not achieve true equilibrium near this peak.

The large peak in inefficiency seen at skews above 10 is rather surprising. It is caused by an anomaly in the state machine at the lowest Qe value, at which the transition is defined to be to the nearest neighbor. This is in violation of our approximate model for estimation, which suggests that the transition should be to the second-nearest neighbor. Indeed, when that transition is defined to be to the second-nearest neighbor, the inefficiency peak is almost completely removed. This can be seen in Figure 14-6.

The inefficiency peak in Figure 14-5 illustrates an interesting aspect of this state machine estimation process—namely, the ease with which

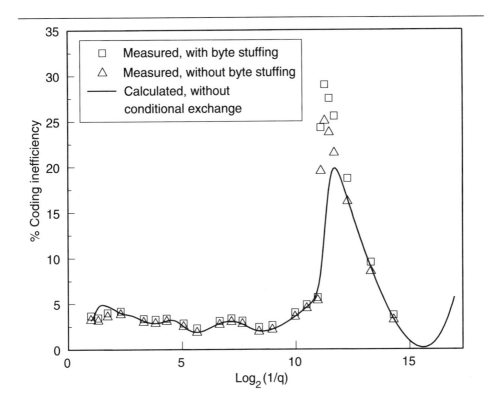

Figure 14-5. Single-context coding inefficiency for QM-coder

"traps" such as that seen in that figure can be constructed.[40] These traps cause an imbalance in the estimate relative to the actual probability, thereby preventing the correct operation of the estimator. A strong form of this trap occurs when the Qe value renormalizes to a value equal to the minimum A-register value and the LPS transition is to the nearest neighbor, and this is is exactly the situation for the smallest Qe value in the QM-coder state machine. When starting from this state, even one MPS event after the LPS transition out of the state causes an immediate MPS renormalization that returns the estimator to the state. The estimator is trapped, and only a pair of LPS without an intervening MPS can take the estimator out of the trap in the LPS direction.

When mixed contexts are coded, the effect of trapping is somewhat moderated by MPS renormalization into contexts that are not trapped. A plot of calculated coding inefficiency computed with the random-interval model is shown in Figure 14-7. As expected, this model does not predict a trapping effect. Note, however, that the effect of the immediate MPS renormalization on the distribution of values of the interval register is not properly incorporated into this model. The predictions of this model are therefore qualitative rather than quantitative.

One other aspect of the data in Figure 14-5 that is of interest is the deviation between the calculated and measured coding efficiencies near

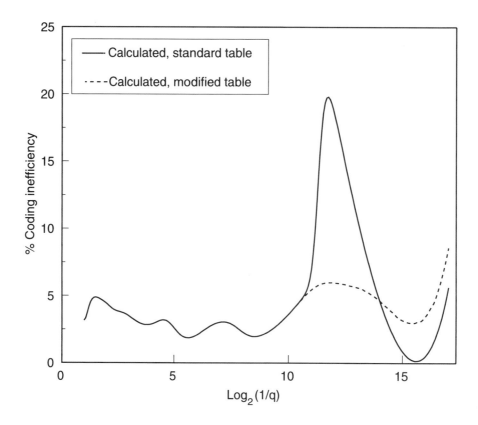

Figure 14-6. Single-context coding inefficiency for revised state machine

$Q = 0.375$. This is caused by conditional exchange, which is not modeled in the calculations. It is interesting that conditional exchange typically improves the coding efficiency by about 0.5%, even in coding of binary data with the JBIG algorithms. This indicates that a relatively high percentage of the coded bits are from contexts that have relatively small skews, even in binary image coding.

Although the coding inefficiency peak in Figure 14-5 is aesthetically displeasing, the problem is more "cosmetic" than real. Although one would prefer a relatively flat coding inefficiency as a function of symbol probability, this preference is based on the notion that a flat coding inefficiency is the best choice in the absence of knowledge about the actual probabilities. Generally, however, if the coding inefficiency is too large for one range of probabilities, it is usually offset by smaller coding inefficiency in some other range of probabilities. A true optimum would require adjusting the coding inefficiency according to the frequency of usage of the different probability estimates.

Shortly after the authors discovered this coding inefficiency peak, W. Equitz[79] measured the compression performance of the modified table

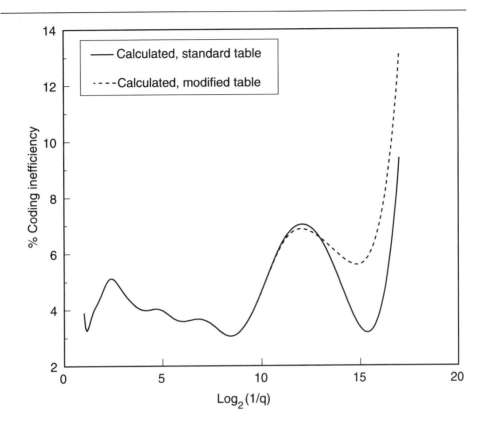

Figure 14-7. Random-internal coding inefficiency for QM-coder

of Figure 14-6 for the test data used in optimizing JBIG. He found no significant difference relative to the standard QM-coder state machine. This could indicate one of three things: (1) high skew states contribute little to the overall bit rate, (2) the QM-coder table is optimized to these data sets, or (3) context mixing effectively suppresses the trapping. Further research will be needed to understand this.

14.9 Initial learning ◑

One aspect of the QM-coder that distinguishes it from the Q-coder is the incorporation of additional states in the estimation state machine that exist solely for the purpose of arriving at the correct probability estimate with a minimum expenditure of coded bits.

The encoder and decoder adapt from an initial state, which, in the absence of any information about the coding decision, is defined for the QM-coder to be at a symbol probability of 0.5 and an MPS sense of 0 for all coding decisions. This initial estimate implies no compression per binary decision, and is therefore presumably far from correct. Some compression is expected (otherwise one would have no reason to do the entropy coding), and therefore a symbol probability will usually be seen that is significantly less than 0.5 and that has either one or the other sense of MPS.

Experimentally, the best setting for the prior information parameter, t, in Equation 14-5 is $t = 0.5$,[46] which indicates a great deal of uncertainty about the initial estimate. At each renormalization the MPS and LPS symbol counts are estimated and used to calculate a probability from

$$Qe = \frac{N_{LPS} + 0.5}{N_{LPS} + N_{MPS} + 1} \qquad [14\text{-}22]$$

Sequences of "learning" states threaded together by MPS renormalizations are defined for each value of N_{LPS}. Within these sequences the values of N_{MPS} are derived from an approximate count estimated at each renormalization.

An approximate value of N_{MPS} can be derived by observing that the MPS renormalization can occur only for any given Qe when the interval register is less than $A - Qe$. However, rather than assuming a flat distribution for A, as was done in the random interval model (see Equation 14-18), we assume (see the JBIG DIS) a $1/A$ probability density for values of A. The probability of the MPS renormalization transition from state k to neighboring state m is then

$$Pm_{k,m} = \frac{1-q}{\ln(2)} \int_{0.75}^{0.75 + Qe} dA/A \quad = \quad (1-q)\log_2(1 + Qe/0.75) \qquad [14\text{-}23]$$

which for small Qe reduces to

$$Pm_{k,m} = \frac{(1-q)}{\ln(2)} \frac{Qe_k}{0.75} \qquad [14\text{-}24]$$

Except for a constant, this is the same as Equation 14-18.

The number of MPS events which produces the MPS renormalization is given by the reciprocal of this probability. Note that as long as N_{LPS} remains zero, each MPS renormalization causes a relatively large change in the estimate and the learning is very rapid. The term "fast attack" has been used to describe this rapid initial learning.[46]

The state machine is initialized to a special state in which the symbol counts are both zero, and the MPS sense is defined arbitrarily to be zero. The first renormalization effectively provides a "guess" as to the sense of the MPS and places the estimator in the fastest attack sequence for that MPS sense. With each MPS renormalization the estimator advances to a state with a smaller Qe, and, as long as no LPS occurs, the Qe value is decreased by a relatively large amount at each MPS renormalization. Whenever an LPS occurs (except when the sense of MPS is switched), the state machine makes a transition to a slower attack sequence until a non-transient state is reached.

14.10 Robustness of estimators versus refinement of models ◑

In the paper on the Q-coder renormalization-driven estimation [42] a second-order model was also discussed in which high correlation between renormalizations was used to increase the estimation rate—the amount by which the estimate would change at each renormalization. A statistically stable system in equilibrium should see a renormalization correlation of slightly less than 0.5, indicating a small preference for the opposite renormalization occurring after a given renormalization. When the renormalization correlation increases, this is an indication that the statistical behavior of the model is unstable; for such models this second-order multirate estimation process can substantially improve the coding efficiency.

The additional cost in terms of storage and state machine size are significant for this second-order model. The original Q-coder used six bits to address the Qe values and MPS sense. The multirate system used a total of 12 bits for each context: seven bits for the Qe value and MPS sense (to give a finer granularity), one bit to "remember" the sense of the last renormalization, and four bits to count renormalization sequences. The QM-coder uses eight bits, including the "fast attack" states.

For statistically stable systems such as the JPEG and (for the most part) JBIG models, this second-order estimation provides little improvement; only for some of the sequential modes in which the JBIG adaptive template is turned off are there any indications of statistical instability. The general consensus in both JPEG and JBIG was that the second-order estimation model did not provide enough performance improvement to justify the additional complexity.

This leads us to a fairly general observation: most of the compression gains of the more robust estimation provided by a multirate estimation system can be achieved and usually exceeded by more sophisticated modelling. Whereas a statistically stable model is not necessarily optimal, the converse appears to be true. If a model is statistically unstable, there is a high likelihood that a better model can be found.

14.11 Other estimation tables ◑

In choosing an estimator table one of the major problems is to define an appropriate granularity. Once all initial learning stages are complete, granularity controls the rate of adaptation. It also determines the coding inefficiency for stationary probabilities, and these two opposing factors must be balanced.

In arriving at the final consensus table for the QM-coder, many variations were explored. One of them was proposed by the authors, and we include it here in Table 14-2 as an interesting example of the kind of trade-offs that were made. The estimator table, which we label JPEG-FA, is really a variant of the Q-coder estimation table with a very simple "fast attack" set of initial learning states. This table can also be found in the JPEG Draft Technical Specification Revision 6 (JPEG 8–R6). It is an example of a coarser-granularity estimation state machine.

Table 14-2. JPEG-FA estimation table

Qe (hex)	Imps	Ilps	Xmps	Index
5601	1	1	1	0
3401	2	6	0	1
1801	3	9	0	2
0AC1	4	12	0	3
0521	5	29	0	4
0221	38	33	0	5
5601	7	6	1	6
5401	8	14	0	7
4801	9	14	0	8
3801	10	14	0	9
3001	11	17	0	10
2401	12	18	0	11
1C01	13	20	0	12
1601	29	21	0	13
5601	15	14	1	14
5401	16	14	0	15
5101	17	15	0	16
4801	18	16	0	17
3801	19	17	0	18
3401	20	18	0	19
3001	21	19	0	20
2801	22	19	0	21
2401	23	20	0	22
2201	24	21	0	23
1C01	25	22	0	24
1801	26	23	0	25
1601	27	24	0	26
1401	28	25	0	27
1201	29	26	0	28
1101	30	27	0	29
0AC1	31	28	0	30
09C1	32	29	0	31
08A1	33	30	0	32
0521	34	31	0	33
0441	35	32	0	34
02A1	36	33	0	35
0221	37	34	0	36
0141	38	35	0	37
0111	39	36	0	38
0085	40	37	0	39
0049	41	38	0	40
0025	42	39	0	41
0015	43	40	0	42
0009	44	41	0	43
0005	45	42	0	44
0001	45	43	0	45

Table 14-3. Comparison of QM-coder and JPEG-FA estimation tables for the JBIG three-line template

B/W Image	AT (X-pos.)	QM-coder (bytes)	JPEG-FA (bytes)	Comparison (FA–QM)/QM
CCITT1	0	14631	14522	–0.8%
CCITT2	0	8500	8459	–0.5%
CCITT3	0	21930	21757	–0.8%
CCITT4	0	54238	53642	–1.1%
CCITT5	0	25803	25608	–0.8%
CCITT6	0	12546	12483	–0.5%
CCITT7	0	56191	56407	+0.4%
CCITT8	0	14204	14210	+0.0%
Total:	0	208043	207088	–0.5%
Budking	0	101247	97394	–4.0%
Budking	3	106429	102727	–3.6%
Budking	4	102332	98610	–3.8%
Budking	5	102757	99582	–3.2%
Budking	6	60988	60679	–0.5%
Budking	7	110546	106544	–3.8%
Boat2	0	16773	16713	–0.4%
Boat2	3	17640	17598	–0.2%
Boat2	4	17041	17040	+0.0%
Boat2	5	17141	17125	–0.1%
Boat2	6	13216	13291	+0.6%
Boat2	7	18218	18137	–0.5%
Jphmesh	0	15280	14804	–3.2%
Jphmesh	3	14829	14447	–2.6%
Jphmesh	4	7929	7966	+0.5%
Jphmesh	5	15307	14868	–3.0%
Jphmesh	6	15365	14902	–3.1%
Jphmesh	7	14962	14544	–2.9%
Jphmesh	8	9466	9414	–0.5%
Jphmesh	9	15309	14861	–3.0%

The relative performance of the QM-coder and the JPEG-FA table for the JBIG binary image sequential compression model is a good vehicle for illustrating the advantages and disadvantages of coarser granularity in the estimator. Table 14-3 provides this comparison.* Looking first at the CCITT test set, we see that JPEG-FA with its faster adaptor has about 0.5%

---

* The high-level review of the JBIG system presented in Chapter 20 provides a description of the JBIG sequential mode, and we refer the reader to that chapter for information regarding the JBIG coding model. For the rest of this section we assume the reader is familiar with the template model and adaptive template described there.

better performance on the average. This performance differential is even greater for the "Budking" halftone, for which the performance difference is quite small only when the adaptive template is positioned correctly for the halftone pattern. This illustrates the additional robustness of the coarser-granularity estimator.

In the other two halftones, however, the best compression is achieved by the QM-coder estimator when the adaptive template is correctly positioned. Note that these are small files in which the more sophisticated initial learning in the QM-coder tables becomes important. This initial learning is also very important in the JBIG progressive mode, in which the initial stages usually produce very small amounts of data.

In fact, the choice of QM-coder estimation table was based entirely on a composite measure of progressive performance that tended to emphasize early stage performance. As such, initial learning strongly affected the performance metric, whereas fast adaptation was de-emphasized. In addition, sequential mode performance was not considered in the performance metric. JPEG performance was considered, but only to the degree that the QM-coder table had to provide compression performance at least as good that provided by the JPEG-FA table. As it turns out, the two tables perform equally well in the JPEG system.

15

COMPRESSION PERFORMANCE

This chapter provides experimental data on the compression performance of the nonhierarchical* JPEG modes of operation. The DCT transformation used was the Chen, Smith, and Fralick algorithm,[21] and the quantization tables are given in Annex K of the JPEG DIS. The data given in this chapter were first published in a book edited by Nier and Courtot.[80]

The measurements were made using the nine JPEG test images. These test images are 8-bit precision YCbCr color images. The Y format is 720 samples per line and 576 lines, whereas the Cb and Cr format is 360 samples per line and 576 lines. This format is sometimes called YYCbCr, and has an average of 16 bits/pixel.

---

* Comparable data for the JPEG hierarchical mode are not available. However, the example given in Chapter 6 indicates that, for a given image quality, the total number of bytes required to complete the final stage of a hierarchical progression is significantly more than is required by either of the DCT-based progressive modes.

Table 15-1. Baseline results

Image	Fixed H. (bytes)	Custom H. (bytes)	Difference
Boats	40854	38955	4.9%
Board	35853	33233	7.9%
Zelda	31526	29373	7.3%
Barbara1	54878	53107	3.3%
Barbara2	56714	55148	2.8%
Hotel	49406	48267	2.4%
Balloons	29617	27489	7.7%
Goldhill	46081	43764	5.3%
Girl	43288	41766	3.6%
Average:	43135	41234	4.6%

For baseline sequential encoding the compressed bytes for the nine test images were measured with the fixed Huffman tables listed in Annex K of the JPEG DIS. These example Huffman tables were designed using a much larger set of test images, including images at other resolutions, and are, therefore, not the best possible tables for these images. The same DCT data were also encoded with Huffman tables customized for each image, with arithmetic coding, with restart markers, with successive approximation only, with spectral selection only, and with a full progression mixing successive approximation and spectral selection. The resulting compressed image file sizes, in bytes, are listed in Tables 15-1 to 15-8. For a given set of DCT coefficient values, these entropy-coding procedures are all lossless. Therefore, the image quality achieved with the DCT-based processes is independent of the entropy-coding procedures. The seven predictors for lossless coding are also compared for these test images.

15.1 Results for baseline sequential DCT ◗

Results for the baseline sequential DCT-based processes are listed in Table 15-1. The baseline process is restricted to two Huffman DC coding tables and two Huffman AC tables. Because the test images are YCbCr data, one set of tables was used to encode the luminance and the other set of tables was used to encode the two chrominance components. The data were interleaved in all sequential mode tests.

The first column lists the images. The first four images were used in the June, 1987, evaluation testing, whereas the last five images were used in the January, 1988, subjective evaluation testing. The second column, labeled "Fixed H.," gives the number of bytes of compressed image data obtained when using the fixed Huffman tables from the JPEG CD. The results obtained with custom Huffman tables for each image are listed in the column labeled "Custom H.." The last column shows the percentage of extra data needed if the fixed tables are used instead of the custom tables.

Table 15-2. Sequential arithmetic-coding results

Image	Arith. (bytes)	Vs. Fixed H.	Vs. Custom H.
Boats	35497	15.1%	9.7%
Board	30528	17.4%	8.9%
Zelda	27599	14.2%	6.4%
Barbara1	49472	10.9%	7.3%
Barbara2	50426	12.5%	9.4%
Hotel	45398	8.8%	6.3%
Balloons	25320	17.0%	8.6%
Goldhill	40083	15.0%	9.2%
Girl	38754	11.7%	7.8%
Average:	38120	13.2%	8.2%

Custom Huffman tables give 2.4%–7.9% better compression than the fixed tables given in the JPEG CD; conversely, using fixed tables increases the storage required for these nine images by an average of 4.6%. The fixed Huffman table compressed image data include 591 bytes of header information, and include a DHT segment for the fixed tables. When customized Huffman tables were used the number of bytes in the header varied from 321 bytes to 347 bytes, with an average of 329 bytes.

15.2 Results for sequential DCT with arithmetic coding ◑

Arithmetic coding consistently achieves higher data compression and has the advantage of being a one-pass adaptive encoder procedure. Table 15-2 lists the results for sequential arithmetic coding. The percentages of additional bytes that would be required if these images were transcoded into a Huffman entropy-coded format, using either the fixed Huffman table suggested in the JPEG committee draft or custom Huffman tables per image, are also given. The Huffman compressed bytes results have already been given in Table 15-1.

Huffman coding with fixed tables required 8.8%–17.4% more bytes than arithmetic coding. Huffman coding with custom tables per image required 6.3%–9.7% more bytes. On the average, for these images, fixed Huffman tables required 13.2% more storage for the compressed data than did arithmetic coding. For custom Huffman tables, only 8.2% more storage was needed. The arithmetic-coding compressed image data byte counts included 183 bytes of header information.

15.3 Results for sequential DCT with restart capability ◑

Some environments will require error recovery or parallel encode or decode capability. To provide this function, JPEG defined a restart capability which allows the image to be segmented into sub-images that can be encoded independently. In Table 15-3 the custom Huffman and

Table 15-3. Sequential results with restart

Image	Custom H. (bytes)	Arith. (bytes)	Difference
Boats	39184	37704	3.8%
Board	33452	32614	2.5%
Zelda	29655	29631	0.1%
Barbara1	53315	52406	1.7%
Barbara2	55243	53394	3.3%
Hotel	48475	47730	1.5%
Balloons	27640	27020	2.2%
Goldhill	43938	42168	4.0%
Girl	41997	41007	2.4%
Average:	41433	40408	2.5%

arithmetic-coding sequential-process performances are compared for a condition where the encoder was restarted after each block row (every eight lines). For Huffman coding, an average of 2.5 bytes were added 89 times to the compressed data (there are 90 block rows). Thus, the compressed data stream expanded by a few hundred bytes. For arithmetic coding, in addition to this overhead, the adaptive statistics were reinitialized in order to achieve coding independence. The loss in coding efficiency was larger, but arithmetic coding still gave fewer bytes than custom Huffman coding. Six extra bytes were added to the previous headers to enable the restart marker capability.

15.4 Results for progressive DCT with arithmetic coding ◑

Table 15-4. Progressive arithmetic-coding results

Image	SS (bytes)	SA (bytes)	Mixed (bytes)
Boats	36291	35020	35587
Board	31609	29964	30512
Zelda	27981	27199	27510
Barbara1	50414	48500	49060
Barbara2	51724	49711	50336
Hotel	46392	44223	44999
Balloons	26208	24968	25366
Goldhill	40373	39601	39852
Girl	39412	38138	38465
Average:	38934	37480	37965

Table 15-5. Arithmetic-coding conditioning

Component	Table	L	U
Y	0	2	3
Cb	1	1	2
Cr	2	1	2

Table 15-4 gives measured compression performance for a few selected examples of progressive coding. Arithmetic coding was used for these experiments. The two methods of progressive coding, spectral selection (SS) and successive approximation (SA), were applied independently and a mixture of the two techniques was also used.

For progressive coding, successive approximation alone gave the best compression, and spectral selection alone gave the worst; a mixed method came out somewhere in between. (Sequential coding also came out between the successive-approximation-only and spectral-selection-only methods.) The progressive coding with only spectral selection had headers with 263 bytes. The header for progressive coding with only successive approximation had 253 bytes. The mixed progressive coding needed 283 bytes of header information.

15.5 Results for lossless mode with arithmetic coding ◑

Figure 6-7 in Chapter 6 illustrates the neighboring samples that are used for prediction in sequential lossless coding. Note that in the test images the luminance component has twice as many samples horizontally as does either of the chrominance components. Therefore, there are, on average, 16 bits for each pixel.

Arithmetic coding with custom conditioning was used in these experiments.* Table 15-5 gives the arithmetic-coding conditioning parameters used for all compression results given in this section. If default conditioning had been used, the compression performance would have been about 1% worse.

The predictor $(a + b)/2$ was used for the lossless coding results listed in Table 15-6, whereas in Table 15-7 the performance achieved with the seven predictors is compared. In the authors' experience, predictor 7 usually was best. However, other predictors sometimes give significantly

---

* In relatively noisy images such as these, Huffman coding typically has about 10% worse compression than arithmetic coding. However, when images compress very well, the Huffman coder becomes less efficient. Each DPCM difference is coded as an independent symbol, and, therefore, the Huffman coding system can never get below 1 bit/sample. The authors have encountered highly compressible medical images where the arithmetic-coding performance is 30% to 50% better than Huffman coding.

Table 15-6. Lossless arithmetic-coding results for selector 7

Image	Arith. (bytes)	Bits/pixel
Boats	369084	7.2
Board	355650	6.9
Zelda	364480	7.0
Barbara1	431372	8.3
Barbara2	449841	8.7
Hotel	422420	8.1
Balloons	296219	5.7
Goldhill	426740	8.2
Girl	376784	7.3
Average:	388066	7.5

better performance when the images are more correlated and the compression is higher.

15.6 Summary of Results ○

Table 15-8 summarizes the results for the different coding processes. In these experiments lossless coding achieved only about 2:1 compression; thus the appeal of the DCT-based lossy coding procedures, which achieved about an order of magnitude more compression. The best compression was achieved with the successive-approximation progressive mode with arithmetic coding.

All lossless compression processes permit, by definition, transcoding from one choice of coding technique or predictor to another. What may not be so obvious is that all of the lossy DCT results can also be transcoded. However, if the decoded data are kept in quantized DCT form, the

Table 15-7. Lossless arithmetic-coding results for nine images

Selector	Predictor	Total (bytes)	Difference (rel. to 7)
1	a	3734235	6.9%
2	b	3543711	1.5%
3	c	3899133	11.6%
4	$a + b - c$	3741431	7.1%
5	$a + (b - c)/2$	3660822	4.8%
6	$b + (a - c)/2$	3555079	1.8%
7	$(a + b)/2$	3492590	0.0%

Table 15-8. Summary of averages for JPEG processes

Process	Average (bytes)	Bits/pixel
Sequential coding:		
Fixed Huffman	43135	0.83
Custom Huffman	41234	0.80
Custom Huffman with restart	41433	0.80
Arithmetic coding	38120	0.74
Arithmetic coding with restart	40408	0.78
Progressive arithmetic coding:		
Spectral selection	38934	0.75
Successive approximation	37480	0.72
Full progression	37965	0.73
Lossless arithmetic coding:		
a	414914	8.00
b	393746	7.60
c	433237	8.36
$a + b - c$	415715	8.02
$a + (b - c)/2$	406758	7.85
$b + (a - c)/2$	395009	7.62
$(a + b)/2$	388066	7.49

DCT-based coding processes are also lossless and, therefore, can be transcoded.

16

JPEG ENHANCEMENTS

In this chapter we discuss some interesting ways in which JPEG performance can be improved through nonstandard processing but within the constraints of a standard JPEG bit stream.

16.1 Removing blocking artifacts with AC prediction ◑

One of the problems that any DCT-based scheme has when extended to low bit rates is blocking artifacts. Basically, when the DCT coefficient quantization step size is above the threshold for visibility, discontinuities in grayscale values become clearly visible at the boundaries between blocks. Of course, the limiting case occurs when only DC coefficients are available, as is the situation after the first scan of a progressive DCT-based frame.

These blocking artifacts can be greatly reduced by predicting the low-frequency AC coefficients from the DC coefficient changes [81] within a 3×3 array of blocks centered on the block in question. Figure 16-1 illustrates this 3×3 array.

A quadratic surface, given by

$$P(x,y) = A_1 x^2 y^2 + A_2 x^2 y + A_3 xy^2 + A_4 x^2 + A_5 xy + A_6 y^2 + A_7 x + A_8 y + A_9 \quad [16-1]$$

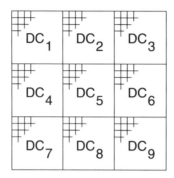

Figure 16-1. 3×3 array of DCT blocks used for AC prediction.

is fitted to the 3×3 array of DC values, and is used to estimate the 8×8 pixel array in the central block. The coefficients $A_1, \ldots, A_9$ are determined by requiring that the mean values computed for the quadratic surface match the DC values, $DC_1, \ldots, DC_9$, with appropriate scaling for the DCT normalization.

If a DCT is computed for the central block, the result is a set of equations relating the AC coefficients required to reproduce the quadratic surface to the DC coefficients which predict the quadratic surface. The equations for the first five coefficients in the zigzag scan are:

$$AC_{01} = (1.13885 / 8)(DC_4 - DC_6) \qquad\qquad \text{[16-2a]}$$

$$AC_{10} = (1.13885 / 8)(DC_2 - DC_8) \qquad\qquad \text{[16-2b]}$$

$$AC_{20} = (0.27881 / 8)(DC_2 + DC_8 - 2DC_5) \qquad\qquad \text{[16-2c]}$$

$$AC_{11} = (0.16213 / 8)((DC_1 - DC_3) - (DC_7 - DC_9)) \qquad\qquad \text{[16-2d]}$$

$$AC_{02} = (0.27881 / 8)(DC_4 + DC_6 - 2DC_5) \qquad\qquad \text{[16-2e]}$$

The subscripts of the AC coefficients indicate the horizontal and vertical position of the coefficient in the DCT array.

These equations predict the unquantized AC coefficients from the unquantized DC values. Note that the DC coefficients are eight times the block mean values, consistent with the DCT normalization defined in Chapter 4. The factors of 8 relating the means to the DC coefficients are left unresolved in the equations to facilitate comparison to equations given in Annex K of the JPEG DIS.

The prediction equations can be mapped to a form in which the quantized DC coefficients are used to predict the quantized AC coefficients. So that integer arithmetic can be used, the constant terms in Equation 16-2a to Equation 16-2e are multiplied by 256 and rounded to the nearest integer. We then have:

$$QAC_{01} = \frac{(RQ_{01} + 36Q_{00}(QDC_4 - QDC_6))}{256Q_{01}} \qquad [16\text{-}3a]$$

$$QAC_{10} = \frac{(RQ_{10} + 36Q_{00}(QDC_2 - QDC_8))}{256Q_{10}} \qquad [16\text{-}3b]$$

$$QAC_{20} = \frac{(RQ_{20} + 9Q_{00}(QDC_2 + QDC_8 - 2QDC_5))}{256Q_{20}} \qquad [16\text{-}3c]$$

$$QAC_{11} = \frac{(RQ_{11} + 9Q_{00}((QDC_1 - QDC_3) - (QDC_7 - QDC_9)))}{256Q_{11}} \qquad [16\text{-}3d]$$

$$QAC_{02} = \frac{(RQ_{02} + 9Q_{00}(QDC_4 + QDC_6 - 2QDC_5))}{256Q_{02}} \qquad [16\text{-}3e]$$

where R is a constant that is +128 for positive numerators and –128 for negative numerators in these equations. QAC_{mn} is the predicted quantized AC coefficient for position mn in the DCT coefficient array and QDC_n is the quantized DC value for block n in the 3×3 block array.

The predicted AC coefficients definitely help to remove blocking artifacts in smooth background areas of the image. However, the predictions often are wrong at the edges, and in any event must be suppressed wherever non-zero AC coefficient values are decoded. The values should be clamped so that they cannot match or exceed the smallest quantized magnitude that the decoder can decode in the current scan. Otherwise, the AC prediction can introduce large artifacts.

Originally AC prediction was proposed as an encoder and decoder "front end," and typically improved compression by about 2%. It was relegated to a decoder option by JPEG in the interests of simplifying the system. However, as a decoder option it can still be used to reduce blocking artifacts in the smooth regions of low-bit rate images. Figure 16-2a is the original of the "Lena" test image, Figure 16-2b is the reconstruction obtained with a standard JPEG decoder, and Figure 16-2c is the reconstruction obtained when blocking artifacts are suppressed by a proprietary variation of AC prediction. Successive approximation with arithmetic coding was used for the experiment.

(a)

Figure 16-2. Removal of blocking artifacts with AC prediction. (a) Original of the "Lena" image.

16.2 Low bitrate VQ enhanced decoding ◐

In a recent paper Wu and Gersho[82] suggested an interesting way of applying vector quantization (VQ) to improve the image quality decoded from a standard JPEG DCT-based mode data stream. This scheme uses VQ concepts to create a new set of basis functions that minimizes the mean square error (MSE). Each quantized DCT coefficient becomes an index into a codebook that provides an incremental output for each pixel in the 8x8 array. The result is much like a Goertzel DCT[66] in which each scaled basis function is added to the pixel values separately, except that the VQ output is not a set of orthogonal basis functions. The net improvement in the image signal-to-noise ratio for typical images (outside the training set) is about 0.5 dB.

16.3 An approximate form of adaptive quantization ◐

Although JPEG made no provision for adaptive quantization, an approximate form is possible. The basic idea is to identify unimportant regions of the image by some reasonable technique, and in these regions selectively set to zero any DCT coefficients with magnitudes below some threshold. This increases the compression in these regions by making zero coefficients more probable and the EOB symbol more likely. If arithmetic coding is used, some additional compression gains can be gotten from clearing low-order bits of the larger non-zero coefficients. This approximate form of

(b)

(c)

Figure   16-2 continued. (b) Standard decoded output for a 0.203 bits/pixel image. (c) 0.203 bits/pixel image when decoded using AC prediction. The decoded images were supplied by D. Garrido.

adaptive quantization can be used with either sequential or progressive mode, and will improve the compression by a significant amount.[*]

16.4 Display-adjusted decoding ○

Display quality can have a major effect on the perceived quality of the decoded images. Indeed, the early evaluations of image quality by JPEG were done on studio-quality monitors, and when the images were later displayed on high-bandwidth monitors artifacts that could not be seen on the lower-bandwidth monitors became easily visible. Only later was it discovered that these problems can be greatly suppressed by adapting the decoder to the display. JPEG uses a uniform quantizer and defines a linear dequantization in which the quantized DCT is simply multiplied by the quantization value. However, experiments have shown that for high-bandwidth displays the image quality can be improved by dequantizing to a value near the lower bound of the quantization interval. (Statistically, this should improve the MSE slightly, too.) Conversely, if the display bandwidth is limited, better quality can be obtained by dequantizing as JPEG specifies, and perhaps even increasing the dequantized DCT coefficient magnitudes modestly. This compensates for some of the loss of response of the display at high frequencies.

---

[*] Adaptive quantization in MPEG improves the bit rate by about 30%. More than half of this improvement is attained with the approximate form possible with JPEG.

17

JPEG APPLICATIONS
AND VENDORS

This chapter gives information about some of the early applications of JPEG, and about vendors selling JPEG products. Much of the material in this chapter is paraphrased (with permission) from the responses provided to the JPEG vendor/application worksheet distributed rather late in the process of writing this book. In some cases it has been supplemented by the authors' own experiences or gleaned from journal articles and vendor data sheets.

The vendor/application worksheet was not distributed uniformly throughout the world, nor were all of the major JPEG players successfully contacted. Thus, the companies profiled here are only a subset of the current JPEG implementers. Figure 17-1 shows the questions asked on the current version of the vendor/application worksheet. Companies wishing to update or provide new information about their JPEG products are invited to fill out a copy of this worksheet.*

---

* Answers to worksheet questions may be mailed to Joan L. Mitchell, IBM T. J. Watson Research Center, P.O. Box 704, Yorktown Heights, NY, U.S.A. 10598.

Material about JPEG products and applications is being solicited for use in a book on JPEG. Even if the material does not go into the book, it is likely to be used in other publications and presentations.

* Please answer at least these items

1. Section heading (what version of your company name would you prefer as the section name?) (e.g., IBM):

2.* Company legal name (e.g., International Business Machines Corporation):

3.* Company short name (e.g., IBM):

4. Brief history (suggestions below—none required):
        4a. Date founded:
        4b. Type of company founded (consulting, SW, HW, chips, boards, an application):
        4c. Original primary product:
        4d. Current primary product:

5.* Repeat 5a, 5b, 5c, 5d, 5e, and 5f for each JPEG/JPEG-like/JPEG-related product/JPEG application/JPEG experience (this is where additional sheets would be appropriate):
        5a.* Complete name (if someone wanted to buy what would he/she ask for?) of JPEG product/JPEG application/etc.:
        5b.* Type (consulting, SW, HW, chips, boards, OEM, applications):
        5c.* Fully JPEG compliant?:
        5d.* Description of amount of JPEG function supported (baseline, sequential, progressive, hierarchical, Huffman, arithmetic, 8-bit/sample, 12-bit/sample, lossless (2–16 bits/sample), vertical and horizontal sampling (if complete now or limited forever)):

5e. Hardware/software required (e.g., 386 PC, OS/2, Scanner):
5f.* Preferred contact:
        a.   Name:
        b.   Title:
        c.*   Company name:
        d.*   Street:
        e.*   City, state, ZIP:
        f.*   Country:
        g.   Telephone:
        h.   Fax:

Figure 17-1. Vendor/Application Worksheet

Companies are described in alphabetical order in the sections of this chapter. Because most of the vendors claim full JPEG compliance and the compliance test data (ISO DIS 10918-2) are not yet available, only the exceptions are noted. The corporate contact person with name, address, telephone number (P:) and fax number (F:) is listed in the references if a name different from that of the JPEG product contact(s) was given.

The authors have not seen all the advertised products and take no responsibility as to the validity of the reported information. However, we do know that many companies, including our own, have been selling pro-

6. Preferred corporate level-contact:
     a.   Name:
     b.   Title:
     c.   Company name:
     d.   Street:
     e.   City, state, ZIP:
     f.   Country:
     g.   Telephone:
     h.   Fax:

7. Company mission statement (one sentence!):

8. Any additional comments/special instructions:

9.a* Name of person to call if there are questions about worksheet responses:
9.b* Telephone:
9.c* Fax:

10. If you have any trademarks included in this material, please list them here and the appropriate words to go with them (e.g., XXX is a registered trademark of YYY Corp.):

Note: There is no guarantee that any of your responses will be used. Include no proprietary or confidential material, please.

Figure   17-1 continued.

ducts using JPEG and successfully exchanging JPEG files, at least on a test basis.

17.1 Adobe Systems Incorporated ○

Adobe Systems Incorporated (Adobe) was founded in 1982 as a software company. Adobe develops, markets, and supports computer software products and technologies that enable users to create, display, print, and communicate electronic documents. The company licenses its technology to major computer and publisher suppliers, and markets a line of type and application software products. Its PostScript Language made Adobe well-known and is still its primary product, with additional support for OEM output devices and professional graphics application packages.[83]

    JPEG function has been designed into all Adobe PostScript™ Language Level 2 products. These OEM products include laser printers, color printers, film recorders, and Display PostScript™ products. Full JPEG baseline and Huffman sequential modes are supported for eight bits/sample per component in the JPEG Interchange format. All sampling combinations are supported. Additionally, JPEG has been extended to support more than 10 blocks in an MCU.* RGB-to-TCC and CMYK-to-YCCK

---

* The JPEG committee discussed extending the number of blocks in the MCU to more than 10 as a possible enhancement to the current JPEG standard. The image

color transforms are optional. JFIF markers are supported. The person to contact about Adobe's PostScript Language 2 products is:

> James C. King, Manager of Advanced Technology
> Adobe Systems Incorporated
> 1585 Charleston Road
> P.O. Box 7900
> Mountain View, CA, U.S.A. 94039-7900
> P: (415)961-4400; F: (415)961-4400

Adobe Photoshop™ photo-editing software product supports JPEG Huffman sequential mode, including baseline for eight bits/sample per component in the JPEG Interchange format.[†] All sampling combinations are supported (even some for more than 10 blocks in the MCU). RGB-to-YCC and CMYK-to-YCCK color transforms are optional. JFIF markers are supported. Adobe Photoshop includes "plug-in modules" for reading and writing JPEG disk files. Further information on this product can be obtained from :

> Jeff Parker, Adobe Photoshop Product Manager
> Adobe Systems Incorporated
> 1585 Charleston Road
> P.O. Box 7900
> Mountain View, CA, U.S.A. 94039-7900
> P: (415)961-4400; F: (415)961-4400

17.2 AT&T Microelectronics ○

AT&T® Microelectronics,[84] a division of American Telephone and Telegraph Co. (AT&T), has implemented JPEG as part of their mission to provide state-of-the-art communications capabilities to their customers through excellence in technology, design, production, and application.[85]

The DSP3210-VMDE is a multimedia software product containing JPEG in conjunction with a digital signal processor (DSP). The AT&T product literature shows "JPEG Image Coder" as part of the "VCOS Multimedia Module Library—Ready-to-Use DSP Modules." The boards and applications that use this DSP are available from third parties.

The JPEG coder is not an "application" in itself, although a sample application source code is provided to developers. The coder could be considered a "JPEG engine" that can be utilized by any number of host applications to perform JPEG compression/decompression through a standard interface.

---

format of 4x4 blocks of luminance for each chrominance block would require 18 blocks in an MCU. At the time the decision was made to limit the MCU to 10, no one suggested this sampling combination as a useful format.

[†] This product is JPEG compliant except that the image size (X and Y) must appear in the frame header because DNL markers are not decoded. All components must be interleaved in one scan.

The complete baseline system is supported. Support is also provided for subsampling and popular forms of color space conversion. The decoder color space conversion coefficients default to YCbCr values but can be overridden by the application for any color space. Support is provided even for YCbCrK conversion for CMYK images. All valid combinations of sampling are directly supported by the decoder. The encoder provides direct support for all valid combinations not requiring fractional subsampling and indirect support for combinations requiring fractional subsampling. The code can be run on any host computer that has a DSP3210 running the AT&T VCOS™ Operating System. For more information about the DSP, contact:

> George Warner, Senior Engineer
> DSP Multimedia
> AT&T
> 555 Union Boulevard, MS 2C-108
> Allentown, PA, U.S.A. 18103
> P: (215)439-7787; F: (215)778-4555

17.3 AutoGraph International ApS ○

AutoGraph International ApS (AGI) was founded in June, 1989, to develop and market high-performance, high-quality software for image applications.[86] AGI produced some of the first commercial very-high-performance JPEG software implementations for developers and OEM.

EasyTech/Codec™ is a developer's tool kit for JPEG compression and decompression. It can handle any digital image format. In the package is an application example with source code, and a high-level interface where the developer simply needs to specify certain facts about the image file in order to compress the image data. Some parameters available at the high-level interface are the level of quality, subsampling, color space, resolution, and interleaving mode.

EasyTech/Codec was among the first commercial products to perform JPEG progressive image compression. In addition, it fully supports the baseline, sequential DCT-based Huffman, progressive Huffman, and lossless Huffman modes of operation. All lossy modes encode eight and 12 bits/sample per component. The lossless Huffman mode supports 2–16 bits/sample per component. All vertical and horizontal sampling factor ratios are implemented according to the JPEG standard.

EasyTech/Codec developer's tool kits exist for UNIX®, PC, and Macintosh® environments. The EasyTech/Codec is available on Digital DecStation, Hewlett-Packard® HP® 9000, IBM RS/6000®, SCO UNIX, Silicon Graphic IRIS, INDIGO, CRIMSON, POWER, SONY NEWS, SUN™ SPARC™, 386/348 PC, and I860.

AGI also sells other image products. EasyCopy/PC Plus™ is a utility for making screen dumps and digital image file printing, file conversion and file viewing on a PC. EasyCopy/X™ is a print utility for making screen dumps from X-Windows and digital image printing to any color printer. More information about AGI's products can be obtained from:

Jens Ole Jensen, Product Manager
AutoGraph International ApS
Gl. Lundtoftevej 1 C
DK 2800 Lyngby
Denmark
P: +45-4593-3399; F: +45-4593-3222

17.4 AutoView ○

AutoView, Inc. (AutoView) was formally founded in April, 1991, although its JPEG-based product has been under development for five years. This value-added reseller and developer is creating a vehicle image data storage system that will allow a storage pool to reproduce high-resolution color images of vehicles for salvage buyers or insurance-claims personnel as part of its mission "to market a product that increases insurance salvage returns and decreases buyers' travel expenses."[87]

AutoView Imaging System is in beta test now. Soon AutoView plans to use a single CD-ROM, express-mailed to each subscriber each week. Then salvage buyers will be able to review and bid on salvage without leaving their offices. Insurance adjustors and appraisers can also view images of vehicles. The AutoView Imaging System includes Hollander parts interchangeability and vehicle identification number (VIN). This development system uses 386PCs with 16- or 32-bit Targa cards and Telephoto's ALICE™ Compression. For more information about AutoView's JPEG-based product, please contact:

Stephen Hite, President
AutoView Inc.
32400 148th Ave SE
Auburn, WA, U.S.A. 98002
P/F: (800)472-7101, (206)939-2747

17.5 Bulletin board systems ○

JPEG is spreading rapidly through the electronic bulletin board system (BBS) community. A student at Williams College, Erik Rutins,[*] created his (single-line) BBS, called Bacchus Plateau, during his sophomore year. His main goals for Bacchus Plateau are to provide a forum for discussion of events that are of global and/or local importance, to support the IBM, Macintosh, and Amiga users on campus, and to provide Williams College students as well as local residents with an enjoyable on-line service that allows them to spend their free time in a fun, relaxing manner.

Right around Easter, 1992, he was introduced to a new format by a user who uploaded a program called Image Alchemy™ (IA). [See section 17.13]. The user claimed that IA could compress files better than anything

---

[*] Erik Rutins is a history major (class of 1993) whose hobby is computer science. His BBS handle is SnowDog.

he had ever seen. Erik become interested after seeing a SAMPLE.JPG file that was a SuperVGA image of stunning quality and that was stored in 65 Kbytes instead of the usual 250-300 Kbytes. He shrank his entire graphics section from 20 Mbytes to 6 MBytes with JPEG. Now he stores all of his images in .JPG format.

According to Rutins, "Bacchus Plateau was the first BBS system [he] knew of to go completely JPEG. The advantages were clear—not only did storage space become less of a premium, but transfer times decreased drastically. After downloading the Image Alchemy archive, users could then download files which were as small as 13 Kbytes, yet yielded 256 color VGA or SVGA images of excellent quality.... Within a month, other BBSs began to support JPEG—Image Alchemy appeared everywhere, followed by a freeware set of GIF2JPG and JPG2GIF converters." Rutins has heard several times in on-line chats during the last year that other bulletin board services "deal exclusively in JPEG."[88]

17.6 California Department of Motor Vehicles ○

The State of California Department of Motor Vehicles (Cal DMV) is one of the first DMVs to utilize imaging technology in the production of driver licenses and identification cards. Since January, 1991, all drivers' pictures have been stored as JPEG compressed images. In the first 18 months over 10 million licenses were processed, and approximately 400,000 more are added each month. The states of New York and Virginia have also awarded contracts for similar processes to begin during late 1992.[89]

According to Gary Nishite from the Cal DMV Office of Technology, "Recognizing the possibility that the images of photographs, signatures, and in some cases fingerprints, will be shared with other jurisdictions including the private sector, we wanted to ensure that unnecessary technology barriers will not result. Through our national organization, the American Association of Motor Vehicle Administrators (AAMVA) a committee of states was established to create a standard for the capture, storage, and retrieval of image data. The result of the committee work was the establishment of a 'Best Practice Standard' selecting options within the proposed JPEG Baseline specification." For more information about the Cal DMV usage of JPEG, please contact:

Gary Nishite, Chief, Office of Technology
California Department of Motor Vehicles
2415 First Avenue (B-239)
Sacramento, CA, U.S.A. 95818
P: (916)657-6723; F: (916)657-5644

17.7 C-Cube Microsystems, Inc. ○

C-Cube Microsystems, Inc.[90] (C-Cube) was founded in August, 1988, to "develop and deliver enabling technologies for digital image and video compression within the computer, communication, and consumer electronics markets."[91] The company pioneered a single-chip implementation of JPEG capable of real-time encoding and decoding of full-motion video at 30 frames per second. Considering that the JPEG committee set a requirement for feasibility to decode at 64 Kbits/sec. in 1986, development of a multi-Mbits/sec. chip just a few years later was not entirely expected.

As part of its mission "to be, and to be perceived as, the world leader in providing compression technology for digital and image solutions,"[91] C-Cube has since developed other JPEG and MPEG compression and decompression chips. In May, 1992, C-Cube shipped the first production MPEG decoder chip.

Currently, C-Cube offers a number of JPEG products. All of these products support the baseline, sequential, Huffman, and eight-bit/sample JPEG mode and permit 2:1 horizontal sampling.* C-Cube has also been active in helping to define file formats for JPEG data streams (see 17.35.5).

The CL550™ JPEG Image Compression Processor is a single-chip, real-time JPEG CODEC that is capable of compressing and decompressing image frames fast enough to permit use in full-motion video. It is available in 10-MHZ, 30-MHZ, MQUAD, and 35-MHZ PGA. It is platform independent. Products are available on ISA bus, Nubus, S-bus, and VME bus, among others. The CL550 supports all sampling factors in its bypass modes.

The CL550 Still Image Board Development Kit is a low-cost JPEG Still Image Compression Board with a Windows Software driver. Implementation requires an IBM compatible PC of series 386SX or higher. The CL550 Still Image Board Development Kit supports all sampling factors in its bypass modes.

The JPEG Video Development Kit is a full-motion, real-time compression/decompression board for capturing live or recorded video on a hard disk and providing playback in a window on a VGA monitor. The Video Development Kit requires an IBM-compatible 386SX or above PC with a minimum of 2 MBytes of RAM and a 40-Mbyte hard drive, as well as Microsoft Windows 3.1 software.

The C-Cube DOS Image Compression Interface is a DOS software library for application and compression utility developers.

For more information about C-Cube's JPEG products, please contact:

Jim Anderson, Director of Product Marketing
C-Cube Microsystems, Inc.
1778 McCarthy Boulevard
Milpitas, CA, U.S.A. 95035
P: (408)944-6300; F: (408)944-6314

17.8 Data Link ○

Data Link Information Solutions, Inc.[92] (Data Link) was founded in 1985. It specializes in multimedia video imaging and identification systems.

---

* C-Cube participated in the baseline validation and has distributed the sequential validation diskette.

Data Link's goal is to provide creative and economic solutions to complex security and information management problems. Data Link installs ID badge-making systems to meet government security requirements, develops integrated management systems for controlled-access areas, produces badges for special events, and develops management information systems to allow instant access to sensitive data.[93]

Data Link's main product, the Electronic Pass and Security System (EPASS®), provides a multimedia information management system for security operations. It uses JPEG baseline Huffman coding at eight bits/sample. No additional hardware or software is needed for this turn-key system, because all required hardware and software are included in the initial system setup. For more information about Data Link's JPEG application, contact:

> Don Stropes, Executive Vice President
> Data Link
> 620 Herndon Parkway, Suite 120
> Herndon, VA, U.S.A. 22070
> P: (703)318-7309; F: (703)318-7309

17.9 Discovery Technologies, Inc. ○

Discovery Technologies, Inc.[94] (DTI) was founded in April, 1987, to "build a successful business by providing high-quality image communication solutions to the medical and industrial markets."[95]

DTI's FilmFAX® is high-resolution teleradiology equipment. It uses the ALICE™ JPEG image compression by Telephoto, and is available for OEM. This turn-key solution requires a digital laser scanner, 386 PC, and proprietary software. For information about DTI's FilmFAX system, please contact:

> Mike Bolser, Vice President of Sales
> Discovery Technologies
> 2405 Trade Centre Drive
> Longmont, CO, U.S.A. 80503
> P: (303)651-6500; F: (303)651-7545

Two more DTI primary products are Tel-Med, a microscope-based image communication system for transmitting high-resolution, full-color medical (pathology) images; and MAT, a microscope-based imaging system for archiving, printing, and transmitting high-resolution, full-color industrial images for quality control. These products also use the ALICE JPEG image compression by Telephoto, and are available for OEM. These turn-key solutions require a digital laser scanner, 386 PC, and proprietary software. For information about the MAT or Tel-Med Microscope Imaging Systems, please contact:

George Leonard, Product Manager
Discovery Technologies
2405 Trade Centre Drive
Longmont, CO, U.S.A. 80503
P: (303)651-6500; F: (303)651-7545

17.10 DSP ○

Digital Sensory Products, Inc. (DSP) is an IBM business partner that provides new technology for the IBM ImagePlus™ offerings and provides customer support for those offerings. DSP also does consulting, custom software, and installation and maintenance support. DSP is a hardware reseller of color devices.[96]

DSP Color Video Drivers for IBM® SAA ImagePlus® Workstations are software drivers and command-language link devices. When the IBM M-Motion adapter or the Video Capture Adapter/A is attached to the IBM SAA ImagePlus Workstation Program/2 (IWP/2), these drivers enable the capture, display and printing of high-quality color video images. DSP software provides a video window for previewing and selecting images before storing them and for displaying retrieved images. IWP/2 is utilized to compress/decompress eight-bit/sample images using the JPEG arithmetic-coding progressive mode. IWP/2 activity can be buffered and overlap the video window activity. For more information about DSP's products, contact:

Michael Cherry, President
Digital Sensory Products, Inc.
14 North Broadway
Tarrytown, NY, U.S.A. 10591
P: (914)332-7070

17.11 Eastman Kodak Company ○

Eastman Kodak Company was founded in 1880 as the Eastman Dry Plate Company. Originally focused on monochrome, still-frame photography, Eastman Kodak Company's history includes development of panoramic camera products, stereoscopic products, and kinetoscope products. Eastman Kodak Company markets over 35,000 products. Traditional photographic products include 35-mm cameras and film, aerospace and reconnaissance films, professional films, high-speed films, motion-picture films, infra-red films, X-ray films, picture films, slide films, photo-CDs, and photo-CD players. Other products run the gamut from image transmission systems, image storage (optical disk) systems, high speed document scanners, very high-resolution film scanners, digital cameras, digital thermal printers, health imaging systems, and blood testing equipment, to vitamins and chemical products. Most Eastman Kodak Company products are based on Imaging and Information Systems.[97]

The Kodak 35mm Rapid Film Scanner™ and Kodak RFS 2035 Scanner™ are not fully JPEG compliant, but are JPEG-like. They handle eight bits/sample, three color components and interface to the Macintosh computer.

The Kodak Professional Digital Camera System™ allows image capture, storage, transmission, and conversion to other standard file formats. It handles eight bit/sample, three color components with some subsampling. JPEG compression is used to increase the storage capacity of all three devices in this line. Compression is also used to decrease image transmission times. Application developers can get access to the JPEG compressed data stream from the Kodak Professional Digital Camera System.[98] Eastman Kodak Co. also supports the use of JPEG compressed data incorporated into an ANPA/IPTC file (see section 17.35.1), which allows images to be easily used with standard photo-journalism applications.

For general information, call (800)242-2424. For purchasing information, call (800)44-KODAK or, for more information by mail, write to:

> Professional Imaging/Electronic Imaging Markets
> 343 State Street
> Rochester, NY, U.S.A. 14650-0406

17.12 Handmade Software ○

Handmade Software, Inc. was founded in June, 1992 "to provide a complete image conversion and processing tool kit which actually works."[99] This software company created Image Alchemy™ (see section 17.5).

"Image Alchemy is a general purpose image processing tool, which does image format conversion, JPEG compression, colour conversion, and image scaling. It supports 51 different image file formats, including TIFF, GIF, PCX, Targa, and Sun Raster. Alchemy has 4 different scaling, 7 different dithering, and 3 different colour conversion algorithms."[99]

Image Alchemy supports baseline, sequential, and (some) progressive Huffman JPEG at eight bits/sample for all vertical and horizontal sampling factors. The arithmetic coding version is written and works, but is not being distributed for reasons relating to patents. There are versions of Image Alchemy 1.6 available for IBM PC/AT, 80386, UNIX™/386, Sun SPARC™, HP9000/700 and 800, SGI Iris and Indigo, and IBM RS/6000®. For more information about Image Alchemy please contact:

> Marcos H. Woehrmann
> Vice President
> Handmade Software, Inc.
> 15951 Los Gatos Boulevard, Suite 17
> Los Gatos, CA, U.S.A. 95032-3488
> P: (408) 358-1292; F: (408) 358-2694
> hsi@netcom.com (e-mail)
> 71330,3136 (CompuServe®)

17.13 IBM ○

International Business Machines Corporation (IBM®) has supported the authors' participation in the JPEG committee since March, 1987. IBM's OS/2 Image Support, a software product, was the first commercial product to use JPEG arithmetic coding. The original product was shipped at the JPEG Technical Specification (JPEG 8) Revision 5 level in June, 1990. It was upgraded to the JPEG CD level in June, 1991.

The IBM ImagePlus® SAA Workstation Program/2® (ImagePlus IWP/2) supports JPEG in software. The primary method of compressing color and grayscale images is with spectral-selection progressive mode with arithmetic coding. Decoding is supported for baseline, extended-sequential Huffman, extended-sequential arithmetic and progressive arithmetic at eight-bits/sample with horizontal subsampling up to 2:1 and no vertical subsampling. Images are wrapped in IBM's Image Object Content Architecture (IOCA) and Mixed Object Document Content Architecture (MOD:CA). ImagePlus IWP/2 runs under OS/2 1.3 or OS/2 2.0.

IBM developers in Austin, Texas, interested in the development of still image and audio subsystems, created an application programming interface (API) for the IBM M-ACPA Card.[100] Parallel processing on the host computer and the M-ACPA card is used to accomplish a full JPEG baseline implementation running on both Family 1 and Family 2 machines under DOS, Windows, and OS/2. Utilities for color conversion and user customization of JPEG parameter streams are included. People desiring more information about IBM's JPEG products should contact an IBM representative.

IBM JPEG Express compression utilities is an implementation of JPEG available to the OEM market. JPEG Express offers the application developer/integrator Huffman or arithmetic entropy coding, lossless compression of 2–16-bit/sample grayscale images, and lossy sequential or progressive (spectral-selection and successive-approximation) compression of eight-bit precision grayscale or color images. Spatial scaling factors of 1/8, 1/4, 3/8, 1/2, 5/8, 3/4, and 7/8 are available to exploit the power of progressive JPEG data streams and allow applications to easily scale down images for display. Support of eight-bit palette color is also provided. Dithering to the OS/2 1.3 8514A and XGA palettes and the OS/2 2.0 default palette from YCrCb, YCbCr, YYCrCb, and YYCbCr color spaces has been integrated into the decoder to speed the display of color and grayscale images. The code is provided as an OS/2, Windows DLL, or DOS Library.[101] For more information about IBM's JPEG OEM products, please contact:

IBM Corporation
OEM Marketing
Dept. B82
472 Wheelers Farms Road
Milford, CT, U.S.A. 06460
T: (203)783-7500

17.14 Identix ○

Identix Incorporated (Identix) was founded in 1982 as a technology manufacturer of biometric identification products. Its mission is to "supply integrated bookings systems to law enforcement for capture and management of fingerprints, mug shots and documents."[102] Its current primary product is I-3 (Integrated Identification Information) Systems. Identix's I-3 Systems, Fingerprint Systems, and Mug Shot Systems use JPEG. These integrated PC-based hardware and software solutions run on 386 and 486 PCs. For more information about Identix's JPEG products, please contact:

> Tim Ruggles, Vice President
> Identix
> 510 N. Pasteria
> Sunnyvale, CA, U.S.A. 94086
> P: (408)739-2000; F: (408)739-3308

17.15 IIT ○

Integrated Information Technology, Inc. (IIT) was founded in 1987. "IIT's mission is to provide innovative solutions for the processing, storage, access, display and communication of electronic information."[103] The company entered the market with the industry's first complete family of 286-and 386-compatible math co-processors. Subsequently, IIT produced the first programmable video compression processor for still-frame and full-motion video. IIT is now focused on developing a series of products for emerging PC graphics standards, still and full-motion video compression, and lossless data compression.

IIT's Vision Controller™(VC™) and Vision Processor™(VP™) chips are programmable chips, supplied with software that performs JPEG encoding and decoding at video rates. The current microcode implements the JPEG baseline system with Huffman, eight bits/sample per component, and unlimited sampling. Future microcode will support all enhanced functions. An evaluation card is available for the IBM PC AT bus.

The VC and VP can compress and decompress not only JPEG but also MPEG and H.261. They can be used in two configurations to implement JPEG. The VC and VP are used together to process JPEG at video rates. For non-video rate applications, the VP can be used as a JPEG accelerator to the system CPU; in this configuration the VP performs all JPEG steps except the Huffman coding.

For more information about IIT's JPEG products, please contact:

> Tim Williams, Product Manager
> Vision Products
> IIT
> 2445 Mission College Boulevard
> Santa Clara, CA, U.S.A. 95054
> P: (408)727-1885; F: (408)980-0432

17.16 Independent JPEG Group ○

The Independent JPEG Group (IJG) is an informal group created to promote JPEG compression and common file formats for JPEG images.[104] They have written highly portable C software for baseline JPEG compression. They make this software freely available for noncommercial or commercial use. The software is called the "Independent JPEG Group's free JPEG software." The software is targeted primarily at Unix and MS-DOS, but it has been used successfully on machines ranging from Apple IIgs to a Cray YMP. Almost any C compiler can be used.

The IJG's software supports baseline and extended-sequential 8-bit and 12-bit lossy compression. All integral sampling ratios are supported. There is no support for progressive, hierarchical, or lossless modes at present. Arithmetic entropy coding is currently not supported due to patent restrictions.

As distributed, the IJG software performs file conversion between JPEG format and several popular image formats; more formats can be added easily. The compression and decompression modules are designed to be reused in other programs, such as image viewers. The software includes a fast, high-quality color quantizer for output to colormapped file formats and displays.*

A key IJG reason for releasing this free software is to help force rapid convergence to *de facto* standards for JPEG file formats. IJG uses and recommends the "JFIF" file format, which is already in use by a number of commercial JPEG vendors. JFIF is a minimal representation; for applications that need to record additional data about an image, IJG recommends the TIFF 6.0 format. IJG intends to support TIFF 6.0 in the future. They hope that these two formats will be sufficient and that other, incompatible JPEG file formats will not proliferate.

---

* The IJG may be contacted by Internet e-mail at "jpeg-info@uunet.uu.net." Note that most commercial nets (CompuServe, AOL, etc.) provide Internet mail links. Readers are asked to refrain from contacting IJG simply to request a copy of the software. IJG makes the software available via a number of electronic networks, and they are NOT set up to mail out diskettes. The official archive site for the IJG software is FTP.UU.NET (Internet address 137.39.1.9). The most recent released version can always be found there in directory graphics/jpeg. If you have Internet mail access but no FTP capability, you can retrieve the software indirectly through an FTP-by-mail server. To get the latest info on FTP-by-mail, read the article "How to find sources" which appears regularly in the Usenet news.answers newsgroup (you can also get it by sending e-mail to "mail-server@rtfm.mit.edu" with message text "send usenet/news.answers/finding-sources"). You can also obtain the IJG software from CompuServe, in the GRAPHSUPPORT forum (GO PICS), library 10. The software is also available on many other archive sites and bulletin board systems, and it has been distributed on several CD-ROMs; of course, copies not obtained directly from FTP.UU.NET may not be the latest version. IJG recommends that copies older than version 3 (March 1992) not be used.

17.17 ITR ○

Information Technologies Research, Inc.[105] (ITR®) was founded as a software-based company in August, 1987. ITR's original purpose was to develop image compression technology based on vector quantization (VQ) and other state-of-the-art methods. The first product (based on ITR's patented VARI Compression method) was targeted for specific applications, such as those found in the medical and government arenas, which require a highly optimized compression method. The company now also sells hardware accelerator boards and does application coding. ITR's objective is to be a recognized leader in the development and marketing of state-of-the-art image compression products.[106]

ITR's current primary product, the ITR ImageEngine™, has JPEG software for DOS and Windows. ITR's products fully support the JPEG interchange format, have baseline function and can handle eight bits/sample for single and multicomponent images in non-interleaved or interleaved format. Current products use Huffman coding, but arithmetic coding is planned for the future. Lossless coding was in beta test at the time this book was being written. Common vertical and horizontal subsampling configurations are supported. Optional color space conversions using YIQ, YUV (analog matrix), YCrCb (also called digital YUV), and YCC (Eastman Kodak Co.) are available. The ITR JPEG products require an 80286-based system, with a high-end 80386-based or an 80486-based system being recommended. Versions are available to run under DOS, Windows, OS/2, and UNIX environments.

ITR offers a range of software, from end-user executable modules up to Developer's Kits, which include linkable libraries and source code. These are all software-based products, but an optional hardware accelerator card is available. ITR has products that support file formats such as JPG, BMP, DIB, TIFF, TGA, and GIF. On-screen scaling, cropping, rotation, image flipping and color hue, saturation, and brightness adjustments capabilities are provided.

ITR's founder and president, Greg Kisor, reports, "We started shipping JPEG product supporting full interchange format in July, 1991. It was initially called the ITR Vision Software. We have successfully swapped files with more than 12 different sources of JPEG files, including files created on non-PC platforms. It appears that the JPEG vendor community has finally achieved a level of interchange."[106]

For more information about ITR's JPEG products, please contact:

Karen Parker, VP/Sales
Information Technologies Research, Inc.
3520 West Hallandale Beach Boulevard
Pembroke Park, FL, U.S.A. 33023
P: (305)962-9961; F: (305)962-6546

17.18 Lewis Siwell, Inc. ○

Lewis Siwell, Inc. is a privately held corporation founded in 1987. Its original product was a police mugshot system based on a proprietary compression algorithm. This sequential Huffman coding based system with adaptive quantization was JPEG-like. The 1989 EV1260/CMO Compression Microcode Option for the Vision Technologies EV1260 co-processor board was based on the JPEG Technical Specification JPEG-8 Revision 5. It was a sequential Huffman coding system for the YIQ color space with 2:1 horizontal and vertical subsampling. It is no longer available. Newer JPEG products help Lewis Siwell, Inc. continue to provide "custom development in areas such as automated inspection systems, printing and image management systems."[107]

The 1991 PUMA/CMO Compression Microcode Option for the Chips & Technologies PUMA chipset is a code generator system able to build specific implementations of JPEG. Multiple color spaces and subsampling factors may be defined. The system supports sequential and progressive modes with Huffman coding for multiple platforms and operating systems.

The 1992 JPEG Application Specific Interface (ASI) is expected to be released third quarter 1992. This code generator system builds specific implementations of JPEG with function similar to the PUMA/CMO Compression Microcode Option.

The 1993 ASI1 Automated Inspection system does sequential Huffman coding 8-bit/sample grayscale images. This hardware and software subsystem is part of a larger inspection and repair station used to provide high-speed inspection of ITO and gold traces on glass.

The 1992 PCBS Police Central Booking System provides sequential Huffman coding with 2:1 horizontal and vertical subsampling of 8-bit/sample YIQ. This system provides for collection and booking and mugshot information on 386/486 platforms using proprietary image processing boards.

For more information about the availability of Lewis Siwell JPEG products, contact:

> William C. Lewis, Marketing Director
> Lewis Siwell, Inc.
> 6170 Cabrillo Court
> Alta Loma, CA, U.S.A. 91701
> P: (909) 987-7708 x 2; F: (909) 987-7708 x 5

17.19 LSI Logic ○

LSI Logic Corporation (LSI) began in 1981 as a semiconductor design and manufacturing company. Its first primary product was an application-specific integrated circuit (ASIC). Since then digital signal processors (DSP) and microprocessors have joined ASIC as primary products. These all help LSI fulfill its mission "to develop, manufacture and bring to market chip level solutions for the capture, storage, display and transmission of digital video, image and audio information."[108]

The JPEG Video Rate Chip Set (L64735, L64745, L64765) consists of hardware chips to encode or decode Huffman baseline or Huffman lossless entropy-coded segments (except for the byte stuffing) with vertical and horizontal sampling. The semi-custom set combines the L64735 DCT processor with the L64745 quantizer plus JPEG Huffman entropy coder.[109]

The chips run with the software provided by LSI to load the Huffman code tables. The chip set includes a statistics mode that allows easy optimization of Huffman code tables. The L64765 does color space conversion and includes gamma correction. It also does the raster-to-block conversion. For more information about LSI's JPEG products, please contact:

> Linda Blauner, Product Line Manager
> LSI Logic Corp.
> 1525 McCarthy Boulevard
> Milpitas, CA, U.S.A. 95035
> P: (408)433-7089; F: (408)954-4855

17.20 Moore Data Management Services ○

Moore Data Management Services Division (Moore Data) was founded in 1953 as a real-estate publishing company. Since then it has expanded in the areas of general real-estate publishing and on-line databases.

Viewpoint® is a real-estate photo-display software package that displays photos of real estate in summary format or full screen. It uses JPEG compression and runs under DOS.[110] For more information about Moore Data's product, contact:

> Mark Weber, Marketing Manager
> Moore Data Management Services
> 1060 S. Highway 100
> Minneapolis, MN, U.S.A. 55416
> P: (612)540-1000; F: (612)540-1002

17.21 NBS Imaging ○

Since 1958, NBS Imaging Systems, Inc. (NBSI) has generated driver licenses and identification (ID) cards. Today NBSI makes both conventional and digital driver-license management systems in order "to provide the best solutions and support to [its] customers."[111] South Dakota, Hawaii, Louisiana, Manitoba, and California all use NBSI systems.

The NBSI Digital Video Workstation and/or retrieval workstations use JPEG compression products for the over-the-counter or central issuance systems for driver licenses and ID cards. NBSI's application supports the early versions of JPEG as well as the latest version of the baseline sequential at eight bits/sample and the lossless coding modes with vertical and horizontal sampling. For more information about the NBSI's use of JPEG, contact:

> Vic Andelin, Direct Technical Support
> NBSI
> 1530 Progress Road
> Fort Wayne, IN, U.S.A. 46808
> P: (219)484-8611; F: (219)482-2428

17.22 NTT Electronics Technology Ltd. ○

NTT Electronics Technology Ltd. (NEL) was founded in 1982 to do hardware, chips, and applications.[112]

The full-color, high-resolution still image CODEC Board (NLP-002) supports JPEG baseline and lossless Huffman processes for eight bits/sample. It plugs into the VME bus interface to workstations.

For more information about NEL's JPEG product, contact:

> Takayoshi Nakashima, General Manager
> NTT Electronics Technology Ltd.
> 1-14-5 Kichijoji-honmachi
> Musashino-shi Tokyo 180
> Japan
> P: +81-422-20-1081; F: +81-422-21-8922

17.23 OPTIBASE® ○

OPTIBASE Inc. (Optibase®) was founded in 1988 as a hardware-oriented company. Its original primary product was a compression algorithm for aerial photography. Its current primary products are JPEG and MPEG compression boards and software.[113, 114]

Optitools 3.1 is a software development tool kit for Microsoft and Borland C developers for DOS or Windows, and for Visual Basic for Windows. It supports baseline and lossless JPEG modes at eight bits/sample for one to four color components with horizontal and vertical subsampling.

Optibase 100 is a JPEG accelerator board with a single Motorola 56001 DSP. It supports baseline, Huffman, eight bits/sample and lossless JPEG modes with vertical and horizontal subsampling. It runs under DOS or Windows on an IBM (or compatible) PC computer.

The Optibase 500 JPEG accelerator board has dual Motorola 56001 DSPs in parallel for increased processing performance. It plugs into the PC AT (ISA).

The JPEG Workshop is an end-user software application package. It supports PCX, TIFF, BMP, and JPG file formats. The Optibase accelerator boards are a compatible option that can be used to boost performance.

For more information about Optibase's JPEG products, contact:

> Ray Harris, Vice President, Marketing
> Optibase, Inc.
> 7800 Deering Avenue
> Canoga Park, CA, U.S.A. 91304
> P: (818)719-6566; F: (818)712-0126

17.24 Optivision, Inc. ○

Optivision, Inc.[115] (Optivision) "develops and manufactures high speed communications products based on innovative technologies in compression and photonics. Founded in 1983, the Company's strategy is to build an R&D base with industrial and government sponsored contracts and then to expand these R&D projects into commercial products."[116] For example,

Optivision develops additional compression algorithms for high-performance image compression, such as recursive block coding and wavelets. This is consistent with its mission "to provide innovative, high quality compression products to the imaging community."[116]

The OPTIPAC™ JPEG Image Compression Accelerator Board provides hardware support for the sequential lossy, Huffman, eight-bit/sample, 12-bit/sample, lossless and variable vertical and horizontal sampling JPEG functions. This OPTIPAC hardware is currently available for the 386, 486, PC AT, XT, or compatible, MicroChannel™ architecture and VME platforms. Currently available operating systems are DOS, Windows, OS/2, and Unix.

The OPTIPAC JPEG Image Compression Software supports the same JPEG function as the OPTIPAC JPEG Image Accelerator Board. OPTIPAC software currently runs under DOS, OS/2, and Windows for PC-based computers and Unix workstations.

Optivision provided JPEG image compression hardware to NBS Imaging Systems, the system contractor for the California Driver's License Program (see sections 17.21 and 17.6). All California Department of Motor Vehicle (DMV) field offices were outfitted with 80386 PC ATs.

Optivision provides JPEG image compression as a newspaper photographic capture service. Using Optivision's image compression hardware, this image capture service provides large, high-quality color images to newspapers, magazines, and news bureaus worldwide. Each transmission site is equipped with a 386 PC AT.

Optivision provides National Imagery Transmission Format Standard JPEG image compression software and custom hardware for PCs and workstations.

For more information on Optivision's JPEG products, please contact:

Kathy Ernest
Optivision, Inc.
1477 Drew Avenue, Suite 102
Davis, CA, U.S.A. 95616
P: (800)562-8934, (916)757-4850; F: (916)756-1309

17.25 Philips Kommunikations Industrie ○

Philips Kommunikations Industrie AG (PKI) was founded to design and market telecommunications products of all kinds: hardware, software, chips, solutions, and applications.

PKI's Picture View is a JPEG still-picture application, consisting of boards, hardware, software, and consulting. It is available for OEM. Picture View supports the baseline, sequential, progressive, Huffman, and 8-bits/sample lossy JPEG modes. The JPEG lossless mode is currently in preparation. Images can be either grayscale or YUV color. Picture View runs on at least a 386 PC, under MS Windows™ 3.X. Picture View is helping to achieve Philips' mission that "Telecommunication from Philips turns high tech into Human tech."[117] For more information on PKI's JPEG product, please contact:

Klaus-Peter Loeser
Philips Kommunications Industrie AG
Thurn-und-Taxis Strasse 10
D-8500 Nuernberg
Germany
P: +49(911)526-6279; F: +49(911)526-3795

17.26 PRISM ○

Prism Interactive Corporation (PRISM) was "founded in November, 1989 by Dr. Richard Bruno with the mission of creating systems for encoding images, audio, and video, and to provide consulting services in the area to an international clientele."[118]

PRISM's primary products do compression/decompression for image (AQIN™), audio (AV100™ and COMO100™), and video (PCE). The AQIN software supports JPEG baseline, sequential and progressive, Huffman and arithmetic coding at eight bits/sample with vertical and horizontal sampling. It runs on a UNIX-based workstation. For more information about PRISM's JPEG product, contact:

Ashok Mathur, Vice President
Prism Interactive Corporation
751 Roosevelt Road, Suite 7-200
Glen Ellyn, IL, U.S.A. 60137
P: (708)469-1215; F: (708)469-1452

17.27 Storm Technology ○

Storm Technology®, Inc.[119] (STORM) was founded in 1990 to create software and hardware accelerator boards for color digital image processing. STORM's mission is "to enhance visual communications by making color digital imaging easy, fast, and exciting."[120]

STORM's PicturePress™ provides JPEG image compression software for the Macintosh computer. It supports baseline, sequential, Huffman, eight bits/sample, 12 bits/sample, and vertical and horizontal subsampling. PicturePress runs on any of the Macintosh II or Quadra family computers as well as a Macintosh SE30. All of these computers must have 32-bit color QuickDraw installed in their systems for PicturePress to run.

The StormCard™ is a programmable hardware accelerator card for the Macintosh that accelerates the JPEG functions of PicturePress and QuickPress™ software, as well as software developed by third-party developers. The StormCard runs on the Macintosh II, IIx, IIfx, IIxc, IIci, IIsi, or Quadra family computers.

QuickPress™ is an operating system extension that programs the StormCard accelerator to provide much faster JPEG image compression and decompression under Apple Computer's Quicktime. QuickPress supports the baseline, sequential, Huffman coding, eight bits/sample and 12 bits/sample with vertical and horizontal sampling. QuickPress runs on any of the Macintosh II or Quadra family computers. All of these computers must have a StormCard, QuickTime software, and 32-bit color QuickDraw installed in their systems for QuickPress to run.

For more information about STORM's JPEG products, contact:

Anne Marie Gunning, Marketing Communications Specialist
Storm Technology, Inc.
1861 Landings Drive
Mountain View, CA, U.S.A. 94043
P: (415)691-6677; F: (415)691-9825
AppleLink address: STORMSALES

17.28 Telephoto Communications ○

Telephoto Communications, Incorporated[121] (Telephoto) "is an outgrowth of Sorrento Valley Associates, an engineering and technical consulting firm established in San Diego during the mid-70's. Sorrento Valley Associates specialized in electronic design and microcomputer based applications.... Telephoto Communications was formed and began development of image processing products in 1986. The primary product focus was on still-frame image compression/expansion products for the IBM® PC/XT, PC/AT and compatible products." Today, Telephoto's mission is "to design, develop and market electronic imaging tools and services."[122]

Telephoto has ALICE™ JPEG Image Compression for MS-DOS®, Windows®, OS/2, Sun Sparcstations and Compatibles, and for the IBM® RS/6000® Workstations. The JPEG baseline, sequential, Huffman, eight-bit/sample and lossless (2–12 bits/sample) modes are supported. All of Telephoto's JPEG products compress with 2:1 horizontal subsampling of the chrominance. The decompression supports 1:1, 2:1 horizontal, and 2:1 horizontal and vertical subsampling. These software utilities or integration libraries include high-quality color mapping, file-format conversion, and autodetect of optional hardware accelerator boards.

For more information about the Telephoto's ALICE products, contact:

Marwan Kadado, National Sales Manager
Telephoto Communications, Inc.
11722-D Sorrento Valley Road
San Diego, CA, U.S.A. 92121-1084
P: (619)452-0903; F: (619)792-0075

Telephoto also provides consulting services for hardware and software development, including the C-Cube and LSI Logic JPEG chips. Telephoto has source code for JPEG compression that can be ported to customers' environments and is used in some of its consulting contracts. For more information about Telephoto's consulting services, contact:

Amy B. Barnhart, Manager of Consulting Services
Telephoto Communications Inc.
11722-D Sorrento Valley Road
San Diego, CA, U.S.A. 92121-1084
P: (619)452-0903; F:(619) 792-0075

17.29 Tribune Publishing Co. ○

Tribune Publishing Co. was established in 1892 as a newspaper. Its primary business continues to be newspaper publishing. However, a separate R&D division, Tribune Solutions, now creates newspaper archiving software applications with a mission to "produce PC-based image archiving products for publishing markets."[123]

Tribune Solutions has an image-archiving software package called PhotoView®. It is compatible with picture desks and is compatible with PC-based prepress color systems. The archives are designed to handle up to 2 billion cutlines and photos. Software products from Telephoto Communications (see section 17.28) are incorporated into the Tribune Solutions system. It runs on 386/486 PCs under DOS 5.0 (or later) or Windows 3.1 (or later). For more information about PhotoView, please contact:

> Glenn Cruickshank, Manager
> Tribune Solutions
> 505 C Street
> Lewiston, ID, U.S.A. 83501
> P: (208)743-9411; F: (208)746-7341

17.30 VideoTelecom ○

VideoTelecom Corporation[124] (VTC) started in 1985 as a manufacturer of VisionPlus™, a desktop audio/video terminal and network. This product had audio/video frequency multiplexing on coax cable. Today VTC manufactures equipment for the video teleconferencing market. VTC's video conferencing system, the MediaMax™, supports both a proprietary mode and a CCITT Px64 standard mode of operation. The MediaMax operates at channel rates of 56 Kbits/sec. to 384 Kbits/sec.

The MediaMax uses the baseline JPEG standard for transmission and storage of still-frame graphics images in its proprietary mode. The still-frame graphics image in the MediaMax has a resolution of 512x480 for NTSC, and 512x576 for PAL. The color space is the same as the CCIR-601 YCbCr color space. Chrominance subsampling ratios may be 2:1 or 1:1, both horizontally and vertically. The MediaMax includes a 386 PC/AT subsystem. This subsystem allows for storage of JPEG encoded images in addition to the transmission of such data. The MediaMax also provides for annotation of graphics images, which can be subsequently stored or transmitted as JPEG encoded data.

David Hein, Chief Scientist at VTC, writes, "In summary, still frame graphics is an important facet of motion video teleconferencing. Standardization is also an important factor, which allows for an increase in connectivity among manufacturers. VideoTelecom strongly feels that JPEG provides the flexibility and bit-rate efficiency required for video teleconferencing."[125]

For more information about VTC's JPEG product, contact:

Dr. David Hein, Chief Scientist
VideoTelecom Corp.
1901 West Braker Lane
Austin, TX, U.S.A. 78758
P: (512)834-9838 ext. 216; F: (512)834-3792

17.31 XImage ○

XImage Corporation[126] (XImage) was founded in 1988 as a systems integrator. It continues to have as its primary product a computerized booking and suspect identification system for the law-enforcement community. "Using leading edge technology, XImage is dedicated to providing cost effective, highest quality Image Database Systems specifically designed to law enforcement and related applications."[127]

XImage's ForceField II$^{®}$ is a digital mug-shot and turn-key booking system. It supports baseline, sequential, Huffman, eight-bit/sample, and lossless JPEG modes. It requires ForceField$^{®}$, Telephoto ALICE JPEG 350 Image Compression, and ALICE Libraries. For more information about XImage's system, contact:

Jag Narasimhan
Systems
XImage Corporation
1050 North Fifth Street
San Jose, CA, U.S.A. 95112
P: (408)288-8800; F: (408)993-1050

17.32 Xing ○

Xing Technology Corporation[128] (Xing) was incorporated in July, 1989, to "provide [the] highest performance software implementations of internationally standardized image handling algorithms on industry-standard computing platforms."[129]

Xing's VT-Compress JPEG Image Compression Software is a JPEG baseline sequential software-only package for JPEG compression or decompression. It provides a 5–8-second response, for disk-to-screen running on a typical 386 platform under Windows 3.1 DOS; Windows 3.0 or 3.1; or OS/2 1.3 or 2.0 on 386/486 PCs.

The VT-Express JPEG Turbo Accelerator typically provides a factor-of-5 software-only speedup for VT-Compress, yielding a 1.5–2.5-second disk-to-screen decompression on a 386 platform and subsecond performance on a 486. It supports Windows MCI and OLE programming interfaces (Windows 3.0 or 3.1; OS/2 1.3 or 2.0).

The VT-Compress JPEG Capture System is the VT-Compress JPEG compression/decompression software bundled with frame-grabber hardware for NTSC/PAL capture, utilities, and unlimited license for duplication of viewing software so as to allow distribution of still images for viewing on other platforms not already possessing JPEG capability (for 386/486 PCs with Windows 3.0 or 3.1).

VT-Compress Software Developers Kit is a software package for 386/486 PCs with Windows 3.0 or 3.1, OS/2 1.3 or 2.0, or DOS. It provides linkable libraries for DOS and DLLs (dynamically linked libraries) for Windows and OS/2, plus appropriate tools and examples to allow program developers to incorporate JPEG compression/decompression into their applications.

For more information about Xing's JPEG products, contact:

Lee Ambrosini, Vice President of Sales
Xing Technology Corporation
456 Carpenter Canyon
Arroyo Grande, CA, U.S.A. 93420
P: (805)473-0145; F: (805)473-0147

Xing has developed the Scalable Compression Architecture (SCA) and its interface to applications (API) in order to allow easy integration of compression into a wide variety of computing systems. SCA supports integration of JPEG, MPEG video, MPEG audio, G3/G4, and binary data compression. For further information about SCA, contact the SCA Administrator at the address and phone numbers above.

17.33 Zoran Corporation ○

Zoran Corporation[130] was founded in 1983 to "create value by developing, manufacturing, and marketing high-quality cost-effective, VLSI solutions for digital image enhancement and compression applications."[131]

The ZR36031 Image Coder/Decoder is a JPEG-like coder/decoder that is not JPEG compliant. It is similar to the baseline algorithm for eight bits/sample. As an extra feature it has bit rate control and supports real-time video (320×240, 30 frames/sec.). A PC evaluation board is available.

Zoran Corporation and Fuji Photo Film Co. Ltd. (Tokyo) jointly developed chip sets for image compression. The first chip set was similar to JPEG with a DCT processor and an image compression coder/decoder chip.[132] These chips are used in the Fujix Memory Card Camera DS-100, with integrated TV playback capabilities. The new chips, introduced in 1992, are JPEG compliant.

The ZR36040 JPEG Image Coder/Decoder Device is a fully JPEG compliant chip that does the baseline JPEG for eight bits/sample. It has bit-rate control and fast reverse, and supports CCIR video compression (720×480, 30 frames/sec.).

For more information about Zoran Corporation's JPEG products, contact:

Steven E. Brook, Senior Product Marketing Engineer
Zoran Corporation
1705 Wyatt Drive
Santa Clara, CA, U.S.A. 95054
P: (408)986-1314 x333; F: (408)986-1240

17.34 3M ○

3M was founded in 1902 in Two Harbors, Minnesota. 3M now designs and manufactures more than 60,000 products, ranging from health-care to industrial tapes and adhesives to high-tech imaging and electronic systems. The company is organized into three sectors: Life Sciences; Industrial and Consumer; and Information, Imaging, and Electronics. 3M's company mission is "Innovation: Meeting the needs of our changing times."[133]

CD Stock is an innovative CD-ROM-based stock-photo-researching tool that has been tailored to meet the needs of art directors and designers. It offers end users a fast, efficient means for making creative decisions and carrying-out image selection and comprehensive layout. CD Stock is a Macintosh-compatible image-searching tool that utilizes Apple QuickTime™ JPEG image compression to facilitate the storage of several (>5) thousand low resolution (72 dpi) images on a single CD-ROM disc.

The CD Stock interface software provides rapid search, selection, and retrieval of images, offering several tools to assist the user in the search for desired images. The user may browse through a catalog of thumbnail images in a manner similar to browsing a printed catalog of images; review images by classification; or search for images by keyword descriptor. Once images have been selected, they can be retrieved, decompressed, and examined, along with accompanying descriptive information. Finally, selected images can be exported for use in graphics or page-layout applications.

For more information about CD Stock, contact:

> Peter Thorp, Manager (Marketing)
> Technology Development and Application
> HESD Laboratory
> 3M
> 260-6A-08
> St. Paul, MN, U.S.A. 55144
> P: (612)733-0997

or

> Martin Kenner, Software Specialist (Technical)
> Technology Development and Application
> HESD Laboratory
> 3M
> 260-6A-08
> St. Paul, MN, U.S.A. 55144
> P: (612)736-3065

17.35 File formats ○

In recognition of the fact that JPEG committee did not specify a file format, several different formats have been developed by applications.

17.35.1 ANPA/IPTC ○

American Newspaper Publisher's Association/International Press Tele-communications Council (ANPA/IPTC) is a world-wide standard for information interchange. This standard includes file headers that go with the image. JPEG is one of the allowable compression formats in the ANPA/IPTC standard.[98]

17.35.2 CCITT and ISO/IEC standards committees ○

CCITT SGVIII application subgroups of facsimile, photovideotex, still image teleconferencing, and audiographics have had preliminary dis-cussions on how to incorporate JPEG. ISO DIS 10918-1 was adopted as CCITT Recommendation T.81 without waiting for the final IS version partly in order to make it easier to reference in other application standards.

Other standards groups, such as ISO/IEC JTC1/SC18 and CCITT ODA/ODIF for "compound documents" and have also indicated strong interest in JPEG.

17.35.3 ETSI ○

The European Telecommunications Standards Institute (ETSI) is stand-ardizing photovideotex in Europe. The JPEG committee introduced inheritance of tables partly at the request of the ETSI project team 22.[134] The ETSI "Videotex Photographic Data Syntax" allows for five different profiles plus a private profile. The first three profiles are for sequential mode with images of increasing size. The next two profiles add progressive modes. The private profile allows access to all of the JPEG modes.[135] The ETSI photovideotex document is being ballotted. The incorporation of JPEG in videotelephony (still image mode) is being worked on.

17.35.4 Image Object Content Architecture ○

IBM ImagePlus uses IBM's Image Object Content Architecture (IOCA) as the JPEG file format. IOCA recognizes JPEG baseline, sequential, pro-gressive, and lossless modes with Huffman and arithmetic coding.

IBM has recently granted the ANSI Accredited Standards Committee X9B the right to use IBM's IOCA as described in IBM publication SC31-6805-01 in the Check Image Interchange standard, provided that IBM's contribution and copyrights are recognized.

17.35.5 JPEG File Interchange Format ○

According to Eric Hamilton of C-Cube Microsystems, Inc., "The JPEG File Interchange Format (JFIF) is a minimal file format which enables JPEG bitstreams to be exchanged between a wide variety of platforms and applications. This file format specifies parameters in addition to those defined by JPEG which allow decoders to interpret the reconstructed image data and provide accurate image presentation on an output device. JFIF does not include any of the advanced features of the TIFF JPEG specifi-cation or any application-specific file format; its primary purpose is to allow the exchange of JPEG compressed images."[136]

The JPEG File Interchange Format uses JPEG interchange-format compressed image format. It uses a standard color space (CCIR 601-1). The APP0 marker is used to identify a JFIF file and includes application-specific information. For additional information about JFIF, contact:

> Eric Hamilton
> C-Cube Microsystems, Inc.
> 1778 McCarthy Boulevard
> Milpitas, CA, U.S.A. 95035
> P: (408)944-6300; F: (408)944-6314

17.35.6 National Imagery Transmission Format ○

In 1984, the federal intelligence community and the Department of Defense (DoD) wanted to take advantage of more imagery possibilities. A task force was started in 1985 to create a National Imagery Transmission Format (NITF) for the exchange of imagery and the associated annotations. Currently, the NITF version 1.1 uses an Adaptive Recursive Interpolative Differential Pulse Code Modulator (ARIDPCM) to compress image data. The next version will include JPEG. In 1991, NITF changed its name to NITFS as part of its conversion to a Department of Defense (DoD) standard.[137]

17.35.7 Tag Image File Format Revision 6.0 ○

Aldus® Corporation published the first version of the Tag Image File Format (TIFF™) specification in the fall of 1986. TIFF has been enhanced to include JPEG support in the extensions section of Revision 6.0. TIFF supports grayscale, RGB, CMYK, and YCbCr color spaces for JPEG baseline and Huffman lossless coding processes. Clever use of the restart intervals allows the TIFF header to point to independently decodable segments of the original image.[138]

18

OVERVIEW OF
CCITT, ISO, AND IEC

This chapter deals with the international standards committees and their procedures that relate to JPEG. The steps for generation of international standards (for the paths followed by JPEG) are documented in the hope that they will help the reader understand how standards are created. Any future changes or enhancements to JPEG would be expected to go through similar procedures. Since standards procedures can change, the latest revisions of the standards manuals should be consulted to get accurate and current information. More detailed information about international standards committees can be found in Wallenstein [139] and McPherson. [140]

JPEG was started in 1986 by cooperative efforts of both the International Organization for Standardization (ISO)* and the International Telegraph and Telephone Consultative Committee (CCITT).† In November, 1987, the International Electrotechnical Commission (IEC) joined with

---

* The three-letter acronym comes from its original name of International Standards Organization.

† The five-letter acronym comes from the French version of its name.

ISO to create a new Joint Technical Committee 1 (JTC1), under which the JPEG committee continued to operate. Thus, these three major international standards organizations are all sponsors of the JPEG committee.

18.1 ISO ○

ISO had its beginnings in World War II, when the United States and its allies needed interface characteristics standardized. This led at first to an *ad hoc* standardization effort. In 1946, however, that standardization effort was upgraded into the International Standards Organization, a non-treaty agency of the United Nations.

ISO is a self-regulating group aimed at setting basic industry standards such as those involving wire gauges, safe voltages, metric units, and screw threads. Seventy-two member bodies participate in ISO, but there are an additional 18 associated nonvoting developing countries. Member national bodies may be either private, voluntary, national agencies supported by participating companies, or government agencies with some industrial participation.

ISO's scope includes "agriculture, nuclear systems, fabrics, documents used in commerce, library science, computer systems, and computer communications."[141] Its programs tend to be user-oriented and generally aim to set standards that are for the "good of the industry." All ISO standards are reviewed every five years.

ISO is organized into technical committees (TC) that are further divided up into study committees (SC). Each SC has working groups (WG), expert groups (EG), *ad hoc* groups, and rapporteur groups (short-term assignments). Study committee members are of two types, primary (P) members and nonvoting observing (O) members. (Note that "member" refers to a national-body standards organization.) Study committees have about 10 to 20 primary members and another two to four observing members. To ensure world-wide interest and need, each working group must have at least five members participating. For each technical committee, study committee, and working group meeting, the member bodies send a delegation and head of delegation with instructions as to the degree of freedom permitted the delegation. Written ballots are used to make most official decisions.

ISO is a decentralized organization. There is a small ISO Central Secretariat in Geneva, the Information Technology Task Force (ITTF), that oversees the operation of the many committees and is responsible for the publication of the International Standards (IS). However, the chairmanship and secretariat for each specialty area are delegated to national member committees that volunteer for the job.

The first step toward creating an ISO standard is getting consensus among the technical experts at the working-group level to create a working draft of the proposed standard. Different revisions of a working draft or technical specification are created to reflect the current tentative agreements as to the contents of the eventual standard. Eventually, a formal Committee Draft (CD) is prepared. The parent study committee must agree that the CD is ready for balloting and approve assignment of a registration number. This approval process can be carried out at an SC meeting (such meetings are held about every 18 months) or by written ballot of the P members.

The three-month CD ballot officially starts two weeks after the SC secretariat mails the document. All SC P members are expected to vote on the CD. Votes can indicate unconditional approval, approval with comments, or disapproval. For disapprovals, mandatory comments must explain the changes necessary to obtain approval. The working group must reply in writing to all comments that address problems more serious than minor typographical errors. The comments are generally of the form "After the third sentence in subclause X change 'yyy' to 'zzz'." Representatives from countries with significant comments need to participate in the resolution-of-comments meeting, in order to facilitate rapid agreement on all changes.

A new document that incorporates all the agreed-upon changes documented in the Disposition of Comments is prepared. This document is then submitted to the SC. When the SC determines that the comments have been responded to satisfactorily, it recommends to its parent TC that the revised CD be promoted to draft international standard (DIS) status. Alternatively, the study committee may require the working group to consider the comments in more detail and require another three-month CD ballot to approve the revised CD.

Extensive editorial changes in the document are allowed between CD and DIS documents, but extensive structural changes are not allowed. Any significant technical changes automatically require another round of CD balloting. Changes are not only possible, but should be expected during the CD stage. This is often the first opportunity for technical review outside the working group.

The DIS document (which is expected to look like the final IS document) goes through another round of national balloting, but this time at the TC level. The six-month ballot begins two weeks after the DIS ballot is mailed. At least two-thirds of the nations voting must agree to the promotion of the DIS to international standard (IS).

After promotion to an IS, the technical-committee secretariat may recommend some editorial changes in style, but all such changes must be approved by the working-group editing committee. The IS is then published and sold through the ISO headquarters and through national standards committees.

18.2 CCITT ○

The need for adequate volume on the European telephone long lines led to the creation of two CCITT predecessor organizations (CCIF and CCIT) in the 1920's. They issued nonbinding recommendations to allow compatibility of telecommunication services across national boundaries. Today the International Telecommunications Union (ITU), through its consultative committees, CCITT and International Radio Consultative Committee (CCIR), continues to issue recommendations that are intended to be international compatibility agreements. Each participating nation voluntarily chooses whether to observe the recommendations inside their national boundaries. There is no international enforcement capability.

However, it is not unusual for administrations of importing countries to cite rigid conformance with CCITT recommendations for their telecommunications equipment. This helps them to avoid dependence upon any particular supplier. Many telecommunications service providers have integrated CCITT performance standards into their internal rule books and require suppliers to meet the apportioned contributions to overall performance. Ignoring such recommendations can lead to a severe competitive disadvantage.

As an intergovernmental treaty organization sponsored by the United Nations, the ITU operates with the rule of "one state, one vote." However, the ITU and its committees, CCITT and CCIR, tend to operate with unanimous agreement as the objective because of the need for end-to-end compatibility. Controversial recommendations may therefore be sent back to the committee for further study rather than passed.

All official CCITT documents are overseen in their preparation, translation, and distribution by the ITU central secretariat, located in Geneva. The CCITT books were originally published in French only. Now they are also issued in English, Spanish, and (after an additional delay) Chinese and Arabic.

Traditionally, all CCITT recommendations were not officially adopted until unanimously approved at a plenary session held once every four years. The recommendations could be considered only if the study group (SG) had reached unanimous consensus sufficiently in advance for the documents to be properly translated and distributed to all member bodies. The final documents were published as a multi-volume book known by the color of its cover (orange for 1980, red for 1984, and blue for 1988). Since 1988, an accelerated procedure called Resolution 2 encourages final approval at any time between plenaries.

An ISO DIS can be approved at a CCITT study-group meeting for accelerated procedures. The document must adhere to the joint CCITT/ISO drafting rules. Only the foreword is changed to reflect its status as a CCITT recommendation. The rest of the document (wherever possible) is left identical.

Translations into Russian, French, and English must be available at least three months in advance of the study-group meeting, at which unanimous approval is needed to request the secretariat to send the CCITT recommendation version of the ISO DIS out for a "three-month default ballot." Any dissenting vote at the SG meeting means that the potential standard must follow the slower, normal process. Since only a two-thirds majority is needed to pass the "default" ballot and the lack of a response is counted as a positive vote, failure to pass this ballot is unusual.

18.3 IEC ○

IEC was started in 1904–1906 to coordinate and unify various national electrotechnical standards. Eventually, it became a nongovernmental standards organization. IEC has been concerned throughout its long history with component measures, information processing systems, safety of data processing, and office machine services, including the connector itself. IEC has more than eighty technical committees (TC).

18.4 Joint coordination ○

The border between telecommunications technology (traditionally CCITT's arena) and computer technology (ISO and IEC's domain) is not easy to define. ISO and IEC solved some of the problems of coordination by creating, in November, 1987, a Joint Technical Committee 1 (JTC1) with the field of information technology as its scope of standardization. An effective collaboration exists among CCITT, ISO, and IEC in data networks and telematic services systems through parallel adoption of recommendations and standards with practically identical text. The 1988 CCITT blue book contained 40 such parallel recommendations. A joint rules committee has been working to create guidelines for identical text for joint recommendations/standards. JPEG is one of these joint collaborations under JTC1. Therefore, the JPEG DIS is known officially as ISO DIS 10918-1 | CCITT Recommendation T.81.

19

HISTORY OF JPEG

This chapter gives the history of JPEG and describes how the JPEG standard passed through the standards process. Readers unfamiliar with the standards procedures, standards committees, and their acronyms are encouraged to read the preceding chapter on international standards committees.

19.1 Formation of JPEG ○

The JPEG Standard is the result of an "anticipatory" standards activity. As early as 1982 researchers interested in photovideotex initiated activity in ISO on a color image data compression standard. A photographic experts group was formed under ISO/TC97/SC2 Working Group 8 (WG8) (Coded Representation of Picture and Audio Information). The group wanted to develop a progressive data compression scheme that would operate at ISDN (64 Kbits/sec.) rates, and produce something that would be recognizable in approximately one second and that would continue to improve until a visually lossless picture was obtained. They were joined in 1986 by the CCITT Study Group VIII's Special Rapporteur Group on New Forms of Image Communication. This CCITT SG had several WGs and rapporteur groups that needed color data compression. At this time joint collaboration between CCITT and ISO was being actively encouraged in order to avoid the creation of competing, independently developed standards. JPEG, the Joint Photographic Experts Group was therefore established for the purpose of developing a still image color data compression standard.[142]

Graham Hudson, of the U.K., was appointed chair of this subgroup of WG8. Hiroshi Yasuda, of Japan, chair of WG8, was instrumental in encouraging the formation of JPEG, and later, JBIG, MPEG, and MHEG. Istvan Sebestyen, of Germany, as special rapporteur of the CCITT group, also actively encouraged the joint collaboration to achieve a data compression standard that would meet the needs of many applications.

Even though the JPEG committee was a subcommittee of an ISO working group, it contained members from both ISO and CCITT. This cooperative effort was so experimental that the rules and procedures were not well defined. Because the procedure to resolve disputes was uncertain, the JPEG committee made all technical decisions on the basis of unanimous agreement, thereby avoiding the problem.

19.2 Original JPEG Goals ○

The JPEG compression standard for still continuous-tone grayscale and color images was intended to cover the widest range of applications consistent with a number of predetermined requirements. This led to a mandatory capability for sequential coding, progressive coding, lossy coding, lossless coding, and feasibility of hardware implementation at 64 Kbits/sec. The JPEG experts wished to achieve nearly state-of-the-art compression performance, so ambitious bits/pixel targets were set for the lossy compression. The hope was that recognizable images could be obtained at 0.25 bit/pixel, excellent-quality images could be achieved at 1.00 bit/pixel (or even at 0.75 bit/pixel), and images indistinguishable from the original image could be obtained at about 4 bits/pixel. The images chosen for the contest had a format appropriate for European video systems: 720 samples/line by 576 lines with 8 bits/sample for the luminance, and 360 samples/line by 576 lines with 8 bits/sample for each chrominance component. However, the stated intent was to have a resolution-independent standard. These goals were a merging of the photovideotex requirements and the CCITT telecommunication requirements.

19.3 Selecting an approach ○

The selection process was carried out by competitive contests. Twelve proposals were registered in March, 1987, as candidates for the compression method. The titles indicate the diversity of approaches: "Generalized Block Truncation Coding,"[143] "Progressive Coding Scheme,"[143] "Adaptive DCT,"[143] "Component VQ,"[143] "Quadtree Extension of Block Truncation Coding,"[144] "Adaptive Discrete Cosine Transform,"[145] "Progressive Recursive Binary Nesting,"[146] "Adaptive Transform and Differential Entropy Coding,"[147] "DCT with Low Block-to-Block Distortion,"[148] "Block List Transform Coding of Images,"[149] "HPC,"[150] and "DPCM using Adaptive Binary Arithmetic Coding."[151]

By the June, 1987, meeting in Copenhagen, 10 proposals were supported with complete documentation and reconstructed images at the various progressive stages. Executable code that could generate the reconstructed images without human interaction was a requirement before a technique could be considered.

After a week of extensive subjective testing using four images, the field was narrowed to three techniques.[152] An Adaptive Discrete Cosine Transform (ADCT)[153] proposed by the ESPRIT PICA group had consistently the best quality at the 0.75 bit/pixel and 1.00 bit/pixel compression rate targets. A proposal[154] based on DPCM (Differential Pulse Code Modulation) with adaptive arithmetic coding had exceptional images at 0.25 bit/pixel and the best lossless compression. A progressive block truncation technique[155] (a Japanese proposal subsequently merged with another proposal from Japan[156]) scored well on many of the required features and had competitive image quality. Development groups were formed to refine and enhance the three methods for further testing. A requirement was adopted that decoding at ISDN (64 Kbits/sec.) data rates should be demonstrated at this meeting. New targets were set for subjective testing at 0.125, 0.25, 0.75, and 3.00 bits/pixel in order to stress the coding techniques. These bit rates were significantly more demanding than the 0.25, 1.00, and 4.00 bits/pixel originally targeted.

At the October, 1987, meeting the development teams shared intermediate results. The bit rate targets were further modified to specify recognizable images at 0.083 bit/pixel, useful images at 0.25 bit/pixel, excellent quality images at 0.75 bit/pixel and indistinguishable images at 2.25 bits/pixel. Five new images were selected at the December, 1987, meeting in England. The three development groups had to submit executable software modules for their refined techniques before receiving these new images.

At the January, 1988, meeting in Copenhagen, extensive subjective evaluation was done to compare the three techniques. The development teams had enhanced all three methods significantly.[157-159] All three techniques met the 64 Kbits/sec. feasibility requirement. Two were demonstrated as software implementations on 80386 machines. Because of the superior image quality achieved by the Adaptive Discrete Cosine Transform and the demonstrated feasibility in both hardware and software, JPEG reached a consensus that the Adaptive Discrete Cosine Transform approach should be the basis for further refinement and enhancement.[160]

19.4 Functional requirements ○

A "CCITT/ISO Photographic Image Data Compression Standard Functionality and Goals" draft document[161] reaffirming the goals and expected functionality of the standard was approved at the January meeting discussed above. (A more detailed revision[162, 163] was unanimously approved at the May, 1988, meeting.) This document was intended to ensure that the final standard would be suitable for the widest range of applications. Since the other two finalists also produced excellent results, JPEG agreed that the other approaches would be reconsidered if the final enhanced DCT technique was found to be deficient in meeting a requirement.

A cost-effectiveness criterion in both software and hardware was added to the functional objectives at January, 1988, meeting. Although initially cost effectiveness in software was a very controversial issue, the adaptive DCT method had by then been demonstrated in both hardware and software at a nominal 64 Kbit/sec. decoding rate. Consequently, the ability to reach this goal was already proven.

The following list of functionality and goals can be useful in separating data compression techniques that have a niche opportunity from general-purpose methods. Although some of the DCT-specific requirements may not apply, any data compression method that might be considered competitive to JPEG needs to be evaluated on the basis of all of the general requirements, not merely on its own strongest features.

Quality: The first goal was that the final enhanced technique should match or exceed the quality shown in Copenhagen at the 0.25-, 0.75- and 2.25-bit/pixel stages. An even lower bit-rate image was expected to be useful, but it was not mandatory that the first image be at the 0.083-bit/pixel testing point.

Progressive: The whole testing methodology was predicated on the need to achieve good progressive compression. The images were judged on the basis not of the best independent 0.25-, 0.75-, 2.25-bit/pixel images, but rather the best quality at each stage having gone through the earlier progressive stages. Because the amount of progressiveness depended upon the application, transmission speeds, and other local considerations, a mandatory number of stages was not set. However, a minimum of five stages to lossless was required. "Graceful" progression was also set as a requirement, meaning that a smooth, visually appealing progression was expected. The ability to guarantee worst-case error limits in multiple stages after "indistinguishable" quality had been reached was also expected.

Reversible: Not only should the standard have a lossless mode of operation, but the compression performance should be close to the best achievable: typically 2:1 compression for images similar to the JPEG test set.

Synchronous: Any data compression technique in which the encoder must take much more time and/or processing power than the decoder is not able to meet this requirement. MPEG in its higher performance modes is an example of such a system. Because teleconferencing was one of the CCITT applications expected to use JPEG, "synchronous operation" was mandatory. Synchronous operation was defined to mean that some of the compressed data could be transmitted after a time relatively short compared to that needed to reach the main stage. The performance of the algorithm was to be evaluated primarily in its progressive synchronous mode. A multipass nonsynchronous mode of encoding showing significant performance advantages was certainly permissible.

Sequential: A sequential mode with performance similar to that of the progressive mode was required. The sequential mode had to be able to operate with minimal buffering (just a few image lines) and be capable of being a single-pass system.

Cost effective in both hardware and software: Computationally efficient implementations for both hardware and software were goals.* For software this was defined to mean decompression at a nominal 64 Kbits/sec. in a microprocessor equivalent to a 16 MHz Intel 80386.

Resolution independence: Arbitrary source image resolutions were to be handled. Ways to handle images whose dimensions were not multiples of eight rows or eight columns were to be specified.

Text images: Grayscale representations of text had to be handled well. Originally, compression of bi-level images was also a requirement, but at the January, 1988, meeting the Joint Bi-level Image Experts Group (JBIG) was formed in order to address that class of images separately. (The JPEG and JBIG groups have cooperated to achieve commonality of the arithmetic coder.)

No absolute bit-rate targets: Even though the competition was held at specific bit rates (0.083, 0.25, 0.75, and 2.25 bits/pixel) in order to compare the different techniques at exactly the same bit rates, the additional complexity involved in creating images at exactly these bit rates was not considered justified. No absolute bit-rate targets were required, but similar aggregate performance was expected. The graceful progression and the ability to modify the quantization table were regarded as adequate mechanisms for reaching specific target bit rates.

Precision: For eight bits/sample, no more than 16-bit precision could be required for internal computations. This decision had significant impact on some software and hardware implementations.

Extensible: The number of lower-resolution images for a hierarchical progression was not to be fixed. More than eight bits/sample should be allowed in hierarchical mode. Up to 16 bits/sample should be compressible. It was recognized that six bits/sample could be handled as eight bits/sample by appending zeros to the input.

Luminance-chrominance sampling: When JPEG thought that it would need to specify image formats and color spaces, it accepted a default image format of YCrCb with horizontal subsampling 2:1 of chrominance relative to luminance. Other samplings of luminance only, equal sampling of all components, and 2:1 subsampling of the chrominance on both axes were expected. As JPEG was generalized to be "colorblind," this was generalized to sampling factors of 1, 2, 3, or 4 both horizontally and vertically.

Luminance-chrominance separability: The ability to recover the luminance-only image from luminance–chrominance encoded images without always having to decode the chrominance indicated a need to use some sort of pointers to separate the luminance from the chrominance compressed data.

---

* Although not formally stated, the intent was to keep the system simple enough to permit single-chip implementations, and many of the design choices JPEG made later were consistent with this unstated goal.

General image models: In order to have a general image model, the DCT quantization was to be controlled solely through the quantization table.

DCT—8×8 blocks: The 8×8 blocks for the DCT were chosen for two reasons, computational complexity and the availability of commercial hardware chips. The 16×16 block sizes were explored and found to not give enough improvement in compression to justify the extra image buffering, precision of internal calculations, and complexity.

DCT coefficients—eight bits maximum: A requirement for the quantized coefficients to fit into eight bits was made in order to reduce the buffering needed to store the coefficients. (This requirement was later modified.)

Quantization matrix (table): Default quantization tables should be specified. (This requirement was later modified as JPEG was made "color blind.")

Specification of IDCT: In a May addendum to the functionality statement[163] a DCT specification was considered. The early seeds for the final compliance tests can be found in that specification. Lossless compression through the DCT-based hierarchical mode required identical IDCTs in the encoder and decoder and, therefore, exact specification of the IDCT. Lossy compression, on the other hand, could tolerate some small differences. Therefore, a relaxed IDCT specification should also be specified as an alternative. Such things as ease of implementation for both hardware and software, simplicity of computation, universality, and amount of mismatch were to be considered before the final specification.

Specification of FDCT: It was recognized that encoders did not need to use exact versions of the FDCT even to ensure lossless decoding, because the encoder's output was in the compressed data. However, the FDCT should be sufficiently accurate that there is no loss in image quality relative to an ideal FDCT.

19.5 Refining the ADCT technique ○

A Transform Technique Enhancement Group was formed within JPEG in 1988 for the purpose of considering refinements and enhancements. One of the areas requiring extensive study was how to achieve the mandatory requirement of lossless coding while not unnecessarily constraining the implementations of the IDCT. Different implementations produced slightly different results after the IDCT step. This made achieving true reversible coding difficult. At a meeting in Rennes, France, in July, 1988, various solutions were proposed for further study.

The ADCT proposal demonstrated at the January, 1988, meeting achieved progression by a hierarchical (pyramidal) scheme. The 0.083 bit/pixel image was really a reduced image (containing 1/16th of the original number of pixels) although it was displayed at full size. The incremental corrections to improve an expanded version of the small image gave good 0.25-bit/pixel images. This process was iterated two more times to achieve the 0.75-bit/pixel and 2.25-bit/pixel images. Each iteration required an IDCT. Since IDCT implementation differences could lead to significant cumulative differences, alternative progressive coding schemes were explored.

Image quality was a major JPEG concern. At the September, 1988, JPEG meeting held in Torino, Italy, and London, England, an *ad hoc* group was formed to study the apparent differences in image quality on different displays and recommend a "front-end" technique for suppression of blocking artifacts and for increasing coding efficiency at low bit rates when using a nonhierarchical progression. At this meeting, Gregory Wallace, of the U.S., served as the new JPEG chair.

Over the next few months the *ad hoc* group on "front-end" techniques determined that the quality differences were due to display bandwidth differences. The *ad hoc* group found, however, that most of the display bandwidth differences could be handled by adjusting the reconstruction level of the AC coefficients to either the bottom or the middle of the quantization interval (or somewhere in between). (See Chapter 16 and JPEG DIS Part 1 section K.9.) This modification is implemented only in the decoder and does not affect the compressed data stream. The *ad hoc* group reached agreement that the front-end approach should be quantized AC prediction of five AC coefficients.[164]

Another area requiring considerable study was the choice of entropy-coding technique. Comparisons were made between adaptive Huffman coding and adaptive arithmetic coding. Work was initiated to compare adaptive Huffman coding with "custom" tables generated for each image to nonadaptive Huffman coding with "fixed" tables generated for a set of "typical" images. In addition, alternative arithmetic coders were proposed and resynchronization of the decoder in the presence of bit errors in the compressed data was investigated.

19.6 Technical specifications ◖

At a meeting in Livingston, New Jersey, in February, 1989, a set of system configurations meeting the requirements was adopted.[165-167] A baseline capability was defined as a lossy DCT-based sequential system with Huffman tables and eight-bit/sample input precision. Decoders had to receive custom Huffman and quantization tables and understand how to handle resynchronization. An extended DCT-based system was also defined that added capabilities such as extended Huffman coding, arithmetic coding, "front-end" processing, progressive coding, hierarchical coding, and reversible coding through differential corrections. The problem of IDCT variations was solved by creating an "independent function" for lossless coding that used the simple reversible coding algorithm defined for encoding of the DC coefficients of the DCT. A 181-page technical compendium of all active proposals was assembled. Where alternative proposals for a procedure were still being considered, they were included as well. This became the first draft of the JPEG Technical Specification (JPEG 8—Revision 0). The committee decided to consider new proposals only with the unanimous consent of the group.

An editing committee was formed consisting of members of JPEG who were native English speakers. The JPEG chair, Greg Wallace, chaired this committee. In the early stages much of the writing was done by the authors of this book. They rewrote the technical specification so that JPEG 8—Revision 1 was available for the next meeting.

At the July, 1989, meeting in Stockholm the coding models were modified to a logarithmic form that allowed higher precision in the DCT coefficients and source data. In addition, the precise form of progressive mode and the choice of the arithmetic coder had not been decided, so all choices had to be documented.* An extensive validation testing process was started in which at least two companies were to verify each part of the JPEG algorithm. The changes to the JPEG specification were incorporated into JPEG 8—Revisions 2 and 3.

JPEG 8—Revision 4 was the first technical specification that attempted to document all technical aspects under consideration. Since the decision about the arithmetic coder had not been made, both the MELCODER and the Q-coder were fully documented.

At a meeting in Japan in October, 1989, a number of simplifications were made to the baseline algorithm. 2-D prediction was eliminated from the DC coefficient coding (but was retained in the lossless coding). Front-end AC prediction was removed from the encoder and was retained only as an optional decoder postprocessing step. This was done in order to allow for transcoding between the progressive and sequential DCT-based modes. The arithmetic coding choice had been resolved by a JPEG *ad hoc* Group on Arithmetic Coding by adding "conditional exchange" to the Q-coder. The number of independent options was significantly reduced so that compatibility would be highly likely for a given functionality. The lossy DCT-based algorithm was now expected to only apply for 8- or 12-bit/component samples. The lossless algorithm (the "independent function") still handled up to 16-bit/component samples.

At this same meeting, JPEG agreed that the next revision of the technical specification (JPEG 8—Revision 5) would be released for external technical peer review by December 15, 1989. The baseline sequential system as documented in JPEG 8—Revision 5 was validated and a validation diskette distributed.

At a March, 1990, meeting in Tampa, Florida, a large number of minor changes were adopted. The basic structure of the JPEG data compression algorithm remained stable, however. Most of the changes dealt with signalling parameter syntax and constraints on system configuration. Much work had also been done looking for default Huffman tables that could be used over a wide range of applications. Finally, JPEG abandoned default Huffman tables and moved updated representative tables to an "examples and guidelines" annex. CCITT and ISO application standards committees may still develop default tables for their own environments.

At its plenary meeting in Washington, D.C., in April, 1990, ISO/IEC JTC1/SC2 restructured WG8 and its subgroups so that JPEG became the responsibility of Working Group 10 (Photographic Image Coding Group). JBIG became the responsibility of Working Group 9 (Bi-level Image Cod-

---

* Two choices for the arithmetic coder, the Q-coder and the MELCODER, were then under consideration.

ing Group). Both WG9 and WG10 continued their strong affiliation with CCITT Study Group VIII by continuing to parent JBIG and JPEG, respectively.

A JPEG delegation attended the CCITT subgroup meetings of Study Group VIII held in Budapest in April, 1990. Extensive editing committee work was accomplished during this meeting. JPEG 8—Revision 6 was written after this meeting. Revision 6 incorporated a new probability estimation table for arithmetic coding, with the fast-attack states for initial learning so crucial to JBIG performance.

Before the next JPEG meeting, extensive efforts were made to find a common JBIG and JPEG arithmetic coder. This work involved comparing proposals and their impact on hardware and software complexity as well as compression performance. The JBIG group made the final refinements in the probability estimation tables based upon performance in the binary arena, but with the constraint that JPEG image compression would not be adversely affected. It was agreed that the arithmetic encoder would resolve any carry-overs before completion of compression. The recommended arithmetic coder was documented in JPEG 8—Revision 7.

At the July, 1990, meeting in Porto, Portugal, this common arithmetic coder for JPEG and JBIG was adopted. A TEM marker code was added to prevent latency problems for high-speed hardware arithmetic encoders. Definition of the hierarchical mode was completed. Preliminary discussions were held on the compliance testing. A successful parsing of the JPEG header into Backus-Naur form was noted.[36] (See section 7.11.) A final revision of the Draft Technical Specification (JPEG 8—Revision 8) was completed in August, 1990, and became the basis for the Committee Draft (CD) Part 1. The committee decided to document the compliance testing in Part 2 in order not to delay balloting on the algorithm specification.

19.7 ISO 10918 Part 1 ○

On the basis of the Draft Technical Specification (JPEG 8—Revision 5), at the April, 1990, meeting of ISO/IEC JTC1/SC2 the study committee approved registration of the JPEG CD as ISO CD 10918, "Digital Compression and Coding of Continuous-Tone Still Images." The actual JPEG CD was to be submitted for SC2 balloting at the JPEG chair's discretion. This eliminated several months of delay that a written ballot among SC2 member nations would have taken. Even after registration, the JPEG Draft Technical Specification continued to be revised until JPEG 8—Revision 8.

An extended JPEG *ad hoc* editing meeting was held in Cambridge, Massachusetts in November, 1990. A plan for preparing the Committee Draft Part 1 was created. From November, 1990, to February, 1991, JPEG members revised and edited the Technical Specification into annexes for the JPEG Committee Draft Part 1. New material was written for the introduction, definitions, and requirements sections. Technical experts from 11 countries (Canada, Denmark, France, Germany, Israel, Italy, Japan, Republic of Korea, The Netherlands, United Kingdom, and the United States) actively participated in this effort. The Committee Draft

Part 1 was submitted to the ISO/IEC JTC1/SC2 secretariat on February 20, 1991.

At a meeting in San Jose, California, in March, 1991, a preliminary draft of Part 2 on compliance testing (JPEG 10—Revision 0) was considered. WG10 members' comments and corrections to Part 1 were also discussed. Further progress on the validation was reported for the progressive mode of operation.

The SC2 members (i.e., national bodies) voted on the three-month CD ballot for JPEG CD Part 1 that closed June 30, 1991. The Peoples Republic of China, Czechoslovakia, Denmark, Germany, Italy, The Netherlands, Switzerland, and Union of Soviet Socialist Republics voted "approved." France, Israel, Japan, United Kingdom, and the United States voted "approved with comments." Canada voted "Disapproved." and Belgium and Brazil abstained. Iran, Republic of Korea, Sweden, Turkey, and Yugoslavia did not vote.

At the JPEG meeting in Santa Clara, California, in August, 1991, the comments on CD Part 1 were handled. The more than 30 pages of comments were addressed by a Disposition of Comments generated at this meeting. The JPEG chair had notified the national bodies in advance that "approved with comments" would be treated as seriously as "disapproved with comments," but that a representative needed to be present at the meeting to help resolve the issues.

Many comments that were in the constructive, clarifying editorial category were adopted with the suggested changes. The committee learned the value of the requirement that comments suggesting change had to contain an acceptable new wording. The JPEG group was able to accept unanimously all of the changes as "editorial" in nature, and the Canadian "disapprove" vote was converted to "approve" when the necessary clarifications were incorporated.*

The principal members of ISO/IEC JTC1 were Australia, Austria, Belgium, Brazil, Canada, People's Republic of China, Cuba, Denmark, Finland, France, Germany, Hungary, Ireland, Italy, Japan, Republic of Korea, The Netherlands, Sweden, Switzerland, Union of Soviet Socialist Republics, United Kingdom, and the U.S. These 22 nations voted on the JPEG Draft International Standard. The SC29 committee agreed that all of its principal member comments had been adequately responded to at its October meeting and so recommended promotion to the DIS stage.

The DIS document was submitted to the JTC1 secretariat in October, 1991. The official ballot was mailed to JTC1 principal members December 17, 1991. The six-month ballot started on January 2, 1992, and ended July 2, 1992.

A new study committee, SC29, was formed under ISO/IEC JTC1 in November, 1991, from SC2/WG8, for the purpose of pursuing standards in the area of coded representations of picture, audio and multimedia/hypermedia information. Study Committee 2 transferred to

---

* Some of those clarifications were originally intended as material for this book, but were incorporated into the CD instead.

this committee its Working Groups 9, 10, 11, 12, and 13. Actions had been taken during the previous year to create such working groups out of the old WG8 subgroups of JBIG, JPEG, MPEG, and MHEG under SC2 in anticipation of the creation of this new committee. Since SC2 had already recommended promotion of the Part 1 DIS to JTC1, the change in structure did not affect Part 1 DIS balloting. The new SC29 confirmed promotion of the ISO CD 10918-1 to DIS.

19.8 JPEG Part 1 DIS ballot results ○

The JPEG Part 1 DIS ballot closed July 2, 1992. Australia, Austria, Belgium, Brazil, Canada, China, Czech and Slovak Federation (observer), Denmark, France, Italy, Japan, Poland, Romania, Russian Federation, Sweden, Switzerland, Turkey (observer), and United Kingdom voted "approved." Israel, Republic of Korea, and the U.S. voted "approved with Comments." Hungary abstained.

At the July, 1992, JPEG meeting in Toronto, the JPEG committee successfully responded to all of the DIS Part 1 Comments. All the changes were minor editorial ones and were incorporated into the DIS document in the process of converting it into the IS document.

19.9 CCITT Recommendation T.81 ○

The DIS document was completed one day before the deadline for the CCITT Study Group VIII proposal submission. Express mail from North America to Europe could not guarantee delivery in time, but, fortunately, faxing the first few pages was sufficient to submit it for consideration for accelerated CCITT procedures as CCITT Recommendation T.81. The rest of the document was delivered by express mail.

The French AFNOR committee worked together with the CCITT secretariat to get a translation of the JPEG Part 1 document. The JPEG 8—Revision 8 document had already been translated into French so it became the starting point for the official translation of the CD.

On April 30, 1992, the SG voted to approve a three-month default ballot on Rec. T.81. Since a failure to respond to a default ballot is counted as an "approved" vote and the ballot is passed with only 66.7% positive votes, most recommendations pass this stage without problems. The default ballot closed September 18, 1992.

Consideration of a revision of Rec. T.81 to achieve a joint common-text version with ISO IS 10918-1 was placed on the agenda for the next meeting.

19.10 ISO 10918 Part 2 ○

ISO rules require a test for all normative requirements. The JPEG group wrestled with how to construct compliance tests and what their purpose should be for the generic JPEG standard.

Eric Hamilton, of the U.S., agreed to chair the editing group of the Part 2 Committee Draft early in 1991. He had prepared the first draft of JPEG 10—Revision 0 for the November, 1990, meeting of JPEG. The

JPEG committee was preoccupied with the Part 1 CD and so did not hold serious discussions.

JPEG 10—Revision 3 was ready for the August, 1991, meeting in Santa Clara. Again, JPEG was too preoccupied with responding to the CD comments on Part 1 to deal with Part 2 in any depth.

At its September, 1991, meeting in La Jolla, California the U.S. ANSI Accredited Standards Committee X3L3, after vigorous debate, finally agreed to recommend strongly to ISO/IEC JTC1/SC2/WG10 that "the goals listed below (in order of importance) should be satisfied by the Part 2 compliance tests:

1. Facilitate image interchange;
2. to assist implementers;
3. to provide a basis for ensuring that the implementations that claim compliance are, in fact, compliant, with regard, for example to, DCT/quantizer accuracy and compressed image syntax;
4. to not constrain implementations unduly;
5. to test computational accuracy." [168]

The U.S. National Body also requested that the "testing procedures be specified so as to determine whether a compressed image data stream is syntactically correct, and that this determination be based upon the compressed image data stream." [168] These resolutions were agreed to at the November, 1991, meeting in Kurihama, Japan, after just a few minutes of discussion. This established a philosophical framework for the decisions about the many details of the Part 2 compliance testing.

A reasonable draft of the Part 2 CD containing most of the technical content and philosophy was successfully completed and approved at the concurrent ISO/IEC JTC1/SC29 meeting held in Tokyo. The editorial committee continued to improve the document until the December 16 deadline for submission to the secretariat. This deadline was the last date on which the comments from the CD ballot could possibly be available in time for the next JPEG meeting.

The Part 2 CD ballot was distributed to the country member bodies on December 16. The official start date for the ballot was December 30, 1991. The results were completed on March 30, 1992. The JPEG chair urged members to have their countries' comments available by March 16 so that they could be dealt with at the Haifa, Israel, meeting the last week of March.

Eight nations voted "approved." France, Israel, Japan, and the U.S. voted "approved with comments." Canada voted "disapproved with comments." The WG10 committee concluded that the large number of comments (more than were received for Part 1) resulted from the fact that Part 2 was written by committee and never revised. A major revision of the document was undertaken. All comments were successfully responded to at this meeting and the Canadian disapprove was converted to an "approved."

JPEG adopted an inverse DCT compliance test in Kurihama with the proviso that it would be modified if it were shown that the test input data had overflow and underflow problems. A contribution from Canada demonstrated that even an ideal IDCT system would fail the test.[169]   This problem was solved by generating encoder reference test data (quantized DCT coefficients from an ideal FDCT transform on the original source image samples) and decoder reference test data (quantized DCT coefficients from an ideal FDCT transform on the reconstructed image samples).

The March, 1992, meeting of WG10 in Haifa was devoted to responding to the Part 2 CD comments. On the basis of the partially rewritten document and the formal disposition of comments, SC29 authorized the JPEG committee to determine whether another CD ballot was needed or to go directly to the DIS ballot. A plan was established to generate the digital compliance test data before the next ballot.

19.11 JPEG Goals Achieved ○

In re-examining the list of functionality and goals, we indicate here how the different modes can be used to achieve them.

Quality: All the DCT-based sequential and progressive modes produce exactly the same final output image, independent of entropy coder and choice of progression (assuming that the progression is completed). Little or no compression penalty is paid for use of a DCT-based progressive mode (indeed, the arithmetic coder typically gets a few percent better compression in the successive-approximation progressive mode).

In these DCT-based modes the quality is controlled solely by the quantization table. A wide range of distortion levels can be achieved, including distortions well below the threshold of visibility. In the authors' experience an all-1's quantizer gives mean-square error on the order of 0.6 or less. However, at this level of distortion the compression is not much better than that achieved with lossless coding.

In lossless modes low levels of distortion may be introduced by means of the point transform. The reader should be aware, however, that the point transform defined by JPEG provides limited control of distortion and is intended only for distortions that are well below the threshold of visibility. If distortions approaching the threshold of visibility are desired, one of the DCT-based modes should be used.

The point transform is actually most useful for PCM coding of differences in the hierarchical mode. After coding the image with one or more DCT-based stages, the final correction can be made using the lossless differential frames. The point transform can be used here to good advantage if a precisely lossless output is not needed or cannot be guaranteed because of IDCT mismatch between encoder and decoder.

Note that when the lossless mode of operation is used for the first stage, a truly lossless progression can be achieved. However, if the first stage is downsampled, the compression performance will be adversely affected.

Progressive: The DCT-based progressive mode allows a graceful progression. An almost unlimited number of stages can be achieved by combining spectral selection and successive approximation. Spectral selection is generally considered to be easier to implement than successive approximation, and for this reason JPEG decided to allow spectral selection as a distinct process in progressive mode. Unfortunately, the image quality achieved at a given bit rate by "pure" spectral selection is generally very much worse than that achieved by successive approximation, and a truly graceful progression is provided only by a full capability for both.

Additional progressive capabilities are provided by the hierarchical mode, but the authors believe that the main use for this mode will probably be for multiresolution output and for correction of DCT-based output to achieve a nearly lossless final stage.

Reversible: The lossless mode is within a few percent of the best achievable with models of similar complexity. For the JPEG test set, the best average compression is 2.14:1 for arithmetic coding.[*]

Synchronous: The sequential mode of operation is synchronous if it uses either pre-defined Huffman tables or arithmetic coding. Several baseline real-time video rate JPEG encoders and decoders are commercially available (see sections 17.7, 17.16, and 17.32). Progressive encoders can be made to be fairly synchronous.

Sequential: The sequential mode has compression performance within a few percent of the DCT-based progressive modes and, although data are limited at present, appears to have significantly better compression performance than the hierarchical mode. The sequential mode is capable of being a single-pass system if predefined Huffman tables are used or if arithmetic coding is used. On-chip storage is minimized by the type of interleaving chosen.

Cost effective in both hardware and software: The number of vendors offering hardware and software indicates the successful accomplishment of this goal.

Resolution independence: Arbitrary source image resolutions can be handled. Images whose dimensions not multiples of eight are padded to multiples of eight internally in the DCT-based modes of operation. Padded data are discarded by the decoder, but some problems may occur in systems in which the decoder immediately writes the decoded image data to a buffer when a DNL marker specifies a line count that is not a multiple of eight.

Text images: JPEG handles grayscale representations of text well. In the authors' experience the block sizes larger than 8×8 tend to exhibit more clipping of character corners.

No absolute bit-rate targets: The bit-rate/quality trade-off is controlled primarily by the quantization table, and custom Huffman tables and arithmetic coding provide some refinement in this trade-off. In addition, the two progressive DCT modes give many choices on how to achieve a variable rate-distortion trade-off, and the hierarchical modes supply additional flexibility in terms of image resolution.

---

[*] Note that at higher compression ratios the lossless JPEG system performs much better with arithmetic coding than with Huffman coding. Because the Huffman codes are defined for each difference independently, the Huffman coder is unable to get below 1 bit/pixel in very smooth regions of the image. The arithmetic coder can get to a tiny fraction of 1 bit/pixel in these regions.

Precision: The JPEG DCT modes of operation are restricted to eight- and 12-bit/sample precision. Other precisions (up to 12 bits/sample) can be handled within this constrained choice by appropriate padding of bits and choice of quantization table. For lossless coding JPEG defines the precision to be from two to 16 bits, although the compression achieved by the JPEG DPCM algorithm below about 4–5 bits/pixel is not as good as can be achieved with JBIG.

At eight bits/sample, 16 bit-internal precision implementations of the FDCT and IDCT are possible. The entire lossless mode can be implemented with 16-bit internal precision, provided that a carry bit is available in the arithmetic logic unit.

Extensible: The system is extensible in a number of ways. There are no meaningful bounds on the number of progressive stages or in the case of hierarchical-progression, lower-resolution stages. For DCT-based modes both eight-bit/sample and 12-bit/sample precisions are allowed; the lossless modes are defined for two to 16 bits/sample. Although for simplicity, interleaves are limited to four components, the JPEG system can handle up to 255 components per image.

Luminance-chrominance sampling: JPEG is "colorblind." Sampling factors of 1, 2, 3, and 4 may be used both horizontally and vertically. As it turns out, however, this range of sampling factors was probably too broad and has given many developers problems. Many of the current JPEG implementations do not yet support the full range of sampling factors.

Luminance-chrominance separability: If the luminance is coded in a separate scan from the chrominance, the scan marker segments identify the component coded in the next entropy-coded segment. This allows the luminance segments to be decoded independently.

General image models: The DCT quantization is controlled solely through the quantization table. All decoders performing the IDCT are required to accept any quantization table that can be specified by the DQT marker segment and that are appropriate for the precision of the data.

DCT—8×8 blocks: The DCT blocks are always 8×8 blocks. One of the primary motivations in making this choice was the wealth of practical hardware already available or in development at the time this parameter was frozen.

DCT coefficients - eight bits maximum: The requirement to store the quantized DCT value in one byte was later relaxed when adverse effects were demonstrated. Experiments with medical images showed that DCT coefficients limited to eight-bit precision could exhibit very visible artifacts under certain conditions.[*]

If all the quantization table entries are eight or greater, as is the case for the suggested luminance and chrominance quantization tables in the DIS, the quantized coefficients will fit into a byte. However, in the authors' experience the suggested quantization tables are not at the threshold for visibility for typical usage with computer displays, and must

---

[*] The images used for these experiments were supplied by H. M. Kroon, M.D., University Hospital, Leiden, The Netherlands.

be divided by a factor of two to achieve nearly visually lossless image quality. This reduces some entries to values less than 8, and quantization values less than 8 require more than eight bits to express completely the quantized DCT coefficients.

Quantization matrix (table): JPEG originally intended to specify a default quantization table. However, as the dependence of quantization upon color space and the desired rate distortion became more obvious, the specification was changed such that the tables originally proposed as defaults were retained only as informative examples of quantization tables for luminance-chrominance images. Abbreviated table specification and abbreviated format compressed image data were introduced to make it easy for applications to specify application-specific default quantization tables.

Specification of IDCT: The exact IDCT specification became unnecessary when a separate "independent function" lossless mode was defined. A pseudo-lossless DCT-based hierarchical mode is still possible in closed systems in which the encoder and decoder IDCT can be made identical. In general, however, truly lossless capability cannot be achieved with DCT-based modes. Part 2 compliance tests require the IDCT output (reconstructed samples) to be within one quantization level of decoder reference test data when re-encoded with an ideal FDCT, and the worst-case deviation between two compliant decoders is therefore potentially quite large.

Specification of FDCT: The compliance tests require the FDCT to generate coefficients within one quantization level of those generated by the ideal FDCT.

20

OTHER IMAGE COMPRESSION STANDARDS

This chapter is devoted primarily to a brief description of two other developing standards for image compression. JBIG (Joint Bi-level Image Experts Group), as its name implies, is expected to become a standard for lossless compression of binary and limited bits/pixel images (ISO DIS 11544).[170] MPEG (Motion Picture Experts Group) is expected to become a standard for image sequence compression (ISO DIS 11172).[171] Like JPEG, JBIG is a joint committee between ISO/IEC and CCITT, and its activities are conducted in ISO/IEC JTC1/SC29/WG9. MPEG is conducted in ISO/IEC JTC1/SC29/WG11. At this point, both the JBIG and MPEG (MPEG-1) documents have been promoted to Draft International Standard status and are currently being balloted.

In this chapter we shall follow our conventions for JPEG, and use the terms "MPEG" and "JBIG" to refer to both the compression standard and the committee.

20.1 CCITT G3 and G4 ○

JBIG was preceded by a very important binary image compression standard known as CCITT Group 3 (G3—Rec. T.4),[172] also known by the acronym MR. "MR" stands for Modified READ, where "READ" is the acronym for Relative Element Address Designate. A follow-on, CCITT Group 4 (G4—Rec. T.6), uses a Modified MR (MMR) compression technique that is a simplified version of G3, which gets better compression by removing some of the error recovery from MR.

"MR" and "MMR" are names that refer to the algorithmic structure of the underlying technique for decomposing the binary image into descriptors. (Recall that "descriptors" is our term for the output of the model to the entropy-coding system.) CCITT G3 and G4 are very important in the world of facsimile and document-storage applications, and are at present the algorithms used for black/white document compression.

20.2 H.261 ○

MPEG also has a precursor, a motion sequence compression standard known as H.261.[173, 174] H.261 is a CCITT standard for low-bandwidth real-time video compression. Also known informally as P×64, this algorithm is intended for use at multiples of 64 Kbits/sec. in teleconferencing applications.

20.3 JBIG ○

The basic structure of the JBIG compression system is the same as described in Chapter 5. However, the model is a null box, because pel values are fed directly to the statistical model. The statistical model compresses binary decisions that are the actual pel values, using binary pel values already coded (and therefore known to both encoder and decoder) to create a "context" for each binary decision. The resulting context and binary decisions (or symbols) are compressed by the arithmetic encoder.

The arithmetic coder defined for JBIG is identical to the arithmetic-coder option in JPEG. We shall assume the reader is already familiar, at least superficially, with this arithmetic coder. It is described in some detail in Chapter 9, and additional details can be found in Chapter 13 and Chapter 14. We remind the reader that the JPEG/JBIG arithmetic coder is an adaptive binary arithmetic coder that dynamically adapts to the statistics for each context.

20.3.1 Neighboring-pel template models ○

The pel values used to create the context are samples selected from immediate-neighborhood pels already coded. Because the model must be causal, the samples must be known to both encoder and decoder. If we are using the JBIG sequential mode, the samples available to us are to the left of the current pel on the current line, or from lines above. If we are using the progressive mode, the samples may also come from values coded in previous stages or "resolution layers" (in JPEG we would say from "previous scans," but this terminology is not used by JBIG.). Figure 20-1 illustrates one of the JBIG sequential-mode templates.

The general idea behind the neighborhood template model derives from a paper by Langdon and Rissanen,[47] which was the first example in the literature where these models were integrated with arithmetic coding. The book chapter by Arps[175] describes earlier work on template models,

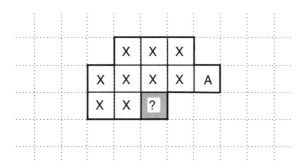

Figure 20-1. A JBIG sequential template. The "?" marks the pel to be coded, and the pels marked *X* or *A* are the template pels.

and a later paper by Arps, Truong, Lu, Pasco, and Friedman describes a seven-pel neighboring template model with the Q-coder arithmetic coder.[44]

20.3.1.1 Arithmetic-coding context ○

A template model such as that shown in Figure 20-1 is a simple and elegant vehicle for presenting the concept of arithmetic-coding context. The pel values from the template, excluding, of course, the pel value being coded, make up the context. We simply concatenate the bits in a given sequence to create a binary representation of the context index. This context index is the index into the table of decision probabilities used by the arithmetic coder and the pel value being coded is the binary decision fed to the encoder and returned by the decoder.

It is highly likely that probabilities for different contexts will be substantially different. Thus, if all the pels in a context are white, the probability value found at that context index is likely to be strongly skewed toward the pel being white. Conversely, if all the pels are black, the probability value is likely to be strongly skewed toward the pel being black.

20.3.1.2 Advantages of template models ○

An excellent standard, MMR, is already available for compressing binary images; one may, therefore, ask why JBIG is needed. One of the primary reasons for JBIG is the improved performance on a class of binary images known as binary halftones. These are binary renditions of continuous-tone images, and have statistical properties which are very different from those of binary text documents.

The Huffman tables used by G3/G4 were tuned to a set of text documents, and the performance of G3/G4 on this class of documents is very good. Generally one finds that if G3 gets 20:1 compression on a simple text document, G4 will get about 30% better compression—basically by stripping out the end-of-line codes and coding the document entirely in 2-D mode. (Depending on resolution, G3 codes every other line or every fourth line in 1-D, and all lines have end-of-line codes.) JBIG improves on the compression of G4 by about another 30%. This is impressive but perhaps not sufficient to justify a new standard.

Unfortunately, the Huffman table in G3/G4 is totally unsuited to binary halftones, and can even give expansion when image data are of this type. However, even if that is taken into account by some form of custom Huffman table, the performance does not match that of a template model in combination with a good adaptive arithmetic coder. JBIG can achieve as much as 8:1 compression on documents that may expand with G3/G4.

20.3.1.3 Adaptive templates ○

In order to get the best performance, a neighborhood template must span the repeat pattern used in the binary halftoning process. Otherwise, as the template is shifted along the halftone pattern the statistical behavior is unstable. When this happens, better compression is achieved with a very coarse granularity probability estimation state machine (or with a true multirate estimator[42]), but coarser granularity compromises the performance elsewhere.

A better solution to this problem is to improve the model by making the template large enough to span the halftone pattern.[176, 177] Another approach is to reformat the document by concatenating lines to get alignment of halftone patterns vertically. Building upon this work, JBIG chose yet a third approach, which is to define one pel in the template as "adaptive."[178] This pel, marked by A in Figure 20-1, can be moved to another position, as illustrated in Figure 20-2.

In principle, this pel can be moved to any sample position where the pel is already coded. However, compression performance will decrease if it is shifted beyond the point where the best match to the repeating pattern is first obtained. The "suggested minimum support" defined by JBIG for decoders for the horizontal and vertical offsets from the current pel are 16 and 0, respectively.

20.3.2 JBIG progressive mode ○

JBIG defines a progressive mode in which a reduced resolution "starting layer" image is followed by progressively higher-resolution "layers" of data. The image resolution reduction is done either by a default algorithm that JBIG has defined or by any other algorithm the user might prefer. The resolution reduction is always by a factor of 2 both vertically and horizontally, and can be done as many times as needed. For example, much of the testing was done with a six-layer system that started at 12.5 pels/inch and finished at 400 pels/inch.

The "starting layer" uses one of the sequential templates, whereas subsequent layers use a template that is a composite of neighboring pels from the preceding layer and pels already coded from the current layer. Figure 20-3 shows the template set for the four phases of high-resolution pel values. These phases are needed because of the different spatial

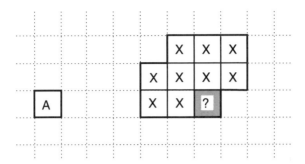

Figure 20-2. The adaptive pel in the JBIG template for sequential coding. The adaptive pel A is usually placed at a position where it matches the repetitive pattern in the halftone.

relationship of each high-resolution pel relative to the lower-resolution pels (marked by P).

The sketch of the JBIG progressive mode in Figure 20-3 does not show two important aspects of the JBIG progressive mode: deterministic prediction and typical prediction. These directly affect the performance and complexity of implementation of progressive mode.

Deterministic prediction [179] is the process by which certain pels can be predicted without error from the data already coded, and therefore do not need to be coded.* Deterministic prediction usually improves the progressive compression by about 5%. Unfortunately, it also adds very significantly to the complexity of implementation. Deterministic prediction must be matched to the resolution reduction algorithm, and therefore JBIG defines a default table for it which is matched to the suggested reduction algorithm. If a different reduction algorithm is used, either deterministic prediction must be disabled or a matched deterministic prediction table must be downloaded.

In sequential-mode typical prediction a single decision is coded at the beginning of the line to tell the decoder that the line above is the same as the current line. The decoder then simply repeats the line. The function is similar in some respects to the speedup mode described in Chapter 12, as both code all-white areas very efficiently. Typical prediction has little impact on compression.

For differential layers, typical prediction decisions are made locally, telling the decoder when matched lower resolution pel values occur within

---

* Unfortunately, the array of pel values needed for deterministic prediction is much larger than the context template. Otherwise, one could simply let the arithmetic coder code the pel value with a very low probability of being wrong.

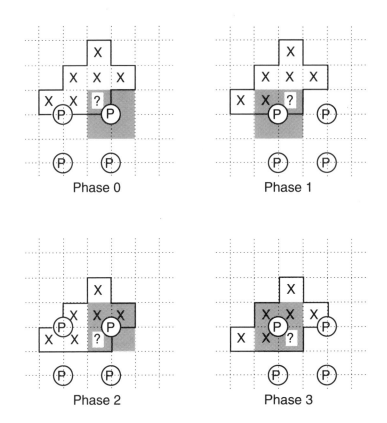

Figure 20-3. **JBIG progressive-mode templates.** Four different arrangements of the template are used that are dependent on the "phase" of the high-resolution pel. The lower-resolution layer data is marked by P and the high-resolution 2×2 block of pixels is shaded in gray.

a region around the higher resolution pel values that have the same color. If this is true for all matched low resolution pel groups for a pair of high resolution lines, a single decision codes them.

Typical prediction and deterministic prediction can be enabled or disabled, allowing some interesting trade-offs between complexity and coding performance.

20.3.3 JBIG sequential mode ○

The JBIG sequential mode is identical to the starting layer of the progressive mode. The adaptive template and typical prediction are options in sequential mode, as is the ability to use either a two-line or a three-line template. The sequential mode usually achieves about 10% better compression than the progressive mode, primarily because the resolution

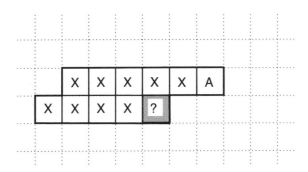

Figure 20-4. The JBIG two-line adaptive pel template for sequential coding. The "?" marks the pel to be coded, and the pels marked X or A are the template pels. Pel A is the adaptive pel for this template.

reduction algorithms are optimized for image quality in early stages rather than overall compression performance.

20.3.3.1 Sequential-mode templates ○

The three-line template defined by JBIG is shown in Figure 20-1 and the two-line template is shown in Figure 20-4. The adaptive pel is marked in each of these illustrations.

These two template choices were defined to allow for a flexibility in trading off implementation complexity against compression performance. According to data compiled by JBIG, the three-line template usually achieves about 5% better compression. Complexity analysis by JBIG members suggests, however, that the third line will decrease software performance (throughput) by 30% or more.

In the authors' experience, some systems designers are more concerned with throughput than with compression performance. Although compression performance is important, sometimes the real measure of a system is the time to decode and display the image. In this respect optimized MMR software is roughly a factor of three faster than similarly optimized software for the two-line template model. Even though this ratio may improve with further work, MMR has such a throughput advantage that it will remain in active use for a long time to come.

20.3.3.2 JBIG compression performance ○

Table 20-1 provides some data on JBIG performance and how it relates to CCITT G4. The authors are grateful to W. Equitz for providing these data.

The data in Table 20-1 illustrate many of the properties of the JBIG sequential mode. We see that JBIG improves upon CCITT-G4 by about 30% for the eight CCITT test images. We also see that for halftones G4

Table 20-1. JBIG sequential-mode compression (bytes)

B/W Image	JBIG–3 line		JBIG–2 line		G4–MMR	Original
	AT off	AT on	AT off	AT on		
Scanned text and line drawings:						
S01a400	13734	13734	14704	14704	17416	1671168
S02a400	15411	15411	16525	16525	20490	1671168
S03a400	146643	146643	155407	155407	186107	1671168
S05a400	34242	34242	36714	36714	46685	1671168
S07a400	24052	24052	25448	25448	35789	1671168
Total:	234082	234082	248798	248798	306487	8355840
Computer-generated line drawings:						
S08a400	6198	6198	7005	7005	25551	1167360
S10a400	13716	13716	18620	18620	56220	2998272
Total:	19914	19914	25625	25625	81771	4165632
Scanned dither halftones images:						
S04a400	130072	77234	90000	79306	1108259	786432
S04b400	157672	89492	130191	85086	636638	786432
S04c400	130592	71323	71751	69026	1522525	786432
S04d400	121237	68195	75832	70964	1329333	786432
S06a400	256145	222282	226604	227032	996253	1671168
S09a400	74289	74289	79101	79101	286708	131072
Total:	870007	602815	673479	610515	5879716	4947968
CCITT images:						
CCITT1	14715	14715	15060	15060	18103	513216
CCITT2	8545	8545	8902	8902	10803	513216
CCITT3	21988	21988	23271	23271	28706	513216
CCITT4	54356	54356	55770	55770	69275	513216
CCITT5	25877	25877	26655	26655	32222	513216
CCITT6	12589	12589	13684	13684	16651	513216
CCITT7	56253	56253	57668	57668	69282	513216
CCITT8	14294	14294	15343	15343	19099	513216
Total:	208617	208617	216353	216353	264141	4105728

Data courtesy of W. Equitz.

actually expands, whereas JBIG gets about 8:1 compression when the adaptive template is active. The effect of the adaptive template is more significant for the three-line template than for the two-line template. The two-line template is wider, and therefore comes closer to spanning the repeating pattern in the halftone. For text images the three-line template outperforms the two-line template by about 5%, except for computer-generated line art, in which the three-line template has significantly better performance.

20.3.4 Bit plane coding of grayscale images ○

The JBIG syntax provides for coding of images with more than one bit plane. Rabbani and Jones[34] provide a review of bit plane encoding. Although no definition is given of the meaning of the data, the JBIG DIS does note that Gray coding, which reduces the number of transitions in most of the bit planes, is superior to direct coding of the binary intensity values.

The compression JBIG achieves with low precision grayscale image data is quite good. Although the JPEG lossless mode (with arithmetic coding) gives modestly higher compression for eight bit precision, JBIG and JPEG performance is about matched for precisions between four and six bits/pixel, depending on image complexity, and JBIG outperforms the JPEG lossless mode at lower precisions.

20.3.5 JBIG implementations ○

To the best of the authors' knowledge, there are no commercially available JBIG implementations at this time. A number of software implementations were created for validation of the JBIG system, however, and it would seem logical that some commercial applications will soon appear. In the authors' opinion the JBIG sequential mode is much simpler than the progressive mode, and will at least initially prove to be the most useful aspect of the JBIG system. This sequential algorithm is likely to be particularly important in the bank-check-processing industry.

20.4 MPEG ○

The MPEG system is intended for image sequence compression. As such, it is designed to take advantage of the correlation in the temporal dimension, and typically achieves about three times better compression than JPEG when measured in bits/pixel per frame.

MPEG development is segmented into two distinct phases, MPEG-1 and MPEG-2. MPEG-1 is intended for image resolutions of approximately 360 pixels × 240 lines and bit rates of about 1.5 Mbits/sec. for both video and a pair of audio channels. MPEG-2 is for higher resolutions (including interlaced video) and higher bit rates (4–10 Mbits/sec. or more). We shall be concerned here primarily with MPEG-1, although much of the MPEG-1 algorithmic structure appears to be carrying over to MPEG-2. However, MPEG-2 is still in active development.

JPEG chose to define both a data stream syntax and the encoder procedures that produce the entropy-coded segments of that data stream. MPEG has taken a somewhat different approach than JPEG, defining the coded data stream and the operations a decoder must perform in order to decode it. MPEG does not define specific algorithms needed to produce a valid data stream.

In MPEG, the encoder can be much more complex than the decoder. Most of the "intelligence" and processing requirements are in the encoder, and the decoder simply "does what it is told to do." Although relatively symmetric MPEG encoder/decoder systems are possible, only with a more powerful encoder can relatively high-quality images be obtained at the low 1.5-Mbit/sec. MPEG-1 bit rate.

The MPEG DIS compressed data stream syntax is documented in a very terse pseudo-C-code format. However, an informational section of the DIS contains an updated version of a comprehensive and very readable review of MPEG by Astle.[180]

One common element between JPEG and MPEG (and H.261 as well) is the 8×8 DCT. Indeed, for all three of these standards the availability of hardware for the 8×8 DCT was an important factor in choosing this size DCT.

The color coordinate system used by MPEG is CCIR YCbCr. A flexible format is supported, consistent with requirements for the 50-Hz and 60-Hz television formats used throughout the world.

20.4.1 Basic structure of the MPEG system ○

An image sequence is divided into the following layers:

1. sequence, made up of groups-of-pictures

2. group-of-pictures, made up of I, P, B or D pictures

3. picture, made up of slices

4. slice, made up of macroblocks

5. macroblock, 4 (2×2) Y blocks + 1 Cb block + 1 Cr block

6. block, 8×8 sample array

Headers are defined for sequences, groups-of-pictures, pictures, slices, and macroblocks. All headers except the macroblock header are byte aligned.

20.4.1.1 Sequences ○

Sequences are independent segments in the MPEG data stream. Each sequence has a header followed by one or more compressed groups-of-pictures. Custom quantization tables can be defined in the sequence header.

20.4.1.2 Pictures and groups-of-pictures ○

In coding an image sequence MPEG classifies the individual pictures into four types:

1. I pictures (intra pictures) are coded without reference to other pictures, and are thus the functional equivalent of a JPEG image.

2. P pictures (predicted pictures) are coded relative to a prediction from a previous frame. The prediction is motion-compensated.

3. B pictures (bidirectional pictures) are coded relative to motion-compensated prediction from a previous and/or a future picture.

4. D pictures (DC pictures) are pictures containing only the DC (8×8 block average) for each block. Support for this category of picture is

optional, and sequences may not contain D pictures intermixed with other types of pictures.

Groups-of-pictures contain I, P, and B pictures, as illustrated in Figure 20-5. The P pictures are always coded relative to the preceding I or P picture, whereas the B pictures can be coded (with motion-compensated prediction) relative to the preceding P or I and the following P or I picture. The prediction of the B picture can be forward or backward relative to a P or I picture or to a simple motion-compensated average of both preceding and following P or I pictures. Referring to the picture numbers Figure 20-5, the coding order for the pictures might be 0, 4, 1, 2, 3, 8, 5, 6, 7, 12, 9, 10, and 11. P and B pictures can make predictions only from pictures already transmitted.

Groups-of-pictures can be either open (predictions may reference another group-of-pictures) or closed (predictions are made solely from other pictures in the same group). Figure 20-5 illustrates a closed group-of-pictures.

20.4.1.3 Slices ○

A slice is a sequence of macroblocks in raster scan order. Both the vertical position and a quantization scale factor are contained in the slice header. More than one slice can be at a given vertical position.

20.4.1.4 Macroblocks ○

Each macroblock consists of four (2×2 array) Y blocks, one Cb block and one Cr block, as shown in Figure 20-6. The MPEG macroblock is equivalent to the JPEG MCU with horizontal and vertical sampling factors of 2, 1, and 1.

Macroblocks are classified into a number of different types, depending on the type of picture. Table 20-2 lists the macroblocks that can occur in each type of picture. Intra-d macroblocks are coded with the currently defined quantization matrix, whereas intra-q macroblocks are coded with a scaled quantization matrix. This, together with the quantization scale defined for the slice, provides a flexible form of adaptive quantization in which the quantization scale factor can be varied from one macroblock to the next to suit the local characteristics of the image.

For P pictures several new classes of macroblock are defined. Note, however, that macroblocks can also be coded independently of any reference data. Macroblocks classified as "pred" are coded relative to a prediction from a prior picture, and the prediction may use motion compensation and/or adaptive quantization. "Pred" macroblocks labelled with "c" require coding of the DCTs of the difference between the input macroblock and the predicted macroblock. For "Pred–m" macroblocks the motion compensated prediction is sufficient to represent the macroblock. Note that a macroblock may be skipped if the block from the predicting frame is good enough as is.

For B pictures the list of macroblock classes becomes even more elaborate, as now there is the possibility of forward (f), backward (b), and interpolated (i) prediction, all with motion compensation and adaptive quantization (q).

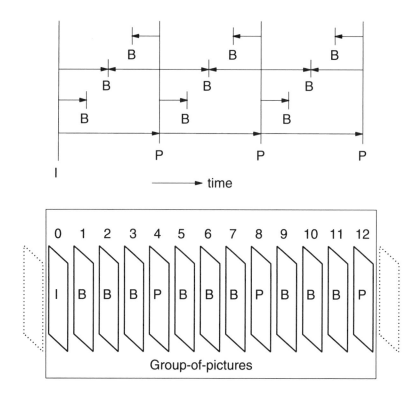

Figure 20-5. An illustration of one possible MPEG group-of-pictures

20.4.2 Huffman coding in MPEG ○

MPEG uses fixed Huffman code tables and there is no provision for downloading custom tables. In I pictures the DC coefficient is always coded with eight-bit precision (the quantizer is always set to 8), and the fixed DC Huffman table has a logarithmic amplitude category structure borrowed from JPEG.

In I pictures the AC coefficients are coded with a model that is similar in many respects to that used by JPEG. The same zigzag scan is used, as are variable-length Huffman codes based on run-amplitude and end-of-block symbols. However, a single Huffman table is used for all blocks, independent of the color component to which they belong, and the code book itself is completely different from that of JPEG. The fixed code table does not allow for AC coefficient precisions exceeding nine bits.

In P pictures the coding becomes more elaborate, with code tables needed for the type of macroblock, the motion vector (relative to the motion vector of the preceding macroblock), and the 8×8 DCTs of the differences between the macroblock samples and the motion compensated

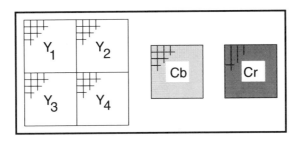

Figure 20-6. MPEG macroblock

predictions. There is no separate DC code table; instead, the differential DC value and AC coefficients are coded together.

In B pictures the number of options is still greater, and the coding is correspondingly more elaborate. In addition to coding of intra macro-blocks, coding of differences relative to forward, backward, and interpolated motion-compensated prediction of macroblocks is needed. All of the variations using motion-compensated prediction may use the motion compensated prediction as is, may code a correction to the motion compensated prediction, and may use adaptive quantization. A new Huffman tables is defined for coding of the macroblock type, whereas the tables already defined for P pictures are used for the motion vectors and the differential DCT corrections.

20.4.3 Adaptive quantization ○

The MPEG process for coding of I pictures is quite similar in many respects to the sequential JPEG mode. While there are significant differences between the two in structure and constraints, the primary difference between MPEG and JPEG from an algorithmic point of view is the provision for adaptive quantization in MPEG. Although this increases complexity—especially for the encoder—significantly better compression can be achieved. MPEG enthusiasts claim that up to 30% better compression can be achieved with adaptive quantization.[181]

Table 20-2. Types of macroblocks in I, P, and B pictures

I pictures	P pictures	B pictures
Intra-d	Intra-d	Intra-d
Intra-q	Intra-q	Intra-q
	Pred-c	Pred-i
	Pred-cq	Pred-ic
	Pred-m	Pred-icq
	Pred-mc	Pred-b
	Pred-mcq	Pred-bc
	Skipped	Pred-bcq
		Pred-f
		Pred-fc
		Pred-fcq
		Skipped

21

POSSIBLE FUTURE
JPEG DIRECTIONS

Chapter 17 provides a compendium of early vendors and applications using JPEG. Many of the implementations are purely baseline, but a few vendors are already moving to the extended-system capabilities. Once the extended systems become more widely known and used, the interest may shift to improving the performance of JPEG. There have already been suggestions within the standards community for possible addenda to the JPEG standard.

21.1 Adaptive quantization ○

In the area of lossy compression the most obvious additional capability to add to JPEG is adaptive quantization. As noted in Chapter 20, for a given image quality, compression can be improved by as much as 30% by adaptive quantization.

Adaptive quantization is not a cost-free option, because the determination of the quantization values that are used with a given DCT block can be a computational burden, and provision for multiple quantization tables or scaling of the quantization table values in the encoder and decoder must be provided. Although the JPEG decision not to include adaptive quantization can be defended on the basis of simplicity, there is no question that adaptive quantization would significantly improve image quality at a given bit rate.

The most obvious way of introducing adaptive quantization would be to encode a binary decision at the beginning of each MCU to tell the decoder when new quantization tables (or modifications of tables) are needed. The arithmetic coder could code this as a single binary decision, perhaps conditioned on the preceding adaptive quantization decision. For the best coding efficiency, the Huffman coder would probably need to run-length encode the sequence of adaptive quantization decisions. Alternatively, a "table switch" code could be defined for selecting from a set of pre-defined tables.[182]

A standard JPEG compressed image data format can be used with adaptive quantization and there are two ways to accomplish this: A crude form of adaptive quantization is possible by "lying" to the decoder and discarding information selectively in regions of less importance. This is described in Chapter 16. This method is completely consistent with standard decoders.

Another way to incorporate adaptive quantization into a standard JPEG compressed data format is through the use of a pseudo-component that contains the adaptive quantization parameters.[183] Each MCU contains a pseudo-component data unit, and the values coded for that data unit tell the decoder how to adapt the quantization. Although a standard decoder would be able to decode the quantized DCT coefficients from the compressed data, the interpretation of the resulting data would be impossible without a knowledge of the format of the pseudo-component information.

21.2 Improvements to lossless coding ◑

The JPEG lossless mode with Huffman coding does not have good compression performance when coding at higher compressions. The problem is introduced by the use of a statistical model in which each DPCM difference is coded with a separate code word. Run-length coding of zero differences is needed to get below one bit/pixel, and this would be a logical extension of the Huffman statistical model. There might even be reason to use a 2-D Huffman table format such as is used for efficient coding of runs of zeros in AC coefficient coding.

Although the compression gains are smaller, the arithmetic coding statistical model can also be improved. Experiments by the authors have shown that when a 7×7 conditioning array is used, the coding performance is improved by about 1%.

Improvements in the DPCM coding model may also lead to significant gains in compression performance. One interesting idea was suggested by Niss[184] as a possible improvement for lossless Huffman coding. However, the concept is general and independent of the entropy entropy-coding technique. Niss found that lossless Huffman compression was improved by about 10% if the data were segmented into N×N blocks and an optimal predictor was selected for each block. The predictors Niss used were the four nearest neighbors, a, b, c (see Figure 10-16), and a fourth neighbor, d, above and to the right. Niss found the optimal block size to be 5×5.

Although Niss experimented with predictor selection based on sample values already transmitted to the decoder, he found that better results were obtained by encoding the predictor selection as part of the compressed data. Coding the predictor selection rather than "predicting" the predictor has definite advantages, in that the decoder is then much simpler. Furthermore, if the complexity of choosing a predictor is too great, the encoder can be simplified by restricting it to a single predictor.

To take advantage of adaptive prediction, encoders must devise a strategy for selecting the predictor. Niss used the predictor which achieved the smallest sum of absolute differences for the block. The selection can also be made on the basis of the best coding performance.

21.3 Other possible addenda ○

A number of other addenda have been suggested for JPEG. The limit of 10 data units in an MCU may be too restrictive, and suggestions have been made to increase this to 18 data units or more. Possible hierarchical-mode enhancements include a new frame for uncompressed "thumbnail" images, inheritance of reference components, and an offset marker capability that would allow tiling of very large images and the selective improvement of portions of an image. There is also a possibility that JBIG data streams may be incorporated within the JPEG syntax. A version number marker has been proposed, and will definitely be needed if any addenda are adopted that modify the meaning of the compressed data stream.

More radical changes have also been proposed. Informal suggestions have been made to allow different DCT block sizes, and even to use wavelets instead of the DCT. Such major changes are not simply enhancements or extensions of the current algorithms, and therefore may require a new work item.*

21.4 Backwards compatibility ○

Image compression is still a very active area of research, and improvements are continually being made. As new concepts are developed and old ones are refined, there will be pressure to migrate JPEG to higher performance levels and new function. However, given the enthusiastic reception that JPEG has received, and the widespread adoption of the current JPEG standard that is now taking place, backwards compatibility with the current standard will almost certainly be maintained. Indeed, JPEG has already defined backwards compatibility to mean that new decoders shall decode old data streams.

---

* A "new work item" is a statement which defines the intended activity. A formal ballot at the Technical Committee level is required for every new work item before serious work can commence on the activities. JPEG has decided that a new work item is needed if the algorithm is changed so fundamentally that backwards compatibility is no longer easily achieved.

APPENDIX A

ISO DIS 10918-1 REQUIREMENTS AND GUIDELINES

The ISO Draft International Standard 10918 Part 1 is reproduced in this appendix with the permission of the ISO secretariat.

The ISO IS 10918-1 incorporates many minor editorial changes to the DIS. The only major change in the document organization was the interchanging of Clause 3 (Introduction) and Clause 4 (Definitions, Abbreviations, and Symbols). Clause 3 was also renamed "General."

In Annex L, a new patent was added as relevant to arithmetic coding [185] and another added as relevant to hierarchical coding with a final lossless stage. [186] Another patent, U.S. Pat. No 4,725,885, [187] that had been mistakenly cited as U.S Pat. No. 4,725,884 was removed as no longer relevant. Some references were added to annex M. [49, 58, 188]

In Annex D including Figures D.8, D.9, D.10, and D.12, the stack counter symbol was changed from SC to ST, in order to avoid confusion with the context-index for coding the correction bits in successive approximation coding (SC). In Annex B the arithmetic conditioning table identifier was changed from Ta to Tb in order to avoid confusion with Ta_j, the AC entropy table selector for the jth component in a scan.

The missing "1"s were inserted into the third column at event count 184 in Table K.7 and event count 222 in Table K.8.

Table A-1. Standards nomenclature

Nomenclature	Equivalent expression
clause	chapter
subclause	section or paragraph
annex	appendix
shall	is required to ...
shall not	is not allowed ...
should	ought to/recommended to ...
should not	ought not to/not recommended to ...
may	is permitted ...
need not	it is not required that ...
can	it is possible to ...
cannot	it is impossible to ...

Other minor changes were made to achieve consistency in the symbols, resulting in the following additions to the list of symbols in subclause 4.2 in the DIS.

Bx	byte modified by a carry-over
CX	conditional exchange
d_{ji}^{k}	d_{ji} for component k
EC	event count
ECS_i	ith entropy-coded segment
$EOB0, EOB1, ..., EOB14$	run length categories for EOB runs
MCU_i	ith MCU
$MPS(S)$	more probable symbol for context-index S
ST	stack counter
Tb	arithmetic conditioning table destination identifier
Tc	Huffman coding or arithmetic coding table class
Th	Huffman table destination identifier
X'values'	values within the quotes are hexadecimal

Table A-1 lists some of the special nomenclature used in standards documents.[189] It is important to notice the distinction that is made between "shall," "may," and "can."

Subclauses labelled as "NOTE" always convey information. They cannot be used to specify requirements, although they may call attention to a requirement specified elsewhere.

Implementers should obtain the IS document rather than relying upon the DIS document. In the U.S. the American National Standards Institute (ANSI) is the distributor.

INTERNATIONAL STANDARD DIS 10918-1

CCITT RECOMMENDATION T.81

DIGITAL COMPRESSION AND CODING OF CONTINUOUS-TONE STILL IMAGES

PART I: REQUIREMENTS AND GUIDELINES

Contents

Page

Foreword .... *iii*

1   Scope 1

2   Normative references 1

3   Introduction 2

4   Definitions, abbreviations, and symbols 14

5   Interchange format requirements 27

6   Encoder requirements 28

7   Decoder requirements 29

Annex A   Mathematical definitions A-1

Annex B   Compressed data formats B-1

Annex C   Huffman table specification C-1

Annex D   Arithmetic coding D-1

Annex E   Encoder and decoder control procedures E-1

Annex F   Sequential DCT-based mode of operation F-1

Annex G   Progressive DCT-based mode of operation G-1

Annex H   Lossless mode of operation H-1

Annex J   Hierarchical mode of operation J-1

Annex K   Examples and guidelines K-1

Annex L   Patents L-1

Annex M   Bibliography M-1

Foreword

This document (Part 1) is at the Draft International Standard (DIS) level of the ISO/IEC JTC1 process for the development of International Standards. After successful balloting and any necessary revision, this DIS will be promoted to International Standard (IS).

This CCITT Recommendation | International Standard, **Digital Compression and Coding of Continuous-tone Still Images**, is published as two parts:

- Part 1: Requirements and guidelines

- Part 2: Compliance testing

This part, Part 1, sets out requirements and implementation guidelines for continuous-tone still image encoding and decoding processes, and for the coded representation of compressed image data for interchange between applications. These processes and representations are intended to be generic, that is, to be applicable to a broad range of applications for colour and grayscale still images within communications and computer systems. Part 2, which is now being drafted and which is expected to enter the CD review process six to nine months after this Part 1, sets out tests for determining whether implementations comply with the requirements for the various encoding and decoding processes specified in Part 1.

The user's attention is called to the possibility that - for some of the coding processes specified herein - compliance with this International Standard may require use of an invention covered by patent rights. See Annex L for further information.

The committee which has prepared this Specification is ISO/IEC JTC1/SC2/WG10, Photographic Image Coding, in collaboration with CCITT SGVIII. Prior to the establishment of WG10 in 1990, the committee existed as an Ad Hoc group, known as the Joint Photographic Experts Group (JPEG), of ISO/IEC JTC1/SC2/WG8. Both the committee and the processes it has developed for standardization continue to be known informally by the name JPEG. The technical content of this Specification is identical (except for various corrections) to that of Revision 8 of the working document "JPEG Draft Technical Specification" (JPEG-8-R8).

The "joint" in JPEG refers to the committee's close but informal collaboration with the Special Rapporteur's committee Q.16 of CCITT SGVIII. In this collaboration, WG10 has performed the work of selecting, developing, documenting, and testing the generic compression processes. CCITT SGVIII has provided the requirements which these processes must satisfy to be useful for specific image communications applications such as facsimile, videotex, and audiographic conferencing. The intent is that the generic processes will be incorporated into the various CCITT Recommendations for terminal equipment for these applications.

Other international organizations which have contributed to the preparation of this Specification include:

- ISO/IEC JTC1/SC18/WG5 ODA Content Architecture and Colour;

- International Press Telecommunications Council (IPTC).

In addition to the applications addressed by the international organizations above, the JPEG committee has developed a compression standard to meet the needs of other applications as well, including desktop publishing, graphic arts, medical imaging, and scientific imaging.

iii

Annexes A, B, C, D, E, F, G, H, and J are normative, and thus form an integral part of this Specification. Annexes K, L and M are informative and thus do not form an integral part of this Specification.

This Specification aims to follow the procedures of CCITT and ISO/IEC JTC1 on "Presentation of CCITT | ISO/IEC Common Text," currently under elaboration by CCITT and ISO/IEC JTC1. The alignment of this Specification to the final version of the above joint drafting rules might result in minor editorial changes to this Specification. It is envisaged that the final CCITT Recommendation | International Standard shall fully conform with the joint drafting rules of CCITT and ISO/IEC JTC1.

INTERNATIONAL STANDARD DIS 10918-1

CCITT RECOMMENDATION T.81

INFORMATION TECHNOLOGY -

DIGITAL COMPRESSION AND CODING OF CONTINUOUS-TONE STILL IMAGES

PART 1: REQUIREMENTS AND GUIDELINES

1 Scope

This CCITT Recommendation I International Standard is applicable to continuous-tone - grayscale or colour - digital still image data. It is applicable to a wide range of applications which require use of compressed images. It is not applicable to bi-level image data.

This Specification

-   specifies processes for converting source image data to compressed image data;

-   specifies processes for converting compressed image data to reconstructed image data;

-   gives guidance on how to implement these processes in practice;

-   specifies coded representations for compressed image data.

NOTE - This Specification does not specify a complete coded image representation. Such representations may include certain parameters, such as aspect ratio and colour space designation, which are application-dependent.

2 Normative references

None

1

3 Introduction

The purpose of this clause is to give an informative overview of the elements specified in this Specification. Another purpose is to introduce many of the terms which are defined in clause 4. These terms are printed in *italics* upon first usage in this clause.

3.1 Elements specified in this Specification

There are three elements specified in this Specification:

1) An *encoder* is an embodiment of an *encoding process*. As shown in Figure 1, an encoder takes as input *digital source image data* and *table specifications*, and by means of a specified set of *procedures* generates as output *compressed image data*.

2) A *decoder* is an embodiment of a *decoding process*. As shown in Figure 2, a decoder takes as input compressed image data and table specifications, and by means of a specified set of procedures generates as output *digital reconstructed image data*.

3) The *interchange format*, shown in Figure 3, is a compressed image data representation which includes all table specifications used in the encoding process. The interchange format is for exchange between *application environments*.

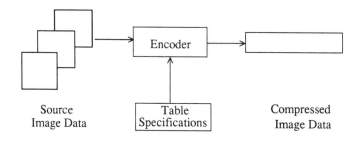

Source
Image Data

Table
Specifications

Compressed
Image Data

Figure 1 – Encoder

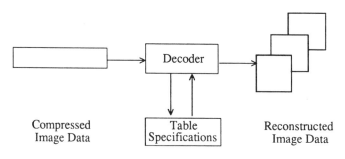

Compressed
Image Data

Table
Specifications

Reconstructed
Image Data

Figure 2 – Decoder

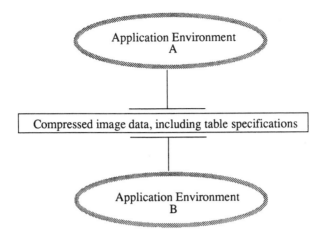

Figure 3 – Interchange format for compressed image data

Figures 1 and 2 illustrate the general case for which the *continuous-tone* source and recon-structed image data consist of multiple *components*. (A *colour* image consists of multiple com-ponents; a *grayscale* image consists only of a single component). A significant portion of this Specification is concerned with how to handle multiple-component images in a flexible, application-independent way.

These figures are also meant to show that the same tables specified for an encoder to use to compress a particular image must be provided to a decoder to reconstruct that image. However, this Specification does not specify how applications should associate tables with compressed image data, nor how they should represent source image data generally within their specific en-vironments.

Consequently, this Specification also specifies the interchange format shown in Figure 3, in which table specifications are included within compressed image data. An image compressed with a specified encoding process within one application environment, A, is passed to a differ-ent environment, B, by means of the interchange format. The interchange format does not spec-ify a complete coded image representation. Application-dependent information, e.g. colour space, is outside the scope of this Specification.

3.2 Lossy and lossless compression

This Specification specifies two *classes* of encoding and decoding processes, *lossy* and *lossless* processes. Those based on the *discrete cosine transform* (DCT) are lossy, thereby allowing substantial compression to be achieved while producing a reconstructed image with high visual fidelity to the encoder's source image.

The simplest DCT-based *coding process* is referred to as the *baseline sequential* process. It provides a capability which is sufficient for many applications. There are additional DCT-based processes which extend the baseline sequential process to a broader range of applications. In any decoder using extended DCT-based decoding processes, the baseline decoding process is required to be present in order to provide a default decoding capability.

3

The second class of coding processes is not based upon the DCT and is provided to meet the needs of applications requiring lossless compression. These lossless encoding and decoding processes are used independently of any of the DCT-based processes.

A table summarizing the relationship among these lossy and lossless coding processes is included in 3.11.

The amount of compression provided by any of the various processes is dependent on the characteristics of the particular image being compressed, as well as on the picture quality desired by the application and the desired speed of compression and decompression.

3.3 DCT-based coding

Figure 4 shows the main procedures for all encoding processes based on the DCT. It illustrates the special case of a single-component image; this is an appropriate simplification for overview purposes, because all processes specified in this Specification operate on each image component independently.

In the encoding process the input component's *samples* are grouped into 8x8 *blocks*, and each block is transformed by the *forward DCT* (FDCT) into a set of 64 values referred to as *DCT coefficients*. One of these values is referred to as the *DC coefficient* and the other 63 as the *AC coefficients*.

Each of the 64 coefficients is then *quantized* using one of 64 corresponding values from a *quantization table* (determined by one of the table specifications shown in Figure 4). No default values for quantization tables are specified in this Specification; applications may specify values which customize picture quality for their particular image characteristics, display devices, and viewing conditions.

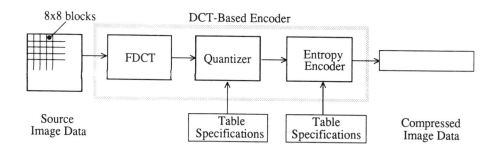

Figure 4 - DCT-based encoder simplified diagram

4

After quantization, the DC coefficient and the 63 AC coefficients are prepared for *entropy encoding*, as shown in Figure 5. The previous quantized DC coefficient is used to predict the current quantized DC coefficient, and the difference is encoded. The 63 quantized AC coefficients undergo no such differential encoding, but are converted into a one-dimensional *zig-zag sequence*, as shown in Figure 5.

The quantized coefficients are then passed to an entropy encoding procedure which compresses the data further. One of two entropy coding procedures can be used, as described in 3.6. If *Huffman encoding* is used, *Huffman table* specifications must be provided to the encoder. If *arithmetic encoding* is used, arithmetic coding *conditioning table* specifications must be provided.

Figure 6 shows the main procedures for all DCT-based decoding processes. Each step shown performs essentially the inverse of its corresponding main procedure within the encoder. The entropy decoder decodes the zig-zag sequence of quantized DCT coefficients. After *dequantization* the DCT coefficients are transformed to an 8x8 block of samples by the *inverse DCT* (IDCT).

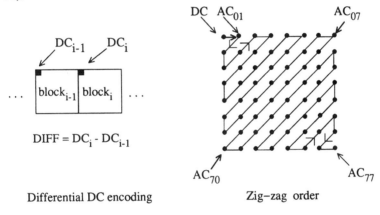

Differential DC encoding   Zig–zag order

Figure 5 - Preparation of quantized coefficients for entropy encoding

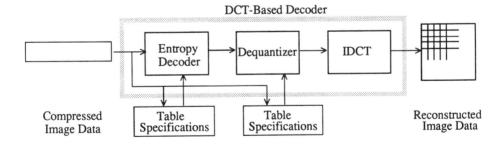

Figure 6 - DCT-based decoder simplified diagram

5

3.4 Lossless coding

Figure 7 shows the main procedures for the lossless encoding processes. A *predictor* combines the reconstructed values of up to three neighborhood samples at positions a, b, and c to form a prediction of the sample at position x as shown in Figure 8. This prediction is then subtracted from the actual value of the sample at position x, and the difference is losslessly entropy-coded by either Huffman or arithmetic coding.

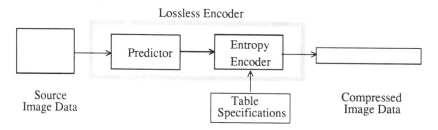

Figure 7 - Lossless encoder simplified diagram

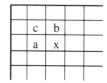

Figure 8 - 3-sample prediction neighborhood

This encoding process may also be used in a slightly modified way, whereby the *precision* of the input samples is reduced by one or more bits prior to the lossless coding. This achieves higher compression than the lossless process (but lower compression than the DCT-based processes for equivalent visual fidelity), and limits the reconstructed image's worst-case sample error to the amount of input precision reduction.

3.5 Modes of operation

There are four distinct *modes of operation* under which the various coding processes are defined: *sequential DCT-based, progressive DCT-based,* lossless, and *hierarchical.* (Implementations are not required to provide all of these). The lossless mode of operation was described in 3.4. The other modes of operation are compared as follows.

For the sequential DCT-based mode, 8x8 sample blocks are typically input block by block from left to right, and *block-row* by block-row from top to bottom. After a block has been quantized and prepared for entropy encoding, all 64 of its quantized DCT coefficients can be immediately entropy encoded and output as part of the compressed image data (as was described in 3.3), thereby minimizing coefficient storage requirements.

For the progressive DCT-based mode, 8x8 blocks are also typically encoded in the same order, but in multiple *scans* through the image. This is accomplished by adding an image-sized coefficient memory buffer (not shown in Figure 4) between the quantizer and the entropy encoder. As each block is quantized, its coefficients are stored in the buffer. The DCT coefficients in the buffer are then partially encoded in each of multiple scans. The typical sequence of image presentation at the output of the decoder for sequential vs. progressive modes of operation is shown in Figure 9.

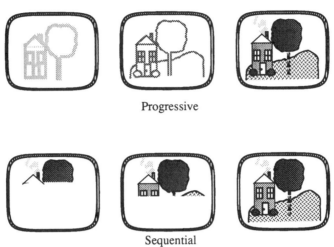

Progressive

Sequential

Figure 9 - Progressive vs. sequential presentation

There are two procedures by which the quantized coefficients in the buffer may be partially encoded within a scan. First, only a specified *band* of coefficients from the zig-zag sequence need be encoded. This procedure is called *spectral selection*, because each band typically contains coefficients which occupy a lower or higher part of the *frequency* spectrum for that 8x8 block. Secondly, the coefficients within the current band need not be encoded to their full (quantized) accuracy within each scan. Upon a coefficient's first encoding, a specified number of most significant bits is encoded first. In subsequent scans, the less significant bits are then encoded. This procedure is called *successive approximation*. Either procedure may used separately, or they may be mixed in flexible combinations.

In hierarchical mode, an image is encoded as a sequence of *frames*. These frames provide *reference reconstructed components* which are usually needed for prediction in subsequent frames. Except for the first frame for a given component, *differential frames* encode the difference between source components and reference reconstructed components. The coding of the differences may be done using only DCT-based processes, only lossless processes, or DCT-based processes with a final lossless process for each component. *Downsampling* and *upsampling filters* may be used to provide a pyramid of spatial resolutions as shown in Figure 10. Alternately, the hierarchical mode can be used to improve the quality of the reconstructed components at a given spatial resolution.

7

Hierarchical mode offers a progressive presentation similar to the progressive DCT-based mode but is useful in environments which have multi-resolution requirements. Hierarchical mode also offers the capability of progressive transmission to a final lossless stage.

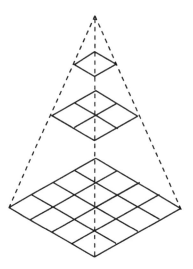

Figure 10 - Hierarchical multi-resolution encoding

3.6 Entropy coding alternatives

Two alternative entropy coding procedures are specified: Huffman coding and arithmetic coding. Huffman coding procedures use Huffman tables, determined by one of the table specifications shown in Figures 1 and 2. Arithmetic coding procedures use arithmetic coding conditioning tables, which may also be determined by a table specification. No default values for Huffman tables are specified, so that applications may choose tables appropriate for their own environments. Default tables are defined for the arithmetic coding conditioning.

The baseline sequential process uses Huffman coding, while the extended DCT-based and lossless processes may use either Huffman or arithmetic coding.

3.7 Sample precision

For DCT-based processes, two alternative sample precisions are specified: either 8 bits or 12 bits per sample. Applications which use samples with other precisions can use either 8-bit or 12-bit precision by shifting their source image samples appropriately. The baseline process uses only 8-bit precision. DCT-based implementations which handle 12-bit source image samples are likely to need greater computational resources than those which handle only 8-bit source images. Consequently in this Specification separate normative requirements are defined for 8-bit and 12-bit DCT-based processes.

For lossless processes the sample precision is specified to be from 2 to 16 bits.

8

3.8 Multiple-component control

3.3 and 3.4 give an overview of one major part of the encoding and decoding processes - those which operate on the sample values in order to achieve compression. There is another major part as well - the procedures which control the order in which the image data from multiple components are processed to create the compressed data, and which ensure that the proper set of table data is applied to the proper *data units* in the image. (A data unit is a sample for lossless processes and an 8x8 block of samples for DCT-based processes).

3.8.1 Interleaving multiple components

Figure 11 shows an example of how an encoding process selects between multiple source image components as well as multiple sets of table data, when performing its encoding procedures. The source image in this example consists of the three components A, B, and C, and there are two sets of table specifications. (This simplified view does not distinguish between the quantization tables and entropy coding tables).

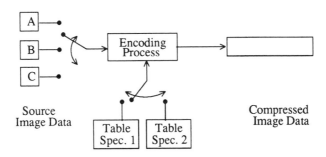

Figure 11 – Component–interleave and table–switching control

In sequential mode, encoding is *non-interleaved* if the encoder compresses all image data units in component A before beginning component B, and then in turn all of B before C. Encoding is *interleaved* if the encoder compresses a data unit from A, a data unit from B, a data unit from C, then back to A, etc. These alternatives are illustrated in Figure 12, which shows a case in which all three image components have identical dimensions: X *columns* by Y rows, for a total of N data units each.

These control procedures are also able to handle cases in which the source image components have different dimensions. Figure 13 shows a case in which two of the components, B and C, have half the number of horizontal samples relative to component A. In this case, two data units from A are interleaved with one each from B and C. Cases in which components of an image have more complex sampling relationships, including subsampling in the vertical dimension, can be handled as well.

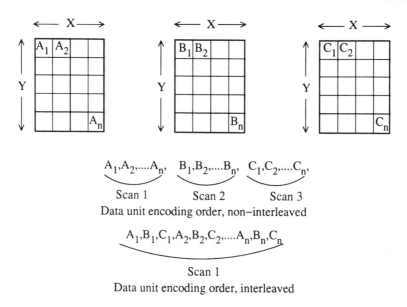

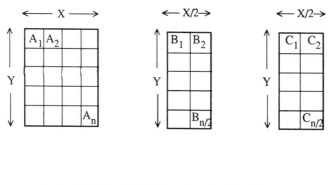

Data unit encoding order, non-interleaved

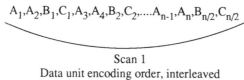

Scan 1

Data unit encoding order, interleaved

Figure 12 - Interleaved vs. non-interleaved encoding order

$A_1,A_2,B_1,C_1,A_3,A_4,B_2,C_2,....A_{n-1},A_n,B_{n/2},C_{n/2}$

Scan 1

Data unit encoding order, interleaved

Figure 13 - Interleaved order for components with different dimensions

10

3.8.2 Minimum coded unit

Related to the concepts of multiple-component interleave is the *minimum coded unit* (MCU). If the compressed image data is non-interleaved, the MCU is defined to be one data unit. For example, in Figure 12 the MCU for the non-interleaved case is a single data unit. If the compressed data is interleaved, the MCU contains one or more data units from each component. For the interleaved case in Figure 12, the (first) MCU consists of the three interleaved data units A_1, B_1, C_1. In the example of Figure 13, the (first) MCU consists of the four data units A_1, A_2, B_1, C_1.

3.9 Structure of compressed data

Figures 1, 2, and 3 all illustrate slightly different views of compressed image data. Figure 1 shows this data as the output of an encoding process, Figure 2 shows it as the input to a decoding process, and Figure 3 shows compressed image data in the interchange format, at the interface between applications.

Compressed image data is described by a uniform structure and set of parameters for both classes of encoding processes (lossy or lossless), and for all modes of operation (sequential, progressive, lossless, and hierarchical).The various parts of the compressed image data are identified by special two-byte codes called *markers*. Some markers are followed by particular sequences of parameters such as table specifications and headers. Others are used without parameters for functions such as marking the start-of-image and end-of-image. When a marker is associated with a particular sequence of parameters, the marker and its parameters comprise a *marker segment*.

The data created by the entropy encoder are also segmented, and one particular marker - the *restart marker* - is used to isolate *entropy-coded data segments*. The encoder outputs the restart markers, intermixed with the entropy-coded data, at regular *restart intervals* of the source image data. Restart markers can be identified without having to decode the compressed data to find them. Because they can be independently decoded, they have application-specific uses, such as parallel encoding or decoding, isolation of data corruptions, and semi-random access of entropy-coded segments.

There are three compressed data formats:

1) the interchange format;

2) the *abbreviated format* for compressed image data;

3) the abbreviated format for table-specification data.

3.9.1 Interchange format

In addition to certain required marker segments and the entropy-coded segments, the interchange format shall include the marker segments for all quantization and entropy-coding table specifications needed by the decoding process. This guarantees that a compressed image can cross the boundary between application environments, regardless of how each environment internally associates tables with compressed image data.

3.9.2 Abbreviated format for compressed image data

The abbreviated format for compressed image data is identical to the interchange format, except that it does not include all tables required for decoding. (It may include some of them.) This format is intended for use within applications where alternative mechanisms are available for supplying some or all of the table-specification data needed for decoding.

11

3.9.3 Abbreviated format for table-specification data

This format contains only table-specification data. It is a means by which the application may install in the decoder the tables required to subsequently reconstruct one or more images.

3.10 Image, frame, and scan

Compressed image data consists of only one image. An image contains only one frame in the cases of sequential and progressive coding processes; an image contains multiple frames for the hierarchical mode.

A frame contains one or more scans. For sequential processes, a scan contains a complete encoding of one or more image components. In Figures 12 and 13, the frame consists of three scans when non-interleaved, and one scan if all three components are interleaved together. The frame could also consist of two scans: one with a non-interleaved component, the other with two components interleaved.

For progressive processes, a scan contains a partial encoding of all data units from one or more image components. Components shall not be interleaved in progressive mode, except for the DC coefficients in the first scan for each component of a progressive frame.

3.11 Summary of Coding Processes

Table 1 provides a summary of the essential characteristics of the distinct coding processes specified in this Specification. The full specification of these processes is contained in Annexes F, G, H and J.

Table 1 - Summary: Essential Characteristics of Coding Processes

Baseline Process (required for all DCT-based decoders)

- DCT-based process
- Source image: 8-bit samples within each component
- Sequential
- Huffman coding: 2 AC and 2 DC tables
- Decoders shall process scans with 1, 2, 3, and 4 components
- Interleaved and non-interleaved scans

Extended DCT-based Processes

- DCT-based process
- Source image: 8-bit or 12-bit samples
- Sequential or progressive
- Huffman or arithmetic coding: 4 AC and 4 DC tables
- Decoders shall process scans with 1, 2, 3, and 4 components
- Interleaved and non-interleaved scans

Lossless Processes

- Predictive process (not DCT-based)
- Source image: N-bit samples ($2 \leq N \leq 16$)
- Sequential
- Huffman or arithmetic coding: 4 DC tables
- Decoders shall process scans with 1, 2, 3, and 4 components
- Interleaved and non-interleaved scans

Hierarchical processes

- Multiple frames (non-differential and differential)
- Uses extended DCT-based or lossless processes
- Decoders shall process scans with 1, 2, 3, and 4 components
- Interleaved and non-interleaved scans

13

4 Definitions, abbreviations, and symbols

4.1 Definitions and abbreviations

For the purposes of this Specification, the following definitions apply.

abbreviated format: A representation of compressed image data which is missing some or all of the table specifications required for decoding.

AC coefficient: Any DCT coefficient for which the frequency is not zero in at least one dimension.

(adaptive) (binary) arithmetic decoding: An entropy decoding procedure which recovers the sequence of symbols from the sequence of bits produced by the arithmetic encoder.

(adaptive) (binary) arithmetic encoding: An entropy encoding procedure which codes by means of a recursive subdivision of the probability of the sequence of symbols coded up to that point.

application environment: The standards for data representation, transmission, and association which have been established for a particular application.

arithmetic decoder: An embodiment of arithmetic decoding procedure.

arithmetic encoder: An embodiment of arithmetic encoding procedure.

baseline (sequential): A particular sequential DCT-based encoding and decoding process specified in this Specification, and which is required for all DCT-based decoding processes.

binary decision: Choice between two alternatives.

bit stream: Partially encoded or decoded sequence of bits comprising an entropy-coded segment.

block-interleaved: The descriptive term applied to the repetitive multiplexing in a specific order of small groups of 8x8 blocks from each component in a scan.

block-row: A sequence of eight contiguous component lines which are partitioned into 8x8 blocks.

byte: Eight-bit octet.

byte stuffing: A procedure in which either the Huffman coder or the arithmetic coder inserts a zero byte into the entropy-coded segment following the generation of an encoded hexadecimal X'FF' byte.

carry bit: A bit in the arithmetic encoder code register which is set if a carry-over in the code register overflows the eight bits reserved for the output byte.

ceiling function: The mathematical procedure in which the greatest integer value of a real number is obtained by selecting the smallest integer value which is greater than or equal to the real number.

class (of coding process): Lossy or lossless coding processes.

code register: The arithmetic encoder register containing the least significant bits of the a partially completed entropy-coded segment. Alternatively, the arithmetic decoder register containing the most significant bits of a partially decoded entropy-coded segment.

14

coding model: A procedure used to convert input data into symbols to be coded.

coding process: A general term for referring to an encoding process, a decoding process, or both.

colour image: A continuous-tone image that has more than one component.

columns: Samples per line in a component.

component: One of the two-dimensional arrays which comprise an image.

compressed data: Either compressed image data or table specification data or both.

compressed image data: A coded representation of an image, as specified in this Specification.

compression: Reduction in the number of bits used to represent source image data.

conditional exchange: The interchange of MPS and LPS probability intervals whenever the size of the LPS interval is greater than the size of the MPS interval (in arithmetic coding).

(conditional) probability estimate: The probability value assigned to the LPS by the probability estimation state machine (in arithmetic coding).

conditioning table: The set of parameters which select one of the defined relationships between prior coding decisions and the conditional probability estimates used in arithmetic coding.

context: The set of previously coded binary decisions which is used to create the index to the probability estimation state machine (in arithmetic coding).

continuous-tone image: An image whose components have more than one bit per sample.

data unit: A block in DCT-based processes; a sample in lossless processes.

DC coefficient: The DCT coefficient for which the frequency is zero in both dimensions.

DC prediction: The procedure used by DCT-based encoders whereby the quantized DC coefficient from the previously encoded 8x8 block of the same component is subtracted from the current quantized DC coefficient.

(DCT) coefficient: The amplitude of a *specific cosine* basis function.

decoder: An embodiment of a decoding process.

decoding process: A process which takes as its input compressed image data and outputs a continuous-tone image.

default conditioning: The values defined for the arithmetic coding conditioning tables at the beginning of coding of an image.

dequantization: The inverse procedure to quantization by which the decoder recovers a representation of the DCT coefficients.

difference image: The two dimensional array of data formed by subtracting a reference image from a source image (in hierarchical mode coding).

differential component: The difference between an input component derived from the source image and the corresponding reference component derived from the preceding frame for that component (in hierarchical mode coding).

differential frame: A frame in a hierarchical process in which differential components are either encoded or decoded.

15

(digital) reconstructed image (data): A continuous-tone image which is the output of any decoder defined in this Specification.

(digital) source image (data): A continuous-tone image used as input to any encoder defined in this Specification.

(digital) (still) image: A set of two-dimensional arrays of data.

discrete cosine transform; DCT: Either the forward discrete cosine transform or the inverse discrete cosine transform.

downsampling (filter): A procedure by which the spatial resolution of an image is reduced (in hierarchical mode coding).

encoder: An embodiment of an encoding process.

encoding process: A process which takes as its input a continuous-tone image and outputs compressed image data.

entropy-coaed (data) segment: An independently decodable sequence of entropy encoded bytes of compressed image data.

(entropy-coded segment) pointer: The variable which points to the most recently placed (or fetched) byte in the entropy encoded segment.

entropy decoder: An embodiment of an entropy decoding procedure.

entropy decoding: A lossless procedure which recovers the sequence of symbols from the sequence of bits produced by the entropy encoder.

entropy encoder: An embodiment of an entropy encoding procedure

entropy encoding: A lossless procedure which converts a sequence of input symbols into a sequence of bits such that the average number of bits per symbol approaches the entropy of the input symbols.

extended (DCT-based) process: A descriptive term for DCT-based encoding and decoding processes in which additional capabilities are added to the baseline sequential process.

forward discrete cosine transform; FDCT: A mathematical transformation using cosine basis functions which converts a block of samples into a corresponding array of basis function amplitudes.

frame: A group of one or more scans (all using the same DCT-based or lossless process) through the data of one or more of the components in an image.

frame header: The start-of-frame marker and frame parameters that are coded at the beginning of a frame.

frequency: A two dimensional index into the two dimensional array of DCT coefficients.

(frequency) band: A contiguous group of coefficients from the zig-zag sequence (in progressive mode coding).

full progression: A process which uses both spectral selection and successive approximation (in progressive mode coding).

grayscale image: A continuous-tone image that has only one component.

16

hierarchical: A mode of operation for coding an image in which the first frame for a given component is followed by frames which code the differences between the source data and the reconstructed data from the previous frame for that component. Resolution changes are allowed between frames.

hierarchical decoder: A sequence of decoder processes in which the first frame for each component is followed by frames which decode an array of differences for each component and adds it to the reconstructed data from the preceding frame for that component.

hierarchical encoder: The mode of operation in which the first frame for each component is followed by frames which encoded the array of differences between the source data and the reconstructed data from the preceding frame for that component.

horizontal sampling factor: The relative number of horizontal data units of a particular component with respect to the number of horizontal data units in the other components.

Huffman decoder: An embodiment of a Huffman decoding procedure

Huffman decoding: An entropy decoding procedure which recovers the symbol from each variable length code produced by the Huffman encoder.

Huffman encoder: An embodiment of a Huffman encoding procedure

Huffman encoding: An entropy encoding procedure which assigns a variable length code to each input symbol.

Huffman table: The set of variable length codes required in a Huffman encoder and Huffman decoder.

image data: Either source image data or reconstructed image data.

interchange format: The representation of compressed image data for exchange between application environments.

interleaved: The descriptive term applied to the repetitive multiplexing of small groups of data units from each component in a scan in a specific order.

inverse discrete cosine transform; IDCT: A mathematical transformation using cosine basis functions which converts an array of basis function amplitudes into a corresponding block of samples.

latent output: Output of the arithmetic encoder which is held, pending resolution of carry-over (in arithmetic coding).

less probable symbol; LPS: For a binary decision, the decision value which has the smaller probability.

level shift: A procedure used by DCT-based encoders and decoders whereby each input sample is either converted from an unsigned representation to a two's complement representation or from a two's complement representation to an unsigned representation.

lossless: A descriptive term for encoding and decoding processes and procedures in which the output of the decoding procedure(s) is identical to the input to the encoding procedure(s).

lossless coding: The mode of operation which refers to any one of the coding processes defined in this Specification in which all of the procedures are lossless (see Annex H).

lossy: A descriptive term for encoding and decoding processes which are not lossless.

marker: A two-byte code in which the first byte is hexadecimal FF (X'FF') and the second byte is a value between 1 and hexadecimal FE (X'FE').

marker segment: A marker and associated set of parameters.

MCU-row: The smallest sequence of MCU which contains at least one row of data units from every component in the scan.

minimum coded unit; MCU: The smallest group of data units that is coded.

modes (of operation): The four main categories of image compression processes defined in this Specification.

more probable symbol; MPS: For a binary decision, the decision value which has the larger probability.

non-differential frame: The first frame for any components in a hierarchical encoder or decoder. The components are encoded or decoded without subtraction from reference components. The term refers also to any frame in modes other than the hierarchical mode.

non-interleaved: The descriptive term applied to the data unit processing sequence when the scan has only one component.

(number of) lines: The number of rows in the component with the largest vertical sampling factor in the image.

parameters: Fixed length integers 4, 8 or 16 bits in length, used in the compressed data formats.

point transform: Scaling of a sample or DCT coefficient.

precision: Number of bits allocated to a particular sample or DCT coefficient.

predictor: A linear combination of previously encoded reconstructed values (in lossless mode coding).

probability estimation state machine: An interlinked table of probability values and indices which is used to estimate the probability of the LPS (in arithmetic coding).

probability interval: The probability of a particular sequence of binary decisions within the ordered set of all possible sequences (in arithmetic coding).

(probability) sub-interval: A portion of a probability interval allocated to either of the two possible binary decision values (in arithmetic coding).

procedure: A set of steps which accomplishes one of the tasks which comprise an encoding or decoding process.

progressive (coding): One of the DCT-based or hierarchical processes defined in this Specification in which each scan typically improves the quality of the reconstructed image.

progressive DCT-based: The mode of operation which refers to any one of the processes defined in Annex G of this Specification.

quantization table: The set of 64 quantization values used to quantize the DCT coefficients.

quantization value: An integer value used in the quantization procedure.

quantize: The act of performing the quantization procedure for a DCT coefficient.

reference (reconstructed) component: Reconstructed component data which is used in a subsequent frame of a hierarchical encoder or decoder process (in hierarchical mode coding).

renormalization: The doubling of the probability interval and the code register value until the probability interval exceeds a fixed minimum value (in arithmetic coding).

18

restart interval: The integer number of MCUs processed as an independent sequence within a scan.

restart marker: The marker that separates two restart intervals in a scan.

run (length): Number of consecutive symbols of the same value.

sample: One element in the two-dimensional array which comprises a component.

sample-interleaved: The descriptive term applied to the repetitive multiplexing of small groups of samples from each component in a scan in a specific order.

scan: A single pass through the data for one or more of the components in an image.

scan header: The start-of-scan marker and scan parameters that are coded at the beginning of a scan.

sequential (coding): One of the lossless or DCT-based coding processes defined in this Specification in which each component of the image is encoded within a single scan.

sequential DCT-based: The mode of operation which refers to any one of the processes defined in Annex F of this Specification.

spectral selection: A progressive coding process in which the zig-zag sequence is divided into bands of one or more contiguous coefficients, and each band is coded in one scan.

stack counter: The count of X'FF' bytes which are held, pending resolution of carry-over in the arithmetic encoder.

statistical bin: The storage location where an index is stored which identifies the value of the conditional probability estimate used for a particular arithmetic coding binary decision.

statistical conditioning: The selection, based on prior coding decisions, of one estimate out of a set of conditional probability estimates (in arithmetic coding).

statistical model: The assignment of a particular conditional probability estimate to each of the binary arithmetic coding decisions.

statistics area: The array of statistics bins required for a coding process which uses arithmetic coding.

successive approximation: A progressive coding process in which the coefficients are coded with reduced precision in the first scan, and precision is increased by one bit with each succeeding scan.

table specification data: The coded representation from which the tables used in the encoder and decoder are generated.

transcoder: A procedure for converting compressed image data of one encoder process to compressed image data of another encoder process.

(uniform) quantization: The procedure by which DCT coefficients are linearly scaled in order to achieve compression.

upsampling (filter): A procedure by which the spatial resolution of an image is increased (in hierarchical mode coding).

vertical sampling factor: The relative number of vertical data units of a particular component with respect to the number of vertical data units in the other components in the frame.

zig-zag sequence: A specific sequential ordering of the DCT coefficients from (approximately) lowest spatial frequency to highest.

(8x8) block: An 8x8 array of samples.

3-sample predictor: A linear combination of the three nearest neighbor reconstructed samples to the left and above (in lossless mode coding).

4.2 Symbols

The symbols used in this Specification are listed below.

A: probability interval

AC: AC DCT coefficient

AC_{ji}: AC coefficient predicted from DC values

Ah: successive approximation bit position, high

Al: successive approximation bit position, low

Ap_i: ith 8-bit parameter in APP_n segment

APP_n: marker reserved for application segments

B: byte of data in entropy-coded segment

B_i: ith statistics bin for coding magnitude bit pattern category

B2: next byte from entropy-coded segment when B=X'FF'

BE: counter for buffered correction bits for Huffman coding in the successive approximation process

BITS: 16 byte list containing number of Huffman codes of each length

BP: pointer to entropy-coded segment

BPST: pointer to byte before start of entropy-coded segment

BR: counter for buffered correction bits for Huffman coding in the successive approximation process

C: value of bit stream in code register

C_i: component identifier for frame

C_u: horizontal frequency dependent scaling factor in DCT

C_v: vertical frequency dependent scaling factor in DCT

C-low: low order 16 bits of the arithmetic decoder code register

Cm_i: ith 8-bit parameter in COM segment

CNT: bit counter in NEXTBYTE

CODE: Huffman code value

20

CODESIZE(V): code size for symbol V

COM: comment marker

Cs: conditioning table value

Cs$_i$: component identifier for scan

CT: renormalization shift counter

Cx: high order 16 bits of arithmetic decoder code register

d$_{ji}$: data unit from horizontal position i, vertical position j

D: decision decoded

Da: in DC coding, the DC difference coded for the previous block from the same component; in lossless coding, the difference coded for the sample immediately to the left

DAC: define-arithmetic-coding conditioning marker

Db: the difference coded for the sample immediately above

DC: DC DCT coefficient

DC$_i$: DC coefficient for ith block in component

DC$_k$: kth DC value used in prediction of AC coefficients

DHP: define hierarchical progression marker

DHT: define-Huffman-tables marker

DIFF: difference between quantized DC and prediction

DNL: define-number-of-lines marker

DQT: define-quantization-table marker

DRI: define restart interval marker

E: exponent in magnitude category upper bound

ECS: abbreviation for entropy-coded segment

Eh: horizontal expansion parameter in EXP segment

EHUFCO: Huffman code table for encoder

EHUFSI: encoder table of Huffman code sizes

EOB: end-of-block for sequential; end-of-band for progressive

EOB$_n$: run length category for EOB runs

EOBx: position of EOB in previous successive approximation scan

EOI: end-of-image marker

Ev: vertical expansion parameter in EXP segment

EXP: expand reference components

21

FREQ(V): frequency of occurrence of symbol V

H_i: horizontal sampling factor for ith component

H_{max}: largest horizontal sampling factor

HUFFCODE: list of Huffman codes corresponding to lengths in HUFFSIZE

HUFFSIZE: list of code lengths

HUFFVAL: list of values assigned to each Huffman code

i: subscript index

I: integer variable

Index(S): index to probability estimation state machine table for context index S

j: subscript index

J: integer variable

JPG: marker reserved for JPEG extensions

JPG_n: marker reserved for JPEG extensions

k: subscript index

K: integer variable

Kmin: index of 1st AC coefficient in band (1 for sequential DCT)

Kx: conditioning parameter for AC arithmetic coding model

L: DC and lossless coding conditioning lower bound parameter

L_i: element in BITS list in DHT segment

$L_i(t)$: element in BITS list in the DHT segment for Huffman table t

La: length of parameters in APP_n segment

LASTK: largest value of K

Lc: length of parameters in COM segment

Ld: length of parameters in DNL segment

Le: length of parameters in EXP segment

Lf: length of frame header parameters

Lh: length of parameters in DHT segment

Lp: length of parameters in DAC segment

LPS: less probable symbol (in arithmetic coding)

Lq: length of parameters in DQT segment

Lr: length of parameters in DRI segment

Ls: length of scan header parameters

LSB: least significant bit

m: modulo 8 counter for RSTm marker

M: bit mask used in coding magnitude of V

MAXCODE: table with maximum value of Huffman code for each code length

MCUR: number of MCU required to make up one row

MINCODE: table with minimum value of Huffman code for each code length

MPS: more probable symbol (in arithmetic coding)

MSB: most significant bit

$\mathbf{m_t}$**:** number of V_{ij} parameters for Huffman table t

M2, M3, M4, ... , M15: designation of context-indices for coding of magnitude bits in the arithmetic coding models

n: integer variable

N: data unit counter for MCU coding

Nf: number of components in frame

Nb: number of data units in MCU

Next_Index_MPS: new value of Index(S) after a MPS renormalization

Next_Index_LPS: new value of Index(S) after a LPS renormalization

NL: number of lines defined in DNL segment

Ns: number of components in scan

OTHERS(V): index to next symbol in chain

P: sample precision

Pq: quantizer precision parameter in DQT segment

Pq(t): quantizer precision parameter in DQT segment for quantization table t

Px: calculated value of sample

PRED: quantized DC coefficient from the most recently coded block of the component

Pt: point transform parameter

$\mathbf{Q_{vu}}$**:** quantization value for DCT coefficient S_{vu}

$\mathbf{Q_{00}}$**:** quantizer value for DC coefficient

$\mathbf{Q_{ji}}$**:** quantizer value for coefficient AC_{ji}

$\mathbf{QAC_{ji}}$**:** quantized AC coefficient predicted from DC values

$\mathbf{QDC_k}$**:** kth quantized DC value used in prediction of AC coefficients

Qe: LPS probability estimate

23

Qe(S): LPS probability estimate for context index S

Qk: kth element of 64 quantization elements in DQT segment

R: length of run of zero amplitude AC coefficients

R_{vu}: dequantized DCT coefficient

Ra: reconstructed sample value

Rb: reconstructed sample value

Rc: reconstructed sample value

Rd: rounding in prediction calculation

RES: reserved markers

Ri: restart interval in DRI segment

RS: composite value used in Huffman coding of AC coefficients

RST_m: restart marker

RRRR: 4-bit value of run length of zero AC coefficients

r_{vu}: reconstructed image sample

S: context index

s_{ji}: sample from horizontal position i, vertical position j in block

S_{vu}: DCT coefficient at horizontal frequency u, vertical frequency v

s_{yx}: reconstructed value from IDCT

SC: context-index for coding of correction bit in successive approximation coding

Se: end of spectral selection band in zigzag sequence

SE: context-index for coding of end-of-block or end-of-band

SI: Huffman code size

SIGN: decoded sense of sign

SIZE: length of a Huffman code

SN: context-index for coding of first magnitude category when V is negative

SOI: start-of-image marker

SOF_0: baseline DCT process frame marker

SOF_1: extended sequential DCT frame marker, Huffman coding

SOF_2: progressive DCT frame marker, Huffman coding

SOF_3: lossless process frame marker, Huffman coding

SOF_5: differential sequential DCT frame marker, Huffman coding

SOF_6: differential progressive DCT frame marker, Huffman coding

24

SOF$_7$: differential lossless process frame marker, Huffman coding

SOF$_9$: sequential DCT frame marker, arithmetic coding

SOF$_{10}$: progressive DCT frame marker, arithmetic coding

SOF$_{11}$: lossless process frame marker, arithmetic coding

SOF$_{13}$: differential sequential DCT frame marker, arithmetic coding

SOF$_{14}$: differential progressive DCT frame marker, arithmetic coding

SOF$_{15}$: differential lossless process frame marker, arithmetic coding

SOS: start-of-scan marker

SP: context-index for coding of first magnitude category when V is positive

Sq$_{vu}$: quantized DCT coefficient

Ss: start of spectral selection band in zigzag sequence

SS: context-index for coding of sign decision

SSSS: 4-bit size catagory of DC difference or AC coefficient amplitude

Switch_MPS: parameter controlling inversion of sense of MPS

Sz: parameter used in coding magnitude of V

S0: context-index for coding of V=0 decision

t: summation index for parameter limits computation

T: temporary variable

Ta: arithmetic conditioning table identifier

Ta$_j$: AC entropy table selector for jth component in scan

Td$_j$: DC entropy table selector for jth component in scan

TEM: temporary marker

Tq$_i$: quantization table for ith component in frame

Tq: quantizer table identifier in DQT segment

U: DC and lossless coding conditioning upper bound parameter

V: Symbol or value being either encoded or decoded

V$_i$: vertical sampling factor for ith component

V$_{ij}$: jth value for length i in HUFFVAL

V$_{max}$: largest vertical sampling factor

VALPTR: list of indices for 1st value in HUFFVAL for each code length

V1: symbol value

V2: symbol value

Vt: temporary variable

X: number of samples per line in component with largest horizontal sampling factor value

x_i: number of columns in ith component

X_i: ith statistics bin for coding magnitude category decision

X1, X2, X3, ... , X15: designation of context-indices for coding of magnitude categories in the arithmetic coding models

XHUFCO: extended Huffman code table

XHUFSI: table of sizes of extended Huffman codes

Y: number of lines in component with largest vertical sampling factor value

y_i: number of rows in ith component

ZRL: value in HUFFVAL assigned to run of 16 zero coefficients

ZZ(k): kth element in zigzag sequence of DCT coefficients

ZZ(0): quantized DC coefficient in zigzag sequence

5 Interchange format requirements

The interchange format is the coded representation of compressed image data for exchange between application environments.

The interchange format requirements are that any compressed image data represented in interchange format shall comply with the syntax and codes assignments appropriate for the decoding process selected, as specified in Annex B.

6 Encoder requirements

An encoding process converts source image data to compressed image data. Each of Annexes F, G, H, and J specifies a number of distinct encoding processes for its particular mode of operation.

An encoder is an embodiment of one (or more) of the encoding processes specified in Annexes F, G, H, or J. In order to comply with this Specification, an encoder shall satisfy at least one of the following two requirements.

An encoder shall:

1) with proper accuracy, convert source image data to compressed image data which complies with the interchange format syntax specified in Annex B for the encoding process(es) embodied by the encoder;

2) with proper accuracy, convert source image data to compressed image data which complies with the abbreviated format syntax for compressed image data specified in Annex B for the encoding process(es) embodied by the encoder.

For each of the encoding processes specified in Annexes F, G, H, and J, the compliance tests for the above requirements are specified in Part 2 of this Specification.

NOTE - There is **no requirement** in this Specification that any encoder which embodies one of the encoding processes specified in Annexes F, G, H, or J shall be able to operate for all ranges of the parameters which are allowed for that process. An encoder is only required to meet the compliance tests specified in Part 2, and to generate the compressed data format according to Annex B for those parameter values which it does use.

7 Decoder requirements

A decoding process converts compressed image data to reconstructed image data. Each of Annexes F, G, H, and J specifies a number of distinct decoding processes for its particular mode of operation.

A decoder is an embodiment of one (or more) of the decoding processes specified in Annexes F, G, H, or J. In order to comply with this Specification, a decoder shall satisfy all three of the following requirements.

A decoder shall:

1) with proper accuracy, convert to reconstructed image data any compressed image data with parameters within the range supported by the application, and which complies with the interchange format syntax specified in Annex B for the decoding process(es) embodied by the decoder;

2) accept and properly store any table-specification data which complies with the abbreviated format syntax for table-specification data specified in Annex B for the decoding process(es) embodied by the decoder;

3) with proper accuracy, convert to reconstructed image data any compressed image data which complies with the abbreviated format syntax for compressed image data specified in Annex B for the decoding process(es) embodied by the decoder, provided that the table-specification data required for decoding the compressed image data has previously been installed into the decoder.

Additionally, any DCT-based decoder, if it embodies any DCT-based decoding process other than baseline sequential, shall also embody the baseline sequential decoding process.

For each of the decoding processes specified in Annexes F, G, H, and J, the compliance tests for the above requirements are specified in Part 2 of this Specification.

29

Annex A (normative)

Mathematical definitions

A.1 Source image

Source images to which the encoding processes specified in this Specification can be applied are defined in this annex.

A.1.1 Dimensions and relative sampling

As shown in Figure A.1, a source image is defined to consist of Nf components. Each component, with unique identifier C_i, is defined to consist of a rectangular array of samples of x_i columns by y_i rows. The image has overall dimensions X samples horizontally by Y samples vertically, where X is the maximum of the x_i values and Y is the maximum of the y_i values for all components in the frame. For each component, sampling factors H_i and V_i are defined relating component dimensions x_i and y_i to overall image dimensions X and Y, according to the following expressions:

$$x_i = \lceil X \times \frac{H_i}{H_{max}} \rceil \quad \text{and} \quad y_i = \lceil Y \times \frac{V_i}{V_{max}} \rceil \,,$$

where H_{max} and V_{max} are the maximum sampling factors for all components in the frame, and $\lceil \; \rceil$ is the ceiling function.

As an example, consider an image having 512 pixels by 512 lines and 3 components sampled according to the following sampling factors:

$$\begin{array}{lll}
\text{Component 0} & H_0 = 4, & V_0 = 1 \\
\text{Component 1} & H_1 = 2, & V_1 = 2 \\
\text{Component 2} & H_2 = 1, & V_2 = 1
\end{array}$$

Then X=512, Y=512, H_{max}=4, V_{max}=2, and x_i and y_i for each component are:

$$\begin{array}{lll}
\textit{Component 0} & x_0 = 512, & y_0 = 256 \\
\text{Component 1} & x_1 = 256, & y_1 = 512 \\
\text{Component 2} & x_2 = 128, & y_2 = 256
\end{array}$$

NOTE - The X, Y, H_i, and V_i parameters are contained in the frame header of the interchange format (see B.2.2), whereas the individual component dimensions x_i and y_i are derived by the decoder. Source images with x_i and y_i dimensions which do not satisfy the expressions above cannot be properly reconstructed.

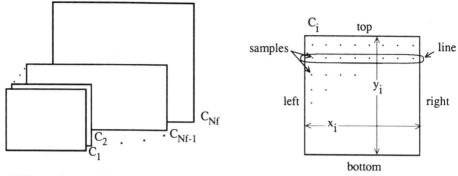

(a) Source image with multiple components   (b) Characteristics of an image component

Figure A.1 - Source image characteristics

A.1.2 Sample precision

A sample is defined to be an integer with precision P bits, with any value in the range $[0, 2^P-1]$. All samples of all components within the same source image shall have the same precision P. Restrictions on the value of P depend on the mode of operation, as specified in B.2 - B.7.

A.1.3 Data unit

A data unit is a sample in lossless processes and an 8x8 block of contiguous samples in DCT-based processes. The left-most 8 samples of each of the top-most 8 rows in the component shall always be the top-left-most block. With this top-left-most block as the reference, the component is partitioned into contiguous data units to the right and to the bottom (as shown in figure A.4).

A.1.4 Orientation

Figure A.1 indicates the orientation of an image component by the terms top, bottom, left, and right. The order by which the data units of an image component are input to the compression encoding procedures is defined to be left-to-right and top-to-bottom within the component. (This ordering is precisely defined in A.2). It is the responsibility of applications to define which edges of a source image shall be considered top, bottom, left, and right.

A.2 Order of source image data encoding

The scan header (see B.2.3) specifies the order by which source image data units shall be encoded and placed within the compressed image data. For a given scan, if the scan header parameter Ns=1, then data from only one source component -- the component specified by parameter Cs_1 -- shall be present within the scan. This data is non-interleaved by definition. If Ns>1, then data from the Ns components Cs_1 through Cs_{N_s} shall be present within the scan. This data shall always be interleaved. The order of components in a scan shall be according to the order specified in the frame header.

The ordering of data units and the construction of minimum coded units (MCU) is defined as follows.

A.2.1 Minimum coded unit (MCU)

For non-interleaved data the MCU is one data unit. For interleaved data the MCU is the sequence of data units defined by the sampling factors of the components in the scan.

A.2.2 Non-interleaved order (Ns=1)

When Ns=1 (where Ns is the number of components in a scan), the order of data units within a scan shall be left-to-right and top-to-bottom, as shown in Figure A.2. This ordering applies whenever Ns=1, regardless of the values of H_1 and V_1.

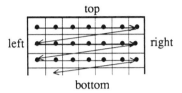

Figure A.2 - Non-interleaved data ordering

A.2.3 Interleaved order (Ns>1)

When Ns>1, each scan component Cs_i is partitioned into small rectangular arrays of H_k horizontal data units by V_k vertical data units. The subscripts k indicate that H_k and V_k are from the position in the frame header component-specification for which $C_k = Cs_i$. Within each H_k by V_k array, data units are ordered from left-to-right and top-to-bottom. The arrays in turn are ordered from left-to-right and top-to-bottom within each component.

As shown in the example of Figure A.3, Ns=4, and MCU_1 consists of data units taken first from the top-left-most region of Cs_1, followed by data units from the same array of Cs_2, then from Cs_3 and then from Cs_4. MCU_2 follows the same ordering for data taken from the next region to the right for the four components

A-3

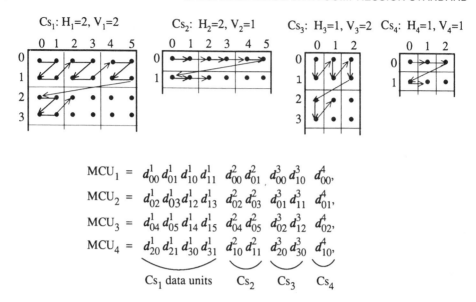

$$MCU_1 = d^1_{00}\, d^1_{01}\, d^1_{10}\, d^1_{11} \quad d^2_{00}\, d^2_{01} \quad d^3_{00}\, d^3_{10} \quad d^4_{00},$$

$$MCU_2 = d^1_{02}\, d^1_{03}\, d^1_{12}\, d^1_{13} \quad d^2_{02}\, d^2_{03} \quad d^3_{01}\, d^3_{11} \quad d^4_{01},$$

$$MCU_3 = d^1_{04}\, d^1_{05}\, d^1_{14}\, d^1_{15} \quad d^2_{04}\, d^2_{05} \quad d^3_{02}\, d^3_{12} \quad d^4_{02},$$

$$MCU_4 = d^1_{20}\, d^1_{21}\, d^1_{30}\, d^1_{31} \quad d^2_{10}\, d^2_{11} \quad d^3_{20}\, d^3_{30} \quad d^4_{10},$$

$$\underbrace{\qquad\qquad}_{Cs_1 \text{ data units}} \quad \underbrace{\quad}_{Cs_2} \quad \underbrace{\quad}_{Cs_3} \quad \underbrace{\quad}_{Cs_4}$$

Figure A.3 - Interleaved data ordering example

A.2.4 Completion of partial MCU

For DCT-based processes the data unit is a block. If x_i is not a multiple of 8, the encoding process shall extend the number of columns to complete the right-most sample blocks. If the component is to be interleaved, the encoding process shall also extend the number of samples by one or more additional blocks, if necessary, so that the number is an integer multiple of H_i. Similarly, if y_i is not a multiple of 8, the encoding process shall extend the number of rows to complete the bottom-most block-row. If the component is to be interleaved, the encoding process shall also extend the number of rows by one or more additional block-rows, if necessary, so that the number is an integer multiple of V_i.

   NOTE: It is recommended that any incomplete MCUs be completed by replication of the right-most column and the bottom row of each component.

For lossless processes the data unit is a sample. If the component is to be interleaved, the encoding process shall extend the number of samples, if necessary, so that the number is a multiple of H_i. Similarly, the encoding process shall extend the number of lines, if necessary, so that the number of lines is a multiple of V_i.

A.3 DCT compression

A.3.1 Level shift

Before a non-differential frame encoding process computes the FDCT for a block of source image samples, the samples shall be level shifted to a signed representation by subtracting 2^{P-1}, where P is the precision parameter specified in B.2.2. Thus, when P=8, the level shift is by 128; when P=12, the level shift is by 2048.

After a non-differential frame decoding process computes the IDCT and produces a block of reconstructed image samples, an inverse level shift shall restore the samples to the unsigned representation.

A.3.2 Orientation of samples for FDCT computation

Figure A.4 shows an image component which has been partitioned into 8x8 blocks for the FDCT computations. Figure A.4 also defines the orientation of the samples within a block by showing the indices used in the FDCT equation of A.3.3.

The definitions of block partitioning and sample orientation also apply to any DCT decoding process and the output reconstructed image. Any extended samples added by an encoding process shall be removed by the decoding process.

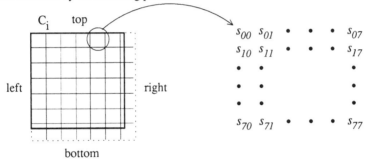

Figure A.4 - Partition and orientation of 8x8 sample blocks

A.3.3 FDCT and IDCT (informative)

The following equations specify the ideal functional definition of the FDCT and the IDCT.

> NOTE - These equations contain terms which cannot be represented with perfect accuracy by any real implementation. The accuracy requirements for the combined FDCT and quantization procedures are specified in Part 2 of this Specification. The accuracy requirements for the combined dequantization and IDCT procedures are also specified in Part 2 of this Specification.

FDCT: $S_{vu} = \frac{1}{4} C_u C_v \sum\limits_{x=0}^{7} \sum\limits_{y=0}^{7} s_{yx} \cos \dfrac{(2x+1)u\pi}{16} \cos \dfrac{(2y+1)v\pi}{16}$

IDCT: $s_{yx} = \frac{1}{4} \sum\limits_{u=0}^{7} \sum\limits_{v=0}^{7} C_u C_v S_{vu} \cos \dfrac{(2x+1)u\pi}{16} \cos \dfrac{(2y+1)v\pi}{16}$

where: $C_u, C_v = 1/\sqrt{2}$ for $u, v = 0$; $C_u, C_v = 1$ otherwise.

A.3.4 DCT coefficient quantization (informative) and dequantization (normative)

After the FDCT is computed for a block, each of the 64 resulting DCT coefficients is quantized by a uniform quantizer. The quantizer step size for each coefficient S_{vu} is the value of the corresponding element Q_{vu} from the quantization table specified by the frame parameter Tq_i (see B.2.2).

The uniform quantizer is defined by the following equation. Rounding is to the nearest integer:

$$Sq_{vu} = round\left(\frac{S_{vu}}{Q_{vu}}\right)$$

Sq_{vu} is the quantized DCT coefficient, normalized by the quantizer step size.

NOTE: This equation contains a term which may not be represented with perfect accuracy by any real implementation. The accuracy requirement for the combined FDCT and quantization procedures are specified in part 2 of this Specification.

At the decoder, this normalization is removed by the following equation, which defines dequantization:

$$R_{vu} = Sq_{vu} \times Q_{vu}$$

The relationship among samples, DCT coefficients, and quantization is illustrated in Figure A.5.

A-6

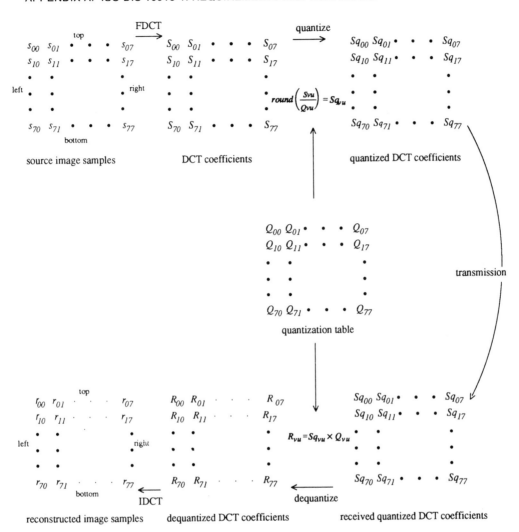

Figure A.5 – Relationship between 8x8–block samples and DCT coefficients

A.3.5 Differential DC encoding

After quantization, and in preparation for entropy encoding, the quantized DC coefficient Sq_{00} is treated separately from the 63 quantized AC coefficients. It is differentially encoded according to the following equation:

$$DIFF = DC_i - PRED$$

where PRED is the quantized DC value (Sq_{00}) of the preceding block of the same component in the interleave.

A.3.6 Zig-zag sequence

After quantization, and in preparation for entropy encoding, the quantized AC coefficients are converted to the zig-zag sequence. The quantized DC coefficient (coefficient zero in the array) is treated separately, as defined in A.3.5. The zig-zag sequence is specified as follows:

0	1	5	6	14	15	27	28
2	4	7	13	16	26	29	42
3	8	12	17	25	30	41	43
9	11	18	24	31	40	44	53
10	19	23	32	39	45	52	54
20	22	33	38	46	51	55	60
21	34	37	47	50	56	59	61
35	36	48	49	57	58	62	63

Figure A.6 – Zig–zag sequence of quantized DCT coefficients

A.4 Point transform

For various procedures data may be optionally divided by a power of 2 by a point transform prior to coding. There are three processes which require a point transform: lossless coding, lossless differential frame coding in the hierarchical mode, and successive approximation coding in the progressive DCT mode.

In the lossless mode of operation the point transform is applied to the input samples. In the difference coding of the hierarchical mode of operation the point transform is applied to the difference between the input component samples and the reference component samples. In both cases the point transform is an integer divide by 2^{Pt}, where Pt is the value of the point transform parameter (see B.2.3).

In successive approximation coding the point transform for the AC coefficients is an integer divide by 2^{Al}, where Al is the successive approximation bit position, low (see B.2.3). The point transform for the DC coefficients is an arithmetic-shift-right by Al bits. This is equivalent to dividing by 2^{Pt} before the level shift (see A.3.1).

The output of the decoder is rescaled by multiplying 2^{Pt}. An example of the point transform is given in K.10.

A.5 Arithmetic procedures in lossless and hierarchical modes of operation

In the lossless mode of operation predictions are calculated with full precision and without clamping of either overflow or underflow beyond the range of values allowed by the precision of the input. However, the division by two which is part of some of the prediction calculations shall be approximated by an arithmetic-shift-right by one bit.

The two's complement differences which are coded in either the lossless mode of operation or the differential frame coding in the hierarchical mode of operation are calculated modulo 65536, thereby restricting the precision of these differences to a maximum of 16 bits. The modulo values are calculated by performing the logical AND operation of the two's complement difference with X'FFFF'. For purposes of coding, the result is still interpreted as a 16 bit two's complement difference. Modulo 65536 arithmetic is also used in the decoder in calculating the output from the sum of the prediction and this two's complement difference.

Annex B (normative)

Compressed data formats

This Annex specifies three compressed data formats:

1) the interchange format, specified in B.2 and B.3;
2) the abbreviated format for compressed image data, specified in B.4;
3) the abbreviated format for table-specification data, specified in B.5.

B.1 describes the constituent parts of these formats. The format specifications in B.2 - B.5 give the conventions for symbols and figures used in the format specifications.

B.1 General aspects of the compressed data format specifications

Structurally, the compressed data formats consist of an ordered collection of parameters, markers, and entropy-coded data segments. Parameters and markers in turn are often organized into marker segments. Because all of these constituent parts are represented with byte-aligned codes, each compressed data format consists of an ordered sequence of 8-bit bytes. For each byte, a most significant bit (MSB) and a least significant bit (LSB) are defined.

B.1.1 Constituent parts

This section gives a general description of each of the constituent parts of the compressed data format.

B.1.1.1 Parameters

Parameters are integers, with values specific to the encoding process, source image characteristics, and other features selectable by the application. Parameters are assigned either 4-bit, 1-byte, or 2-byte codes. Except for certain optional groups of parameters, parameters encode critical information without which the decoding process cannot properly reconstruct the image.

The code assignment for a parameter shall be an unsigned integer of the specified length in bits with the particular value of the parameter.

For parameters which are 2 bytes (16 bits) in length, the most significant byte shall come first in the interchange format's ordered sequence of bytes. Parameters which are 4 bits in length always come in pairs, and the pair shall always be encoded in a single byte. The first 4-bit parameter of the pair shall occupy the most significant 4 bits of the byte. Within any 16, 8 or 4 bit parameter, the MSB shall come first and LSB shall come last.

B.1.1.2 Markers

Markers serve to identify the various structural parts of the compressed data formats. Most markers start marker segments containing a related group of parameters; some markers stand alone. All markers are assigned two-byte codes: an X'FF' byte followed by a byte which is not equal to 0 or X'FF' (see Table B.1). Any marker may optionally be preceded by any number of fill bytes, which are bytes assigned code X'FF'.

> NOTE - Because of this special code-assignment structure, markers make it possible for a decoder to parse the interchange format and locate its various parts without having to decode other segments of image data.

B-1

B.1.1.3 Marker assignments

All markers shall be assigned two-byte codes: a X'FF' byte followed by a second byte which is not equal to 0 or X'FF'. The second byte is specified in Table B.1 for each defined marker. An asterisk (*) indicates a marker which stands alone, that is, which is not the start of a marker segment.

Table B.1 - Marker code assignments

Code Assignment	Symbol	Description

Start Of Frame markers, non-differential Huffman coding:

X'FFC0'	SOF_0	Baseline DCT
X'FFC1'	SOF_1	Extended sequential DCT
X'FFC2'	SOF_2	Progressive DCT
X'FFC3'	SOF_3	Spatial (sequential) lossless

Start Of Frame markers, differential Huffman coding:

X'FFC5'	SOF_5	Differential sequential DCT
X'FFC6'	SOF_6	Differential progressive DCT
X'FFC7'	SOF_7	Differential spatial

Start Of Frame markers, non-differential arithmetic coding:

X'FFC8'	JPG	Reserved for JPEG extensions
X'FFC9'	SOF_9	Extended sequential DCT
X'FFCA'	SOF_{10}	Progressive DCT
X'FFCB'	SOF_{11}	Spatial (sequential) lossless

Start Of Frame markers, differential arithmetic coding:

X'FFCD'	SOF_{13}	Differential sequential DCT
X'FFCE'	SOF_{14}	Differential progressive DCT
X'FFCF'	SOF_{15}	Differential spatial

Huffman table specification:

X'FFC4'	DHT	Define Huffman table(s)

Arithmetic coding conditioning specification:

X'FFCC'	DAC	Define arithmetic coding conditioning(s)

Restart interval termination:

'FFD0'-X'FFD7'	RST_m *	Restart with modulo 8 count "m".

Other markers:

X'FFD8'	SOI *	Start of image
X'FFD9'	EOI *	End of image
X'FFDA'	SOS	Start of scan
X'FFDB'	DQT	Define quantization table(s)
X'FFDC'	DNL	Define number of lines
X'FFDD'	DRI	Define restart interval
X'FFDE'	DHP	Define hierarchical progression
X'FFDF'	EXP	Expand reference component(s)
X'FFE0'-X'FFEF'	APP_n	Reserved for application segments
X'FFF0'-X'FFFD'	JPG_n	Reserved for JPEG extensions
X'FFFE'	COM	Comment

Reserved markers:

X'FF01'	TEM *	For temporary private use in arithmetic coding
X'FF02'-X'FFBF'	RES	Reserved

B.1.1.4 Marker segments

A marker segment consists of a marker followed by a sequence of related parameters. The first parameter in a marker segment is the two-byte length parameter. This length parameter encodes the number of bytes in the marker segment, including the length parameter and excluding the two byte marker. The marker segments identified by the SOF and SOS marker codes are referred to as headers: the frame header and the scan header respectively.

B.1.1.5 Entropy-coded data segments

An entropy-coded data segment contains the output of an entropy-coding procedure. It consists of an integer number of bytes, whether the entropy-coding procedure used is Huffman or arithmetic.

NOTES

1. Making entropy-coded segments an integer number of bytes is achieved as follows: for Huffman coding, 1-bits are used, if necessary, to pad the end of the compressed data to complete the final byte of a segment. For arithmetic coding, byte alignment is achieved in the procedure which terminates the entropy-coded segment.

2. In order to ensure that a marker does not occur within an entropy-coded segment, any X'FF' byte generated by either a Huffman or arithmetic encoder is followed by a "stuffed" zero byte.

B.1.2 Syntax

In B.2 and B.3 the interchange format syntax is specified. For the purposes of this Specification, the syntax specification consists of:

- the required ordering of markers, parameters, and entropy-coded segments;

- identification of optional or conditional constituent parts;

- the name, symbol, and definition of each marker and parameter;

- the allowed values of each parameter;

- any restrictions on the above which are specific to the various coding processes.

The ordering of constituent parts and the identification of which are optional or conditional is specified by the syntax figures in B.2 and B.3. Names, symbols, definitions, allowed values, and restrictions are specified immediately below each syntax figure.

B.1.3 Conventions for syntax figures

The syntax figures in B.2 and B.3 are a part of the interchange format specification. The following conventions, illustrated in Figure B.1, apply to these figures:

- parameter/marker indicator: a thin-lined box encloses either a marker or a single parameter;
- segment indicator: a thick-lined box encloses either a marker segment, an entropy-coded data segment, or combinations of these;
- parameter length indicator: the width of a thin-lined box is proportional to the parameter length (4, 8, or 16 bits, shown as E, B, and D respectively in Figure B.1) of the marker or parameter it encloses; the width of thick-lined boxes is not meaningful;
- optional/conditional indicator: square brackets indicate that a marker or marker segment is only optionally or conditionally present in the compressed image data;
- ordering: in the interchange format a parameter or marker shown in a figure precedes all of those shown to its right, and follows all of those shown to its left;
- entropy-coded data indicator: angled brackets indicate that the quantity enclosed has been entropy encoded.

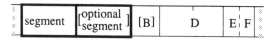

Figure B.1 – Syntax notation conventions

B.1.4 Conventions for symbols, code lengths, and values

Following each syntax figure in B.2 and B.3, the symbol, name, and definition for each marker and parameter shown in the figure are specified. For each parameter, the length and allowed values are also specified in tabular form.

The following conventions apply to symbols for markers and parameters:

- all marker symbols have three upper-case letters, and some also have a subscript. Examples: SOI, SOF_n;
- all parameter symbols have one upper-case letter; some also have one lower-case letter and some have subscripts. Examples: Y, Nf, H_i, Tq_i.

B.2 General sequential and progressive syntax

This section specifies the interchange format syntax which applies to all coding processes for sequential DCT-based, progressive DCT-based, and lossless modes of operation.

B.2.1 High-level syntax

Figure B.2 specifies the order of the high-level constituent parts of the interchange format for all non-hierarchical encoding processes specified in this Specification.

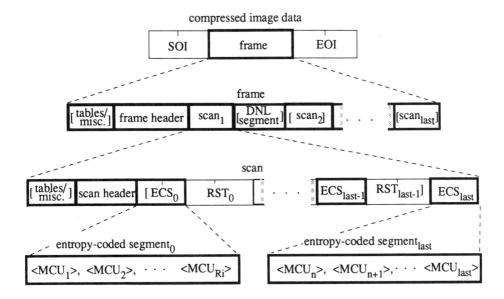

Figure B.2 - Syntax for sequential DCT-based, progressive DCT-based,
and lossless modes of operation

The three markers shown in Figure B.2 are defined as follows:

SOI: start of image marker: marks the start of a compressed image represented in the interchange format.

EOI: end of image marker: marks the end of a compressed image represented in the interchange format.

RST$_m$: restart marker: an optional marker which is placed between entropy-coded segment only if restart is enabled. There are 8 unique restart markers (m=0-7) which repeat in sequence from 0 to 7 to provide a modulo 8 restart interval count.

The top level of Figure B.2 specifies that the non-hierarchical interchange format shall begin with an SOI marker, shall contain one frame, and shall end with an EOI marker.

The second level of Figure B.2 specifies that a frame shall begin with a frame header and shall contain one or more scans. A frame header may be preceded by one or more table-specification or miscellaneous marker segments. If a DNL segment (see B.2.5) is present, it shall immediately follow the first scan.

For sequential DCT-based and lossless processes each scan shall contain from one to four image components. If two to four components are contained within a scan, they shall be interleaved within the scan. For progressive DCT-based processes each image component is only partially contained within any one scan. Only the first scan(s) for the components (which contain only DC coefficient data) may be interleaved.

B-5

The third level of Figure B.2 specifies that a scan shall begin with a scan header and shall contain one or more entropy-coded data segments. Each scan header may be preceded by one or more table-specification or miscellaneous marker segments. If restart is not enabled, there shall be only one entropy-coded segment (the one labeled "last"), and no restart markers shall be present. If restart is enabled, the number of entropy-coded segments is defined by the size of the image and the defined restart interval. In this case, a restart marker shall follow each entropy-coded segment except the last one.

The fourth level of Figure B.2 specifies that each entropy-coded segment is comprised of a sequence of entropy-coded MCUs. If restart is enabled and the restart interval is defined to be Ri, each entropy-coded segment except the last one shall contain Ri MCUs. The last one shall contain whatever number of MCUs completes the scan.

Figure B.2 specifies the locations where table-specification segments **may** be present. However, this Specification hereby specifies that the interchange format **shall** contain all table-specification data necessary for decoding the compressed image. Consequently, the required table-specification data **shall** be present at one or more of the allowed locations.

B.2.2 Frame header syntax

Figure B.3 specifies the frame header which shall be present at the start of a frame. This header specifies the source image characteristics (see A.1), the components in the frame, and the sampling factors for each component, and selects the quantization table to be used with each component.

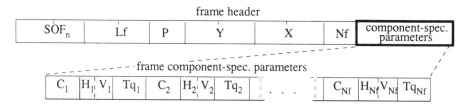

Figure B.3 - Frame header syntax

The markers and parameters shown in Figure B.3 are defined below. The size and allowed values of each parameter are given in Table B.2. In Table B.2 (and similar tables which follow), value choices are separated by commas (e.g., 8, 12) and inclusive bounds are separated by dashes (e.g., 0 - 3).

SOF_n: Start of frame marker; marks the beginning of the frame parameters. The subscript n identifies whether the encoding process is baseline sequential, extended sequential, progressive, or lossless, as well as which entropy encoding procedure is used.

SOF_0 - Baseline DCT sequential
SOF_1 - Extended DCT sequential, Huffman coding
SOF_2 - Progressive DCT, Huffman coding
SOF_3 - Lossless (sequential), Huffman coding
SOF_9 - Extended DCT sequential, arithmetic coding
SOF_{10} - Progressive DCT, arithmetic coding
SOF_{11} - Lossless (sequential), arithmetic coding

Lf: frame header length; specifies the length of the frame header shown in Figure B.3 (see B.1.1.4).

P: sample precision; specifies the precision in bits for the samples of the components in the frame.

Y: number of lines; specifies the number of lines in the source image. This shall be equal to the number of lines in the component with the maximum number of vertical samples (see A.1.1). Value 0 indicates that the number of lines shall be defined by the DNL marker and parameters at the end of the first scan (see B.2.5).

X: number of samples per line; specifies the number of samples per line in the source image. This shall be equal to the number of samples per line in the component with the maximum number of horizontal samples (see A.1.1).

Nf: number of image components in frame; specifies the number of source image components in the frame. The value of Nf shall be equal to the number of sets of frame component specification parameters (C_i, H_i, V_i, and Tq_i) present in the frame.

C_i: component identifier; assigns a unique label to the ith component in the sequence of frame component specification parameters. These values shall be used in the scan headers to identify the components in the scan. The value of C_i shall be different from the values of C_1 through C_{i-1}.

H_i: horizontal sampling factor; specifies the number of horizontal data units of component C_i in each MCU, when more than one component is encoded in a scan.

V_i: vertical sampling factor; specifies the number of vertical data units of component C_i in each MCU, when more than one component is encoded in a scan.

Tq_i: quantization table selector; selects one of four possible quantization tables to use for dequantization of DCT coefficients of component C_i. If the decoding process uses the dequantization procedure, this table shall have been specified by the time the decoder is ready to decode the scan(s) containing component C_i, and shall not be re-specified until all scans containing C_i have been completed.

Table B.2 - Frame header parameter sizes and values

parameter	size (bits)	Values			
		sequential DCT baseline	extended	progressive DCT	lossless
Lf	16	8 + 3 × Nf			
P	8	8	8, 12	8, 12	2-16
Y	16	0-65535			
X	16	1-65535			
Nf	8	1-255	1-255	1-4	1-255
C_i	8	0-255			
H_i	4	1-4			
V_i	4	1-4			
Tq_i	8	0-3	0-3	0-3	0

B.2.3 Scan header syntax

Figure B.4 specifies the scan header which shall be present at the start of a scan. This header specifies which component(s) are contained in the scan, the selection of the entropy coding tables used for each component in the scan, and (for the progressive DCT) which part of the DCT quantized coefficient data is contained in the scan. For lossless processes the scan parameters specify the predictor and the point transform.

scan header

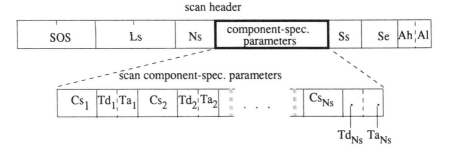

Figure B.4 - Scan header syntax

The marker and parameters shown in Figure B.4 are defined below. The size and allowed values of each parameter are given in Table B.3.

SOS: start of scan marker; marks the beginning of the scan parameters.

Ls: scan header length; specifies the length of the scan header shown in Figure B.4 (see B.1.1.4).

Ns: number of image components in scan; specifies the number of source image components in the scan. The value of Ns shall be equal to the number of sets of scan component specification parameters (Cs_j, Td_j, and Ta_j) present in the scan.

Cs_j: scan component selector; selects which of the Nf image components specified in the frame parameters shall be the jth component in the scan. Each Cs_j shall match one of the C_i values specified in the frame header, and the ordering in the scan header shall follow the ordering in the frame header. If Ns>1, the order of interleaved components in the MCU is Cs_1 first, Cs_2 second, etc.. If Ns>1, the following restriction shall be placed on the image components contained in the scan:

$$\sum_{j=1}^{Ns} H_j \times V_j \le 10,$$

where H_j and V_j are the horizontal and vertical sampling factors for scan component j. These sampling factors are specified in the frame header for component i, where i is the frame component specification index for which frame component identifier C_i matches scan component selector Cs_j.

As an example, consider an image having 512 pixels by 512 lines and 3 components sampled according to the following sampling factors:

Component 0	$H_0 = 4$,	$V_0 = 1$
Component 1	$H_1 = 1$,	$V_1 = 2$
Component 2	$H_2 = 2$,	$V_2 = 2$

Then the summation of $H_j \times V_j$ is $(4 \times 1) + (1 \times 2) + (2 \times 2) = 10$.

Td_j: DC entropy coding table selector; selects one of four possible DC entropy coding tables needed for decoding of the DC coefficients of component Cs_j. The DC entropy table selected shall have been specified (see B.2.4.2) by the time the decoder is ready to decode the current scan. This parameter selects the entropy coding tables for the lossless processes.

Ta_j: AC entropy coding table selector; selects one of four possible AC entropy coding tables needed for decoding of the AC coefficients of component Cs_j. The AC entropy table selected shall have been specified (see B.2.4.2) by the time the decoder is ready to decode the current scan. This parameter is zero for the lossless processes.

Ss: start of spectral or predictor selection; In the DCT modes of operation, this parameter specifies the first DCT coefficient in each block which shall be coded in the scan. This parameter shall be set to zero for the sequential DCT processes. In the lossless mode of operations this parameter is used to select the predictor.

Se: end of spectral selection; specifies the last DCT coefficient in each block which shall be coded in the scan. This parameter shall be set to 63 for the sequential DCT processes.

In the lossless mode of operations this parameter has no meaning. It shall be set to zero.

Ah: successive approximation bit position high; this parameter specifies the point transform used in the preceding scan (i.e. successive approximation bit position low in the preceding scan) for the band of coefficients specified by Ss and Se. This parameter shall be set to zero for the first scan of each band of coefficients. In the lossless mode of operations this parameter has no meaning. It shall be set to zero.

Al: successive approximation bit position low or point transform; in the DCT modes of operation this parameter specifies the point transform, i.e. bit position low, used before coding the band of coefficients specified by Ss and Se. This parameter shall be set to zero for the sequential DCT processes. In the lossless mode of operations, this parameter specifies the point transform, Pt.

Table B.3 - Scan header parameter sizes and values

parameter	size (bits)	Values			
		sequential DCT		progressive DCT	lossless
		baseline	extended		
Ls	16	$6 + 2 \times Ns$			
Ns	8	1-4			
Cs_j	8	0-255*			
Td_j	4	0-1	0-3	0-3	0-3
Ta_j	4	0-1	0-3	0-3	0
Ss	8	0	0	0-63	1-7**
Se	8	63	63	Ss-63***	0
Ah	4	0	0	0-13	0
Al	4	0	0	0-13	0-15

* Cs_j shall be a member of the set of C_i specified in the frame header

** 0 for lossless differential frames in the hierarchical mode

*** 0 if Ss equals zero

The entropy coding table selectors, Td_j and Ta_j, select either Huffman tables (in frames using Huffman coding) or arithmetic coding tables (in frames using arithmetic coding). In the latter case the entropy coding table selector selects both an arithmetic coding conditioning table and an associated statistics area.

B.2.4 Table-specification and miscellaneous marker segment syntax

Figure B.5 specifies that, at the places indicated in Figure B.2, any of the table-specification segments or miscellaneous marker segments specified in B.2.4.1 - B.2.4.6 may be present in any order and with no limit on the number of segments.

tables or miscellaneous marker segment

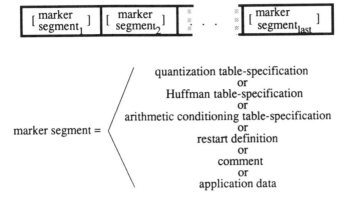

Figure B.5 – Table/miscellaneous marker segment syntax

If any table specifications occur in the compressed image data, they shall replace any previous specifications, and shall be used whenever the tables are required in the remaining scans in the frame. If a table specification for a given table occurs more than once in the compressed image data, each specification shall replace the previous specification. The quantization table specification shall not be altered between progressive DCT scans of a given component.

B.2.4.1 Quantization table-specification syntax

Figure B.6 specifies the marker segment which defines one or more quantization tables.

define quantization table segment

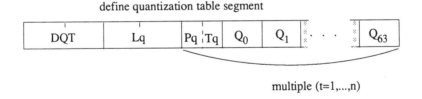

multiple (t=1,...,n)

Figure B.6 – Quantization table syntax

The markers and parameters shown in Figure B.6 are defined below. The size and allowed values of each parameter are given in Table B.4.

DQT: define quantization table marker; marks the beginning of quantization table-specification parameters.

Lq: quantization table definition length; specifies the length of all quantization table parameters shown in Figure B.6 (see B.1.1.4).

Pq: quantization table element precision; specifies the precision of the Q_k values. Value 0 indicates 8-bit Q_k values; value 1 indicates 16-bit Q_k values. Pq shall be zero for 8-bit sample precision P (see B.2.2).

Tq: quantization table identifier; specifies one of four possible destinations at the decoder into which the quantization table shall be installed.

Q_k: quantization table element; specifies the kth element out of 64 elements, where k is the index in the zig-zag ordering of the DCT coefficients. The quantization elements shall be specified in zig-zag scan order.

Table B.4 - Quantization table-specification parameter sizes and values

parameter	size (bits)	Values			lossless
		sequential DCT baseline	sequential DCT extended	progressive DCT	
Lq	16	$2 + \sum_{t=1}^{n} (65 + 64 \times Pq(t))$			undefined
Pq	4	0	0, 1	0, 1	undefined
Tq	4	0-3			undefined
Q_k	8, 16	1-255 , 1-65535			undefined

The value n in Table B.4 is the number of quantization tables specified in the DQT marker segment.

Once a quantization table has been defined, it may be used for subsequent images. If a table has never been defined, the results are unpredictable.

An 8 bit DCT-based process shall not use a 16 bit precision quantization table.

B.2.4.2 Huffman table-specification syntax

Figure B.7 specifies the marker segment which defines one or more Huffman table specifications.

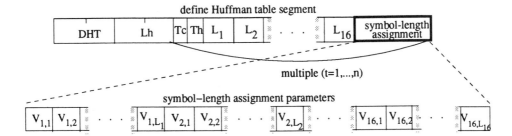

Figure B.7 – Huffman table syntax

The markers and parameters shown in Figure B.7 are defined below. The size and allowed values of each parameter are given in Table B.5.

DHT: define Huffman table marker; marks the beginning of Huffman table definition parameters.

Lh: Huffman table definition length; specifies the length of all Huffman table parameters shown in Figure B.7 (see B.1.1.4).

Tc: table class; 0=DC table or lossless table; 1=AC table.

Th: Huffman table identifier; specifies one of four possible destinations at the decoder into which the Huffman table shall be installed.

L_i: number of Huffman codes of length i; specifies the number of Huffman codes for each of the 16 possible lengths allowed by this Specification. L_i's are the elements of the list BITS.

$V_{i,j}$: value associated with each Huffman code; specifies, for each i, the value associated with each Huffman code of length i. The meaning of each value is determined by the Huffman coding model. The V_{ij}'s are the elements of the list HUFFVAL.

Table B.5 - Huffman table specification parameter sizes and values

parameter	size (bits)	Values			
		sequential DCT baseline	sequential DCT extended	progressive DCT	lossless
Lh	16	$2 + \sum\limits_{t=1}^{n}(17 + m_t)$			
Tc	4	0,1			0
Th	4	0,1	0 - 3		
L_i	8	0 - 255			
$V_{i,j}$	8	0 - 255			

The value n in Table B.5 is the number of Huffman tables specified in the DHT marker segment. The value m_t is the number of parameters which follow the 16 $L_i(t)$ parameters for Huffman table t, and is given by:

$$m_t = \sum_{i=1}^{16} L_i(t)$$

In general, m_t is different for each table.

Once a Huffman table has been defined, it may be used for subsequent images. If a table has never been defined, the results are unpredictable.

B.2.4.3 Arithmetic conditioning table-specification syntax

Figure B.8 specifies the marker segment which defines one or more arithmetic coding conditioning table specifications. These replace the default arithmetic coding conditioning tables established by the SOI marker.

define arithmetic conditioning segment

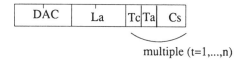

multiple (t=1,...,n)

Figure B.8 - Arithmetic conditioning table-specification syntax

The markers and parameters shown in Figure B.8 are defined below. The size and allowed values of each parameter are given in Table B.6.

DAC: define arithmetic coding conditioning marker; marks the beginning of the definition of arithmetic coding conditioning parameters.

La: arithmetic coding conditioning definition length; specifies the length of all arithmetic coding conditioning parameters shown in Figure B.8 (see B.1.1.4).

Tc: table class; 0 = DC table or lossless table, 1 = AC table.

Ta: arithmetic coding conditioning table identifier; specifies one of four possible destinations at the decoder into which the arithmetic coding conditioning table shall be installed.

Cs: conditioning table value; value in either the AC or the DC (and lossless) conditioning table. A single value of Cs shall follow each value of Ta. For AC conditioning tables Tc shall be one and Cs shall contain a value of Kx in the range $1 \leq Kx \leq 63$. For DC (and lossless) conditioning tables Tc shall be zero and Cs shall contain two 4-bit parameters, U and L. U and L shall be in the range $0 \leq L \leq U \leq 15$ and the value of Cs shall be $L + 16 \times U$.

Table B.6 - Arithmetic coding conditioning table-specification parameter sizes and values

parameter	size (bits)	Values			
		sequential DCT baseline	extended	progressive DCT	lossless
La	16	undefined	$2 + 2 \times n$		
Tc	4	undefined	0,1		0
Ta	4	undefined	0 - 3		
Cs	8	undefined	0-255 (Tc=0), 1-63 (Tc=1)		0-255

The value n in Table B.6 is the number of arithmetic coding conditioning tables specified in the DAC marker segment. The parameters L and U are the lower and upper conditioning bounds used in the arithmetic coding procedures defined for DC coefficient coding and lossless coding. The separate value range 1-63 listed for DCT coding is the Kx, conditioning used in AC coefficient coding.

B.2.4.4 Restart interval definition syntax

Figure B.9 specifies the marker segment which defines the restart interval.

define restart interval segment

DRI	Lr	Ri

Figure B.9 - Restart interval definition syntax

The markers and parameters shown in Figure B.9 are defined below. The size and allowed values of each parameter are given in Table B.7.

DRI: define restart interval marker; marks the beginning of the parameters which define the restart interval.

Lr: define restart interval segment length; specifies the length of the parameters in the DRI segment shown in Figure B.9 (see B.1.1.4).

Ri: restart interval; specifies the number of MCU in the restart interval.

Table B.7 - Define restart interval segment parameter sizes and values

parameter	size (bits)	Values sequential DCT baseline	extended	progressive DCT	lossless
Lr	16	4			
Ri	16	0-65535			$n \times MCUR$

In Table B.7 the value n is the number of rows of MCU in the restart interval. The value MCUR is the number of MCU required to make up one row of samples of each component in the scan. The SOI marker disables the restart intervals. A DRI marker segment with Ri nonzero shall be present to enable restart interval processing for the following scans. A DRI marker segment with Ri equal to zero shall disable restart intervals for the following scans.

B.2.4.5 Comment syntax

Figure B.10 specifies the marker segment structure for a comment segment.

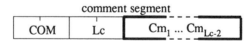

comment segment

COM	Lc	$Cm_1 \dots Cm_{Lc-2}$

Figure B.10 - Comment segment syntax

The markers and parameters shown in Figure B.10 are defined below. The size and allowed values of each parameter are given in Table B.8.

COM: comment marker; marks the beginning of a comment.

Lc: comment segment length; specifies the length of the comment segment shown in Figure B.10 (see B.1.1.4).

Cm$_i$: comment byte; the interpretation is left to the application.

Table B.8 - Comment segment parameter sizes and values

parameter	size (bits)	Values			
		baseline	sequential DCT extended	progressive DCT	lossless
Lc	16	2-65535			
Cm$_i$	8	0-255			

B.2.4.6 Application data syntax

Figure B.11 specifies the marker segment structure for an application data segment.

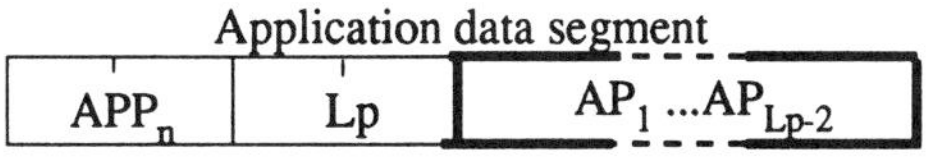

Figure B.11 - Application data syntax

The markers and parameters shown in Figure B.11 are defined below. The size and allowed values of each parameter are given in Table B.9.

APPn: application data marker; marks the beginning of an application data segment.

Lp: application data segment length; specifies the length of the application data segment shown in Figure B.11(see B.1.1.4).

Ap$_i$: application data byte; any 8-bit value.

Table B.9 - Application data segment parameter sizes and values

parameter	size (bits)	Values			
		baseline	sequential DCT extended	progressive DCT	lossless
Lp	16	2-65535			
Ap_i	8	0-255			

The APPn (Application) segments are reserved for application use. Since these segments may be defined differently for different applications, they should be removed when the data are exchanged between application environments.

B.2.5 Define number of lines syntax

Figure B.12 specifies the marker segment for defining the number of lines. The DNL (Define Number of Lines) segment provides a mechanism for defining or redefining the number of lines in the frame (the Y parameter in the frame header) at the end of the first scan. The value specified shall be consistent with the number of MCU-rows encoded in the first scan. This segment, if used, shall only occur at the end of the first scan, and only after coding of an integer number of MCU rows.

Define number of lines segment

DNL	Ld	NL

Figure B.12 - Define number of lines syntax

The markers and parameters shown in Figure B.12 are defined below. The size and allowed values of each parameter are given in Table B.10.

DNL: define number of lines marker; marks the beginning of the define number of lines segment.

Ld: define number of lines segment length; specifies the length of the define number of lines segment shown in Figure B.12 (see B.1.1.4).

NL: number of lines; specifies the number of lines in the frame (see definition of Y in B.2.2).

Table B.10 - Define number of lines segment parameter sizes and values

parameter	size (bits)	Values			
		sequential DCT baseline	extended	progressive DCT	lossless
Ld	16	4			
NL	16	1-65535*			

* The value specified shall be consistent with the number of lines coded at the point where the DNL segment terminates the compressed data segment.

B.3 Hierarchical syntax

B.3.1 High level hierarchical mode syntax

Figure B.13 specifies the order of the high level constituent parts of the interchange format for hierarchical encoding processes.

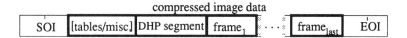

Figure B.13 - Syntax for the hierarchical mode of operation

The hierarchical mode normally uses a sequence of frames including differential frames, and needs two additional markers, DHP and EXP. Frame structure is identical to the frame in non-hierarchical mode.

The non-differential frames in the hierarchical sequence shall use one of the coding processes specified for SOF markers SOF_0, SOF_1, SOF_2, SOF_3, SOF_9, SOF_{10}, and SOF_{11}. The differential frames shall use one of the processes specified for SOF_5, SOF_6, SOF_7, SOF_{13}, SOF_{14}, and SOF_{15}.

If the non-differential frames use DCT-based processes, all differential frames except the final frame shall use DCT-based processes. The final differential frames for a component may use a spatial process.

If the non-differential frames use spatial processes, all differential frames shall use a spatial process.

B.3.2 DHP segment syntax

The DHP segment defines the image components, size and sampling factors for the completed hierarchical sequence of frames. The DHP segment shall precede the first frame; a single DHP

B-19

segment shall occur in the compressed image data.

The DHP segment structure is identical to the frame header syntax, except that the DHP marker is used instead of the SOF$_n$ marker. The figures and description of B.2.2 then apply, except that the quantization table selector parameter shall be set to zero in the DHP segment.

B.3.3 EXP segment syntax

Figure B.14 specifies the marker segment structure for the EXP segment. The EXP segment shall be present if (and only if) expansion of the reference components is required either horizontally or vertically. The EXP segment parameters apply only to the next frame (which shall be a differential frame) in the image. If required, the EXP segment shall be one of the table-specification segments or miscellaneous marker segments preceding the frame header; the EXP segment shall not be one of the table specification segments or miscellaneous segments preceding a scan header.

expand segment

| EXP | Le | Eh ¦ Ev |

Figure B.14 – Syntax of the expand segment

The parameters shown in Figure B.14 are defined below. The size and allowed values of each parameter are given in Table B.11.

EXP: expand reference components marker: marks the beginning of the expand reference components segment.

Le: EXP segment length; specifies the length of the EXP segment (see B.1.1.4).

Eh: expand horizontally; if one, the reference components shall be expanded horizontally by a factor of two. If horizontal expansion is not required, the value shall be zero.

Ev: expand vertically; if one, the reference components shall be expanded vertically by a factor of two. If vertical expansion is not required, the value shall be zero.

Table B.11 - Expand segment parameter sizes and values

parameter	size (bits)	Values			
		sequential DCT baseline	extended	progressive DCT	lossless
Le	16	3			
Eh	4	0,1			
Ev	4	0,1			

Both Eh and Ev shall be one if expansion is required both horizontally and vertically.

B.4 Abbreviated format for compressed image data

Figure B.2 shows the high-level constituent parts of the interchange format. This format includes all table specifications required for decoding. If an application environment provides methods for table specification other than by means of the compressed image data, some or all of the table specifications may be omitted. Compressed image data which is missing any table specification data required for decoding has the abbreviated format.

B.5 Abbreviated format for table-specification data

Figure B.2 shows the high-level constituent parts of the interchange format. If no frames are present in the compressed image data, the only purpose of the compressed image data is to convey table specifications or miscellaneous marker segments defined in B.2.4.1, B.2.4.2, B.2.4.5 and B.2.4.6. In this case the compressed image data has the abbreviated format for table specification data.

compressed image data

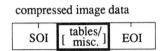

Figure B.15 − Abbreviated format for table-specification data syntax

B.6 Summary

The order of the constituent parts of interchange format and all marker segment structures is summarized in Figures B.16 and B.17. Note that in Figure B.16 double-lined boxes enclose marker segments. In Figures B.16 and B.17 thick-lined boxes encloses only markers.

The EXP segment can be mixed with the other table/misc. marker segments preceding the frame header but not with the table/misc. marker segments preceding the DHP segment or the scan header.

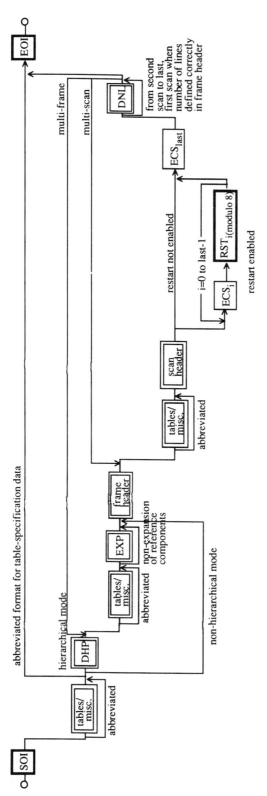

Figure B.16 - Flow of compressed data syntax

Figure B.17 - Flow of marker segment

Annex C (normative)

Huffman table specification

A Huffman coding procedure may be used for entropy coding in any of the coding processes. Coding models for Huffman encoding are defined in Annexes F, G, and H. In this Annex, the Huffman table specification is defined.

Huffman tables are specified in the interchange format in terms of a 16-byte list (BITS) giving the number of codes for each code length from 1 to 16. This is followed by a list of the 8-bit symbol values (HUFFVAL), each of which is assigned a Huffman code. The symbol values are placed in the list in order of increasing code length. Code lengths greater than 16 bits are not allowed. In addition, the codes shall be generated such that the all-1-bits code word of any length is reserved as a prefix for longer code words.

> NOTE: The order of the symbol values within HUFFVAL is determined only by code length. Within a given code length the ordering of the symbol values is arbitrary.

This annex specifies the procedure by which the Huffman table (of Huffman code words and their corresponding 8-bit symbol values) are derived from the two lists (BITS and HUFFVAL) in the interchange format. However, the way in which these lists are generated is not specified. The lists should be generated in a manner which is consistent with the rules for Huffman coding, and it shall observe the constraints discussed in the previous paragraph. Annex K contains an example of a procedure for generating lists of Huffman code lengths and values which are in accord with these rules.

> NOTE - There is **no requirement** in this Specification that any encoder or decoder shall implement the procedures in precisely the manner specified by the flow charts in this Annex. It is necessary only that an encoder or decoder implement the **function** specified in this Annex. The sole criterion for an encoder or decoder to be considered in compliance with this Specification is that it satisfy the requirements given in clause 6 (for encoders) or clause 7 (for decoders), as determined by the compliance tests specified in Part 2.

C.1 Marker segments for Huffman table specification

The DHT marker identifies the start of Huffman table definitions within the compressed image data. B.2.4.2 specifies the syntax for Huffman table specification.

C.2 Conversion of Huffman tables specified in interchange format to tables of codes and code lengths

Given a list BITS (1..16) containing the number of codes of each size, and a list HUFFVAL containing the symbol values to be associated with those codes as described above, two tables are generated. The HUFFSIZE table contains a list of code lengths; the HUFFCODE table contains the Huffman codes corresponding to those lengths.

C- 1

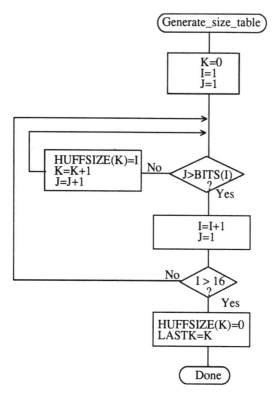

Figure C.1 - Generation of table of Huffman code sizes

Note that the variable LASTK is set to the index of the last entry in the table.

A Huffman code table, HUFFCODE, containing a code for each size in HUFFSIZE is generated by the procedure in figure C.2. The notation "SLL CODE 1" in this figure indicates a
shift-left-logical of CODE by one bit position.

C- 2

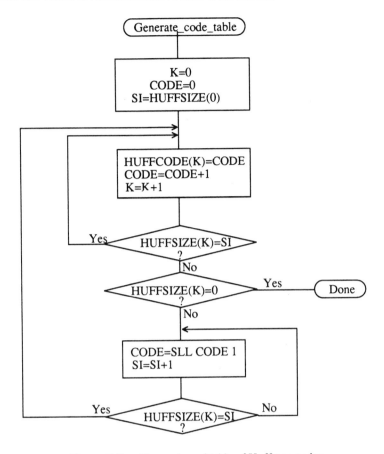

Figure C.2 - Generation of table of Huffman codes

Two tables, HUFFCODE, and HUFFSIZE, have now been initialized. The entries in the tables are ordered according to increasing Huffman code numeric value and length.

The encoding procedure code tables, EHUFCO and EHUFSI, are created by reordering the codes specified by HUFFCODE and HUFFSIZE according to the symbol values assigned to each code in HUFFVAL.

Figure C.3 illustrates this ordering procedure.

C- 3

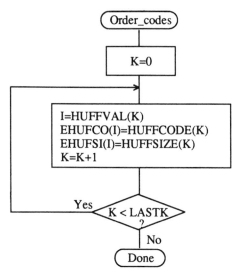

Figure C.3 - Ordering procedure for encoding procedure code tables

C.3 Bit ordering within bytes

The root of a Huffman code is place toward the MSB (most-significant-bit) of the byte, and successive bits are placed in the direction MSB to LSB (least-significant-bit) of the byte. Remaining bits, if any, go into the next byte following the same rules.

Integers associated with Huffman codes are appended with the MSB adjacent to the LSB of the preceding Huffman code.

Annex D (normative)

Arithmetic coding

An adaptive binary arithmetic coding procedure may be used for entropy coding in any of the coding processes except the baseline sequential process. Coding models for adaptive binary arithmetic coding are defined in Annexes F, G, and H. In this Annex the arithmetic encoding and decoding procedures used in those models are defined.

In K.4 a simple test example is given which should be helpful in determining if a given implementation is correct.

> NOTE - There is **no requirement** in this Specification that any encoder or decoder shall implement the procedures in precisely the manner specified by the flow charts in this Annex. It is necessary only that an encoder or decoder implement the **function** specified in this Annex. The sole criterion for an encoder or decoder to be considered in compliance with this Specification is that it satisfy the requirements given in clause 6 (for encoders) or clause 7 (for decoders), as determined by the compliance tests specified in Part 2.

D.1 Arithmetic encoding procedures

Four arithmetic encoding procedures are required in a system with arithmetic coding. They are:

Table D.1. Procedures for binary arithmetic encoding

Procedure	Purpose
Code_0(S)	Code a "0" binary decision with context-index S
Code_1(S)	Code a "1" binary decision with context-index S
Initenc	Initialize the encoder
Flush	Terminate entropy-coded segment

The "Code_0(S)" and "Code_1(S)" procedures code the 0-decision and 1-decision respectively; S is a context-index which identifies a particular conditional probability estimate used in coding the binary decision. The "Initenc" procedure initializes the arithmetic coding entropy encoder. The "Flush" procedure terminates the entropy-coded segment in preparation for the marker which follows.

D.1.1 Binary arithmetic encoding principles

The arithmetic coder encodes a series of binary symbols, zeros and ones, each symbol representing one possible result of a binary decision.

Each "binary decision" provides a choice between two alternatives. The binary decision might be between positive and negative signs, a magnitude being zero or nonzero, or a particular bit in a sequence of binary digits being zero or one.

The output bit stream (entropy-coded data segment) represents a binary fraction which increases in precision as bytes are appended by the encoding process.

D-1

D.1.1.1 Recursive interval subdivision

Recursive probability interval subdivision is the basis for the binary arithmetic encoding proce-dures. With each binary decision the current probability interval is subdivided into two sub-intervals, and the bit stream is modified (if necessary) so that it points to the base (the lower bound) of the probability sub-interval assigned to the symbol which occurred.

In the partitioning of the current probability interval into two sub-intervals, the subinterval for the less probable symbol (LPS) and the sub-interval for the more probable symbol (MPS) are ordered such that usually the MPS sub-interval is closer to zero. Therefore, when the LPS is coded, the MPS sub-interval size is added to the bit stream. This coding convention requires that symbols be recognized as either MPS or LPS rather than 0 or 1. Consequently, the size of the LPS sub-interval and the sense of the MPS for each decision must be known in order to encode that decision.

The subdivision of the current probability interval would ideally require a multiplication of the interval by the probability estimate for the LPS. Because this subdivision is done approxi-mately, it is possible for the LPS sub-interval to be larger than the MPS sub-interval. When that happens a "conditional exchange" interchanges the assignment of the sub-intervals such that the MPS is given the larger sub-interval.

Since the encoding procedure involves addition of binary fractions rather than concatenation of integer code words, the more probable binary decisions can sometimes be coded at a cost of much less than one bit per decision.

D.1.1.2 Conditioning of probability estimates

An adaptive binary arithmetic coder requires a statistical model - a model for selecting condi-tional probability estimates to be used in the coding of each binary decision. When a given bi-nary decision probability estimate is dependent on a particular feature or features (the context) already coded, it is "conditioned" on that feature. The conditioning of probability estimates on previously coded decisions must be identical in encoder and decoder, and therefore can use only information known to both.

Each conditional probability estimate required by the statistical model is kept in a separate stor-age location or "bin" identified by a unique context-index S. The arithmetic coder is adaptive, which means that the probability estimates at each context-index are developed and maintained by the arithmetic coding system on the basis of prior coding decisions for that context-index.

D.1.2 Encoding conventions and approximations

The encoding procedures use fixed precision integer arithmetic and an integer representation of fractional values in which X'8000' can be regarded as the decimal value 0.75. The probability interval, A, is kept in the integer range $X'8000' <= A < X'10000'$ by doubling it whenever its integer value falls below X'8000'. This is equivalent to keeping A in the decimal range 0.75 $<=A < 1.5$. This doubling procedure is called renormalization.

The code register, C, contains the trailing bits of the bit stream. C is also doubled each time A is doubled. Periodically - to keep C from overflowing - a byte of data is removed from the high order bits of the C-register and placed in the entropy-coded segment.

Carry-over into the entropy-coded segment is limited by delaying X'FF' output bytes until the carry-over is resolved. Zero bytes are stuffed after each X'FF' byte in the entropy-coded seg-ment in order to avoid the accidental generation of markers in the entropy-coded segment.

Keeping A in the range 0.75 <= A < 1.5 allows a simple arithmetic approximation to be used in the probability interval subdivision. Normally, if the current estimate of the LPS probability for context-index S is Qe(S), precise calculation of the sub-intervals would require:

Qe(S)×A	= probability sub-interval for the LPS
A-(Qe(S)×A)	= probability sub-interval for the MPS

Because the decimal value of A is of order unity, these can be approximated by

Qe(S)	= probability sub-interval for the LPS
A-Qe(S)	= probability sub-interval for the MPS

Whenever the LPS is coded, the value of A-Qe(S) is added to the code register and the probability interval is reduced to Qe(S). Whenever the MPS is coded, the code register is left unchanged and the interval is reduced to A-Qe(S). The precision range required for A is then restored, if necessary, by renormalization of both A and C.

With the procedure described above, the approximations in the probability interval subdivision process can sometimes make the LPS sub-interval larger than the MPS sub-interval. If, for example, the value of Qe(S) is 0.5 and A is at the minimum allowed value of 0.75, the approximate scaling gives one third of the probability interval to the MPS and two thirds to the LPS. To avoid this size inversion, conditional exchange is used. The probability interval is subdivided using the simple approximation, but the MPS and LPS sub-interval assignments are exchanged whenever the LPS sub-interval is larger than the MPS sub-interval. This MPS/LPS conditional exchange can only occur when a renormalization will be needed.

Each binary decision uses a context. A context is the set of prior coding decisions which determine the context-index, S, identifying the probability estimate used in coding the decision.

Whenever a renormalization occurs, a probability estimation procedure is invoked which determines a new probability estimate for the context currently being coded. No explicit symbol counts are needed for the estimation. The relative probabilities of renormalization after coding of LPS and MPS provide, by means of a table-based probability estimation state machine, a direct estimate of the probabilities.

D.1.3 Encoder code register conventions

The flow charts in this annex assume the following register structures for the encoder:

	MSB			LSB
C-register	0000cbbb,	bbbbbsss,	xxxxxxxx,	xxxxxxxx
A-register	00000000,	00000000,	aaaaaaaa,	aaaaaaaa

The "a" bits are the fractional bits in the A-register (the current probability interval value) and the "x" bits are the fractional bits in the code register. The "s" bits are optional spacer bits which provide useful constraints on carry-over, and the "b" bits indicate the bit positions from which the completed bytes of data are removed from the C-register. The "c" bit is a carry bit. Except at the time of initialization, bit 15 of the A-register is always set and bit 16 is always clear (the LSB is bit 0).

These register conventions illustrate one possible implementation. However, any register conventions which allow resolution of carry-over in the encoder and which produce the same entropy-coded segment may be used. The handling of carry-over and the byte stuffing following X'FF' will be described in a later part of this section.

D.1.4 Code_1(S) and Code_0(S) procedures

When a given binary decision is coded, one of two possibilities occurs - either a 1-decision or a 0-decision is coded. Code_1(S) and Code_0(S) are shown in Figures D.1 and D.2. The Code_1(S) and Code_0(S) procedures use probability estimates with a context-index S. The context-index S is determined by the statistical model and is, in general, a function of the previous coding decisions; each value of S identifies a particular conditional probability estimate which is used in encoding the binary decision.

The context-index S selects a storage location which contains Index(S), an index to the tables which make up the probability estimation state machine. When coding a binary decision, the symbol being coded is either the more probable symbol or the less probable symbol. Therefore, additional information is stored at each context-index identifying the sense of the more probable symbol, MPS(S).

For simplicity, the flow charts in this section assume that the context storage for each context-index S has an additional storage field for Qe(S) containing the value of Qe(Index(S)). If only the value of Index(S) and MPS(S) are stored, all references to Qe(S) should be replaced by Qe(Index(S)).

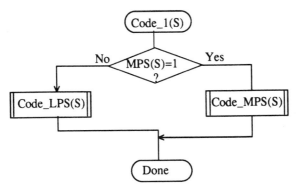

Figure D.1 - Code_1(S) procedure

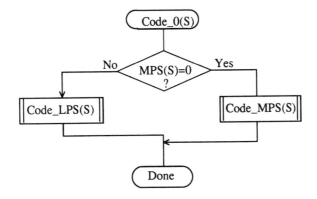

Figure D.2 - Code_0(S) procedure

The Code_LPS(S) procedure normally consists of the addition of the MPS sub-interval A-Qe(S) to the bit stream and a scaling of the interval to the sub-interval, Qe(S). It is always followed by the procedures for obtaining a new LPS probability estimate (Estimate_Qe(S)_after_LPS), and renormalization (Renorm_e).

However, in the event that the LPS sub-interval is larger than the MPS sub-interval, the conditional MPS/LPS exchange occurs and the MPS sub-interval is coded.

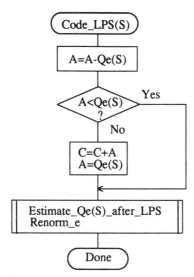

Figure D.3 - Code_LPS(S) procedure with conditional MPS/LPS exchange

The Code_MPS(S) procedure normally reduces the size of the probability interval to the MPS sub-interval. However, if the LPS sub-interval is larger than the MPS sub-interval, the conditional exchange occurs and the LPS sub-interval is coded instead. Note that conditional exchange cannot occur unless the procedures for obtaining a new LPS probability estimate (Estimate_Qe(S)_after_MPS) and renormalization (Renorm_e) are required after the coding of the symbol

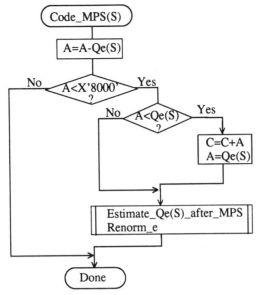

Figure D.4 - Code_MPS(S) procedure with conditional MPS/LPS exchange

D.1.5 Probability estimation in the encoder

D.1.5.1 Probability estimation state machine

The probability estimation state machine consists of a number of sequences of probability estimates. These sequences are interlinked in a manner which provides probability estimates based on approximate symbol counts derived from the arithmetic coder renormalization. Some of these sequences are used during the initial "learning" stages of probability estimation; the rest are used for "steady state" estimation.

Each entry in the probability estimation state machine is assigned an index, and each index has associated with it a Qe value and two Next_Index values. The Next_Index_MPS gives the index to the new probability estimate after an MPS renormalization; the Next_Index_LPS gives the index to the new probability estimate after an LPS renormalization. Note that both the index to the estimation state machine and the sense of the MPS are kept for each context-index S. The sense of the MPS is changed whenever the entry in the Switch_MPS is one.

The probability estimation state machine is given in Table D.2. Initialization of the arithmetic coder is always with an MPS sense of zero and a Qe index of zero in Table D.2.

The Qe values listed in Table D.2 are expressed as hexadecimal integers. To convert the 15 bit integer representation of Qe to a decimal probability, divide the Qe values by $(4/3) \times (X'8000')$.

Table D.2 Qe values and probability estimation state machine

Index	Qe _Value	Next_ _LPS	Index _MPS	Switch _MPS	Index	Qe _Value	Next_Index _LPS	_MPS	Switch _MPS
0	X'5A1D'	1	1	1	57	X'01A4'	55	58	0
1	X'2586'	14	2	0	58	X'0160'	56	59	0
2	X'1114'	16	3	0	59	X'0125'	57	60	0
3	X'080B'	18	4	0	60	X'00F6'	58	61	0
4	X'03D8'	20	5	0	61	X'00CB'	59	62	0
5	X'01DA'	23	6	0	62	X'00AB'	61	63	0
6	X'00E5'	25	7	0	63	X'008F'	61	32	0
7	X'006F'	28	8	0	64	X'5B12'	65	65	1
8	X'0036'	30	9	0	65	X'4D04'	80	66	0
9	X'001A'	33	10	0	66	X'412C'	81	67	0
10	X'000D'	35	11	0	67	X'37D8'	82	68	0
11	X'0006'	9	12	0	68	X'2FE8'	83	69	0
12	X'0003'	10	13	0	69	X'293C'	84	70	0
13	X'0001'	12	13	0	70	X'2379'	86	71	0
14	X'5A7F'	15	15	1	71	X'1EDF'	87	72	0
15	X'3F25'	36	16	0	72	X'1AA9'	87	73	0
16	X'2CF2'	38	17	0	73	X'174E'	72	74	0
17	X'207C'	39	18	0	74	X'1424'	72	75	0
18	X'17B9'	40	19	0	75	X'119C'	74	76	0
19	X'1182'	42	20	0	76	X'0F6B'	74	77	0
20	X'0CEF'	43	21	0	77	X'0D51'	75	78	0
21	X'09A1'	45	22	0	78	X'0BB6'	77	79	0
22	X'072F'	46	23	0	79	X'0A40'	77	48	0
23	X'055C'	48	24	0	80	X'5832'	80	81	1
24	X'0406'	49	25	0	81	X'4D1C'	88	82	0
25	X'0303'	51	26	0	82	X'438E'	89	83	0
26	X'0240'	52	27	0	83	X'3BDD'	90	84	0
27	X'01B1'	54	28	0	84	X'34EE'	91	85	0
28	X'0144'	56	29	0	85	X'2EAE'	92	86	0
29	X'00F5'	57	30	0	86	X'299A'	93	87	0
30	X'00B7'	59	31	0	87	X'2516'	86	71	0
31	X'008A'	60	32	0	88	X'5570'	88	89	1
32	X'0068'	62	33	0	89	X'4CA9'	95	90	0
33	X'004E'	63	34	0	90	X'44D9'	96	91	0
34	X'003B'	32	35	0	91	X'3E22'	97	92	0
35	X'002C'	33	9	0	92	X'3824'	99	93	0
36	X'5AE1'	37	37	1	93	X'32B4'	99	94	0
37	X'484C'	64	38	0	94	X'2E17'	93	86	0
38	X'3A0D'	65	39	0	95	X'56A8'	95	96	1
39	X'2EF1'	67	40	0	96	X'4F46'	101	97	0

D-8

40	X'261F'	68	41	0	97	X'47E5'	102	98	0
41	X'1F33'	69	42	0	98	X'41CF'	103	99	0
42	X'19A8'	70	43	0	99	X'3C3D'	104	100	0
43	X'1518'	72	44	0	100	X'375E'	99	93	0
44	X'1177'	73	45	0	101	X'5231'	105	102	0
45	X'0E74'	74	46	0	102	X'4C0F'	106	103	0
46	X'0BFB'	75	47	0	103	X'4639'	107	104	0
47	X'09F8'	77	48	0	104	X'415E'	103	99	0
48	X'0861'	78	49	0	105	X'5627'	105	106	1
49	X'0706'	79	50	0	106	X'50E7'	108	107	0
50	X'05CD'	48	51	0	107	X'4B85'	109	103	0
51	X'04DE'	50	52	0	108	X'5597'	110	109	0
52	X'040F'	50	53	0	109	X'504F'	111	107	0
53	X'0363'	51	54	0	110	X'5A10'	110	111	1
54	X'02D4'	52	55	0	111	X'5522'	112	109	0
55	X'025C'	53	56	0	112	X'59EB'	112	111	1
56	X'01F8'	54	57	0					

D.1.5.2 Renormalization driven estimation

The change in state in Table D.2 occurs only when the arithmetic coder interval register is renormalized. This must always be done after coding an LPS, and whenever the probability interval register is less than X'8000' (0.75 in decimal notation) after coding an MPS.

When the LPS renormalization is required, Next_Index_LPS gives the new index for the LPS probability estimate. When the MPS renormalization is required, Next_Index_MPS gives the new index for the LPS probability estimate. If Switch_MPS is 1 for the old index, the MPS symbol sense must be inverted after an LPS.

D.1.5.3 Estimation following renormalization after MPS

The procedure for estimating the probability on the MPS renormalization path is given in Figure D.5. Index(S) is part of the information stored for context-index S. The new value of Index(S) is obtained from Table D.2 from the column labeled Next_Index_MPS, as that is the next index after an MPS renormalization. This next index is stored as the new value of Index(S) in the context storage at context-index S, and the value of Qe at this new Index(S) becomes the new Qe(S). MPS(S) does not change.

D-9

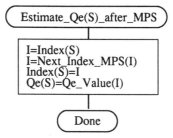

Figure D.5 - Probability estimation on MPS renormalization path

D.1.5.4 Estimation following renormalization after LPS

The procedure for estimating the probability on the LPS renormalization path is shown in Figure D.6. The procedure is similar to that of Figure D.5 except that when Switch_MPS(I) is 1, the sense of MPS(S) must be inverted.

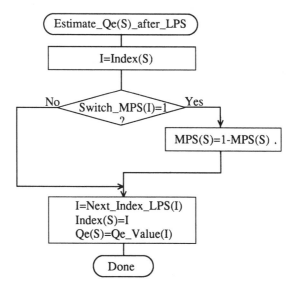

Figure D.6 - Probability estimation on LPS renormalization path

D.1.6 Renormalization in the encoder

The Renorm_e procedure for the encoder renormalization is shown in Figure D.7. Both the probability interval register A and the code register C are shifted, one bit at a time. The number of shifts is counted in the counter CT; when CT is zero, a byte of compressed data is removed from C by the procedure Byte_out and CT is reset to 8. Renormalization continues until A is no longer less than X'8000'.

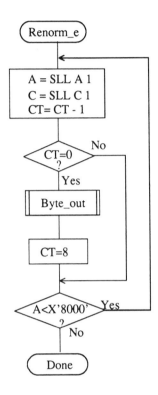

Figure D.7 - Encoder renormalization procedure

The Byte_out procedure used in Renorm_e is shown in Figure D.8. This procedure uses byte-stuffing procedures which prevent accidental generation of markers by the arithmetic encoding procedures. It also includes an example of a procedure for resolving carry-over. For simplicity of exposition, the buffer holding the entropy-coded segment is assumed to be large enough to contain the entire segment.

In Figure D.8 BP is the entropy-coded segment pointer and B is the compressed data byte pointed to by BP. T in Byte_out is a temporary variable which is used to hold the output byte and carry bit. SC is the stack counter which is used to count X'FF' output bytes until any carry-

D-11

over through the X'FF' sequence has been resolved. The value of SC rarely exceeds 3. However, since the upper limit for the value of SC is bounded only by the total entropy-coded segment size, a precision of 32 bits is recommended for SC.

Since large values of SC represent a latent output of compressed data, the following procedure may be needed in high speed synchronous encoding systems for handling the burst of output data which occurs when the carry is resolved:

When the stack count reaches an upper bound determined by output channel capacity, the stack is emptied and the stacked X'FF' bytes (and stuffed zero bytes) are added to the compressed data before the carry-over is resolved. If a carry-over then occurs, the carry is added to the final stuffed zero, thereby converting the final X'FF00' sequence to the X'FF01' temporary private marker. The entropy-coded segment must then be post-processed to resolve the carry-over and remove the temporary marker code. For any reasonable bound on SC this post processing is very unlikely.

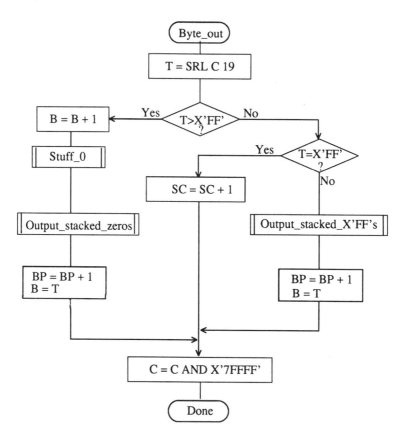

Figure D.8 - Byte_out procedure for encoder

Referring to Figure D.8 the shift of the code register by 19 bits aligns the output bits with the low order bits of T. The first test then determines if a carry-over has occurred. If so, the carry must be added to the previous output byte before advancing the segment pointer BP. The Stuff_0 procedure stuffs a zero byte whenever the addition of the carry to the data already in the entropy-coded segments creates a X'FF' byte. Any stacked output bytes - converted to zeros by the carry-over - are then placed in the entropy encoder segment. Note that when the output byte is later transferred from T to the entropy-coded segment (to byte B), the carry bit is ignored if it is set.

If a carry has not occurred, the output byte is tested to see if it is X'FF'. If so, the stack count SC is incremented, as the output must be delayed until the carry-over is resolved. If not, the carry-over has been resolved, and any stacked X'FF' bytes must then be placed in the entropy-coded segment. Note that a zero byte is stuffed following each X'FF'.

The procedures used by Byte_out are defined in Figures D.9 to D.11.

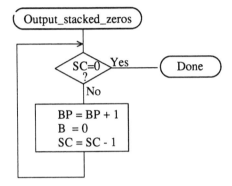

Figure D.9 - Output_stacked_zeros procedure for encoder

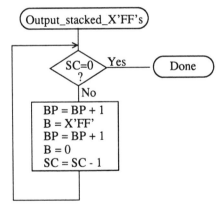

Figure D.10 - Output_stacked_X'FF's procedure for encoder

D-13

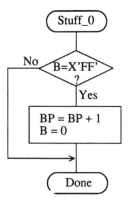

Figure D.11 - Stuff_0 procedure for encoder

D.1.7 Initialization of the encoder

The Initenc procedure is used to start the arithmetic coder. The basic steps are shown in Figure D.12.

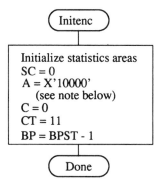

Figure D.12 - Initialization of the encoder

The probability estimation tables are defined by Table D.2. The statistics areas are initialized to an MPS sense of 0 and a Qe index of zero as defined by Table D.2. The stack count (SC) is cleared, the code register (C) is cleared, and the interval register is set to X'10000'. The counter (CT) is set to 11, reflecting the fact that when A is initialized to X'10000' three spacer bits plus eight output bits in C must be filled before the first byte is removed. Note that BP is initialized to point to the byte before the start of the entropy-coded segment (which is at BPST). Note also that the statistics areas are initialized for all values of context-index S to MPS(S)=0 and Index(S)=0.

> NOTE - Although the probability interval is initialized to X'10000' in both Initenc and Initdec, the precision of the probability interval register can still be limited to 16 bits. When the precision of the interval register is 16 bits, it is initialized to zero.

D-14

D.1.8 Termination of encoding

The Flush procedure is used to terminate the arithmetic encoding procedures and prepare the entropy-coded segment for the addition of the X'FF' prefix of the marker which follows the arithmetically coded data. Figure D.13 shows this flush procedure. The first step in the procedure is to set as many low order bits of the code register to zero as possible without pointing outside of the final interval. Then, the output byte is aligned by shifting it left by CT bits; Byte_out then removes it from C. C is then shifted left by 8 bits to align the second output byte and Byte_out is used a second time. The remaining low order bits in C are guaranteed to be zero, and these trailing zero bits shall not be written to the entropy-coded segment.

Any trailing zero bytes already written to the entropy-coded segment and not preceded by a X'FF' may, optionally, be discarded. This is done in the Discard_final_zeros procedure. Stuffed zero bytes shall not be discarded.

Entropy coded segments are always followed by a marker. For this reason, the final zero bits needed to complete decoding shall not be included in the entropy coded segment. Instead, when the decoder encounters a marker, zero bits shall be supplied to the decoding procedure until decoding is complete. This convention guarantees that when a DNL marker is used, the decoder will intercept it in time to correctly terminate the decoding procedure.

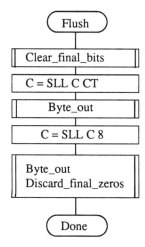

Figure D.13 - Flush procedure

D-15

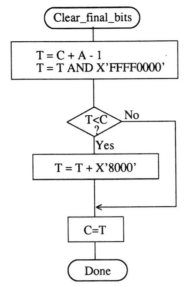

Figure D.14 - Clear_final_bits procedure in Flush

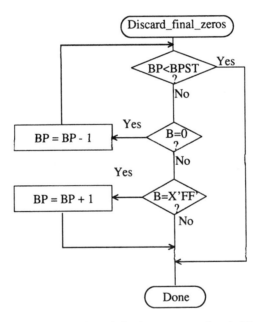

Figure D.15 - Discard_final_zeros procedure in Flush

D-16

D.2 Arithmetic decoding procedures

Three arithmetic decoding procedures are used for arithmetic decoding. They are:

Table D.3 Procedures for binary arithmetic decoding

Procedure	Purpose
Decode(S)	Decode a binary decision
Initdec	Initialize the decoder

The "Decode(S)" procedure decodes the binary decision for a given context-index S and returns a value of either 0 or 1. It is the inverse of the "Code_0(S)" and "Code_1(S)" procedures described in D.1. "Initdec" initializes the arithmetic coding entropy decoder.

D.2.1 Binary arithmetic decoding principles

The probability interval subdivision and sub-interval ordering defined for the arithmetic encoding procedures also apply to the arithmetic decoding procedures.

Since the bit stream always points within the current probability interval, the decoding process is a matter of determining, for each decision, which sub-interval is pointed to by the bit stream. This is done recursively, using the same probability interval sub-division process as in the encoder. Each time a decision is decoded, the decoder subtracts from the bit stream any interval the encoder added to the bit stream. Therefore, the code register in the decoder is a pointer into the current probability interval relative to the base of the interval.

If the size of the sub-interval allocated to the LPS is larger than the sub-interval allocated to the MPS, the encoder invokes the conditional exchange procedure. When the interval sizes are inverted in the decoder, the sense of the symbol decoded must be inverted.

D.2.2 Decoding conventions and approximations

The approximations and integer arithmetic defined for the probability interval subdivision in the encoder must also be used in the decoder. However, where the encoder would have added to the code register, the decoder subtracts from the code register.

D.2.3 Decoder code register conventions

The flow charts given in this section assume the following register structures for the decoder:

	MSB	LSB
Cx register	xxxxxxxx,	xxxxxxxx
C-low	bbbbbbbb,	00000000
A-register	aaaaaaaa,	aaaaaaaa

Cx and C-low can be regarded as one 32 bit C-register, in that renormalization of C shifts a bit of new data from bit 15 of C-low to bit 0 of Cx. However, the decoding comparisons use Cx alone. New data are inserted into the "b" bits of C-low one byte at a time.

NOTE: The comparisons shown in the various procedures use arithmetic comparisons, and therefore assume precisions greater than 16 bits for the variables. Unsigned (logical) comparisons should be used in 16 bit precision implementations.

D.2.4 The decode procedure

The decoder decodes one binary decision at a time. After decoding the decision, the decoder subtracts any amount from the code register that the encoder added. The amount left in the code register is the offset from the base of the current probability interval to the sub-interval allocated to the binary decisions not yet decoded. In the first test in the decode procedure shown in Figure D.16 the code register is compared to the size of the MPS sub-interval. Unless a conditional exchange is needed, this test determines whether the MPS or LPS for context-index S is decoded. Note that the LPS for context-index S is given by 1-MPS(S).

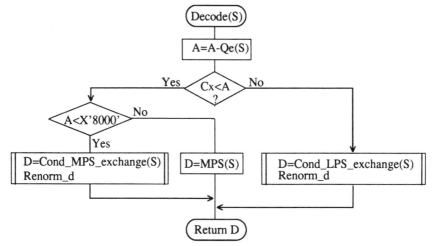

Figure D.16 - Decode(S) procedure

When a renormalization is needed, the MPS/LPS conditional exchange may also be needed. For the LPS path the conditional exchange procedure is shown in Figure D.17. Note that the probability estimation in the decoder is identical to the probability estimation in the encoder (Figures D.5 and D.6).

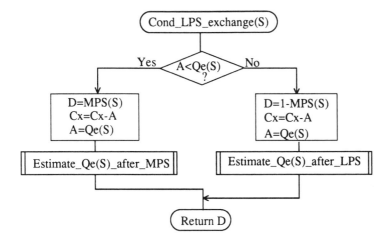

Figure D.17 - Decoder LPS path conditional exchange procedure

For the MPS path of the decoder the conditional exchange procedure is given in Figure D.18.

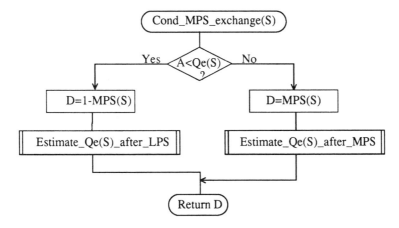

Figure D.18 - Decoder MPS path conditional exchange procedure

D.2.5 Probability estimation in the decoder

The procedures defined for obtaining a new LPS probability estimate in the encoder are also used in the decoder.

D.2.6 Renormalization in the decoder

The Renorm_d procedure for the decoder renormalization is shown in Figure D.19. CT is a counter which keeps track of the number of compressed bits in the C-low section of the C-register. When CT is zero, a new byte is inserted into C-low by the procedure Byte_in and CT is reset to 8.

Both the probability interval register A and the code register C are shifted, one bit at a time, until A is no longer less than X'8000'.

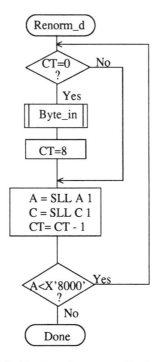

Figure D.19 - Decoder renormalization procedure

The Byte_in procedure used in Renorm_d is shown in Figure D.20. This procedure fetches one byte of data, compensating for the stuffed zero byte which follows any X'FF' byte. It also detects the marker which must follow the entropy-coded segment. The C-register in this procedure is the concatenation of the Cx and C-low registers. For simplicity of exposition, the buffer holding the entropy-coded segment is assumed to be large enough to contain the entire segment.

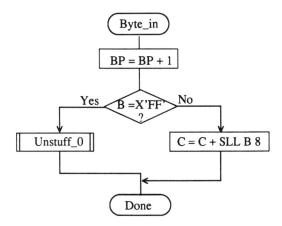

Figure D.20 - Byte_in procedure for decoder

B is the byte pointed to by the entropy-coded segment pointer BP. BP is first incremented. If the new value of B is not a X'FF', it is inserted into the high order 8 bits of C-low.

The Unstuff_0 procedure is shown in Figure D.21. If the new value of B is X'FF', BP is incremented to point to the next byte and this next B is tested to see if it is zero. If so, B contains a stuffed byte which must be skipped. The zero B is ignored, and the X"FF" B value which preceded it is inserted in the C-register.

If the value of B after a X'FF' byte is not zero, then a marker has been detected. The marker is interpreted as required and the entropy-coded segment pointer is adjusted ("Adjust BP" in Figure D.21) so that 0-bytes will be fed to the decoder until decoding is complete. One way of accomplishing this is to point BP to the byte preceding the marker which follows the entropy-coded segment.

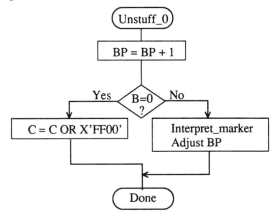

Figure D.21 - Unstuff_0 procedure for decoder

D-21

D.2.7 Initialization of the decoder

The Initdec procedure is used to start the arithmetic decoder. The basic steps are shown in Figure D.22.

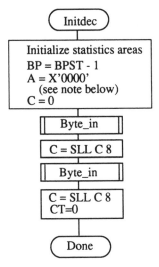

Figure D.22 - Initialization of the decoder

The estimation tables are defined by Table D.2. The statistics areas are initialized to an MPS sense of 0 and a Qe index of zero as defined by Table D.2. BP, the pointer to the entropy-coded segment, is then initialized to point to the byte before the start of the entropy-coded segment at BPST, and the interval register is set to the same starting value as in the encoder. The first byte of compressed data is fetched and shifted into Cx. The second byte is then fetched and shifted into Cx. The count is set to zero, so that a new byte of data will be fetched by Renorm_d.

> NOTE - Although the probability interval is initialized to X'10000' in both Initenc and Initdec, the precision of the probability interval register can still be limited to 16 bits. When the precision of the interval register is 16 bits, it is initialized to zero.

D.3 Bit ordering within bytes

The arithmetically encoded entropy-coded segment is an integer of variable length. Therefore, the ordering of bytes and the bit ordering within bytes is the same as for parameters (see B.1.1.1).

Annex E   (normative)

Encoder and decoder control procedures

This annex describes the encoder and decoder control procedures for the sequential, progressive, and lossless modes of operation.

> NOTE - There is **no requirement** in this Specification that any encoder or decoder shall implement the procedures in precisely the manner specified by the flow charts in this Annex. It is necessary only that an encoder or decoder implement the **function** specified in this Annex. The sole criterion for an encoder or decoder to be considered in compliance with this Specification is that it satisfy the requirements given in clause 6 (for encoders) or clause 7 (for decoders), as determined by the compliance tests specified in Part 2.

E.1 Encoder control procedures

E.1.1 Control procedure for encoding an image

The encoder control procedure for encoding an image is shown in Figure E.1.

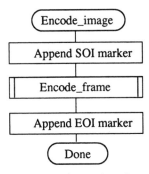

Figure E.1 - Control procedure for encoding an image

E.1.2 Control procedure for encoding a frame

In all cases where markers are appended to the compressed data, optional X'FF' fill bytes may precede the marker.

The control procedure for encoding a frame is oriented around the scans in the frame. The frame header is first appended, and then the scans are coded. Table specifications and other marker segments may precede the SOF marker, as indicated by [tables/misc.] in Figure E.2.

Figure E.2 shows the encoding process frame control procedure.

E-1

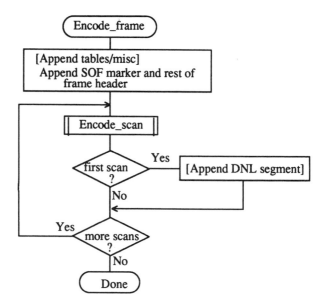

Figure E.2 - Control procedure for encoding a frame

E.1.3 Control procedure for encoding a scan

A scan consists of a single pass through the data of each component in the scan. Table specifications and other marker segments may precede the SOS marker. If more than one component is coded in the scan, the data are interleaved. If restart is enabled, the data are segmented into restart intervals; the encoding process is reset at the start of each restart interval.

If restart is enabled, a RSTm marker is placed in the coded data between each restart interval. If restart is disabled, the control procedure is the same, except that the entire scan contains a single restart interval. The compressed image data generated by a scan is always followed by a marker.

Figure E.3 shows the encoding process scan control procedure. The loop is terminated when the encoding process has coded the number of restart intervals which make up the scan. Note that the number of minimum coded units (MCU) in the final restart interval must be adjusted to match the number of MCU in the scan. The number of MCU is calculated from the frame and scan parameters (see Annex B). "m" is the restart interval modulo counter needed for the RSTm marker. The modulo arithmetic for this counter is shown after the "Append RSTm marker" procedure.

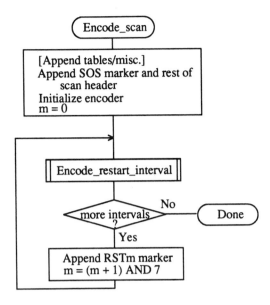

Figure E.3 - Control procedure for encoding a scan

E.1.4 Control procedure for encoding a restart interval

Figure E.4 shows the encoding process control procedure for a restart interval. The loop is terminated either when the encoding process has coded the number of minimum coded units
(MCU) in the restart interval or when it has completed the image scan.

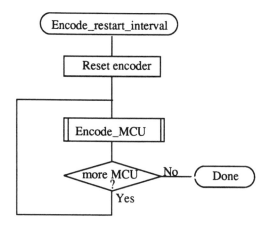

Figure E.4 - Control procedure for encoding a restart interval

If the encoding process is reset, the DC predictions are set to defaults. If arithmetic coding is used, the statistics areas are also reset. If the number of lines must be set or reset, the DNL marker segment shall precede the marker which would normally follow first scan in the frame. If the DNL marker is used, no more restart intervals shall be coded in the scan.

E.1.5 Control procedure for encoding a minimum coded unit (MCU)

The minimum coded unit is defined in A.2. Within a given MCU the data units are coded in the order in which they occur in the MCU. The control procedure for encoding a MCU is shown Figure E.5.

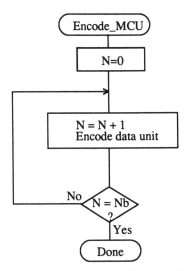

Figure E.5 - Control procedure for encoding a minimum coded unit (MCU)

In Figure E.5, Nb refers to the number of data units in the MCU. The order in which data units occur in the MCU is defined in A.2. The data unit is an 8x8 block for DCT-based processes, and a single sample for lossless processes.

E.2 Decoder control procedure

E.2.1 Control procedure for decoding an image

Figure E.6 shows the decoding process control for an image.

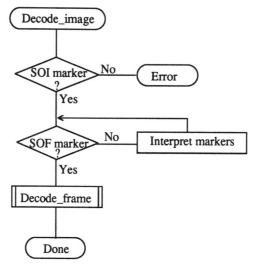

Figure E.6 - Control procedure for decoding an image

Decoding control centers around identification of various markers. The first marker must be the SOI (Start Of Image) marker. The next marker is normally a SOFn (Start Of Frame) marker; if this is not found, one of the marker segments listed in Table E.1 has been received.

Table E.1 Markers recognized by "Interpret markers"

Marker	Purpose
DHT	Define Huffman Tables
DAC	Define Arithmetic Conditioning
DQT	Define Quantization Tables
DRI	Define Restart Interval
APPn	Application defined marker
COM	Comment

Note that optional X'FF' fill bytes which may precede any marker shall be discarded before determining which marker is present

The additional logic to interpret these various markers is contained in the box labeled "Interpret markers". DHT markers shall be interpreted by processes using Huffman coding. DAC markers shall be interpreted by processes using arithmetic coding. DQT markers shall be interpreted

E-6

by DCT-based decoders. DRI markers shall be interpreted by all decoders. APPn and COM markers shall be interpreted only to the extent that they do not interfere with the decoding.

Decoding is terminated when the decoder finds the EOI marker. By definition, the procedures in "Interpret markers" leave the system at the next marker. Note that if the expected SOI marker is missing at the start of the compressed image data, an error condition has occurred. The techniques for detecting and managing error conditions can be as elaborate or as simple as desired.

E.2.2 Control procedure for decoding a frame

Figure E.7 shows the control procedure for the decoding of a frame.

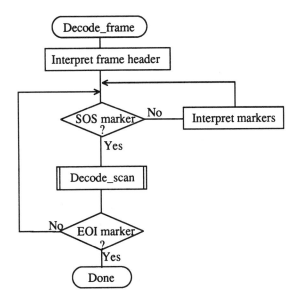

Figure E.7 - Control procedure for decoding a frame

The loop is terminated if the EOI marker is found at the end of the scan.

The markers recognized by "Interpret markers" are listed in Table E.1. E.2.1 describes the extent to which the various markers shall be interpreted.

E-7

E.2.3 Control procedure for decoding a scan

Figure E.8 shows the decoding of a scan.

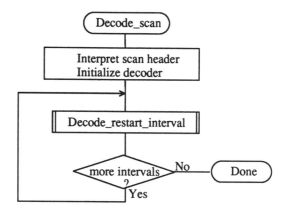

Figure E.8 - Control procedure for decoding a scan

The loop is terminated when the expected number of restart intervals has been decoded. Note that the final restart interval may be smaller than the size specified by the DRI marker segment, as it includes only the number of MCUs remaining in the scan. The decoder is reset at the start of each restart interval.

E.2.4 Control procedure for decoding a restart interval

The procedure for decoding a restart interval is shown in Figure E.9. At the end of the restart interval, the next marker is located. If a problem is detected in locating this marker, error handling procedures may be invoked. While such procedures are optional, the decoder shall be able to correctly recognize RST markers in the compressed data and reset the decoder when they are encountered. The decoder shall also be able to recognize the DNL marker and set the number of lines defined in the DNL segment.

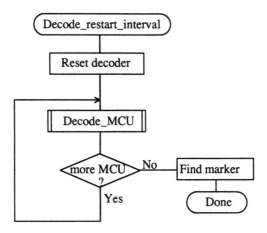

Figure E.9 - Control procedure for decoding a restart interval

E.2.5 Control procedure for decoding a minimum coded unit (MCU)

The procedure for decoding a minimum coded unit (MCU) is shown in Figure E.10.

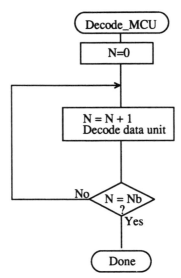

Figure E.10 - Control procedure for decoding a minimum coded unit (MCU)

In Figure E.10 Nb is the number of data units in a MCU.

E-10

Annex F (normative)

Sequential DCT-based mode of operation

This Annex provides a **functional specification** of the following coding processes for the sequential DCT-based mode of operation:

1) baseline sequential;

2) extended sequential, Huffman coding, 8-bit sample precision;

3) extended sequential, arithmetic coding, 8-bit sample precision;

4) extended sequential, Huffman coding, 12-bit sample precision;

5) extended sequential, arithmetic coding, 12-bit sample precision.

For each of these, the encoding process is specified in F.1, and the decoding process is specified in F.2. The functional specification is presented by means of specific flow charts for the various procedures which comprise these coding processes.

> NOTE - There is **no requirement** in this Specification that any encoder or decoder which embodies one of the above-named processes shall implement the procedures in precisely the manner specified by the flow charts in this Annex. It is necessary only that an encoder or decoder implement the **function** specified in this Annex. The sole criterion for an encoder or decoder to be considered in compliance with this Specification is that it satisfy the requirements given in clause 6 (for encoders) or clause 7 (for decoders), as determined by the compliance tests specified in Part 2.

F.1 Sequential DCT-based encoding processes

F.1.1 Sequential DCT-based control procedures and coding models

F.1.1.1 Control procedures for sequential DCT-based encoders

The control procedures for encoding an image and its constituent parts - the frame, scan, restart interval and MCU - are given in Figures E.1 to E.5. The procedure for encoding a MCU (Figure E.5) repetitively calls the procedure for encoding a data unit. For DCT-based encoders the data unit is an 8x8 block of samples.

F.1.1.2 Procedure for encoding an 8x8 block data unit

In the sequential DCT-based encoding process encoding a 8x8 block data unit consists of the following procedures:

(1) Calculate forward 8x8 DCT and quantize using table assigned in frame header.
(2) Encode DC coefficient for 8x8 block using DC table assigned in scan header.
(3) Encode AC coefficients for 8x8 block using AC table assigned in scan header.

F.1.1.3 Forward DCT (FDCT)

The mathematical definition of the FDCT is given in A.3.3.

Prior to computing the FDCT the input data are level shifted to a signed two's complement representation as described in Annex A. For 8-bit input precision the level shift is achieved by subtracting 128. For 12-bit input precision the level shift is achieved by subtracting 2048.

F.1.1.4 Quantization of the FDCT

The uniform quantization procedure described in Annex A is used to quantize the DCT coefficients. One of four quantization tables may be used by the encoder. No default quantization tables are specified in this Specification. However, some typical quantization tables are given in Annex K.

The quantized DCT values are signed, two's complement integers with 11-bit precision for 8-bit input precision and 15-bit precision for 12-bit input precision.

F.1.1.5 Encoding models for the sequential DCT procedures

The two dimensional array of quantized DCT coefficients is rearranged in a zig-zag sequence order defined in A.6. The zig-zag order coefficients are denoted $ZZ(0)$ through $ZZ(63)$ with:

$$ZZ(0) = Sq_{00}$$
$$ZZ(1) = Sq_{01}$$
$$ZZ(2) = Sq_{10}$$
$$\bullet$$
$$\bullet$$
$$\bullet$$
$$ZZ(63) = Sq_{77}$$

Sq_{ij} are defined in Figure A.6.

Two coding procedures are used, one for the DC coefficient $ZZ(0)$ and the other for the AC coefficients $ZZ(1)..ZZ(63)$. The coefficients are encoded in the order in which they occur in ZZ, starting with the DC coefficient. The coefficients are represented as two's complement integers.

F.1.1.5.1 Encoding model for DC coefficients

The DC coefficients are coded differentially, using a one-dimensional predictor, PRED, which is the quantized DC value from the most recently coded 8x8 block from the same component. The difference, DIFF, is obtained from

DIFF=ZZ(0)-PRED.

At the beginning of the scan and at the beginning of each restart interval, the prediction for the DC coefficient prediction is initialized to 0. (Recall that the input data have been level shifted to two's complement representation.)

F.1.1.5.2 Encoding model for AC coefficients

Since many coefficients are zero, runs of zeros are identified and coded efficiently. In addition, if the last part of ZZ is entirely zero, this is coded explicitly as an end-of-block (EOB).

F.1.2 Baseline Huffman encoding procedures

The baseline encoding procedure is for 8-bit sample precision. The encoder may employ up to two DC and two AC Huffman tables within one scan.

F.1.2.1. Huffman encoding of DC coefficients

F.1.2.1.1 Structure of DC code table

The DC code table consists of a set of Huffman codes (maximum length 16 bits) and appended additional bits (in most cases) which can code any possible value of DIFF, the difference between the current DC coefficient and the prediction. The Huffman codes for the difference categories are generated in such a way that no code consists entirely of 1-bits (X'FF' prefix marker code avoided).

The two's complement difference magnitudes are grouped into 12 categories, SSSS, and a Huffman code is created for each of the 12 difference magnitude categories.

Table F.1 Difference magnitude categories for DC coding

SSSS	DIFF values
0	0
1	-1,1
2	-3,-2,2,3
3	-7..-4,4..7
4	-15..-8,8..15
5	-31..-16,16..31
6	-63..-32,32..63
7	-127..-64,64..127
8	-255..-128,128..255
9	-511..-256,256..511
10	-1023..-512,512..1023
11	-2047..-1024,1024..2047

For each category, except SSSS=0, an additional bits field is appended to the code word to uniquely identify which difference in that category actually occurred. The number of extra bits is given by SSSS; the extra bits are appended to the LSB of the preceding Huffman code, most significant bit first. When DIFF is positive, the SSSS low order bits of DIFF are transmitted. When DIFF is negative, the SSSS low order bits of (DIFF-1) are transmitted. Note that the most significant bit of the appended bit sequence is 0 for negative differences and 1 for positive differences.

F.1.2.1.2 Defining Huffman tables for the DC coefficients

The syntax for specifying the Huffman tables is given in Annex B. The procedure for creating a code table from this information is described in Annex C. No more than two Huffman tables may be defined for coding of DC coefficients. Two examples of Huffman tables for coding of DC coefficients are provided in Annex K.

F.1.2.1.3 Huffman encoding procedures for DC coefficients

The encoding procedure is defined in terms of a set of extended tables, XHUFCO and XHUFSI, which contain the complete set of Huffman codes and sizes for all possible difference values. For full 12 bit precision the tables are relatively large. For the baseline system, however, the precision of the differences may be small enough to make this description practical.

XHUFCO and XHUFSI are generated from the encoder tables EHUFCO and EHUFSI (see Annex C) by appending to the Huffman codes for each difference category the additional bits that completely define the difference. By definition, XHUFCO and XHUFSI have entries for each possible difference value. XHUFCO contains the concatenated bit pattern of the Huffman code and the additional bits field; XHUFSI contains the total length in bits of this concatenated bit pattern. Both are indexed by DIFF, the difference between the DC coefficient and the prediction.

The Huffman encoding procedure for the DC difference, DIFF, is:

> SIZE=XHUFSI(DIFF)
> CODE=XHUFCO(DIFF)
> code SIZE bits of CODE

where DC is the quantized DC coefficient value and PRED is the predicted quantized DC value. The Huffman code (CODE) (including any additional bits) is obtained from XHUFCO and SIZE (length of the code including additional bits) is obtained from XHUFSI, using DIFF as the index to the two tables.

F.1.2.2 Huffman encoding of AC coefficients

F.1.2.2.1 Structure of AC code table

Each nonzero AC coefficient in ZZ is described by a composite 8-bit value, RS, of the form

> RS = binary 'RRRRSSSS'

The 4 least significant bits, 'SSSS', define a category for the amplitude of the next nonzero coefficient in ZZ, and the 4 most significant bits, 'RRRR', give the position of the coefficient in ZZ relative to the previous nonzero coefficient (i.e. the run-length of zero coefficients between nonzero coefficients). Since the run length of zero coefficients may exceed 15, the value 'RRRRSSSS'=X'F0' is defined to represent a run length of 15 zero coefficients followed by a coefficient of zero amplitude. (This can be interpreted as a run length of 16 zero coefficients.) In addition, a special value 'RRRRSSSS'='00000000' is used to code the end-of-block (EOB), when all remaining coefficients in the block are zero.

The general structure of the code table is illustrated in Figure F.1.

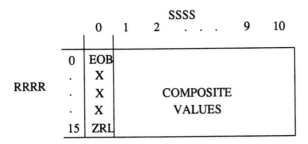

Figure F.1 - Two-dimensional value array for Huffman coding

The entries marked 'X' are undefined for the baseline procedure.

The magnitude ranges assigned to each value of SSSS are defined in table F.2.

Table F.2. Categories assigned to coefficient values

SSSS	AC coefficients
1	-1,1
2	-3,-2,2,3
3	-7..-4,4..7
4	-15..-8,8..15
5	-31..-16,16..31
6	-63..-32,32..63
7	-127..-64,64..127
8	-255..-128,128..255
9	-511..-256,256..511
10	-1023..-512,512..1023

The composite value, RRRRSSSS, is Huffman coded and each Huffman code is followed by additional bits (assumed to be randomly distributed and thus uncoded) which specify the sign and exact amplitude of the coefficient.

The AC code table consists of one Huffman code (maximum length 16 bits, not including additional bits) for each possible composite value. The Huffman codes for the 8-bit composite values are generated in such a way that no code consists entirely of 1-bits.

The format for the additional bits is the same as in the coding of the DC coefficients. ZZ(K) is the Kth coefficient in the zig-zag sequence of coefficients being coded. The value of SSSS gives the number of additional bits required to specify the sign and precise amplitude of the coefficient. The additional bits are either the low-order SSSS bits of ZZ(K) when ZZ(K) is positive or the low-order SSSS bits of ZZ(K)-1 when ZZ(K) is negative.

F.1.2.2.2.2 Defining Huffman tables for the AC coefficients

The syntax for specifying the Huffman tables is given in Annex B. The procedure for creating a code table from this information is described in Annex C.

In the baseline system no more than two Huffman tables may be defined for coding of AC coefficients. Two examples of Huffman tables for coding of AC coefficients are provided in Annex K.

F.1.2.2.3 Huffman encoding procedures for AC coefficients

As defined in Annex C, the Huffman code table is assumed to be available as a pair of vectors, EHUFCO (containing the code bits) and EHUFSI (containing the length of each code in bits), both indexed by the composite value defined above.

The procedure for encoding the AC coefficients in a block is shown in Figures F.2 and F.3.

F-6

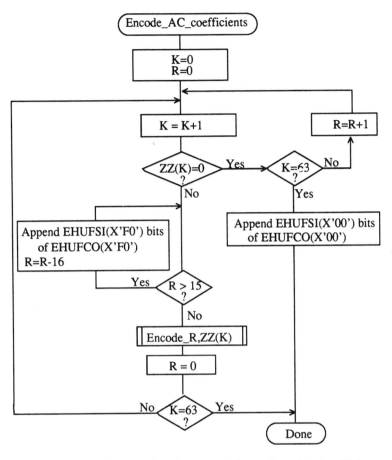

Figure F.2 - Procedure for sequential encoding of AC coefficients
with Huffman coding

In this figure K is the index to the zig-zag scan position and R is the run length of zero coefficients.

F-7

The procedure "Code EHUFSI(240) bits of EHUFCO(240)" codes a run of 16 zero coefficients (ZRL code of Figure F.2). The procedure "Code EHUFSI(0) bits of EHUFCO(0)" codes the end-of-block (EOB code). If the last coefficient (K=63) is not zero, the EOB code is bypassed.

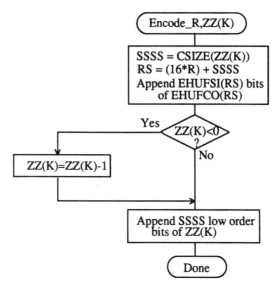

Figure F.3 - Sequential encoding of a nonzero AC coefficient

CSIZE is a procedure which maps an AC coefficient to the SSSS value as defined in Table F.2.

F.1.2.3 Byte stuffing

In order to provide code space for marker codes which can be located in the compressed image data without decoding, byte stuffing is used.

Whenever, in the course of normal encoding, the byte value X'FF' is created in the code string, a X'00' byte is stuffed into the code string.

If a X'00' byte is detected after a X'FF' byte, the decoder must discard it. If the byte is not zero, a marker has been detected, and shall be interpreted to the extent needed to complete the decoding of the scan.

Byte alignment of markers is achieved by padding incomplete bytes with 1-bits. If padding with 1-bits creates a X'FF' value, a zero byte is stuffed before adding the marker.

F.1.3   Sequential DCT encoding process with 8-bit precision extended to four sets of Huffman tables

This process is identical to the Baseline encoding process described in F.1.2, with the exception that the number of sets of Huffman tables which may be used within the same scan is increased to four. Four DC and four AC Huffman tables is the maximum allowed by this Specification.

F.1.4. Sequential DCT encoding process with arithmetic coding

This subclause describes the use of arithmetic coding procedures in the sequential DCT-based encoding process.

The arithmetic coding extensions have the same DCT model as the Baseline DCT encoder. Therefore, Annex F.1.1 also applies to arithmetic coding. As with the Huffman coding technique, the binary arithmetic coding technique is lossless. It is possible to transcode between the two systems without either FDCT or IDCT computations, and without modification of the reconstructed image.

The basic principles of adaptive binary arithmetic coding are described in Annex D. Up to four DC and four AC conditioning tables and associated statistics areas may be used within one scan.

The arithmetic encoding procedures for encoding binary decisions, initializing the statistics area, initializing the encoder, terminating the code string, and adding restart markers are listed in Table D.1 of Annex D.

Some of the procedures in Table D.1 are used in the higher level control structure for scans and restart intervals described in Annex E. At the beginning of scans and restart intervals, the probability estimates used in the arithmetic coder are reset to the standard initial value as part of the Initenc procedure which restarts the arithmetic coder. At the end of scans and restart intervals, the Flush procedure is invoked to empty the code register before the next marker is appended.

F.1.4.1 Arithmetic encoding of DC coefficients

The basic structure of the decision sequence for encoding a DC difference value, DIFF, is shown in Figure F.4.

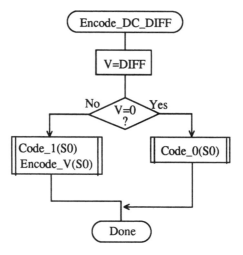

Figure F.4 - Coding model for arithmetic coding of DC difference

The context-index S0 and other context-indices used in the DC coding procedures are defined in Table F.4 (see F.1.4.4.1.3). A 0-decision is coded if the difference value is zero and a 1-decision is coded if the difference is not zero. If the difference is not zero, the sign and magnitude are coded using the procedure Encode_V(S0), which is described in F.1.4.3.1.

F.1.4.2 Arithmetic encoding of AC coefficients

The AC coefficients are coded in the order in which they occur in the zig-zag sequence ZZ(1,...,63). An end-of-block (EOB) binary decision is coded before coding the first AC coefficient in ZZ, and after each nonzero coefficient. If the EOB occurs, all remaining coefficients in ZZ are zero. Figure F.5 illustrates the decision sequence. The equivalent procedure for the Huffman coder is found in Figure F.2.

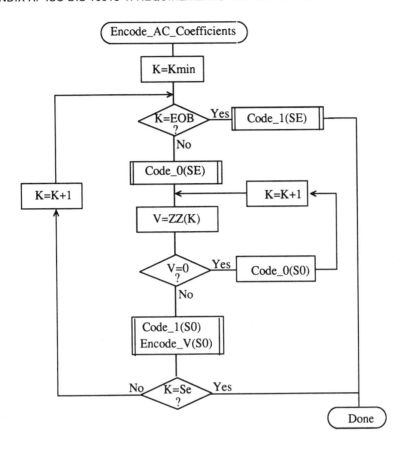

Figure F.5 - AC coding model for arithmetic coding

The context-indices SE and S0 used in the AC coding procedures are defined in Table F.5 (see F.1.4.4.2). In Figure F.5 K is the index to the zig-zag sequence position. For the sequential scan, Kmin is 1 and Se is 63. The V=0 decision is part of a loop which codes runs of zero coefficients. Whenever the coefficient is nonzero, "Encode_V(S0)" codes the sign and magnitude of the coefficient. Each time a nonzero coefficient is coded, it is followed by an EOB decision. If the EOB occurs, a 1-decision is coded to indicate that the coding of the block is complete. Note that if the coefficient for K=Se is not zero, the EOB decision is skipped.

F.1.4.3 Encoding the binary decision sequence for nonzero DC differences and AC coefficients

Both the DC difference and the AC coefficients are represented as signed two's complement 16 bit integer values. The decomposition of these signed integer values into a binary decision tree is done in the same way for both the DC and AC coding models.

Although the binary decision trees for this section of the DC and AC coding models are the same, the statistical models for assigning statistics bins to the binary decisions in the tree are quite different.

F.1.4.3.1 Structure of the encoding decision sequence

The encoding sequence can be separated into three procedures, a procedure which encodes the sign, a second procedure which identifies the magnitude category, and a third procedure which identifies precisely which magnitude occurred within the category identified in the second procedure.

At the point where the binary decision sequence in Encode_V(S0) starts, the coefficient or difference has already been determined to be nonzero. That determination was made in the procedures in Figures F.4 and F.5.

Denoting either DC differences (DIFF) or AC coefficients as V, the nonzero signed integer value of V is encoded by the sequence shown in Figure F.6. This sequence first codes the sign of V. It then (after converting V to a magnitude and decrementing it by 1 to give Sz) codes the magnitude category of Sz (code_log2_Sz), and then codes the low order magnitude bits (code_Sz_bits) to identify the exact magnitude value.

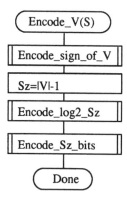

Figure F.6 - Sequence of procedures in encoding nonzero values of V

There are two significant differences between this sequence and the similar set of operations described in F.1.2 for Huffman coding. First, the sign is encoded before the magnitude category is identified, and second, the magnitude is decremented by 1 before the magnitude category is identified.

F.1.4.3.1.1 Encoding the sign

The sign is encoded by coding a 0-decision when the sign is positive and a 1-decision when the sign is negative.

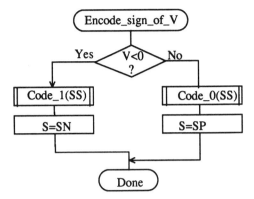

Figure F.7 - Encoding the sign of V

The context-indices SS, SN and SP are defined for DC coding in Table F.4 and for AC coding in Table F.5. After the sign is coded, the context-index S is set to either SN or SP, establishing an initial value for Encode_log2_Sz.

F.1.4.3.1.2 Encoding the magnitude category

The magnitude category is determined by a sequence of binary decisions which compares Sz against an exponentially increasing bound (which is a power of 2) in order to determine the position of the leading 1-bit. This establishes the magnitude category in much the same way that the Huffman encoder transmits a code for the value associated with the difference category. The flow chart for this procedure is shown in Figure F.8.

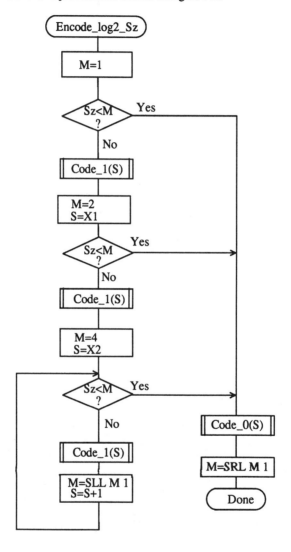

Figure F.8 - Decision sequence to establish the magnitude category

The starting value of the context-index S is determined in Encode_sign_of_V, and the context-index values X1 and X2 are defined for DC coding in Table F.4 and for AC coding in Table F.5. In this figure M is the exclusive upper bound for the magnitude and the abbreviations "SLL" and "SRL" refer to the shift-left-logical and shift-right-logical operations - in this case by one bit position. The SRL operation at the completion of the procedure aligns M with the most significant bit of Sz.

Table F.3   Categories for each maximum bound

exclusive upper bound (M)	Sz range	number of low order magnitude bits
1	0	0
2	1	0
4	2,3	1
8	4,...,7	2
16	8,...,15	3
32	16,...,31	4
64	32,...,63	5
128	64,...,127	6
256	128,...,255	7
512	256,...,511	8
1024	512,...,1023	9
2048	1024,...,2047	10
4096	2048,...,4095	11
8192	4096,...,8191	12
16384	8192,...,16383	13
32768	16384,...,32767	14

The highest precision allowed for the DCT is 15 bits. Therefore, the highest precision required for the coding decision tree is 16 bits for the DC coefficient difference and 15 bits for the AC coefficients, including the sign bit.

F.1.4.3.1.3 Encoding the exact value of the magnitude

After the magnitude category is encoded, the low order magnitude bits are encoded. These bits are encoded in order of decreasing bit significance. The procedure is shown in Figure F.9. The abbreviation "SRL" indicates the shift-right-logical operation, and M is the exclusive bound established in Figure F.8; note that M has only one bit set; shifting M right converts it into a bit mask for the logical "AND" operation.

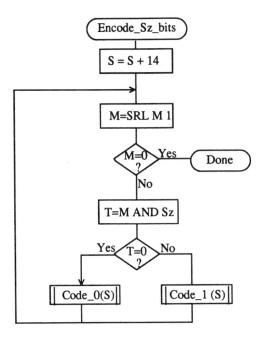

Figure F.9 - Decision sequence to code the magnitude bit pattern

The starting value of the context-index S is determined in Encode_log2_Sz. The increment of S by 14 at the beginning of this procedure sets the context-index to the value required in Tables F.4 and F.5.

F.1.4.4 Statistical models

An adaptive binary arithmetic coder requires a statistical model. The statistics model defines the contexts which are used to select the conditional probability estimates used in the encoding and decoding procedures.

Each decision in the binary decision trees is associated with one or more contexts. These contexts identify the sense of the MPS and the index in Table D.2 of the conditional probability estimate Qe which is used to encode and decode the binary decision.

The arithmetic, coder is adaptive, which means that the probability estimates for each context are developed and maintained by the arithmetic coding system on the basis of prior coding decisions for that context.

F.1.4.4.1 Statistical model for coding DC prediction differences

The statistical model for coding the DC difference conditions some of the probability estimates for the binary decisions on previous DC coding decisions.

F.1.4.4.1.1 Statistical conditioning on sign

In coding the DC coefficients, four separate statistics bins (probability estimates) are used in coding the zero/not-zero (V=0) decision, the sign decision and the first magnitude category decision. Two of these bins are used to code the V=0 decision and the sign decision. The other two bins are used in coding the first magnitude decision, Sz<1; one of these bins is used when the sign is positive, and the other is used when the sign is negative. Thus, the first magnitude decision probability estimate is conditioned on the sign of V.

F.1.4.4.1.2 Statistical conditioning on DC difference in previous block

The probability estimates for these first three decisions are also conditioned on Da, the difference value coded for the previous DCT block of the same component. The differences are classified into five groups: zero, small positive, small negative, large positive and large negative. The relationship between the default classification and the quantization scale is shown in Figure F.10.

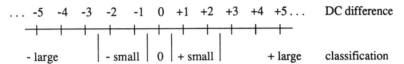

Figure F.10 - Conditioning classification of difference values

The bounds for the "small" difference category determine the classification. Defining L and U as integers in the range 0 to 15 inclusive, the lower bound (exclusive) for difference magnitudes classified as "small" is zero for L=0, and is 2^{L-1} for L>0.

The upper bound (inclusive) for difference magnitudes classified as "small" is 2^{U}.

L shall be less than or equal to U.

These bounds for the conditioning category provide a segmentation which is identical to that listed in table F.3.

F.1.4.4.1.3 Assignment of statistical bins to the DC binary decision tree

As shown in Table F.4, each statistics area for DC coding consists of a contiguous set of 49 statistics bins. The first 20 bins consist of five sets of four bins selected by a context-index S0. The value of S0 is given by DC_Context(Da), which provides a value of 0, 4, 8, 12 or 16, depending on the difference classification of Da (see F.1.4.4.1.2). The remaining 29 bins, X1,...,X15,M2,...,M15, are used to code magnitude category decisions and magnitude bits.

Table F.4. Statistical model for DC coefficient coding

Context-index	Value	Coding decision
S0	DC_Context(Da)	V=0
SS	S0+1	sign of V
SP	S0+2	Sz<1 if V>0
SN	S0+3	Sz<1 it V<0
X1	20	Sz<2
X2	X1+1	Sz<4
X3	X1+2	Sz<8
.	.	.
.	.	.
.	.	.
X15	X1+14	$Sz<2^{15}$
M2	X2+14	magnitude bits if Sz<4
M3	X3+14	magnitude bits if Sz<8
.	.	.
.	.	.
.	.	.
M15	X15+14	magnitude bits if $Sz<2^{15}$

F.1.4.4.1.4 Default conditioning for DC statistical model

The bounds, L and U, for determining the conditioning category have the default values L=0 and U=1. Other bounds may be set using the DAC (Define Arithmetic coding Conditioning) marker segment, as described in Annex B.

F.1.4.4.1.5 Initial conditions for DC statistical model

At the start of a scan and at the beginning of each restart interval, the difference for the previous DC value is defined to be zero in determining the conditioning state.

F.1.4.4.2 Statistical model for coding the AC coefficients

As shown in Table F.5, each statistics area for AC coding consists of a contiguous set of 245 statistics bins. Three bins are used for each value of the zig-zag index K, and two sets of 28 additional bins X2,...,X15,M2,...,M15 are used for coding the magnitude category and magnitude bits.

The value of SE (and also S0, SP and SN) is determined by the zig-zag index K. Since K is in the range 1 to 63, the lowest value for SE is 0 and the largest value for SP is 188. SS is not assigned a value in AC coefficient coding, as the signs of the coefficients are coded with a fixed probability value of 0.5 (Qe=X'5A1D', mps=0).

Table F.5 Statistical model for AC coefficient coding

Context-index	Value	Coding decision
SE	$3 \times (K-1)$	K=EOB
S0	SE+1	V=0
SS	fixed estimate	sign of V
SN,SP	S0+1	Sz<1
X1	S0+1	Sz<2
X2	AC_Context(K)	Sz<4
X3	X2+1	Sz<8
.	.	.
.	.	.
.	.	.
X15	X2+13	$Sz<2^{15}$
M2	X2+14	magnitude bits if Sz<4
M3	X3+14	magnitude bits if Sz<8
.	.	.
.	.	.
M15	X15+14	magnitude bits if $Sz<2^{15}$

The value of X2 is given by AC_Context(K). This gives X2=189 when K<=Kx and X2=217 when K>Kx, where Kx is defined using the DAC marker segment (see B.2.4.3)

Note that a X1 statistics bin is not used in this sequence. Instead, the 63x1 array of statistics bins for the magnitude category is used for two decisions. Once the magnitude bound has been determined - at statistics bin Xn, for example - a single statistics bin, Bn, is used to code the magnitude bit sequence for that bound.

F.1.4.4.2.1 Default conditioning for AC coefficient coding

The default value of Kx is 5. This may be modified using the DAC marker segment, as described in Annex B.

F.1.4.4.2.2 Initial conditions for AC statistical model

At the start of a scan and at each restart, all statistics bins are re-initialized to the standard default value described in Annex D.

F.1.5 Sequential DCT encoding process with Huffman coding and 12-bit precision

This process is identical to the sequential DCT process for 8-bit precision extended to four Huffman tables as documented in F.1.3, with the following changes.

F.1.5.1 Structure of DC code table for 12-bit input precision

The two's complement difference magnitudes are grouped into 16 categories, SSSS, and a Huffman code is created for each of the 16 difference magnitude categories.

The Huffman table for DC coding is extended as shown in Table F.6

F-19

Table F.6 Difference magnitude categories for DC coding

SSSS	Difference values
12	-4095..-2048,2048..4095
13	-8191..-4096,4096..8191
14	-16383..-8192,8192..16383
15	-32767..-16384,16384..32767

F.1.5.2 Structure of AC code table for 12-bit input precision

The general structure of the code table is extended as illustrated in Figure F.11.

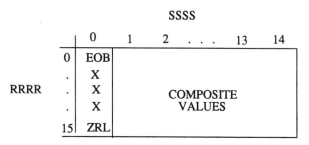

Figure F.11 - Two-dimensional value array for Huffman coding

The Huffman table for AC coding is extended as shown in Table F.7.

Table F.7 Values assigned to coefficient amplitude ranges.

SSSS	AC coefficients
11	-2047..-1024,1024..2047
12	-4095..-2048,2048..4095
13	-8191..-4096,4096..8191
14	-16383..-8192,8192..16383

F.1.6 Sequential DCT encoding process with arithmetic coding and 12-bit precision

The process is identical to the sequential DCT process for 8-bit precision except for changes in the precision of the FDCT computation.

The structure of the encoding procedure in F.1.4 is already defined for a 12-bit input precision.

F.2 Sequential DCT-based decoding processes

F.2.1 Sequential DCT-based control procedures and coding models

F.2.1.1 Control procedures for sequential DCT-based decoders

The control procedures for decoding an image and its constituent parts - the frame, scan, restart interval and MCU - are given in Figures E.6 to E.10. The procedure for decoding a MCU (Figure E.10) repetitively calls the procedure for decoding a data unit. For DCT-based decoders the data unit is an 8x8 block of samples.

F.2.1.2 Procedure for decoding an 8x8 block data unit

In the sequential DCT-based decoding process decoding an 8x8 block data unit consists of the following procedures:

(1) Decode DC coefficient for 8x8 block using the DC table assigned in the scan header.
(2) Decode AC coefficients for 8x8 block using the AC table assigned in the scan header.
(3) Dequantize using table assigned in the frame header and calculate the inverse 8x8 DCT.

F.2.1.3 Decoding models for the sequential DCT procedures

Two decoding procedures are used, one for the DC coefficient $ZZ(0)$ and the other for the AC coefficients $ZZ(1)...ZZ(63)$. The coefficients are decoded in the order in which they occur in ZZ, starting with the DC coefficient. The coefficients are represented as two's complement integers.

F.2.1.3.1 Decoding model for DC coefficients

The decoded difference, DIFF, is added to PRED, the DC value from the most recently decoded 8x8 block from the same component. Thus $ZZ(0) = PRED + DIFF$.

At the beginning of the scan and at the beginning of each restart interval, the prediction for the DC coefficient is initialized to 0.

F.2.1.3.2 Decoding model for AC coefficients

The AC coefficients are decoded in the order in which they occur in ZZ. When the EOB is decoded, all remaining coefficients in ZZ are set to zero.

F.2.1.4 Dequantization of the quantized DCT coefficients

The dequantization of the quantized DCT coefficients as described in Annex A, is accomplished by multiplying each quantized coefficient value by the quantization table value for that coefficient. The decoder shall be able to use up to four quantization tables.

F.2.1.5 Inverse DCT (IDCT)

The mathematical definition of the IDCT is given in A.3.3.

After computation of the IDCT, the signed output samples are level-shifted, as described in Annex A, converting the output to an unsigned representation. If necessary, the output samples should be clamped to stay within the range appropriate for the precision.

F.2.2 Baseline Huffman Decoding procedures

The baseline decoding procedure is for 8-bit sample precision. The decoder shall be capable of using up to two DC and two AC Huffman tables within one scan.

F.2.2.1 Huffman decoding of DC coefficients

The decoding procedure for the DC difference, DIFF, is:

>   T=DECODE
>   DIFF=RECEIVE(T)
>   DIFF=EXTEND(DIFF,T)

where DECODE is a procedure which returns the 8 bit value associated with the next Huffman code in the compressed image data (see F.2.2.3) and RECEIVE(T) is a procedure which places the next T bits of the serial bit string into the low order bits of DIFF, msb first. If T is zero, DIFF is set to zero. EXTEND is a procedure which converts the partially decoded DIFF value of precision T to the full precision difference. EXTEND is shown in Figure F.12.

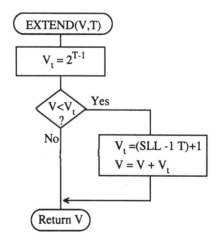

Figure F.12 - Extending the sign bit of a decoded value in V

F.2.2.2 Decoding procedure for AC coefficients

The decoding procedure for AC coefficients is shown in Figures F.13 and F.14.

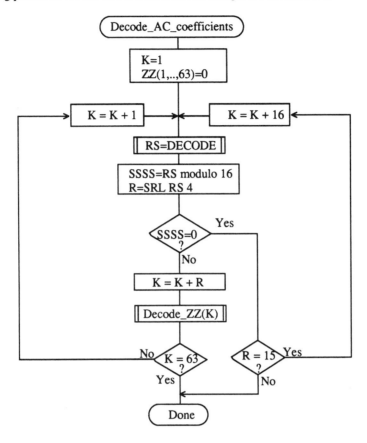

Figure F.13 - Huffman decoding procedure for AC coefficients

The decoding of the amplitude and sign of the nonzero coefficient is done in the procedure "Decode_ZZ(K)", shown in Figure F.14.

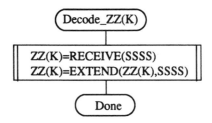

Figure F.14 - Decoding a nonzero AC coefficient

DECODE is a procedure which returns the value, RS, associated with the next Huffman code in the code stream (see F.2.2.3). The values SSSS and R are derived from RS. The value of SSSS is the four low order bits of the composite value and R contains the value of RRRR (the four high order bits of the composite value). The interpretation of these values is described in F.1.2.2. EXTEND is shown in Figure F.12.

F.2.2.3 The DECODE procedure

The DECODE procedure decodes an 8-bit value which, for the DC coefficient, determines the difference magnitude category. For the AC coefficient this 8-bit value determines the zero run length and nonzero coefficient category.

Three tables, HUFFVAL, HUFFCODE, and HUFFSIZE, have been defined in Annex C. This particular implementation of DECODE makes use of the ordering of the Huffman codes in HUFFCODE according to both value and code size. Many other implementations of DECODE are possible.

> NOTE: The values in HUFFVAL are assigned to each code in HUFFCODE and HUFFSIZE in sequence. There are no ordering requirements for the values in HUFFVAL which have assigned codes of the same length.

The implementation of DECODE described in this subclause uses three tables, MINCODE, MAXCODE and VALPTR, to decode a pointer to the HUFFVAL table. MINCODE, MAX-CODE and VALPTR each have 16 entries, one for each possible code size. MINCODE(I) contains the smallest code value for a given length I, MAXCODE(I) contains the largest code value for a given length I, and VALPTR(I) contains the index to the start of the list of values in HUFFVAL which are decoded by code words of length I. The values in MINCODE and MAXCODE are signed 16 bit integers; therefore, a value of -1 sets all of the bits.

The procedure for generating these tables is shown in Figure F.15.

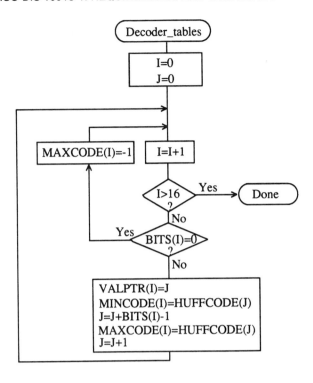

Figure F.15 - Decoder table generation

The procedure for DECODE is shown in Figure F.16. Note that the 8-bit "VALUE" is returned to the procedure which invokes DECODE.

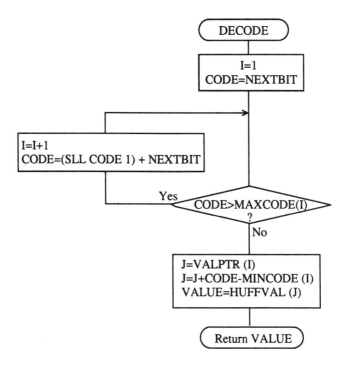

Figure F.16 - Procedure for DECODE

F.2.2.4 The RECEIVE procedure

RECEIVE(SSSS) is a procedure which places the next SSSS bits of the entropy-coded segment into the low order bits of DIFF, MSB first. It calls NEXTBIT and it returns the value of DIFF to the calling procedure.

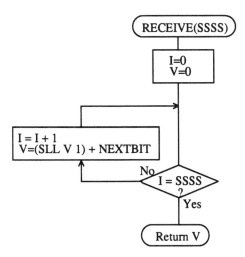

Figure F.17 - Procedure for RECEIVE(SSSS)

F.2.2.5 The NEXTBIT procedure

NEXTBIT reads the next bit of compressed data and passes it to higher level routines. It also intercepts and removes stuff bytes and detects markers. NEXTBIT reads the bits of a byte starting with the MSB.

Before starting the decoding of a scan, and after processing a RST marker, CNT is cleared. The compressed data are read one byte at a time, using the procedure NEXTBYTE. Each time a byte, B, is read, CNT is set to 8.

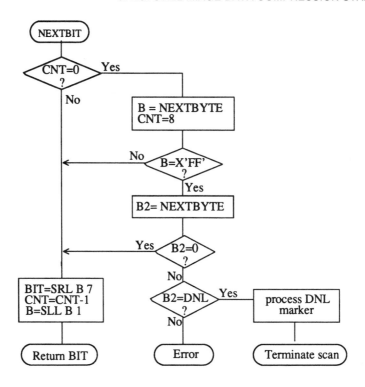

Figure F.18 - Procedure for fetching the next bit of compressed data

The only valid marker which may occur within the Huffman coded data is the RSTm marker. Other than the EOI or markers which may occur at or before the start of a scan, the only marker which can occur at the end of the scan is the DNL (define-number-of-lines).

Normally, the decoder will terminate the decoding at the end of the final restart interval before the terminating marker is intercepted. If the DNL marker is encountered, the current line count is set to the value specified by that marker. Since the DNL marker can only be used at the end of the first scan, the scan decode procedure must be terminated when it is encountered.

F.2.3   Sequential DCT decoding process with 8-bit precision extended to four sets of Huffman tables

This process is identical to the Baseline decoding process described in F.2.2, with the exception that the decoder shall be capable of using up to four DC and four AC Huffman tables within one scan. Four DC and four AC Huffman tables is the maximum allowed by this Specification.

F.2.4 Sequential DCT decoding process with arithmetic coding

This subclause describes the sequential DCT decoding process with arithmetic decoding.

The arithmetic decoding procedures for decoding binary decisions, initializing the statistics model, initializing the decoder, and resynchronizing the decoder are listed in Table D.3 of Annex D.

Some of the procedures in Table D.3 are used in the higher level control structure for scans and restart intervals described in F.2. At the beginning of scans and restart intervals, the probability estimates used in the arithmetic decoder are reset to the standard initial value as part of the Initdec procedure which restarts the arithmetic coder.

The statistical models defined in F.1.4.4 also apply to this decoding process.

The decoder shall be capable of using up to four DC and four AC conditioning tables and associated statistics areas within one scan.

F.2.4.1 Arithmetic decoding of DC coefficients

The basic structure of the decision sequence for decoding a DC difference value, DIFF, is shown in Figure F.19 The equivalent structure for the encoder is found in Figure F.4.

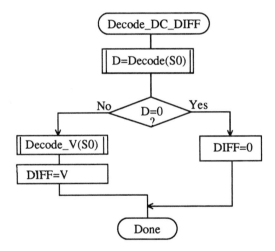

Figure F.19 - Arithmetic decoding of DC difference

The context-indices used in the DC decoding procedures are defined in Table F.4 (see F.1.4.4.1.3).

The "Decode" procedure returns the value "D" of the binary decision. If the value is not zero, the sign and magnitude of the nonzero DIFF must be decoded by the procedure "Decode_V(S0)".

F-29

F.2.4.2 Arithmetic Decoding of AC coefficients

The AC coefficients are decoded in the order that they occur in ZZ(1,...,63). The encoder procedure for the coding process is found in Figure F.5. Figure F.20 illustrates the decoding sequence.

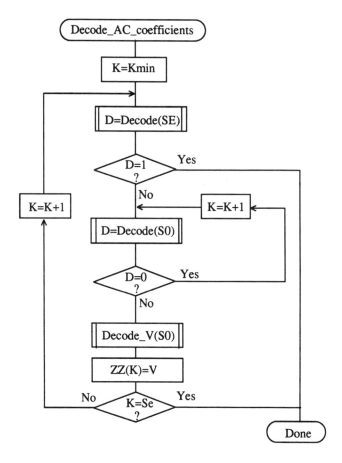

Figure F.20 - Procedure for decoding the AC coefficients

The context-indices used in the AC decoding procedures are defined in Table F.5 (see F.1.4.4.2).

In this Figure K is the index to the zig-zag sequence position. For the sequential scan, Kmin=1 and Se=63. The decision at the top of the loop is the EOB decision. If the EOB occurs (D=1), the remaining coefficients in the block are set to zero. The inner loop just below the EOB decoding decodes runs of zero coefficients. Whenever the coefficient is nonzero, "Decode_V" decodes the sign and magnitude of the coefficient. After each nonzero coefficient is decoded, the EOB decision is again decoded unless K=Se.

F.2.4.3 Decoding the binary decision sequence for nonzero DC differences and AC coefficients

Both the DC difference and the AC coefficients are represented as signed twos complement 16 bit integer values. The decoding decision tree for these signed integer values is the same for both the DC and AC coding models. Note, however, that the statistical models are not the same.

F.2.4.3.1 Arithmetic decoding of nonzero values

Denoting either DC differences or AC coefficients as V, the nonzero signed integer value of V is decoded by the sequence shown in Figure F.21. This sequence first decodes the sign of V. It then decodes the magnitude category of V (Decode_log2_Sz), and then decodes the low order magnitude bits (Decode_Sz_bits). Note that the value decoded for Sz must be incremented by 1 to get the actual coefficient magnitude.

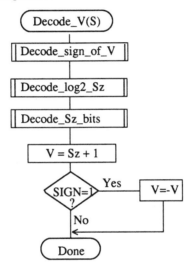

Figure F.21 - Sequence of procedures in decoding nonzero values of V

F.2.4.3.1.1 Decoding the sign

The sign is decoded by the following procedure:

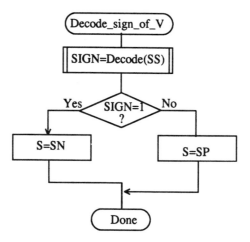

Figure F.22 - Decoding the sign of V

The context-indices are defined for DC decoding in Table F.4 and AC decoding in Table F.5.

If SIGN=0, the sign of the coefficient is positive; if SIGN=1, the sign of the coefficient is negative.

F.2.4.3.1.2 Decoding the magnitude category

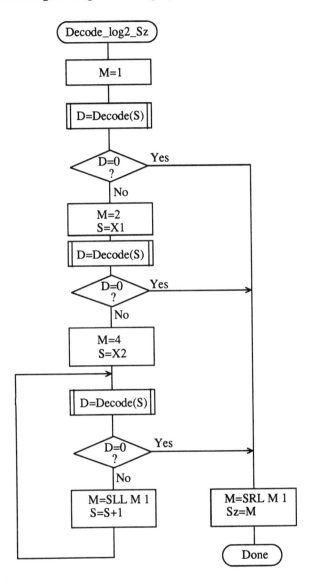

Figure F.23 - Decoding procedure to establish the magnitude category

F-33

The context-index S is set in Decode_sign_of_V and the context-index values X1 and X2 are defined for DC coding in Table F.4 and for AC coding in Table F.5.

In this figure, M set to the upper bound for the magnitude and shifted left until the decoded decision is zero. It is then shifted right by 1 to become the leading bit of the magnitude of Sz.

F.2.4.3.1.3 Decoding the exact value of the magnitude

After the magnitude category is decoded, the low order magnitude bits are decoded. These bits are decoded in order of decreasing bit significance. The procedure is shown in Figure F.24.

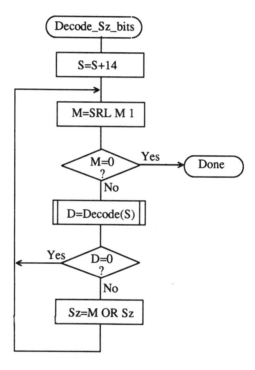

Figure F.24 - Decision sequence to decode the magnitude bit pattern

The context-index S is set in Decode_log2_Sz.

F.2.4.4 Decoder restart

The RSTm markers which are added to the compressed data between each restart interval have a two byte value which cannot be generated by the coding procedures. These two byte sequences can be located without decoding, and can therefore be used to resynchronize the decoder. RSTm markers can therefore be used for error recovery.

Before error recovery procedures can be invoked, the error condition must first be detected. Errors during decoding can show up in two places.

1) The decoder fails to find the expected marker at the point where it is expecting resynchronization.

2) Physically impossible data are decoded. For example, decoding a magnitude beyond the range of values allowed by the model is quite likely when the compressed data are corrupted by errors. For arithmetic decoders this error condition is extremely important to detect, as otherwise the decoder may reach a condition where it uses the compressed data very slowly.

> Note that some errors will not cause the decoder to lose synchronization. In addition, recovery is not possible for all errors; for example, errors in the headers are likely to be catastrophic. The two error conditions listed above, however, almost always cause the decoder to lose synchronization in a way which permits recovery.

In regaining synchronization, the decoder can make use of the modulo 8 coding restart interval number in the low order bits of the RSTm marker. By comparing the expected restart interval number to the value in the next RSTm marker in the compressed image data, the decoder can usually recover synchronization. It then fills in missing lines in the output data by replication or some other suitable procedure, and continues decoding. Of course, the reconstructed image will usually be highly corrupted for at least a part of the restart interval where the error occurred.

F.2.5 Sequential DCT decoding process with Huffman coding and 12-bit precision

This process is identical to the sequential DCT process defined for 8-bit sample precision and extended to four Huffman tables, as documented in F.2.3, but with the following changes.

F.2.5.1 Structure of DC Huffman decode table

The general structure of the DC Huffman decode table is extended as described in F.1.5.1

F.2.5.2 Structure of AC Huffman decode table

The general structure of the AC Huffman decode table is extended as described in F.1.5.2

F.2.6 Sequential DCT decoding process with arithmetic coding and 12-bit precision

The process is identical to the sequential DCT process for 8-bit precision except for changes in the precision of the IDCT computation.

The structure of the decoding procedure in F.2.4 is already defined for a 12-bit input precision.

Annex G (normative)

Progressive DCT-based mode of operation

This Annex provides a **functional specification** of the following coding processes for the progressive DCT-based mode of operation:

1) spectral-selection only, Huffman coding, 8-bit sample precision;

2) spectral-selection only, arithmetic coding, 8-bit sample precision;

3) full progression, Huffman coding, 8-bit sample precision;

4) full progression, arithmetic coding, 8-bit sample precision;

5) spectral-selection only, Huffman coding, 12-bit sample precision;

6) spectral-selection only, arithmetic coding, 12-bit sample precision;

7) full progression, Huffman coding, 12-bit sample precision;

8) full progression, arithmetic coding, 12-bit sample precision.

For each of these, the encoding process is specified in G.1, and the decoding process is specified in G.2. The functional specification is presented by means of specific flow charts for the various procedures which comprise these coding processes.

> NOTE - There is **no requirement** in this Specification that any encoder or decoder which embodies one of the above-named processes shall implement the procedures in precisely the manner specified by the flow charts in this Annex. It is necessary only that an encoder or decoder implement the **function** specified in this Annex. The sole criterion for an encoder or decoder to be considered in compliance with this Specification is that it satisfy the requirements given in clause 6 (for encoders) or clause 7 (for decoders), as determined by the compliance tests specified in Part 2.

The number of Huffman or arithmetic conditioning tables which may be used within the same scan is four.

Two complementary progressive procedures are defined, spectral selection and successive approximation.

In spectral selection the DCT coefficients of each block are segmented into frequency bands. The bands are coded in separate scans.

In successive approximation the DCT coefficients are divided by a power of two before coding. In the decoder the coefficients are multiplied by that same power of two before computing the IDCT. In the succeeding scans the precision of the coefficients is increased by one bit in each scan until full precision is reached.

An encoder or decoder implementing a full progression uses spectral selection within successive approximation. An allowed subset is spectral selection alone.

Figure G.1 illustrates the spectral selection and successive approximation progressive processes.

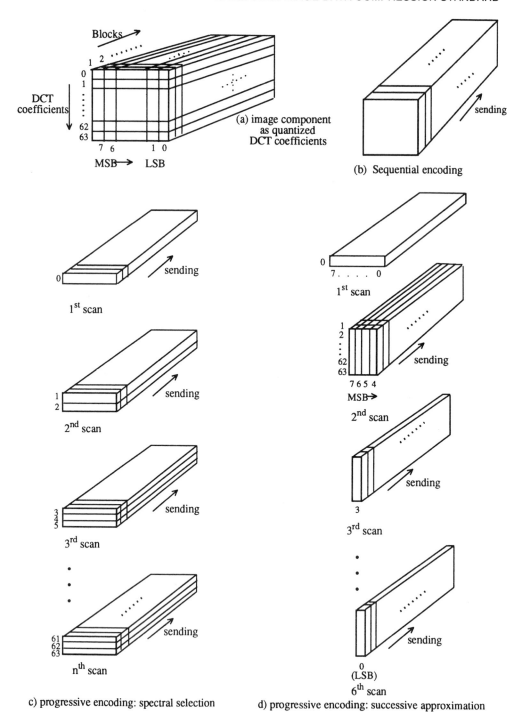

Figure G.1 - Spectral selection and successive approximation progressive processes

G.1 Progressive DCT-based encoding processes

G.1.1 Control procedures and coding models for progressive DCT-based procedures

G.1.1.1 Control procedures for progressive DCT-based encoders

The control procedures for encoding an image and its constituent parts - the frame, scan, restart interval and MCU - are given in Figures E.1 to E.5.

The control structure for encoding a frame is the same as for the sequential procedures. However, it is convenient to calculate the FDCT for the entire set of components in a frame before starting the scans. A buffer which is large enough to store all of the DCT coefficients may be used for this progressive mode of operation.

The number of scans is determined by the progression defined; the number of scans may be much larger than the number of components in the frame.

The procedure for encoding a MCU (Figure E.5) repetitively invokes the procedure for coding a data unit. For DCT-based encoders the data unit is an 8x8 block of samples.

Only a portion of each 8x8 block is coded in each scan, the portion being determined by the scan header parameters Ss, Se, Ah, and Al (see B.2.3). The procedures used to code portions of each 8x8 block are described in this annex. Note, however, that where these procedures are identical to those used in the sequential DCT-based mode of operation, the sequential procedures are simply referenced.

G.1.1.1.1 Spectral selection control

In spectral selection the zig-zag sequence of DCT coefficients is segmented into bands. A band is defined in the scan header by specifying the starting and ending indices in the zig-zag sequence. One band is coded in a given scan of the progression. DC coefficients are always coded separately from AC coefficients, and only scans which code DC coefficients may have interleaved blocks from more than one component. All other scans shall have only one component. With the exception of the first DC scans for the components, the sequence of bands defined in the scans need not follow the zig-zag ordering. For each component, a first DC scan shall precede any AC scans.

G.1.1.1.2 Successive approximation control

If successive approximation is used, the DCT coefficients are reduced in precision by the point transform (see A.4) defined in the scan header (see B.2.3). The successive approximation bit position parameter Al specifies the actual point transform, and the high four bits (Ah) - if there are preceding scans for the band - contain the value of the point transform used in those preceding scans. If there are no preceding scans for the band, Ah is zero.

Each scan which follows the first scan for a given band progressively improves the precision of the coefficients by one bit, until full precision is reached.

G-3

G.1.1.2 Coding models for progressive DCT-based encoders

If successive approximation is used, the DCT coefficients are reduced in precision by the point transform (see A.4) defined in the scan header (see B.2.3). These models also apply to the progressive DCT-based encoders, but with the following changes:

G.1.1.2.1 Progressive encoding model for DC coefficients

If Al is not zero, the point transform for DC coefficients shall be used to reduce the precision of the DC coefficients. If Ah is zero, the coefficient values (as modified by the point transform) shall be coded, using the procedure described in Annex F. If Ah is not zero, the least significant bit of the point transformed DC coefficients shall be coded, using the procedures described in this Annex.

G.1.1.2.2 Progressive encoding model for AC coefficients

If Al is not zero, the point transform for AC coefficients shall be used to reduce the precision of the AC coefficients. It Ah is zero, the coefficient values (as modified by the point transform) shall be coded using modifications of the procedures described in Annex F. These modifications are described in this Annex. If Ah is not zero, the precision of the coefficients shall be improved using the procedures described in this Annex.

G.1.2 Progressive encoding procedures with Huffman

G.1.2.1 Progressive encoding of DC coefficients with Huffman coding

The first scan for a given component shall encode the DC coefficient values using the procedures described in F.1.2.1. If the successive approximation bit position parameter Al is not zero, the coefficient values shall be reduced in precision by the point transform described in Annex A before coding.

In subsequent scans using successive approximation the least significant bits are appended to the compressed bit stream without compression or modification (see G.1.2.3), except for byte stuffing.

G.1.2.2 Progressive encoding of AC coefficients with Huffman coding

In spectral selection and in the first scan of successive approximation for a component, the AC coefficient coding model is similar to that used by the sequential procedures. However, the Huffman code tables are extended to include coding of runs of End-Of-Bands (EOBs).

The end-of-band run structure allows efficient coding of blocks which have only zero coefficients. An EOB run of length 5 means that the current block and the next four blocks have an end-of-band with no intervening nonzero coefficients. The EOB run length is limited only by the restart interval.

G-4

The extension of the code table is illustrated in Figure G.2.

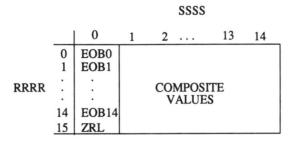

Figure G.2 - Two-dimensional value array for Huffman coding

The EOBn code sequence is defined as follows: Each EOBn code is followed by an extension field similar to the extension field for the coefficient amplitudes (but with positive numbers only). The number of bits appended to the EOBn code is the minimum number required to specify the run length.

Table G.1 EOBn code run length extensions

EOBn code	run length
EOB0	1
EOB1	2,3
EOB2	4..7
EOB3	8..15
EOB4	16 .31
EOB5	32..63
EOB6	64..127
EOB7	128..255
EOB8	256..511
EOB9	512..1023
EOB10	1024..2047
EOB11	2048..4095
EOB12	4096..8191
EOB13	8192..16383
EOB14	16384..32767

If an EOB run is greater than 32767, it is coded as a sequence of EOB runs of length 32767 followed by a final EOB run sufficient to complete the run.

At the beginning of each restart interval the EOB run count, EOBRUN, is set to zero. At the end of each restart interval any remaining EOB run is coded.

The Huffman encoding procedure for AC coefficients in spectral selection and in the first scan of successive approximation is illustrated in Figures G.3 and G.4.

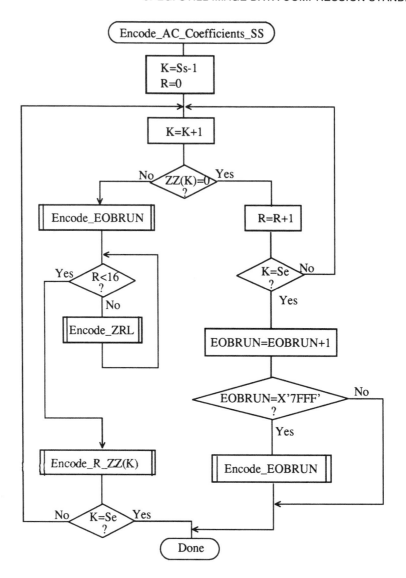

Figure G.3 - Procedure for progressive encoding of AC coefficients
with Huffman coding

In the preceding figure, Ss is the start of spectral selection, Se is the end of spectral selection, K is the index into the list of coefficients stored in the zig-zag sequence ZZ, R is the run length of zero coefficients, and EOBRUN is the run length of EOBs. EOBRUN is set to zero at the start of each restart interval.

If the scan header parameter Al (successive approximation bit position low) is not zero, the DCT coefficient values ZZ(K) in Figure G.3 and figures which follow in this annex shall be replaced by the point transformed values ZZ'(K), where ZZ'(K) is defined by:

$$ZZ'(K) = \frac{ZZ(K)}{2^{Al}}$$

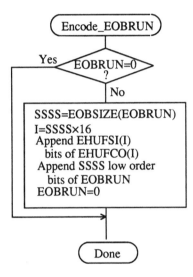

Figure G.4 - Progressive encoding of a nonzero AC coefficient

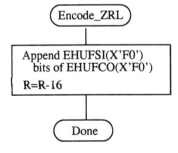

Figure G.5 - Encoding of the run of zero coefficients

G-7

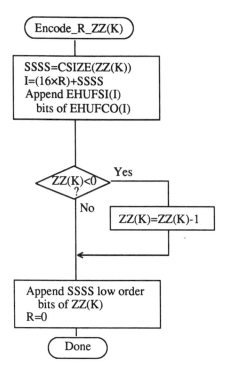

Figure G.6 - Encoding of the zero run and nonzero coefficient

EOBSIZE is a procedure which returns the size of the EOB extension field given the EOB run length as input. CSIZE is a procedure which maps an AC coefficient to the SSSS value defined in the subclauses on sequential encoding (F.1.1 and F.1.3).

G.1.2.3 Coding model for subsequent scans of successive approximation

The Huffman coding structure of the subsequent scans of successive approximation for a given component is similar to the coding structure of the first scan of that component.

The structure of the AC code table is identical to the structure described in sub-clause G.1.2.2. Each nonzero point transformed coefficient that has a zero history (i.e. that has a value ± 1, and therefore has not been coded in a previous scan) is defined by a composite 8-bit run length-magnitude value of the form:

RRRRSSSS

The four most significant bits, RRRR, give the number of zero coefficients that are between the current coefficient and the previously coded coefficient (or the start of band). Coefficients with nonzero history (a nonzero value coded in a previous scan) are skipped over when counting the zero coefficients. The four least significant bits, SSSS, provide the magnitude category of the nonzero coefficient; for a given component the value of SSSS can only be one.

The run length-magnitude composite value is Huffman coded and each Huffman code is followed by additional bits:

1) One bit codes the sign of the newly nonzero coefficient. A 0-bit codes a negative sign; a 1-bit codes a positive sign.

2) For each coefficient with a nonzero history, one bit is used to code the correction. A 0-bit means no correction and a 1-bit means that one shall be added to the (scaled) decoded magnitude of the coefficient.

Nonzero coefficients with zero history are coded with a composite code of the form:

$$HUFFCO(RRRRSSSS) \; + \; \text{additional bit (rule 1)} \; + \; \text{correction bit (rule 2)}$$

In addition whenever zero runs are coded with ZRL or EOBn codes, correction bits for those coefficients with nonzero history contained within the zero run are appended according to rule 2 above.

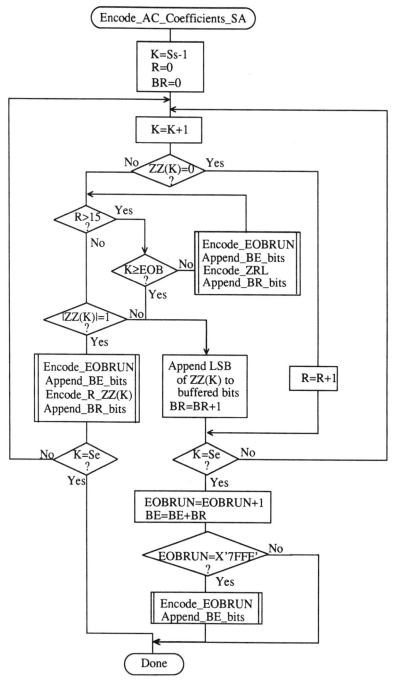

Figure G.7 - Successive approximation coding of AC coefficients using Huffman coding
G-10

For the Huffman coding version of Encode_AC_Coefficients_SA the EOB is defined to be the position of the last point transformed coefficient of magnitude 1 in the band. If there are no co-efficients of magnitude 1, the EOB is defined to be zero.

Note: the definition of EOB is different for Huffman and arithmetic coding procedures.

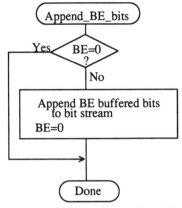

Figure G.8 - Transferring BE buffered bits from buffer to bit stream

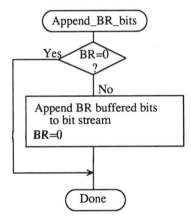

Figure G.9 - Transferring BR buffered bits from buffer to bit stream

In Figures G.7 and G.8 BE is the count of buffered correction bits at the start of coding of the block. BE is initialized to zero at the start of each restart interval. At the end of each restart interval any remaining buffered bits are appended to the bit stream following the last EOBn Huffman code and associated appended bits.

G-11

In Figures G.7 and G.9 BR is the count of buffered correction bits which are appended to the bit stream according to rule 2. BR is set to zero at the beginning of each Encode_AC_Coefficients_SA. At the end of each restart interval any remaining buffered bits are appended to the bit stream following the last Huffman code and associated appended bits.

G.1.3 Progressive encoding procedures with arithmetic coding

G.1.3.1 Progressive encoding of DC coefficients with arithmetic coding

The first scan for a given component shall encode the DC coefficient values using the procedures described in F.1.3.1. If the successive approximation bit position parameter is not zero, the coefficient values shall be reduced in precision by the point transform described in Annex A before coding.

In subsequent scans using successive approximation the least significant bits shall be coded as binary decisions using a fixed probability estimate of 0.5 (Qe=X'5A1D', MPS=0).

G.1.3.2 Progressive encoding of AC coefficients with arithmetic coding

Except for the point transform scaling of the DCT coefficients and the grouping the coefficients into bands, the first scan(s) of successive approximation is identical to the sequential encoding procedure described in F.1.4. If Kmin is equated to Ss, the index of the first AC coefficient index in the band, the flow chart shown in Figure F.4 applies. The EOB decision in that figure refers to the "end-of-band" rather than the "end-of-block". For the arithmetic coding version of Encode_AC_Coefficients_SA (and all other AC coefficient coding procedures) the EOB is defined to be the position following the last nonzero coefficient in the band.

Note: the definition of EOB is different for Huffman and arithmetic coding procedures.

The statistical model described in F.1.4 also holds. For this model the default value of Kx is 5. Other values of Kx may be specified using the DAC marker code (Annex B). The following calculation for Kx has proven to give good results for 8 bit precision samples:

$$Kx = Kmin + SRL \ (8 + Se - Kmin) \ 4$$

This expression reduces to the default of Kx=5 when the band is from index 1 to index 63.

G.1.3.3 Coding model for subsequent scans of successive approximation

The procedure "Encode_AC_Coefficient_SA" shown in Figure G.10 increases the precision of the AC coefficient values in the band by one bit.

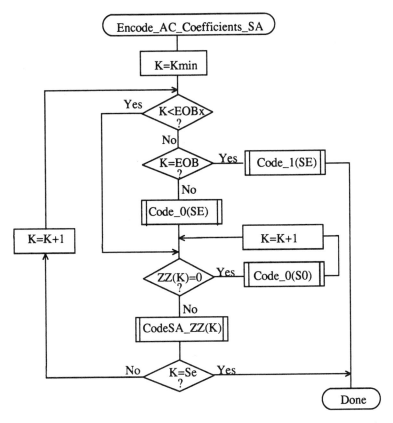

Figure G.10 - Subsequent successive approximation scans for coding of
AC coefficients using arithmetic coding

As in the first scan of successive approximation for a component, an EOB decision is coded at the start of the band and after each nonzero coefficient.

However, since the end-of-band index of the previous successive approximation scan for a given component, EOBx, is known from the data coded in the prior scan of that component, this decision is bypassed whenever the current index, K, is less than EOBx. As in the first scan(s), the EOB decision is also bypassed whenever the last coefficient in the band is not zero. The decision ZZ(K)=0 decodes runs of zero coefficients. If the decoder is at this step of the procedure, at least one nonzero coefficient remains in the band of the block being coded. If ZZ(K) is not zero, the procedure in Figure G.11 is followed to code the value.

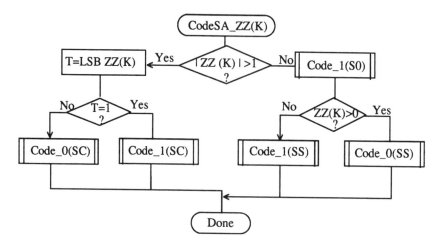

Figure G.11 - Coding nonzero coefficients for subsequent successive approximation scans

The context-indices in Figures G.10 and G.11 are defined in Table G.2 (see G.1.3.3.1). The signs of coefficients with magnitude of one are coded with a fixed probability value of 0.5 (Qe=X'5A1D', mps=0).

G.1.3.3.1 Statistical model for subsequent successive approximation scans

As shown in Table G.2, each statistics area for subsequent successive approximation scans of AC coefficients consists of a contiguous set of 189 statistics bins. The signs of coefficients with magnitude of one are coded with a fixed probability value of 0.5 (Qe=X'5A1D', mps=0).

Table G.2 Statistical model for subsequent scans of successive approximation coding of AC coefficients

Context-index	AC coding	Coding decision
SE	$3 \times (K-1)$	K=EOB
S0	SE+1	V=0
SS	fixed estimate	sign
SC	S0+1	LSB ZZ(K)=1

G.2 Progressive decoding of the DCT

The description of the computation of the IDCT and the dequantization procedure contained in A.3.3 and A.3.4 apply to the progressive operation.

Progressive decoding processes must be able to decompress compressed image data which requires up to four sets of Huffman or arithmetic coder conditioning tables within a scan.

In order to avoid repetition, detailed flow diagrams of progressive decoder operation are not included. Decoder operation is defined by reversing the function of each step described in the encoder flow charts, and performing the steps in reverse order.

Annex H (normative)

Lossless mode of operation

This Annex provides a **functional specification** of the following coding processes for the lossless mode of operation:

1) lossless processes with Huffman coding;

2) lossless processes with arithmetic coding.

For each of these, the encoding process is specified in H.1, and the decoding process is specified in H.2. The functional specification is presented by means of specific procedures which comprise these coding processes.

> NOTE - There is **no requirement** in this Specification that any encoder or decoder which embodies one of the above-named processes shall implement the procedures in precisely the manner specified in this Annex. It is necessary only that an encoder or decoder implement the **function** specified in this Annex. The sole criterion for an encoder or decoder to be considered in compliance with this Specification is that it satisfy the requirements given in clause 6 (for encoders) or clause 7 (for decoders), as determined by the compliance tests specified in Part 2.

The processes which provide for sequential lossless encoding and decoding are not based on the DCT. The processes used are spatial processes based on the coding model developed for the DC coefficients of the DCT. However, the model is extended by incorporating a set of selectable one and two dimensional predictors, and for interleaved data the ordering of samples for the one-dimensional predictor can be different from that used in the DCT-based processes.

Either Huffman coding or arithmetic coding entropy coding may be employed for these lossless encoding and decoding processes. The Huffman code table structure is extended to allow up to 16-bit precision for the input data. The arithmetic coder statistical model is extended to a two-dimensional form.

H.1 Lossless encoder processes

H.1.1 Lossless encoder control procedures

E.1 contains the encoder control procedures. In applying these procedures to the lossless encoder, the data unit is one sample.

Input data precision may be from 2 to 16 bits/sample. If the input data path has different precision from the input data, the data should always be aligned with the least significant bits of the input data path. Input data is represented as unsigned integers and is not level shifted prior to coding.

When the encoder is reset in the restart interval control procedure (see E.1.4), the prediction is reset to a default value. If arithmetic coding is used, the statistics are also reset.

For the lossless processes the restart interval shall be an integer multiple of the number of MCU in an MCU row, and for interleaved data the ordering of samples for the one-dimensional predictor can be different.

H.1.2 Coding model for lossless encoding

The coding model developed for encoding the DC coefficients of the DCT is extended to allow a selection from a set of seven one-dimensional and two-dimensional predictors . The predictor is selected in the scan header (see annex B). Each component in the scan is modeled independently, using predictions derived from neighboring samples of that component.

H.1.2.1 Prediction

Figure H.1 shows the relationship between the positions (a, b, c) of the reconstructed neighboring samples used for prediction and the position of x, the sample being coded.

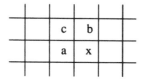

Figure H.1 - Relationship between sample and prediction samples

Define Px to be the prediction and Ra, Rb, and Rc to be the reconstructed samples immediately to the left, immediately above, and diagonally to the left of the current sample. The allowed predictors, one of which is selected in the scan header, are listed in Table H.1

Table H.1 Predictors for lossless coding

Selection-value	Prediction
0	no prediction (See Annex J)
1	Px = Ra
2	Px = Rb
3	Px = Rc
4	Px = Ra+Rb-Rc
5	Px = Ra+((Rb-Rc)/2)
6	Px = Rb+((Ra-Rc)/2)
7	Px = (Ra+Rb)/2

Selection-value 0 shall only be used for differential coding in the hierarchical mode of operation. Selections 1, 2 and 3 are one-dimensional predictors and selections 4, 5, 6, and 7 are two-dimensional predictors.

The one-dimensional horizontal predictor (prediction sample Ra) is used for the first line of samples at the start of the scan and at the beginning of each restart interval. The selected predictor is used for all other lines. The sample from the line above (prediction sample Rb) is used at the start of each line, except for the first line. At the beginning of the first line and at the beginning of each restart interval the prediction value of 2^{P-1} is used, where P is the input precision.

If the point transformation parameter (see A.4) is nonzero, the prediction value at the beginning of the first lines and the beginning of each restart interval is 2^{P-Pt-1}, where Pt is the value of the point transformation parameter.

Each prediction is calculated with full integer arithmetic precision, and without clamping of either underflow or overflow beyond the input precision bounds. For example, if Ra and Rb are both 16-bit integers, the sum is a 17-bit integer. After dividing the sum by 2 (predictor 7), the prediction is a 16-bit integer.

For simplicity of implementation, the divide by 2 in the prediction selections 5 and 6 of Table H.1 is done by an arithmetic-right-shift of the integer values.

The difference between the prediction value and the input is calculated modulo 2^{16}. In the decoder the difference is decoded and added, modulo 2^{16}, to the prediction.

H.1.2.2 Huffman coding of the modulo difference

The Huffman coding procedures defined in Annex F for coding the DC coefficients are used to code the modulo 2^{16} differences. The table for DC coding contained in Tables F.1 and F.6 is extended by one additional entry. No extra bits are appended after SSSS=16 is encoded.

Table H.2. Difference categories for lossless Huffman coding

SSSS	Difference values
0	0
1	-1,1
2	-3,-2,2,3
3	-7..-4,4..7
4	-15..-8,8..15
5	-31..-16,16..31
6	-63..-32,32..63
7	-127..-64,64..127
8	-255..-128,128..255
9	-511..-256,256..511
10	-1023..-512,512..1023
11	-2047..-1024,1024..2047
12	-4095..-2048,2048..4095
13	-8191..-4096,4096..8191
14	-16383..-8192,8192..16383
15	-32767..-16384,16384..32767
16	32768

H.1.2.3 Arithmetic coding of the modulo difference

The statistical model defined for the DC coefficient arithmetic coding model (see F.1.4.4.1) is generalized to a two-dimensional form in which differences coded for the sample to the left and for the line above are used for conditioning.

H.1.2.3.1 Two-dimensional statistical model

The binary decisions are conditioned on the differences coded for the neighboring samples immediately above and immediately to the left from the same component. As in the coding of the DC coefficients, the differences are classified into 5 categories: zero(0), small positive (+S), small negative (-S), large positive (+L) and large negative (-L). The two independent difference categories combine to give 25 different conditioning states. Figure H.2 shows the two-dimensional array of conditioning indices. For each of the 25 conditioning states probability estimates for four binary decisions are kept.

difference above (position b)

		0	+S	-S	+L	-L
	0	0	4	8	12	16
difference	+S	20	24	28	32	36
to left	-S	40	44	48	52	56
(position a)	+L	60	64	68	72	76
	-L	80	84	88	92	96

Figure H.2 - 5x5 Conditioning array for two-dimensional statistical model

At the beginning of the scan and each restart interval the conditioning derived from the line above is set to zero for the first line of each component.

H.1.2.3.2 Assignment of statistical bins to the DC binary decision tree

Each statistics area for lossless coding consists of a contiguous set of 158 statistics bins. The first 100 bins consist of 25 sets of four bins selected by a context-index S0. The value of S0 is given by L_Context(Da,Db), which provides a value of 0, 4, ... , 92 or 96, depending on the difference classifications of Da and Db (see H.1.2.3.1.). The value for S0 provided by L_Context(Da,Db) is from the array in Figure H.2.

The remaining 58 bins consist of two sets of 29 bins, X1,...,X15,M2,...,M15, which are used to code magnitude category decisions and magnitude bits. The value of X1 is given by X1_Context(Db), which provides a value of 100 when Db is in the zero, small positive or small negative categories and a value of 129 when Db is in the large positive or large negative categories.

H-4

The assignment of statistical bins to the binary decision tree used for coding the difference is given in Table H.3.

Table H.3. Statistical model for lossless coding

Context-index	Value	Coding decision
S0	L_Context(Da,Db)	V=0
SS	S0+1	sign
SP	S0+2	Sz<1 if V>0
SN	S0+3	Sz<1 if V<0
X1	X1_Context(Db)	Sz<2
X2	X1+1	Sz<4
X3	X1+2	Sz<8
.	.	.
X15	X1+14	$Sz<2^{15}$
M2	X2+14	magnitude bits if Sz<4
M3	X3+14	magnitude bits if Sz<8
.	.	.
M15	X15+14	magnitude bits if $Sz<2^{15}$

H.1.2.3.3 Default conditioning bounds

The bounds, L and U, for determining the conditioning category have the default values L=0 and U=1. Other bounds may be set using the DAC (Define-Arithmetic-Conditioning) marker segment, as described in Annex B.

H.1.2.3.4 Initial conditions for statistical model

At the start of a scan and at each restart, all statistics bins are re-initialized to the standard default value described in Annex D.

H.2 Lossless decoder processes

Lossless decoders may employ either Huffman decoding or arithmetic decoding. They shall be capable of using up to four tables in a scan. Lossless decoders shall be able to decode encoded image source data with any input precision from 2 to 16 bits per sample.

H.2.1 Lossless decoder control procedures

Annex E.2 contains the decoder control procedures. In applying these procedures to the lossless decoder the data unit is one sample.

When the decoder is reset in the restart interval control procedure (see E.2.4) the prediction is reset to the same value used in the encoder (see H.1.2.1). If arithmetic coding is used, the statistics are also reset.

Restrictions on the restart interval are specified in H.1.1.

H.2.2 Coding model for lossless decoding

The predictor calculations defined in H.1.2 also apply to the lossless decoder processes.

The lossless decoders, decode the differences and add them, modules 2^{16}, to the predictions to create the output. The lossless decoders shall be able to interpret the point transform parameter, and if non zero, multiply the output of the lossless decoder by 2^{Pt}.

In order to avoid repetition, detailed flow charts of the lossless decoding procedures are omitted.

Annex J (normative)

Hierarchical mode of operation

This Annex provides a **functional specification** of the following coding processes for the hierarchical mode of operation:

For the first 12 processes listed in Table J.1 below, the last frame for a component may either be coded using the process defined for the previous frames in the image, or with the lossless sequential differential process (defined in this annex) which matches the entropy coding method of the previous frames in the image.

Table J.1: Coding processes for hierarchical mode

P	Type	Ec	Precision	Annex
1	Extended sequential	H	8	F, process 2
2	Extended sequential	A	8	F, process 3
3	Extended sequential	H	12	F, process 4
4	Extended sequential	A	12	F, process 5
5	Spectral selection only	H	8	G, process 1
6	Spectral selection only	A	8	G, process 2
7	Full progression	H	8	G, process 3
8	Full progression	A	8	G, process 4
9	Spectral selection only	H	12	G, process 5
10	Spectral selection only	A	12	G, process 6
11	Full progression	H	12	G, process 7
12	Full progression	A	12	G, process 8
13	Lossless (DPCM)	H	2 - 16	H, process 1
14	Lossless (DPCM)	A	2 - 16	H, process 2

P=Process, Ec=Entropy coding, H=Huffman coding, A=Arithmetic coding.

NOTE: Baseline sequential is a subset of the extended sequential process. Therefore, it is not excluded from use in any of the above processes that specify extended sequential Huffman.

For each of these, the encoding processes is specified in J.1, and the decoding process is specified in J.2. The functional specification is presented by means of specific flow charts for the various procedures which comprise these coding processes.

NOTE - There is **no requirement** in this Specification that any encoder or decoder which embodies one of the above-named processes shall implement the procedures in precisely the manner specified by the flow charts in this Annex. It is necessary only that an encoder or decoder implement the **function** specified in this Annex. The sole criterion for an encoder or decoder to be considered in compliance with this Specification is that it satisfy the requirements given in clause 6 (for encoders) or clause 7 (for decoders), as determined by the compliance tests specified in Part 2.

In the hierarchical mode of operation each component is encoded or decoded in a non-differential frame followed by a sequence of differential frames. A non-differential frame shall use the procedures defined in Annexes F, G, and H. Differential frame procedures are defined in this Annex.

J.1 Hierarchical encoding

J.1.1 Hierarchical control procedure for encoding an image

The control structure for encoding of an image using the hierarchical mode is given in Figure J.1.

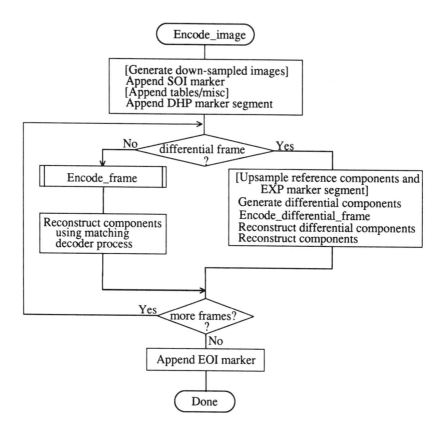

Figure J.1 - Hierarchical control procedure for encoding an image

In Figure J.1 procedures in brackets shall be performed whenever the particular hierarchical encoding sequence being followed requires them.

In the hierarchical mode the define-hierarchical-process (DHP) marker segment shall be placed in the compressed image data before the first start-of-frame. The DHP segment is used to signal the size of the image components of the completed image. The syntax of the DHP segment is specified in Annex B.

The first frame for each component or group of components in a hierarchical process shall be encoded by a non-differential frame. Differential frames shall then be used to encode the two's complement differences between source input components (possibly downsampled) and the reference components (possibly upsampled). The reference components are reconstruct components created by previous frames in the hierarchical process. For either differential or non-differential frames, reconstructions of the components shall be generated if needed as reference components for a subsequent frame in the hierarchical process.

Resolution changes may occur between hierarchical frames in a hierarchical process. These changes occur if down-sampling filters are used to reduce the spatial resolution of some or all of the components of the source image. When the resolution of a reference component does not match the resolution of the component input to a differential frame, an upsampling filter shall be used to increase the spatial resolution of the reference component. The EXP marker segment shall be added to the compressed image data before the start-of-frame whenever upsampling of a reference component is required. No more than one EXP marker segment shall precede a given frame.

Any of the markers segments allowed before a start-of-frame for the encoder process selected may be used before either non-differential or differential frames.

For 16 bit input precision (lossless encoder), the differential components which are input to a differential frame are calculated modulo 2^{16}. The reconstructed components calculated from the reconstructed differential components are also calculated modulo 2^{16}.

If a hierarchical encoder process uses a DCT encoder process for the first frame, all frames in the hierarchical process except for the final frame for each component shall use the DCT encoder processes defined in either Annex F or Annex G, or the modified DCT encoder processes defined in this Annex. The final frame may use a modified lossless process defined in this Annex.

If a hierarchical encoder process uses a lossless encoded process for the first frame, all frames in the hierarchical process shall use a lossless encoder process defined in Annex H, or a modified lossless process defined in this Annex.

J.1.1.1 Downsampling filter

The downsampled components are generated using a downsampling filter that is not specified in this Specification. This filter should, however, be consistent with the upsampling filter. An example of a downsampling filter is provided in K.5.

J.1.1.2 Upsampling filter

The upsampling filter increases the spatial resolution by a factor of two horizontally, vertically, or both. Bi-linear interpolation is used for the upsampling filter, as illustrated in Figure J.2.

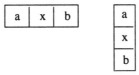

Figure J.2 - Diagram of sample positions for upsampling rules

The rule for calculating the interpolated value is;

$$Px = (Ra+Rb)/2$$

where Ra and Rb are sample values from adjacent positions a and b of the lower resolution image and Px is the interpolated value. The division indicates truncation, not rounding. The left-most column of the upsampled image matches the left-most column of the lower resolution image. The top row of the upsampled image matches the top row of the lower resolution image. The right column and the bottom row of the lower resolution image are replicated to provide the values required for the right column edge and bottom row interpolations. The upsampling process always doubles the line length or the number of rows.

If both horizontal and vertical expansions are signaled, they are done in sequence - first the horizontal expansion and then the vertical.

J.1.2 Control procedure for encoding a differential frame

The control procedures in Annex E for frames, scans, restart intervals and MCU also apply to the encoding of differential frames, and the scans, restart intervals and MCU from which the differential frame is constructed. The differential frames differ from the frames of Annexes F, G, and H only at the coding model level.

J.1.3 Encoder coding models for differential frames

The coding models defined in Annexes F, G, and H are modified to allow them to be used for coding of two's complement differences.

J.1.3.1 Modifications to encoder DCT encoding models for differential frames

Two modifications are made to the DCT coding models to allow them to be used in differential frames. First, the FDCT of the differential input is calculated without the level shift. Second, the DC coefficient of the DCT is coded directly - without prediction.

J.1.3.2 Modifications to lossless encoding models for differential frames

One modification is made to the lossless coding models. The difference is coded directly - without prediction. The prediction selection parameter in the scan header shall be set to zero. The point transform which may be applied to the differential inputs is defined in Annex A.

J.1.4 Modifications to the entropy encoders for differential frames

The coding of two's complement differences requires one extra bit of precision for the Huffman coding of AC coefficients. The extension to Tables F.1 and F.7 is given in Table J.2

Table J.2 Modifications to table of AC coefficient amplitude ranges

SSSS	AC coefficients
15	-32767..-16384, 16384..32767

The arithmetic coding models are already defined for the precision needed in differential frames.

J.2 Hierarchical decoding

J.2.1 Hierarchical control procedure for decoding an image

The control structure for decoding an image using the hierarchical mode is given in Figure J.3.

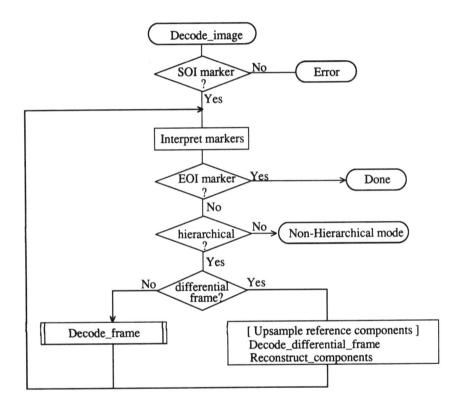

Figure J.3 - Hierarchical control procedure for decoding an image

The Interpret markers procedure shall decode the markers which may precede the SOF marker, continuing this decoding until either a SOF or EOI marker is found. If the DHP marker is encountered before the first frame, a flag is set which selects the hierarchical decoder at the "hierarchical?" decision point. In addition to the DHP marker (which shall precede any SOF) and the EXP marker (which shall precede any differential SOF requiring resolution changes in the reference components), any other markers which may precede a SOF shall be interpreted to the extent required for decoding of the compressed image data.

If a differential SOF marker is found, the differential frame path is followed. If the EXP was encountered in the Interpret markers procedure, the reference components for the frame shall be upsampled as required by the parameters in the EXP segment. The upsampling procedure described in J.1.1.2 shall be followed.

The Decode_differential_frame procedure generates a set of differential components. These differential components shall be added, modulo 2^{16}, to the upsampled reference components in the Reconstruct_components procedure. This creates a new set of reference components which shall be used when required in subsequent frames of the hierarchical process.

J.2.2　Control procedure for decoding a differential frame

The control procedures in Annex E for frames, scans, restart intervals and MCU also apply to the decoding of differential frames and the scans, restart intervals and MCU from which the differential frame is constructed. The differential frame differs from the frames of Annexes F, G, and H only at the decoder coding model level.

J.2.3　Decoder coding models for differential frames

The decoding models described in Annexes F, G, and H are modified to allow them to be used for decoding of two's complement differential components.

J.2.3.1　Modifications to the differential frame decoder DCT coding model

Two modifications are made to the decoder DCT coding models to allow them to code differential frames. First, the IDCT of the differential output is calculated without the level shift. Second, the DC coefficient of the DCT is decoded directly - without prediction.

J.2.3.2　Modifications to the differential frame decoded lossless coding model

One modification is made to the lossless decoder coding model. The difference is decoded directly - without prediction. If the point transformation parameter in the scan header is not zero, the point transform, defined in Annex A, shall be applied to the differential output.

J.2.4　Modifications to the entropy decoders for differential frames

The decoding of two's complement differences requires one extra bit of precision in the Huffman code table. This is described in J.1.4. The arithmetic coding models are already defined for the precision needed in differential frames.

Annex K (informative)

Examples and guidelines

This annex provides examples of various tables, procedures, and other guidelines.

K.1 Quantization tables for luminance and chrominance components

Two examples of quantization tables are given in tables K.1 and K.2. These are based on psychovisual thresholding and are derived empirically using luminance and chrominance and 2:1 horizontal subsampling. These tables are provided as examples only and are not necessarily suitable for any particular application. These quantization values have been used with good results on 8 bit per sample luminance and chrominance images of the format illustrated in Figure 13. Note that these quantization values are appropriate for the DCT normalization defined in A.3.3.

Table K.1 Luminance quantization table

16	11	10	16	24	40	51	61
12	12	14	19	26	58	60	55
14	13	16	24	40	57	69	56
14	17	22	29	51	87	80	62
18	22	37	56	68	109	103	77
24	35	55	64	81	104	113	92
49	64	78	87	103	121	120	101
72	92	95	98	112	100	103	99

Table K.2 Chrominance quantization table

17	18	24	47	99	99	99	99
18	21	26	66	99	99	99	99
24	26	56	99	99	99	99	99
47	66	99	99	99	99	99	99
99	99	99	99	99	99	99	99
99	99	99	99	99	99	99	99
99	99	99	99	99	99	99	99
99	99	99	99	99	99	99	99

If these quantization values are divided by 2, the resulting reconstructed image is usually nearly indistinguishable from the source image.

K.2 A procedure for generating the lists which specify a Huffman code table

A Huffman table is generated from a collection of statistics in two steps. The first step is the generation of the list of lengths and values which are in accord with the rules for generating the Huffman code tables. The second step is the generation of the Huffman code table from the list of lengths and values.

K-1

The first step, the topic of this section, is needed only for custom Huffman table generation and is done only in the encoder. In this step the statistics are used to create a table associating each value to be coded with the size (in bits) of the corresponding Huffman code. This table is sorted by code size.

A procedure for creating a Huffman table for a set of up to 256 symbols is shown in Figure K.1. Three vectors are defined for this procedure:

FREQ(V)	frequency of occurrence of symbol V.
CODESIZE(V)	code size of symbol V.
OTHERS(V)	index to next symbol in chain of all symbols in current branch of code tree.

where V goes from 0 to 256.

Before starting the procedure, the values of FREQ are collected for V=0 to 255 and the FREQ value for V=256 is set to 1 to reserve one code point. FREQ values for unused symbols are defined to be zero. In addition, the entries in CODESIZE are all set to 0, and the indices in OTHERS are set to -1, the value which terminates a chain of indices. Reserving one code point guarantees that no code word can ever be all '1' bits.

The search for the entry with the least value of FREQ(V) selects the largest value of V with the least value of FREQ(V) greater than zero.

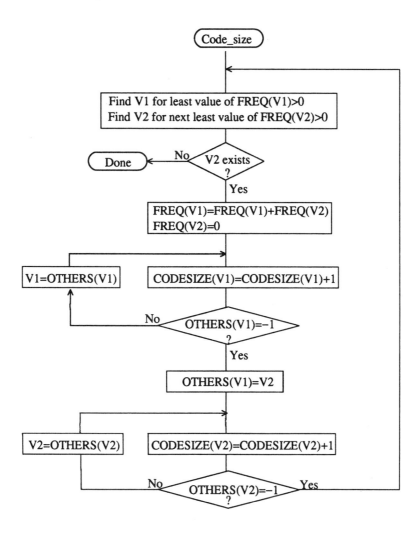

Figure K.1 – Procedure to find Huffman code sizes

The procedure "Find V1 for least value of FREQ(V1)>0" always selects the value with the largest value of V1 when more than one V1 with the same frequency occurs. The reserved code point is then guaranteed to be in the longest code word category.

Once the code lengths for each symbol have been obtained, the number of codes of each length is obtained using the procedure in Figure K.2. The count for each size is contained in the list, BITS. The counts in BITS are zero at the start of the procedure. The procedure assumes that the probabilities are large enough that code lengths greater than 32 bits never occur. Note that until the final Adjust_BITS procedure is complete, BITS may have more than the 16 entries required in the table specification (Annex C)

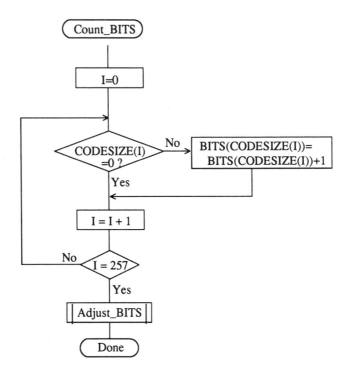

Figure K. 2 – Procedure to find the number of codes of each size

Figure K.3 gives the procedure for adjusting the BITS list so that no code is longer than 16 bits. Since symbols are paired for the longest Huffman code, the symbols are removed from this length category two at a time. The prefix for the pair (which is one bit shorter) is allocated to one of the pair; then (skipping the BITS entry for that prefix length) a code word from the next shortest nonzero BITS entry is converted into a prefix for two code words one bit longer. After the BITS list is reduced to a maximum code length of 16 bits, the last step removes the reserved code point from the code length count.

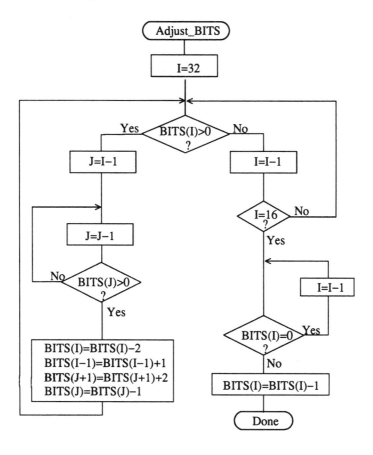

Figure K.3 − Procedure for limiting code lengths to 16 bits

K-5

The input values are sorted according to code size as shown in Figure K.4. HUFFVAL is the
list containing the input values associated with each code word, in order of increasing code
length.

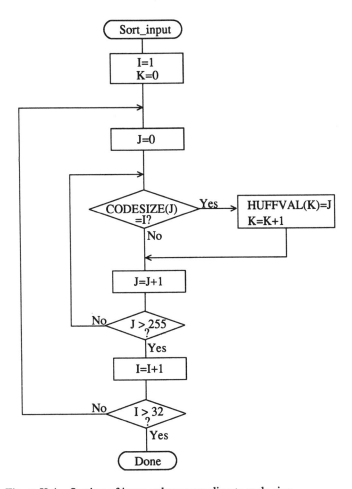

Figure K.4 – Sorting of input values according to code size

At this point, the list of code lengths (BITS) and the list of values (HUFFVAL) can be used to
generate the code tables. These procedures are described in Annex C.

K.3 Typical Huffman tables for 8-bit precision luminance and chrominance

Huffman table-specification syntax is specified in B.2.4.2.

K.3.1 Typical Huffman tables for the DC coefficients

Tables K.3 and K.4 give Huffman tables which have been developed from the average statistics of a large set of video images with 8 bit precision. Table K.3 is appropriate for luminance components and Table K.4 is appropriate for chrominance components. Although there are no default tables, these tables may prove to be useful for many applications.

Table K.3 Table for luminance DC difference

Category	Code length	Code word
0	2	00
1	3	010
2	3	011
3	3	100
4	3	101
5	3	110
6	4	1110
7	5	11110
8	6	111110
9	7	1111110
10	8	11111110
11	9	111111110

Table K.4 Table for chrominance DC difference

Category	Code length	Code word
0	2	00
1	2	01
2	2	10
3	3	110
4	4	1110
5	5	11110
6	6	111110
7	7	1111110
8	8	11111110
9	9	111111110
10	10	1111111110
11	11	11111111110

K.3.2 Typical Huffman tables for the AC coefficients

Tables K.5 and K.6 give Huffman tables for the AC coefficients which have been developed from the average statistics of a large set of images with 8 bit precision. Table K.5 is appropriate for luminance components and Table K.6 is appropriate for chrominance components. Although there are no default tables, these tables may prove to be useful for many applications.

K-7

Table K.5 Table for luminance AC coefficients

Run/Size	Code length	Code word
0/0 (EOB)	4	1010
0/1	2	00
0/2	2	01
0/3	3	100
0/4	4	1011
0/5	5	11010
0/6	7	1111000
0/7	8	11111000
0/8	10	1111110110
0/9	16	1111111110000010
0/A	16	1111111110000011
1/1	4	1100
1/2	5	11011
1/3	7	1111001
1/4	9	111110110
1/5	11	11111110110
1/6	16	1111111110000100
1/7	16	1111111110000101
1/8	16	1111111110000110
1/9	16	1111111110000111
1/A	16	1111111110001000
2/1	5	11100
2/2	8	11111001
2/3	10	1111110111
2/4	12	111111110100
2/5	16	1111111110001001
2/6	16	1111111110001010
2/7	16	1111111110001011
2/8	16	1111111110001100
2/9	16	1111111110001101
2/A	16	1111111110001110
3/1	6	111010
3/2	9	111110111
3/3	12	111111110101
3/4	16	1111111110001111
3/5	16	1111111110010000
3/6	16	1111111110010001
3/7	16	1111111110010010
3/8	16	1111111110010011
3/9	16	1111111110010100
3/A	16	1111111110010101
4/1	6	111011

K-8

4/2	10	1111111000
4/3	16	1111111110010110
4/4	16	1111111110010111
4/5	16	1111111110011000
4/6	16	1111111110011001
4/7	16	1111111110011010
4/8	16	1111111110011011
4/9	16	1111111110011100
4/A	16	1111111110011101
5/1	7	1111010
5/2	11	11111110111
5/3	16	1111111110011110
5/4	16	1111111110011111
5/5	16	1111111110100000
5/6	16	1111111110100001
5/7	16	1111111110100010
5/8	16	1111111110100011
5/9	16	1111111110100100
5/A	16	1111111110100101
6/1	7	1111011
6/2	12	111111110110
6/3	16	1111111110100110
6/4	16	1111111110100111
6/5	16	1111111110101000
6/6	16	1111111110101001
6/7	16	1111111110101010
6/8	16	1111111110101011
6/9	16	1111111110101100
6/A	16	1111111110101101
7/1	8	11111010
7/2	12	111111110111
7/3	16	1111111110101110
7/4	16	1111111110101111
7/5	16	1111111110110000
7/6	16	1111111110110001
7/7	16	1111111110110010
7/8	16	1111111110110011
7/9	16	1111111110110100
7/A	16	1111111110110101
8/1	9	111111000
8/2	15	111111111000000
8/3	16	1111111110110110
8/4	16	1111111110110111
8/5	16	1111111110111000
8/6	16	1111111110111001
8/7	16	1111111110111010

K-9

8/8	16	1111111110111011
8/9	16	1111111110111100
8/A	16	1111111110111101
9/1	9	111111001
9/2	16	1111111110111110
9/3	16	1111111110111111
9/4	16	1111111111000000
9/5	16	1111111111000001
9/6	16	1111111111000010
9/7	16	1111111111000011
9/8	16	1111111111000100
9/9	16	1111111111000101
9/A	16	1111111111000110
A/1	9	111111010
A/2	16	1111111111000111
A/3	16	1111111111001000
A/4	16	1111111111001001
A/5	16	1111111111001010
A/6	16	1111111111001011
A/7	16	1111111111001100
A/8	16	1111111111001101
A/9	16	1111111111001110
A/A	16	1111111111001111
B/1	10	1111111001
B/2	16	1111111111010000
B/3	16	1111111111010001
B/4	16	1111111111010010
B/5	16	1111111111010011
B/6	16	1111111111010100
B/7	16	1111111111010101
B/8	16	1111111111010110
B/9	16	1111111111010111
B/A	16	1111111111011000
C/1	10	1111111010
C/2	16	1111111111011001
C/3	16	1111111111011010
C/4	16	1111111111011011
C/5	16	1111111111011100
C/6	16	1111111111011101
C/7	16	1111111111011110
C/8	16	1111111111011111
C/9	16	1111111111100000
C/A	16	1111111111100001
D/1	11	11111111000
D/2	16	1111111111100010
D/3	16	1111111111100011

K-10

D/4	16	1111111111100100
D/5	16	1111111111100101
D/6	16	1111111111100110
D/7	16	1111111111100111
D/8	16	1111111111101000
D/9	16	1111111111101001
D/A	16	1111111111101010
E/1	16	1111111111101011
E/2	16	1111111111101100
E/3	16	1111111111101101
E/4	16	1111111111101110
E/5	16	1111111111101111
E/6	16	1111111111110000
E/7	16	1111111111110001
E/8	16	1111111111110010
E/9	16	1111111111110011
E/A	16	1111111111110100
F/0 (ZRL)	11	11111111001
F/1	16	1111111111110101
F/2	16	1111111111110110
F/3	16	1111111111110111
F/4	16	1111111111111000
F/5	16	1111111111111001
F/6	16	1111111111111010
F/7	16	1111111111111011
F/8	16	1111111111111100
F/9	16	1111111111111101
F/A	16	1111111111111110

Table K.6 Table for chrominance AC coefficients

Run/Size	Code length	Code word
0/0 (EOB)	2	00
0/1	2	01
0/2	3	100
0/3	4	1010
0/4	5	11000
0/5	5	11001
0/6	6	111000
0/7	7	1111000
0/8	9	111110100
0/9	10	1111110110
0/A	12	111111110100
1/1	4	1011

K-11

1/2	6	111001
1/3	8	11110110
1/4	9	111110101
1/5	11	11111110110
1/6	12	111111110101
1/7	16	1111111110001000
1/8	16	1111111110001001
1/9	16	1111111110001010
1/A	16	1111111110001011
2/1	5	11010
2/2	8	11110111
2/3	10	1111110111
2/4	12	111111110110
2/5	15	111111111000010
2/6	16	1111111110001100
2/7	16	1111111110001101
2/8	16	1111111110001110
2/9	16	1111111110001111
2/A	16	1111111110010000
3/1	5	11011
3/2	8	11111000
3/3	10	1111111000
3/4	12	111111110111
3/5	16	1111111110010001
3/6	16	1111111110010010
3/7	16	1111111110010011
3/8	16	1111111110010100
3/9	16	1111111110010101
3/A	16	1111111110010110
4/1	6	111010
4/2	9	111110110
4/3	16	1111111110010111
4/4	16	1111111110011000
4/5	16	1111111110011001
4/6	16	1111111110011010
4/7	16	1111111110011011
4/8	16	1111111110011100
4/9	16	1111111110011101
4/A	16	1111111110011110
5/1	6	111011
5/2	10	1111111001
5/3	16	1111111110011111
5/4	16	1111111110100000
5/5	16	1111111110100001
5/6	16	1111111110100010
5/7	16	1111111110100011

K-12

5/8	16	1111111110100100
5/9	16	1111111110100101
5/A	16	1111111110100110
6/1	7	1111001
6/2	11	11111110111
6/3	16	1111111110100111
6/4	16	1111111110101000
6/5	16	1111111110101001
6/6	16	1111111110101010
6/7	16	1111111110101011
6/8	16	1111111110101100
6/9	16	1111111110101101
6/A	16	1111111110101110
7/1	7	1111010
7/2	11	11111111000
7/3	16	1111111110101111
7/4	16	1111111110110000
7/5	16	1111111110110001
7/6	16	1111111110110010
7/7	16	1111111110110011
7/8	16	1111111110110100
7/9	16	1111111110110101
7/A	16	1111111110110110
8/1	8	11111001
8/2	16	1111111110110111
8/3	16	1111111110111000
8/4	16	1111111110111001
8/5	16	1111111110111010
8/6	16	1111111110111011
8/7	16	1111111110111100
8/8	16	1111111110111101
8/9	16	1111111110111110
8/A	16	1111111110111111
9/1	9	111110111
9/2	16	1111111111000000
9/3	16	1111111111000001
9/4	16	1111111111000010
9/5	16	1111111111000011
9/6	16	1111111111000100
9/7	16	1111111111000101
9/8	16	1111111111000110
9/9	16	1111111111000111
9/A	16	1111111111001000
A/1	9	111111000
A/2	16	1111111111001001
A/3	16	1111111111001010

K-13

A/4	16	1111111111001011
A/5	16	1111111111001100
A/6	16	1111111111001101
A/7	16	1111111111001110
A/8	16	1111111111001111
A/9	16	1111111111010000
A/A	16	1111111111010001
B/1	9	111111001
B/2	16	1111111111010010
B/3	16	1111111111010011
B/4	16	1111111111010100
B/5	16	1111111111010101
B/6	16	1111111111010110
B/7	16	1111111111010111
B/8	16	1111111111011000
B/9	16	1111111111011001
B/A	16	1111111111011010
C/1	9	111111010
C/2	16	1111111111011011
C/3	16	1111111111011100
C/4	16	1111111111011101
C/5	16	1111111111011110
C/6	16	1111111111011111
C/7	16	1111111111100000
C/8	16	1111111111100001
C/9	16	1111111111100010
C/A	16	1111111111100011
D/1	11	11111111001
D/2	16	1111111111100100
D/3	16	1111111111100101
D/4	16	1111111111100110
D/5	16	1111111111100111
D/6	16	1111111111101000
D/7	16	1111111111101001
D/8	16	1111111111101010
D/9	16	1111111111101011
D/A	16	1111111111101100
E/1	14	11111111100000
E/2	16	1111111111101101
E/3	16	1111111111101110
E/4	16	1111111111101111
E/5	16	1111111111110000
E/6	16	1111111111110001
E/7	16	1111111111110010
E/8	16	1111111111110011
E/9	16	1111111111110100

K-14

E/A	16	1111111111110101
F/0 (ZRL)	10	1111111010
F/1	15	111111111000011
F/2	16	1111111111110110
F/3	16	1111111111110111
F/4	16	1111111111111000
F/5	16	1111111111111001
F/6	16	1111111111111010
F/7	16	1111111111111011
F/8	16	1111111111111100
F/9	16	1111111111111101
F/A	16	1111111111111110

K.3.3 Huffman table-specification examples

K.3.3.1 Specification of typical tables for DC difference coding

A set of typical tables for DC component coding is given in K.3.1. The specification of these table is as follows:

For Table K.3 (for luminance DC coefficients), the 16 bytes which specify the list of code lengths for the table are:

X'00 01 05 01 01 01 01 01 01 00 00 00 00 00 00 00'

The set of values following this list is:

X'00 01 02 03 04 05 06 07 08 09 0A 0B'

For Table K.4 (for chrominance DC coefficients), the 16 bytes which specify the list of code lengths for the table are:

X'00 03 01 01 01 01 01 01 01 01 01 00 00 00 00 00'

The set of values following this list is:

X'00 01 02 03 04 05 06 07 08 09 0A 0B'

K.3.3.2 Specification of typical tables for AC coefficient coding

A set of typical tables for AC component coding is given in K.3.2. The specification of these tables is as follows:

For Table K.5 (for luminance AC coefficients), the 16 bytes which specify the list of code lengths for the table are:

X'00 02 01 03 03 02 04 03 05 05 04 04 00 00 01 7D'

K-15

The set of values which follows this list is:

```
X'01  02  03  00  04  11  05  12  21  31  41  06  13  51  61  07
      22  71  14  32  81  91  A1  08  23  42  B1  C1  15  52  D1  F0
      24  33  62  72  82  09  0A  16  17  18  19  1A  25  26  27  28
      29  2A  34  35  36  37  38  39  3A  43  44  45  46  47  48  49
      4A  53  54  55  56  57  58  59  5A  63  64  65  66  67  68  69
      6A  73  74  75  76  77  78  79  7A  83  84  85  86  87  88  89
      8A  92  93  94  95  96  97  98  99  9A  A2  A3  A4  A5  A6  A7
      A8  A9  AA  B2  B3  B4  B5  B6  B7  B8  B9  BA  C2  C3  C4  C5
      C6  C7  C8  C9  CA  D2  D3  D4  D5  D6  D7  D8  D9  DA  E1  E2
      E3  E4  E5  E6  E7  E8  E9  EA  F1  F2  F3  F4  F5  F6  F7  F8
      F9  FA'
```

For Table K.6 (for chrominance AC coefficients), the 16 bytes which specify the list of code lengths for the table are:

```
X'00  02  01  02  04  04  03  04  07  05  04  04  00  01  02  77'
```

The set of values which follows this list is:

```
X'00  01  02  03  11  04  05  21  31  06  12  41  51  07  61  71
      13  22  32  81  08  14  42  91  A1  B1  C1  09  23  33  52  F0
      15  62  72  D1  0A  16  24  34  E1  25  F1  17  18  19  1A  26
      27  28  29  2A  35  36  37  38  39  3A  43  44  45  46  47  48
      49  4A  53  54  55  56  57  58  59  5A  63  64  65  66  67  68
      69  6A  73  74  75  76  77  78  79  7A  82  83  84  85  86  87
      88  89  8A  92  93  94  95  96  97  98  99  9A  A2  A3  A4  A5
      A6  A7  A8  A9  AA  B2  B3  B4  B5  B6  B7  B8  B9  BA  C2  C3
      C4  C5  C6  C7  C8  C9  CA  D2  D3  D4  D5  D6  D7  D8  D9  DA
      E2  E3  E4  E5  E6  E7  E8  E9  EA  F2  F3  F4  F5  F6  F7  F8
      F9  FA'
```

K.4 Additional information on arithmetic coding

K.4.1 Test sequence for a small data set for the arithmetic coder

The following 256 bit test sequence (in hexadecimal form) is structured to test many of the encoder and decoder paths:

   X'00020051 000000C0 0352872A AAAAAAAA 82C02000 FCD79EF6 74EAABF7 697EE74C'

Tables K.7 and K.8 provide a symbol-by-symbol list of the arithmetic encoder and decoder operation. In these tables the event count, EC, is listed first, followed by the value of Qe used in encoding and decoding that event. The decision D to be encoded (and decoded) is listed next. The column labeled MPS contains the sense of the MPS, and if it is followed by an X (in the

K-16

"X" column), the conditional MPS/LPS exchange occurs when encoding and decoding the decision (see Figures D.3, D.4 and D.17). The contents of the A and C registers are the values before the event is encoded and decoded. SC is the number of X'FF' bytes stacked in the encoder waiting for a resolution of the carry-over. Note that the A register is always greater than X'7FFF'. (The starting value has an implied value of X'10000'.)

In the encoder test, the code bytes (B) are listed if they were completed during the coding of the preceding event. If additional bytes follow, they were also completed during the coding of the preceding event. If a byte is listed in the Bx column, the preceding byte in column B was modified by a carry-over.

In the decoder the code bytes are listed if they were placed in the code register just prior to the event EC.

For this file the coded bit count is 240, including the overhead to flush the final data from the C register. When the marker X'FFD9' is appended, a total of 256 bits are output. The actual compressed data sequence for the encoder is (in hexadecimal form):

X'655B5144 F7969D51 7855BFFF 00FC5184 C7CEF939 00287D46 708ECBC0
   F6FFD900'

Table K.7 Encoder test sequence

EC	D	MPS	X	Qe	A	C	CT	SC	Bx	B
1	0	0		5A1D	0000	00000000	11	0		
2	0	0	X	5A1D	A5E3	00000000	11	0		
3	0	0		2586	B43A	0000978C	10	0		
4	0	0		2586	8EB4	0000978C	10	0		
5	0	0		1114	D25C	00012F18	9	0		
6	0	0		1114	C148	00012F18	9	0		
7	0	0		1114	B034	00012F18	9	0		
8	0	0		1114	9F20	00012F18	9	0		
9	0	0		1114	8E0C	00012F18	9	0		
10	0	0		080B	F9F0	00025E30	8	0		
11	0	0		080B	F1E5	00025E30	8	0		
12	0	0		080B	E9DA	00025E30	8	0		
13	0	0		080B	E1CF	00025E30	8	0		
14	0	0		080B	D9C4	00025E30	8	0		
15	1	0		080B	D1B9	00025E30	8	0		
16	0	0		17B9	80B0	00327DE0	4	0		
17	0	0		1182	D1EE	0064FBC0	3	0		
18	0	0		1182	C06C	0064FBC0	3	0		
19	0	0		1182	AEEA	0064FBC0	3	0		
20	0	0		1182	9D68	0064FBC0	3	0		
21	0	0		1182	8BE6	0064FBC0	3	0		
22	0	0		0CEF	F4C8	00C9F780	2	0		
23	0	0		0CEF	E7D9	00C9F780	2	0		

K-17

24	0	0	0CEF	DAEA	00C9F780	2	0	
25	0	0	0CEF	CDFB	00C9F780	2	0	
26	1	0	0CEF	C10C	00C9F780	2	0	
27	0	0	1518	CEF0	000AB9D0	6	0	65
28	1	0	1518	B9D8	000AB9D0	6	0	
29	0	0	1AA9	A8C0	005AF480	3	0	
30	0	0	1AA9	8E17	005AF480	3	0	
31	0	0	174E	E6DC	00B5E900	2	0	
32	1	0	174E	CF8E	00B5E900	2	0	
33	0	0	1AA9	BA70	00050A00	7	0	5B
34	0	0	1AA9	9FC7	00050A00	7	0	
35	0	0	1AA9	851E	00050A00	7	0	
36	0	0	174E	D4EA	000A1400	6	0	
37	0	0	174E	BD9C	000A1400	6	0	
38	0	0	174E	A64E	000A1400	6	0	
39	0	0	174E	8F00	000A1400	6	0	
40	0	0	1424	EF64	00142800	5	0	
41	0	0	1424	DB40	00142800	5	0	
42	0	0	1424	C71C	00142800	5	0	
43	0	0	1424	B2F8	00142800	5	0	
44	0	0	1424	9ED4	00142800	5	0	
45	0	0	1424	8AB0	00142800	5	0	
46	0	0	119C	ED18	00285000	4	0	
47	0	0	119C	DB7C	00285000	4	0	
48	0	0	119C	C9E0	00285000	4	0	
49	0	0	119C	B844	00285000	4	0	
50	0	0	119C	A6A8	00285000	4	0	
51	0	0	119C	950C	00285000	4	0	
52	0	0	119C	8370	00285000	4	0	
53	0	0	0F6B	E3A8	0050A000	3	0	
54	0	0	0F6B	D43D	0050A000	3	0	
55	0	0	0F6B	C4D2	0050A000	3	0	
56	0	0	0F6B	B567	0050A000	3	0	
57	1	0	0F6B	A5FC	0050A000	3	0	
58	1	0	1424	F6B0	00036910	7	0	51
59	0	0	1AA9	A120	00225CE0	4	0	
60	0	0	1AA9	8677	00225CE0	4	0	
61	0	0	174E	D79C	0044B9C0	3	0	
62	0	0	174E	C04E	0044B9C0	3	0	
63	0	0	174E	A900	0044B9C0	3	0	
64	0	0	174E	91B2	0044B9C0	3	0	
65	0	0	1424	F4C8	00897380	2	0	
66	0	0	1424	E0A4	00897380	2	0	
67	0	0	1424	CC80	00897380	2	0	
68	0	0	1424	B85C	00897380	2	0	
69	0	0	1424	A438	00897380	2	0	

70	0	0		1424	9014	00897380	2	0	
71	1	0		119C	F7E0	0112E700	1	0	
72	1	0		1424	8CE0	001E6A20	6	0	44
73	0	0		1AA9	A120	00F716E0	3	0	
74	1	0		1AA9	8677	00F716E0	3	0	
75	0	0		2516	D548	00041570	8	0	F7
76	1	0		2516	B032	00041570	8	0	
77	0	0		299A	9458	00128230	6	0	
78	0	0		2516	D57C	00250460	5	0	
79	1	0		2516	B066	00250460	5	0	
80	0	0		299A	9458	00963EC0	3	0	
81	1	0		2516	D57C	012C7D80	2	0	
82	0	0		299A	9458	0004B798	8	0	96
83	0	0		2516	D57C	00096F30	7	0	
84	0	0		2516	B066	00096F30	7	0	
85	0	0		2516	8B50	00096F30	7	0	
86	1	0		1EDF	CC74	0012DE60	6	0	
87	1	0		2516	F6F8	009C5FA8	3	0	
88	1	0		299A	9458	0274C628	1	0	
89	0	0		32B4	A668	0004C398	7	0	9D
90	0	0		2E17	E768	00098730	6	0	
91	1	0		2E17	B951	00098730	6	0	
92	0	0		32B4	B85C	002849A8	4	0	
93	1	0		32B4	85A8	002849A8	4	0	
94	0	0		3C3D	CAD0	00A27270	2	0	
95	1	0		3C3D	8E93	00A27270	2	0	
96	0	0		415E	F0F4	00031318	8	0	51
97	1	0		415E	AF96	00031318	8	0	
98	0	0	X	4639	82BC	000702A0	7	0	
99	1	0		415E	8C72	000E7E46	6	0	
100	0	0	X	4639	82BC	001D92B4	5	0	
101	1	0		415E	8C72	003B9E6E	4	0	
102	0	0	X	4639	82BC	0077D304	3	0	
103	1	0		415E	8C72	00F01F0E	2	0	
104	0	0	X	4639	82BC	01E0D444	1	0	
105	1	0		415E	8C72	0002218E	8	0	78
106	0	0	X	4639	82BC	0004D944	7	0	
107	1	0		415E	8C72	000A2B8E	6	0	
108	0	0	X	4639	82BC	0014ED44	5	0	
109	1	0		415E	8C72	002A538E	4	0	
110	0	0	X	4639	82BC	00553D44	3	0	
111	1	0		415E	8C72	00AAF38E	2	0	
112	0	0	X	4639	82BC	01567D44	1	0	
113	1	0		415E	8C72	0005738E	8	0	55
114	0	0	X	4639	82BC	000B7D44	7	0	
115	1	0		415E	8C72	0017738E	6	0	

K-19

116	0	0	X	4639	82BC	002F7D44	5	0	
117	1	0		415E	8C72	005F738E	4	0	
118	0	0	X	4639	82BC	00BF7D44	3	0	
119	1	0		415E	8C72	017F738E	2	0	
120	0	0	X	4639	82BC	02FF7D44	1	0	
121	1	0		415E	8C72	0007738E	8	0	BF
122	0	0	X	4639	82BC	000F7D44	7	0	
123	1	0		415E	8C72	001F738E	6	0	
124	0	0	X	4639	82BC	003F7D44	5	0	
125	1	0		415E	8C72	007F738E	4	0	
126	0	0	X	4639	82BC	00FF7D44	3	0	
127	1	0		415E	8C72	01FF738E	2	0	
128	0	0	X	4639	82BC	03FF7D44	1	0	
129	1	0		415E	8C72	0007738E	8	1	
130	0	0	X	4639	82BC	000F7D44	7	1	
131	0	0		415E	8C72	001F738E	6	1	
132	0	0		3C3D	9628	003EE71C	5	1	
133	0	0		375E	B3D6	007DCE38	4	1	
134	0	0		32B4	F8F0	00FB9C70	3	1	
135	1	0		32B4	C63C	00FB9C70	3	1	
136	0	0		3C3D	CAD0	03F0BFE0	1	1	
137	1	0		3C3D	8E93	03F0BFE0	1	1	
138	1	0		415E	F0F4	000448D8	7	0	FF00FC
139	0	0	X	4639	82BC	0009F0DC	6	0	
140	0	0		415E	8C72	00145ABE	5	0	
141	0	0		3C3D	9628	0028B57C	4	0	
142	0	0		375E	B3D6	00516AF8	3	0	
143	0	0		32B4	F8F0	00A2D5F0	2	0	
144	0	0		32B4	C63C	00A2D5F0	2	0	
145	0	0		32B4	9388	00A2D5F0	2	0	
146	0	0		2E17	C1A8	0145ABE0	1	0	
147	1	0		2E17	9391	0145ABE0	1	0	
148	0	0		32B4	B85C	00084568	7	0	51
149	0	0		32B4	85A8	00084568	7	0	
150	0	0		2E17	A5E8	00108AD0	6	0	
151	0	0		299A	EFA2	002115A0	5	0	
152	0	0		299A	C608	002115A0	5	0	
153	0	0		299A	9C6E	002115A0	5	0	
154	0	0		2516	E5A8	00422B40	4	0	
155	0	0		2516	C092	00422B40	4	0	
156	0	0		2516	9B7C	00422B40	4	0	
157	0	0		1EDF	ECCC	00845680	3	0	
158	0	0		1EDF	CDED	00845680	3	0	
159	0	0		1EDF	AF0E	00845680	3	0	
160	0	0		1EDF	902F	00845680	3	0	
161	1	0		1AA9	E2A0	0108AD00	2	0	

162	1	0		2516	D548	000BA7B8	7	0	84
163	1	0		299A	9458	00315FA8	5	0	
164	1	0		32B4	A668	00C72998	3	0	
165	1	0		3C3D	CAD0	031E7530	1	0	
166	1	0		415E	F0F4	000C0F0C	7	0	C7
167	0	0	X	4639	82BC	00197D44	6	0	
168	0	0		415E	8C72	0033738E	5	0	
169	1	0		3C3D	9628	0066E71C	4	0	
170	1	0		415E	F0F4	019D041C	2	0	
171	0	0	X	4639	82BC	033B6764	1	0	
172	1	0		415E	8C72	000747CE	8	0	CE
173	0	0	X	4639	82BC	000F25C4	7	0	
174	1	0		415E	8C72	001EC48E	6	0	
175	1	0	X	4639	82BC	003E1F44	5	0	
176	1	0		4B85	F20C	00F87D10	3	0	
177	1	0	X	504F	970A	01F2472E	2	0	
178	0	0	X	5522	8D76	03E48E5C	1	0	
179	0	0		504F	AA44	00018D60	8	0	F9
180	1	0		4B85	B3EA	00031AC0	7	0	
181	1	0	X	504F	970A	0007064A	6	0	
182	1	0	X	5522	8D76	000E0C94	5	0	
183	1	0		59EB	E150	00383250	3	0	
184	0			59EB	B3D6	0071736A	2	0	
185	1	0		59EB	B3D6	00E39AAA	1	0	
186	1	1		59EB	B3D6	0007E92A	8	0	38
187	1	1		5522	B3D6	000FD254	7	0	
188	1	1		504F	BD68	001FA4A8	6	0	
189	0	1		4B85	DA32	003F4950	5	0	
190	1	1	X	504F	970A	007FAFFA	4	0	
191	1	1		4B85	A09E	00FFED6A	3	0	
192	0	1		4639	AA32	01FFDAD4	2	0	
193	0	1	X	4B85	8C72	04007D9A	1	0	
194	1	1	X	504F	81DA	0000FB34	8	0	39 00
195	1	1		4B85	A09E	0002597E	7	0	
196	1	1		4639	AA32	0004B2FC	6	0	
197	0	1		415E	C7F2	000965F8	5	0	
198	1	1	X	4639	82BC	0013D918	4	0	
199	0	1		415E	8C72	00282B36	3	0	
200	0	1	X	4639	82BC	0050EC94	2	0	
201	1	1		4B85	F20C	0003B250	8	0	28
202	1	1		4B85	A687	0003B250	8	0	
203	1	1		4639	B604	000764A0	7	0	
204	0	1		415E	DF96	000EC940	6	0	
205	1	1	X	4639	82BC	001ECEF0	5	0	
206	0	1		415E	8C72	003E16E6	4	0	
207	1	1	X	4639	82BC	007CC3F4	3	0	

K-21

208	0	1		415E	8C72	00FA00EE	2	0	
209	1	1	X	4639	82BC	01F49804	1	0	
210	0	1		415E	8C72	0001A90E	8	0	7D
211	1	1	X	4639	82BC	0003E844	7	0	
212	0	1		415E	8C72	0008498E	6	0	
213	1	1	X	4639	82BC	00112944	5	0	
214	0	1		415E	8C72	0022CB8E	4	0	
215	1	1	X	4639	82BC	00462D44	3	0	
216	1	1		415E	8C72	008CD38E	2	0	
217	1	1		3C3D	9628	0119A71C	1	0	
218	1	1		375E	B3D6	00034E38	8	0	46
219	1	1		32B4	F8F0	00069C70	7	0	
220	1	1		32B4	C63C	00069C70	7	0	
221	0	1		32B4	9388	00069C70	7	0	
222	1	1		3C3D	CAD0	001BF510	5	0	
223	1	1		3C3D	8E93	001BF510	5	0	
224	1	1		375E	A4AC	0037EA20	4	0	
225	0	1		32B4	DA9C	006FD440	3	0	
226	1	1		3C3D	CAD0	01C1F0A0	1	0	
227	1	1		3C3D	8E93	01C1F0A0	1	0	
228	0	1		375E	A4AC	0003E140	8	0	70
229	1	1		3C3D	DD78	00113A38	6	0	
230	0	1		3C3D	A13B	00113A38	6	0	
231	0	1		415E	F0F4	00467CD8	4	0	
232	1	1	X	4639	82BC	008E58DC	3	0	
233	0	1		415E	8C72	011D2ABE	2	0	
234	1	1	X	4639	82BC	023AEBA4	1	0	
235	1	1		415E	8C72	0006504E	8	0	8E
236	1	1		3C3D	9628	000CA09C	7	0	
237	1	1		375E	B3D6	00194138	6	0	
238	1	1		32B4	F8F0	00328270	5	0	
239	1	1		32B4	C63C	00328270	5	0	
240	0	1		32B4	9388	00328270	5	0	
241	1	1		3C3D	CAD0	00CB8D10	3	0	
242	1	1		3C3D	8E93	00CB8D10	3	0	
243	1	1		375E	A4AC	01971A20	2	0	
244	0	1		32B4	DA9C	032E3440	1	0	
245	0	1		3C3D	CAD0	000B70A0	7	0	CB
246	1	1		415E	F0F4	002FFCCC	5	0	
247	1	1		415E	AF96	002FFCCC	5	0	
248	1	1		3C3D	DC70	005FF998	4	0	
249	0	1		3C3D	A033	005FF998	4	0	
250	1	1		415E	F0F4	01817638	2	0	
251	0	1		415E	AF96	01817638	2	0	
252	0	1	X	4639	82BC	0303C8E0	1	0	
253	1	1		4B85	F20C	000F2380	7	0	C0

254	1	1		4B85	A687	000F2380	7	0		
255	0	1		4639	B604	001E4700	6	0		
256	0	1	X	4B85	8C72	003D6D96	5	0		
flush:					81DA	007ADB2C	4	0	F6	FFD9

Table K.8 Decoder test sequence

EC	D	MPS	X	Qe	A	C	CT	B
1	0	0		5A1D	0000	655B0000	0	65 5B
2	0	0	X	5A1D	A5E3	655B0000	0	
3	0	0		2586	B43A	332AA200	7	51
4	0	0		2586	8EB4	332AA200	7	
5	0	0		1114	D25C	66554400	6	
6	0	0		1114	C148	66554400	6	
7	0	0		1114	B034	66554400	6	
8	0	0		1114	9F20	66554400	6	
9	0	0		1114	8E0C	66554400	6	
10	0	0		080B	F9F0	CCAA8800	5	
11	0	0		080B	F1E5	CCAA8800	5	
12	0	0		080B	E9DA	CCAA8800	5	
13	0	0		080B	E1CF	CCAA8800	5	
14	0	0		080B	D9C4	CCAA8800	5	
15	1	0		080B	D1B9	CCAA8800	5	
16	0	0		17B9	80B0	2FC88000	1	
17	0	0		1182	D1EE	5F910000	0	
18	0	0		1182	C06C	5F910000	0	
19	0	0		1182	AEEA	5F910000	0	
20	0	0		1182	9D68	5F910000	0	
21	0	0		1182	8BE6	5F910000	0	
22	0	0		0CEF	F4C8	BF228800	7	44
23	0	0		0CEF	E7D9	BF228800	7	
24	0	0		0CEF	DAEA	BF228800	7	
25	0	0		0CEF	CDFB	BF228800	7	
26	1	0		0CEF	C10C	BF228800	7	
27	0	0		1518	CEF0	B0588000	3	
28	1	0		1518	B9D8	B0588000	3	
29	0	0		1AA9	A8C0	5CC40000	0	
30	0	0		1AA9	8E17	5CC40000	0	
31	0	0		174E	E6DC	B989EE00	7	F7
32	1	0		174E	CF8E	B989EE00	7	
33	0	0		1AA9	BA70	0A4F7000	4	
34	0	0		1AA9	9FC7	0A4F7000	4	
35	0	0		1AA9	851E	0A4F7000	4	
36	0	0		174E	D4EA	149EE000	3	
37	0	0		174E	BD9C	149EE000	3	

K-23

38	0	0	174E	A64E	149EE000	3	
39	0	0	174E	8F00	149EE000	3	
40	0	0	1424	EF64	293DC000	2	
41	0	0	1424	DB40	293DC000	2	
42	0	0	1424	C71C	293DC000	2	
43	0	0	1424	B2F8	293DC000	2	
44	0	0	1424	9ED4	293DC000	2	
45	0	0	1424	8AB0	293DC000	2	
46	0	0	119C	ED18	527B8000	1	
47	0	0	119C	DB7C	527B8000	1	
48	0	0	119C	C9E0	527B8000	1	
49	0	0	119C	B844	527B8000	1	
50	0	0	119C	A6A8	527B8000	1	
51	0	0	119C	950C	527B8000	1	
52	0	0	119C	8370	527B8000	1	
53	0	0	0F6B	E3A8	A4F70000	0	
54	0	0	0F6B	D43D	A4F70000	0	
55	0	0	0F6B	C4D2	A4F70000	0	
56	0	0	0F6B	B567	A4F70000	0	
57	1	0	0F6B	A5FC	A4F70000	0	
58	1	0	1424	F6B0	E6696000	4	96
59	0	0	1AA9	A120	1EEB0000	1	
60	0	0	1AA9	8677	1EEB0000	1	
61	0	0	174E	D79C	3DD60000	0	
62	0	0	174E	C04E	3DD60000	0	
63	0	0	174E	A900	3DD60000	0	
64	0	0	174E	91B2	3DD60000	0	
65	0	0	1424	F4C8	7BAD3A00	7	9D
66	0	0	1424	E0A4	7BAD3A00	7	
67	0	0	1424	CC80	7BAD3A00	7	
68	0	0	1424	B85C	7BAD3A00	7	
69	0	0	1424	A438	7BAD3A00	7	
70	0	0	1424	9014	7BAD3A00	7	
71	1	0	119C	F7E0	F75A7400	6	
72	1	0	1424	8CE0	88B3A000	3	
73	0	0	1AA9	A120	7FBD0000	0	
74	1	0	1AA9	8677	7FBD0000	0	
75	0	0	2516	D548	9F7A8800	5	51
76	1	0	2516	B032	9F7A8800	5	
77	0	0	299A	9458	517A2000	3	
78	0	0	2516	D57C	A2F44000	2	
79	1	0	2516	B066	A2F44000	2	
80	0	0	299A	9458	5E910000	0	
81	1	0	2516	D57C	BD22F000	7	78
82	0	0	299A	9458	32F3C000	5	
83	0	0	2516	D57C	65E78000	4	

K-24

84	0	0		2516	B066	65E78000	4	
85	0	0		2516	8B50	65E78000	4	
86	1	0		1EDF	CC74	CBCF0000	3	
87	1	0		2516	F6F8	F1D00000	0	
88	1	0		299A	9458	7FB95400	6	55
89	0	0		32B4	A668	53ED5000	4	
90	0	0		2E17	E768	A7DAA000	3	
91	1	0		2E17	B951	A7DAA000	3	
92	0	0		32B4	B85C	72828000	1	
93	1	0		32B4	85A8	72828000	1	
94	0	0		3C3D	CAD0	7E3B7E00	7	BF
95	1	0		3C3D	8E93	7E3B7E00	7	
96	0	0		415E	F0F4	AF95F800	5	
97	1	0		415E	AF96	AF95F800	5	
98	0	0	X	4639	82BC	82BBF000	4	
99	1	0		415E	8C72	8C71E000	3	
100	0	0	X	4639	82BC	82BBC000	2	
101	1	0		415E	8C72	8C718000	1	
102	0	0	X	4639	82BC	82BB0000	0	
103	1	0		415E	8C72	8C71FE00	7	FF 00
104	0	0	X	4639	82BC	82BBFC00	6	
105	1	0		415E	8C72	8C71F800	5	
106	0	0	X	4639	82BC	82BBF000	4	
107	1	0		415E	8C72	8C71E000	3	
108	0	0	X	4639	82BC	82BBC000	2	
109	1	0		415E	8C72	8C718000	1	
110	0	0	X	4639	82BC	82BB0000	0	
111	1	0		415E	8C72	8C71F800	7	FC
112	0	0	X	4639	82BC	82BBF000	6	
113	1	0		415E	8C72	8C71E000	5	
114	0	0	X	4639	82BC	82BBC000	4	
115	1	0		415E	8C72	8C718000	3	
116	0	0	X	4639	82BC	82BB0000	2	
117	1	0		415E	8C72	8C700000	1	
118	0	0	X	4639	82BC	82B80000	0	
119	1	0		415E	8C72	8C6AA200	7	51
120	0	0	X	4639	82BC	82AD4400	6	
121	1	0		415E	8C72	8C548800	5	
122	0	0	X	4639	82BC	82811000	4	
123	1	0		415E	8C72	8BFC2000	3	
124	0	0	X	4639	82BC	81D04000	2	
125	1	0		415E	8C72	8A9A8000	1	
126	0	0	X	4639	82BC	7F0D0000	0	
127	1	0		415E	8C72	85150800	7	84
128	0	0	X	4639	82BC	74021000	6	
129	1	0		415E	8C72	6EFE2000	5	

K-25

130	0	0	X	4639	82BC	47D44000	4	
131	0	0		415E	8C72	16A28000	3	
132	0	0		3C3D	9628	2D450000	2	
133	0	0		375E	B3D6	5A8A0000	1	
134	0	0		32B4	F8F0	B5140000	0	
135	1	0		32B4	C63C	B5140000	0	
136	0	0		3C3D	CAD0	86331C00	6	C7
137	1	0		3C3D	8E93	86331C00	6	
138	1	0		415E	F0F4	CF747000	4	
139	0	0	X	4639	82BC	3FBCE000	3	
140	0	0		415E	8C72	0673C000	2	
141	0	0		3C3D	9628	0CE78000	1	
142	0	0		375E	B3D6	19CF0000	0	
143	0	0		32B4	F8F0	339F9C00	7	CE
144	0	0		32B4	C63C	339F9C00	7	
145	0	0		32B4	9388	339F9C00	7	
146	0	0		2E17	C1A8	673F3800	6	
147	1	0		2E17	9391	673F3800	6	
148	0	0		32B4	B85C	0714E000	4	
149	0	0		32B4	85A8	0714E000	4	
150	0	0		2E17	A5E8	0E29C000	3	
151	0	0		299A	EFA2	1C538000	2	
152	0	0		299A	C608	1C538000	2	
153	0	0		299A	9C6E	1C538000	2	
154	0	0		2516	E5A8	38A70000	1	
155	0	0		2516	C092	38A70000	1	
156	0	0		2516	9B7C	38A70000	1	
157	0	0		1EDF	ECCC	714E0000	0	
158	0	0		1EDF	CDED	714E0000	0	
159	0	0		1EDF	AF0E	714E0000	0	
160	0	0		1EDF	902F	714E0000	0	
161	1	0		1AA9	E2A0	E29DF200	7	F9
162	1	0		2516	D548	D5379000	4	
163	1	0		299A	9458	94164000	2	
164	1	0		32B4	A668	A5610000	0	
165	1	0		3C3D	CAD0	C6B4E400	6	39
166	1	0		415E	F0F4	E0879000	4	
167	0	0	X	4639	82BC	61E32000	3	
168	0	0		415E	8C72	4AC04000	2	
169	1	0		3C3D	9628	95808000	1	
170	1	0		415E	F0F4	EE560000	7	00
171	0	0	X	4639	82BC	7D800000	6	
172	1	0		415E	8C72	81FA0000	5	
173	0	0	X	4639	82BC	6DCC0000	4	
174	1	0		415E	8C72	62920000	3	
175	1	0	X	4639	82BC	2EFC0000	2	

176	1	0		4B85	F20C	BBF00000	0	
177	1	0	X	504F	970A	2AD25000	7	28
178	0	0	X	5522	8D76	55A4A000	6	
179	0	0		504F	AA44	3AA14000	5	
180	1	0		4B85	B3EA	75428000	4	
181	1	0	X	504F	970A	19BB0000	3	
182	1	0	X	5522	8D76	33760000	2	
183	1	0		59EB	E150	CDD80000	0	
184	0	1		59EB	B3D6	8CE6FA00	7	7D
185	1	0		59EB	B3D6	65F7F400	6	
186	1	1		59EB	B3D6	1819E800	5	
187	1	1		5522	B3D6	3033D000	4	
188	1	1		504F	BD68	6067A000	3	
189	0	1		4B85	DA32	C0CF4000	2	
190	1	1	X	504F	970A	64448000	1	
191	1	1		4B85	A09E	3B130000	0	
192	0	1		4639	AA32	76268C00	7	46
193	0	1	X	4B85	8C72	245B1800	6	
194	1	1	X	504F	81DA	48B63000	5	
195	1	1		4B85	A09E	2E566000	4	
196	1	1		4639	AA32	5CACC000	3	
197	0	1		415E	C7F2	B9598000	2	
198	1	1	X	4639	82BC	658B0000	1	
199	0	1		415E	8C72	52100000	0	
200	0	1	X	4639	82BC	0DF8E000	7	70
201	1	1		4B85	F20C	37E38000	5	
202	1	1		4B85	A687	37E38000	5	
203	1	1		4639	B604	6FC70000	4	
204	0	1		415E	DF96	DF8E0000	3	
205	1	1	X	4639	82BC	82AC0000	2	
206	0	1		415E	8C72	8C520000	1	
207	1	1	X	4639	82BC	827C0000	0	
208	0	1		415E	8C72	8BF31C00	7	8E
209	1	1	X	4639	82BC	81BE3800	6	
210	0	1		415E	8C72	8A767000	5	
211	1	1	X	4639	82BC	7EC4E000	4	
212	0	1		415E	8C72	8483C000	3	
213	1	1	X	4639	82BC	72DF8000	2	
214	0	1		415E	8C72	6CB90000	1	
215	1	1	X	4639	82BC	434A0000	0	
216	1	1		415E	8C72	0D8F9600	7	CB
217	1	1		3C3D	9628	1B1F2C00	6	
218	1	1		375E	B3D6	363E5800	5	
219	1	1		32B4	F8F0	6C7CB000	4	
220	1	1		32B4	C63C	6C7CB000	4	
221	0	1		32B4	9388	6C7CB000	4	

222	1			3C3D	CAD0	2EA2C000	2	
223	1	1		3C3D	8E93	2EA2C000	2	
224	1	1		375E	A4AC	5D458000	1	
225	0	1		32B4	DA9C	BA8B0000	0	
226	1	1		3C3D	CAD0	4A8F0000	6	C0
227	1	1		3C3D	8E93	4A8F0000	6	
228	0	1		375E	A4AC	951E0000	5	
229	1	1		3C3D	DD78	9F400000	3	
230	0	1		3C3D	A13B	9F400000	3	
231	0	1		415E	F0F4	E9080000	1	
232	1	1	X	4639	82BC	72E40000	0	
233	0	1		415E	8C72	6CC3EC00	7	F6
234	1	1	X	4639	82BC	435FD800	6	
235	1	1		415E	8C72	0DB9B000	5	
236	1	1		3C3D	9628	1B736000	4	
237	1	1		375E	B3D6	36E6C000	3	
238	1	1		32B4	F8F0	6DCD8000	2	
239	1	1		32B4	C63C	6DCD8000	2	
240	0	1		32B4	9388	6DCD8000	2	
241	1	1		3C3D	CAD0	33E60000	0	
242	1	1		3C3D	8E93	33E60000	0	

Marker detected: zero byte fed to decoder

243	1	1		375E	A4AC	67CC0000	7
244	0	1		32B4	DA9C	CF980000	6
245	0	1		3C3D	CAD0	9EC00000	4
246	1	1		415E	F0F4	40B40000	2
247	1	1		415E	AF96	40B40000	2
248	1	1		3C3D	DC70	81680000	1
249	0	1		3C3D	A033	81680000	1

Marker detected: zero byte fed to decoder

250	1	1		415E	F0F4	75C80000	7
251	0	1		415E	AF96	75C80000	7
252	0	1	X	4639	82BC	0F200000	6
253	1	1		4B85	F20C	3C800000	4
254	1	1		4B85	A687	3C800000	4
255	0	1		4639	B604	79000000	3
256	0	1	X	4B85	8C72	126A0000	2

K.5 Low-pass downsampling filters for hierarchical coding

In this section simple examples are given of downsampling filters which are compatible with the upsampling filter defined in J.1.1.2.

Figure K.5 shows the weighting of neighboring samples for simple one-dimensional horizontal and vertical low-pass filters. The output of the filter must be normalized by the sum of the neighborhood weights.

Figure K.5 - Low-pass filter example

The center sample in Figure K.5 should be aligned with the left column or top row of the high resolution image when calculating the left column or top row of the low resolution image. Sample values which are situated outside of the image boundary are replicated from the sample values at the boundary to provide missing edge values.

If the image being downsampled has an odd width or length, the odd dimension is increased by 1 by sample replication on the right edge or bottom row before downsampling.

K.6 Domain of applicability of DCT and spatial coding techniques

The DCT coder is intended for lossy coding in a range from quite visible loss to distortion well below the threshold for visibility. However in general, DCT-based processes cannot be used for true lossless coding.

The lossless coder is intended for completely lossless coding. The lossless coding process is significantly less effective than the DCT-based processes for distortions near and above the threshold of visibility.

The point transform of the input to the lossless coder permits a very restricted form of lossy coding with the "lossless" coder. (The coder is still lossless after the input point transform.) Since the DCT is intended for lossy coding, there may be some confusion about when this alternative lossy technique should be used.

Lossless coding with a point transformed input is intended for applications which cannot be addressed by DCT coding techniques. Among these are:

1. True lossless coding to a specified precision.

2. Lossy coding with precisely defined error bounds.

3. Hierarchical progression to a truly lossless final stage.

If lossless coding with a point transformed input is used in applications which can be met effectively by DCT coding, the results will be significantly less satisfactory. For example, distortion in the form of visible contours usually appears when precision of the luminance component is reduced to about six bits. For normal image data, this occurs at bit rates well above those for which the DCT gives outputs which are visually indistinguishable from the source.

K-29

K.7 Domain of applicability of the progressive coding modes of operation

Two very different progressive coding modes of operation have been defined, progressive coding of the DCT coefficients and hierarchical progression. Progressive coding of the DCT coefficients has two complementary procedures, spectral selection and successive approximation. Because of this diversity of choices, there may be some confusion as to which method of progression to use for a given application.

K.7.1 Progressive coding of the DCT

In progressive coding of the DCT coefficients two complementary procedures are defined for decomposing the 8x8 DCT coefficient array, spectral selection and successive approximation. Spectral selection partitions zig-zag array of DCT coefficients into "bands", one band being coded in each scan. Successive approximation codes the coefficients with reduced precision in the first scan; in each subsequent scan the precision is increased by one bit.

A single forward DCT is calculated for these procedures. When all coefficients are coded to full precision, the DCT is the same as in the sequential mode. Therefore, like the sequential DCT coding, progressive coding of DCT coefficients is intended for applications which need very good compression for a given level of visual distortion.

The simplest progressive coding technique is spectral selection; indeed, because of this simplicity, some applications may choose - despite the limited progression that can be achieved - to use only spectral selection. Note, however, that the absence of high frequency bands typically leads - for a given bit rate - to a significantly lower image quality in the intermediate stages than can be achieved with the more general progressions. The net coding efficiency at the completion of the final stage is typically comparable to or slightly less than that achieved with the sequential DCT

A much more flexible progressive system is attained at some increase in complexity when successive approximation is added to the spectral selection progression. For a given bit rate, this system typically provides significantly better image quality than spectral selection alone. The net coding efficiency at the completion of the final stage is typically comparable to or slightly better than that achieved with the sequential DCT.

K.7.2 Hierarchical progression

Hierarchical progression permits a sequence of outputs of increasing spatial resolution, and also allows refinement of image quality at a given spatial resolution. Both DCT and spatial versions of the hierarchical progression are allowed, and progressive coding of DCT coefficients may be used in a frame of the DCT hierarchical progression.

The DCT hierarchical progression is intended for applications which need very good compression for a given level of visual distortion; the spatial hierarchical progression is intended applications which need a simple progression with a truly lossless final stage. Figure K.6 Illistrates examples of these two basic hierarchical progressions.

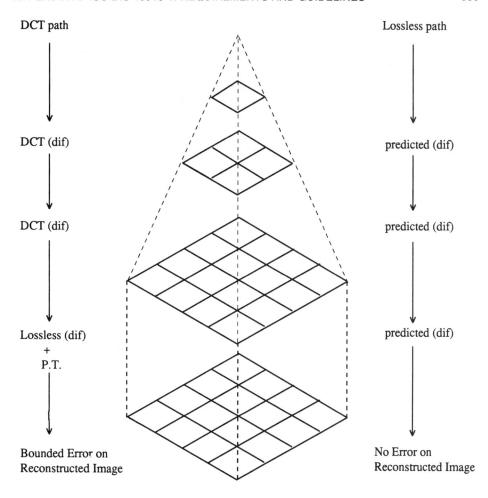

DCT path

DCT (dif)

DCT (dif)

Lossless (dif)
+
P.T.

Bounded Error on
Reconstructed Image

Lossless path

predicted (dif)

predicted (dif)

predicted (dif)

No Error on
Reconstructed Image

Figure K.6 - Sketch of the basic operations of the hierarchical mode

K-31

K.7.2.1 DCT Hierarchical progression

If a DCT hierarchical progression uses reduced spatial resolution, the early stages of the progression can have better image quality for a given bit rate than the early stages of non-hierarchical progressive coding of the DCT coefficients. However, at the point where the distortion between source and output becomes indistinguishable, the coding efficiency achieved with a DCT hierarchical progression is typically significantly lower than the coding efficiency achieved with a non-hierarchical progressive coding of the DCT coefficients.

While the hierarchical DCT progression is intended for lossy progressive coding, a final spatial differential coding stage can be used. When this final stage is used, the output can be almost lossless, limited only by the difference between the encoder and decoder IDCT implementations. Since IDCT implementations can differ significantly, truly lossless coding after a DCT hierarchical progression cannot be guaranteed. An important alternative, therefore, is to use the input point transform of the spatial stage to reduce the precision of the differential input. This allows a bounding of the difference between source and output at a significantly lower cost in coded bits than coding of the full precision spatial difference would require.

K.7.2.2 Spatial hierarchical progression

If lossless progression is required, a very simple hierarchical progression may be used in which the spatial lossless coder with point transformed input is used as a first stage. This first stage is followed by one or more spatial differential coding stages. The first stage should be nearly lossless, such that the low order bits which are truncated by the point transform are essentially random - otherwise the compression efficiency will be degraded relative to nonprogressive lossless coding.

K.8 Suppression of block-to-block discontinuities in decoded images

A simple technique is available for suppressing the block-to-block discontinuities which can occur in images compressed by DCT techniques.

The first few (five in this example) low frequency DCT coefficients are predicted from the nine DC values of the block and the eight nearest-neighbor blocks, and the predicted values are used to suppress blocking artifacts in smooth areas of the image.

The prediction equations for the first five AC coefficients in the zig-zag sequence are obtained as follows:

K.8.1 AC prediction

The sample field in a 3 by 3 array of blocks (each block containing an 8x8 array of samples) is modeled by a two-dimensional second degree polynomial of the form:

$$P(x,y) = A1(x^2y^2)+A2(x^2y)+A3(xy^2)+A4(x^2)+A5(xy)+A6(y^2)+A7(x)+A8(y)+A9$$

The nine coefficients A1 through A9 are uniquely determined by imposing the constraint that the mean of $P(x,y)$ over each of the nine blocks must yield the correct DC-values.

Applying the DCT to the quadratic field predicting the samples in the central block gives a prediction of the low frequency AC coefficients depicted in Figure K.7.

```
DC x  x  .  .  .  .  .
   x  x  .  .  .  .  .  .
   x  .  .  .  .  .  .  .
   .  .  .  .  .  .  .  .
   .  .  .  .  .  .  .  .
   .  .  .  .  .  .  .  .
   .  .  .  .  .  .  .  .
   .  .  .  .  .  .  .  .
```

Figure K.7 - DCT array positions of predicted AC coefficients

The prediction equations derived in this manner are as follows:

For the two dimensional array of DC values

DC1	DC2	DC3
DC4	DC5	DC6
DC7	DC8	DC9

the unquantized prediction equations are:

$$AC01 = 1.13885 \, (DC4 - DC6)$$
$$AC10 = 1.13885 \, (DC2 - DC8)$$
$$AC20 = 0.27881 \, (DC2 + DC8 - 2 \times DC5)$$
$$AC11 = 0.16213 \, ((DC1 - DC3) - (DC7 - DC9))$$
$$AC02 = 0.27881 \, (DC4 + DC6 - 2 \times DC5)$$

The scaling of the predicted AC coefficients is consistent with the DCT normalization defined in A.3.3.

K.8.2 Quantized AC prediction

The prediction equations can be mapped to a form which uses quantized values of the DC coefficients and which computes quantized AC coefficients using integer arithmetic. The quantized DC coefficients need to be scaled, however, such that the predicted coefficients have fractional bit precision.

First, the prediction equation coefficients are scaled by 32 and rounded to the nearest integer. Thus,

$$1.13885 \times 32 = 36$$
$$0.27881 \times 32 = 9$$
$$0.16213 \times 32 = 5$$

K-33

The multiplicative factors are then scaled by the ratio of the DC and AC quantization factors and rounded appropriately. The normalization defined for the DCT introduces another factor of 8 in the unquantized DC values. Therefore, in terms of the quantized DC values, the predicted quantized AC coefficients are given by the equations below. Note that if (for example) the DC values are scaled by a factor of 4, the AC predictions will have 2 fractional bits of precision relative to the quantized DCT coefficients.

$$QAC01 = ((Rd \times Q01) + (36 \times Q00 \times (QDC4-QDC6)))/(256 \times Q01)$$
$$QAC10 = ((Rd \times Q10) + (36 \times Q00 \times (QDC2-QDC8)))/(256 \times Q10)$$
$$QAC20 = ((Rd \times Q20) + (9 \times Q00 \times (QDC2+QDC8-2\times QDC5)))/(256 \times Q20)$$
$$QAC11 = ((Rd \times Q11) + (5 \times Q00 \times ((QDC1-QDC3)-(QDC7-QDC9))))/(256 \times Q11)$$
$$QAC02 = ((Rd \times Q02) + (9 \times Q00 \times (QDC4+QDC6-2\times QDC5)))/(256 \times Q02)$$

where QDCx and QACxy are the quantized and scaled DC and AC coefficient values. The constant Rd is added to get a correct rounding in the division. Rd is 128 for positive numerators, and -128 for negative numerators.

Predicted values should not override transmitted values. Therefore, predicted values for coefficients which are already nonzero should be set to zero. Predictions should be clamped if they exceed a value which would be quantized to a nonzero value for the current precision in the successive approximation.

K.9 Modification of dequantization to improve displayed image quality

For a progression where the first stage successive approximation bit, Al, is set to 3, uniform quantization of the DCT gives the following quantization and dequantization levels for a sequence of successive approximation scans:

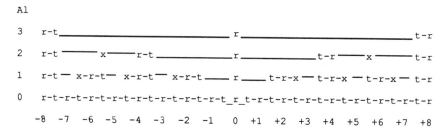

Quantized DCT coefficient value

Figure K.8 − Illustration of two reconstruction strategies

The column to the left labeled 'Al' gives the bit position specified in the scan header. The quantized DCT coefficient magnitudes are therefore divided by 2^{Al} during that scan.

Referring to the final scan (Al=0), the points marked with 't' are the threshold values, while the points marked with 'r' are the reconstruction values. The unquantized output is obtained by multiplying the horizontal scale in Figure K.8 by the quantization value.

K-34

The quantization interval for a coefficient value of zero is indicated by the depressed interval of the line. As the bit position Al is increased, a "fat zero" quantization interval develops around the zero DCT coefficient value. In the limit where the scaling factor is very large, the zero interval is twice as large as the rest of the quantization intervals.

Two different reconstruction strategies are shown. The points marked 'r' are the reconstruction obtained using the normal rounding rules for the DCT for the complete full precision output. This rule seems to give better image quality when high bandwidth displays are used. The points marked 'x' are an alternative reconstruction which tends to give better images on lower bandwidth displays. 'x' and 'r' are the same for slice 0. The system designer must determine which strategy is best for the display system being used.

K.10 Example of point transform

The difference between the arithmetic-shift-right by Pt and divide by 2^{Pt} can be seen from the following:

After the level shift the DC has values from +127 to -128. Consider values near zero (after the level shift), and the case where Pt=1:

Before level shift	Before point transform	After divide by 2	After shift-right-arithmetic 1
131	+3	+1	+1
130	+2	+1	+1
129	+1	0	˙0
128	0	0	0
127	-1	0	-1
126	-2	-1	-1
125	-3	-1	-2
124	-4	-2	-2
123	-5	-2	-3

The key difference is in the truncation of precision. The divide truncates the magnitude; the arithmetic shift truncates the LSB. With a divide by 2 we would get non-uniform quantization of the DC values; therefore we use the shift-right-arithmetic operation.

For positive values, the divide by 2 and the shift-right-arithmetic by 1 operations are the same. Therefore, the shift-right-arithmetic by 1 operation effectively is a divide by 2 when the point transform is done before the level shift.

K-35

Annex L (informative)

Patents

L.1 Introductory remarks

The user's attention is called to the possibility that - for some of the coding processes specified in Annexes F, G, H, and J - compliance with this Specification may require use of an invention covered by patent rights.

By publication of this Specification, no position is taken with respect to the validity of this claim or of any patent rights in connection therewith. However, for each patent listed in this Annex, the patent holder has filed with the ITTF a statement of willingness to grant a license under these rights on reasonable and non-discriminatory terms and conditions to applicants desiring to obtain such a license.

The criteria for including patents in this Annex are:

1. The patent has been identified by someone who is familiar with the technical fields relevant to this Specification, and who believes use of the invention covered by the patent is *required* for implementation of one or more of the coding processes specified in Annexes F, G, H, and J.

2. The patent-holder has written a letter to the ITTF, stating willingness to grant a license to an unlimited number of applicants throughout the world under reasonable terms and conditions that are demonstrably free of any unfair discrimination.

This list of patents shall be updated during preparation and maintenance of this Specification. During preparation, the list shall be updated upon publication of the DIS and any subsequent revisions, and upon publication of the IS. During maintenance, the list shall be updated, if necessary, upon publication of any revisions to the IS.

L.2 List of patents

The following patents may be required for implementation of any one of the processes specified in Annexes F, G, H, and J which uses arithmetic coding:

US 4,633,490, December 30, 1986, J.L. Mitchell and G. Goertzel, "Symmetrical Adaptive Data Compression/Decompression System".

US 4,652,856, February 4, 1986, K.M. Mohiuddin and J.J. Rissanen, "A Multiplication-free Multi-Alphabet Arithmetic Code".

US 4,369,463, January 18, 1983, D. Anastassiou and J.L. Mitchell, "Grey Scale Image Compression with Code Words a Function of Image History".

US 4,725,884, February 16, 1988, C.A. Gonzales, J.L. Mitchell, and W.B. Pennebaker, "Adaptive Graylevel Image Compression System".

L-1

US 4,749,983, June 7, 1988 G. Langdon, "Compression of Multilevel Signals".

US 4,935,882, June 19, 1990, W.B. Pennebaker and J.L. Mitchell, "Probability Adaption for Arithmetic Coders".

US 4,905,297, February 27, 1990, G.G. Langdon, Jr., J.L. Mitchell, W.B. Pennebaker, and F.F. Rissanen, "Arithmetic Coding Encoder and Decoder System".

US 4,973,961, November 27, 1990, C. Chamzas, D.L. Duttweiler, "Method and Apparatus for Carry-over Control in Arithmetic Entropy Coding".

US 5,025,258, June 18, 1991, D.L. Duttweiler, "Adaptive Probability Estimator for Entropy Encoding/Decoding".

Japanese Patent Application 2-46275, February 26, 1990, F. Ono, T. Kimura, M. Yoshida, and S. Kino, "Coding System"

No other patents required for implementation of any of the other processes specified in Annexes F, G, H, or J had been identified at the time of publication of this Specification.

Annex M (informative)

Bibliography

M.1 General references

1. A. Leger, T. Omachi, and G.K. Wallace, "JPEG Still Picture Compression Algorithm," Optical Engineering, Vol. 30, No. 7, pp. 947-954 (July 1991).

2. M. Rabbani and P. Jones, "Digital Image Compression Techniques", Tutorial Texts in Optical Engineering, vol. TT7, SPIE Press, 1991.

3. G. Hudson, H. Yasuda, and I. Sebestyen, "The International Standardization of a Still Picture Compression Technique," Proc. of IEEE Global Telecommunications Conference, pp. 1016-1021 (Nov. 1988).

4. A. Leger, J. Mitchell, and Y. Yamazaki, "Still Picture Compression Algorithm Evaluated for International Standardization," Proc. of the IEEE Global Telecommunications Conference, pp. 1028-1032 (Nov. 1988).

5. G. Wallace, R. Vivian, and H. Poulsen, "Subjective Testing Results for Still Picture Compression Algorithms for International Standardization," Proc. of the IEEE Global Telecommunications Conference, pp. 1022-1027 (Nov. 1988).

6. J.L. Mitchell and W.B. Pennebaker, "Evolving JPEG Color Data Compression Standard," *Standards for Electronic Imaging systems*, M. Nier, M.E. Courtot, Editors, SPIE Vol. CR37, pp. 68-97 (1991).

7. G.K. Wallace, "The JPEG Still Picture Compression Standard," Communications of the ACM, Vol. 34, No. 4, pp. 31-44 (April 1991).

8. A.N. Netravali and B.G. Haskell, *Digital Pictures: Representation and Compression*, Plenum Press, New York, 1988.

M.2 DCT references

1. W. Chen, C.H. Smith, and S.C. Fralick, "A Fast Computational Algorithm for the Discrete Cosine Transform", IEEE Trans. on Comm., COM-25, pp. 1004-1009, Sept. 1977.

2. N. Ahmed, T. Natarajan, and K. R. Rao, "Discrete Cosine Transform," IEEE Trans. on Computers, vol. C-23, pp. 90-93 (Jan. 1974).

3. N. J. Narasinha and A. M. Peterson, "On the Computation of the Discrete Cosine Transform," IEEE Trans. Communications, Vol. COM-26, No. 6, pp. 966-968 (Oct. 1978).

M-1

4. P. Duhamel and C. Guillemot, "Polynomial Transform Computation of the 2-D DCT," Proc. IEEE ICASSP-90, pp. 1515-1518, Albuquerque, New Mexico, 1990.

5. E. Feig, "A Fast Scaled DCT Algorithm," in *Image Processing Algorithms and Techniques*, Proc. SPIE, Vol. 1244, K.S. Pennington and R. J. Moorhead II, Editors, pp. 2-13, Santa Clara, California, February 11-16, 1990.

6 .H. S. Hou, "A Fast Recursive Algorithm for Computing the Discrete Cosine Transform," IEEE Trans. Acoust. Speech and Signal Processing, Vol. ASSP-35, No. 10, pp. 1455-1461.

7. B.G. Lee, "A New Algorithm to Compute the Discrete Cosine Transform," IEEE Trans. on Acoust., Speech and Signal Processing, Vol. ASSP-32, No. 6, pp. 1243-1245 (Dec. 1984).

8. E. N. Linzer and E. Feig, "New DCT and Scaled DCT Algorithms for Fused Multiply/Add Architectures," Proc. IEEE ICASSP-91, pp. 2201-2204, Toronto, Canada, May 1991.

9. M. Vetterli and H.J. Nussbaumer, "Simple FFT and DCT Algorithms with Reduced Number of Operations," Signal Processing, Aug. 1984.

10. M. Vetterli, "Fast 2-D Discrete Cosine Transform," Proc. IEEE ICASSP-85, pp. 1538-1541, Tampa, Florida, March 1985.

11. Y. Arai, T. Agui, and M. Nakajima, "A Fast DCT-SQ Scheme for Images," Trans. of IEICE, Vol. E 71, No. 11, pp. 1095-1097 (Nov. 1988).

12. N. Suehiro and M. Hatori, "Fast Algorithms for the DFT and other Sinusoidal Transforms," IEEE Trans. on Acoust., Speech and Signal Processing, Vol ASSP-34, No. 3, pp. 642-644 (June 1986).

M.3 Quantization and human visual model references

1. W.H. Chen and W.K. Pratt, "Scene adaptive coder", IEEE Transactions on Communications, vol. COM-32, pp. 225-232, March 1984.

2. D.J. Granrath, "The role of human visual models in image processing", Proceedings of the IEEE, vol. 67 pp. 552-561, May 1981.

3. H. Lohscheller, "Vision adapted progressive image transmission", Proceedings of EUSIPCO-83, pp. 191-194, September 1983.

4. H. Lohscheller and U. Franke, "Color picture coding - Algorithm optimization and technical realization", Frequenze, vol. 41, pp. 291-299, 1987.

5. H. Lohscheller, "A subjectively adapted image communication system", IEEE Transactions on Communications, vol. COM-32, pp. 1316-1322, December 1984.

6. H.A. Peterson, et al., "Quantization of color image components in the DCT domain", SPIE/IS&T 1991 Symposium on Electronic Imaging Science and Technology, Feb. 1991.

M.4 Arithmetic coding references

1. G. Langdon, "An Introduction to Arithmetic Coding", IBM J. Res. Develop. 28, pp. 135149, March 1984.

2. W.B. Pennebaker, J.L. Mitchell, G. Langdon, Jr., and R.B. Arps, "An Overview of the Basic Principles of the Q-Coder Binary Arithmetic Coder," IBM J. Res. Develop. 32, No. 6, pp. 717-726, November 1988.

3. J.L. Mitchell and W.B. Pennebaker, "Optimal Hardware and Software Arithmetic Coding Procedures for the Q-Coder Binary Arithmetic Coder", IBM J. Res. Develop.32, No. 6, pp. 727-736, November 1988.

4. W.B. Pennebaker and J.L. Mitchell, "Probability Estimation for the Q-Coder", IBM J. Res. Develop.32, No. 6, pp. 737-752, November 1988.

5. J.L. Mitchell and W.B. Pennebaker, "Software Implementations of the Q-Coder," IBM J. Res. Develop.32, No. 6, pp. 753-774, November 1988.

6. R. Arps, T.K. Truong, D.J. Lu, R.C. Pasco, and T.D. Reiedman, "A Multi-Purpose VLSI Chip for Adaptive Data Compression of Bilevel Images", IBM J. Res. Develop.32, No. 6, pp. 775-795, November 1988.

7. F. Ono, M. Yoshida, T. Kimura, and S. Kino, "Subtraction-type Arithmetic Coding with MPS/LPS Conditional Exchange", Annual Spring Converence of IECED, Japan, D-288, 1990.

8. C. Chamzas and D. Duttweiler, "Probability Estimation in Arithmetic Coders," in preparation.

M.5 Huffman coding references

1. D.A. Huffman, "A Method for the Construction of Minimum Redundancy codes", Proc. IRE 40, pp. 1098-1101, September 1952.

APPENDIX B

DRAFT
ISO DIS 10918-2
COMPLIANCE TESTING

This appendix contains an unapproved draft of ISO DIS 10918 Part 2. It is also an unapproved draft of CCITT recommendation T.83. Unlike ISO DIS 10918 Part 1, Part 2 is not yet an approved CCITT recommendation. This draft is reproduced with the permission of the ISO secretariat.

A later draft with many minor editorial changes has been sent to the JPEG committee for review and approval. Once enough of the compliance test data is generated, the DIS will be submitted to the SC29 secretariat for the DIS ballot. Since this is an unapproved draft, Part 2 should be regarded as likely to change. However, the authors believe that it is still useful for implementers. A sampling of the differences between this draft Part 2 and a later draft are given below.

Foreword
   History of JPEG revised. "for interchange between applications" deleted.
Clause 3
   A new definition is added: "orthogonal representation: The 2-dimensional row-column format illustrated in figure A.5 in Part 1 of this Specification." In the definition of the reference DCT-based decoder "the reference dequantizer" is changed to "a

dequantizer." "(64-bit)" is added after "double precision" in the reference forward/inverse discrete cosine transform definitions. The definition of source image test data is changed to: "source image test data: The data sets to be used as input to the encoder compliance tests. This data is a sequence of pseudo-random numbers generated with uniform distribution over the range from 0 to 255. The algorithm used to generate this data is described in Annex 1 of CCITT Recommendation H.261. (This data is distributed as part of the compliance test data.)" At the end of some of the symbol definitions, a phrase is added to explain where they appear. The upper-case "O" symbol is changed to a lower-case "o" symbol.

Clause 4

Remove "pixel."

Clause 5

After the first sentence of clause 5 is added: "These test procedures utilize the common additional procedures in 5.4." An "EOI?" test is added after the Check_frame procedure in Figure 1.

Clause 7

The second sentence of subclause 7.1 is changed to: "A decoder is found to be compliant if the resulting test data, for all the tests specified for a particular process in subclause 7.4, 7.5, 7.7.1 or 7.7.2, meet the requirements on accuracy specified in A.1.4." The second sentence of subclause 7.2 is changed to: "A decoder is found to be compliant if the resulting test data, for all the tests specified for a particular process in subclause 7.6 or 7.7.3, exactly match the decoder reference test data." Before the final sentence on page 29 is added: "Each of the two compressed image test data streams is tested separately."

Annex A

The third sentence in A.1 is changed to: "Processes with 8-bit sample precision and the corresponding processes with 12-bit sample precision are tested using identical test procedures. However, DCT-based encoders are tested for compliance with processes having 12-bit sample precision by first left-justifying the 8-bit source image test data within the 12-bit samples." Item 1 in A.1.2 is changed to: "1. With the supplied source image test data as input and using the quantization tables specified in Annex B, generate compressed image data using the encoder under test. The source image test data has four components designated A to D. The dimensions of the four components are listed in Annex C."

Annex B

B.1 header is removed. Subclause B.2 is removed.

Annex C

In the Annex C title, "generic" is inserted in front of "decoder." "Ta" is changed to "Tb" in the arithmetic coding compressed test data streams to match the corresponding change in Part 1.

INTERNATIONAL STANDARD DIS 10918-2

CCITT RECOMMENDATION T.83

DIGITAL COMPRESSION AND CODING OF CONTINUOUS-TONE STILL IMAGES

PART 2: COMPLIANCE TESTING

Contents   *Page*

Foreword .... *iii*

1 Scope 1

2 Normative references 2

3 Definitions, abbreviations, and symbols 3

4 General 8

5 Compressed data format compliance tests 12

6 Encoder compliance tests 27

7 Decoder compliance tests 28

Annex A Procedures for determining generic encoder and decoder compliance A-1

Annex B Quantization tables for generic compliance testing of DCT-based processes ... B-1

Annex C Compressed test data stream structure for decoder compliance tests C-1

Annex D Construction of application-specific compliance tests D-1

Annex E Compliance test data for testing of greater computational accuracy E-1

Annex F Specification of supported parameter ranges F-1

Annex G Test data to assist with validation of implementations G-1

Annex H Examples and guidelines H-1

Foreword

This document (Part 2) is at the Draft International Standard (DIS) level of the ISO/IEC JTC1 process for the development of International Standards. After successful balloting and any necessary revision, this DIS will be promoted to International Standard (IS).

This CCITT Recommendation | International Standard, Digital Compression and Coding of Continuous-tone Still Images, is published as two parts:

- Part 1: Requirements and guidelines

- Part 2: Compliance testing

Part 1 sets out requirements and implementation guidelines for continuous-tone still image encoding and decoding processes, and for the coded representation of compressed image data for interchange between applications. These processes and representations are intended to be generic, that is, to be applicable to a broad range of applications for colour and grayscale still images within communications and computer systems.

This part, Part 2, sets out tests for determining whether implementations comply with the requirements for the various encoding and decoding processes specified in Part 1. Part 2 also specifies tests for determining whether any specific instance of compressed image data complies with the Part 1 specification for the compressed image data interchange format.

The committee which has prepared this Specification is ISO/IEC JTC1/SC29/WG10, Coded Representation of Digital Continuous-tone Still Pictures (JPEG). Prior to the establishment of SC29/WG10 in November 1991, the committee existed as ISO/IEC JTC1/SC2/WG10. Prior to the establishment of SC2/WG10 in 1990, the committee existed as an Ad Hoc group, known as the Joint Photographic Experts Group (JPEG), of ISO/IEC JTC1/SC2/WG8. Both the committee and the Part 1 and Part 2 Specifications which it is developing continue to be known informally by the name JPEG.

The "joint" in JPEG refers to the committee's close but informal collaboration with the CCITT SGVIII Special Rapporteur's committee Q.16. In this collaboration, WG10 has performed the work of selecting, developing, documenting, and testing the generic compression processes. CCITT SGVIII has provided the requirements which these processes must satisfy to be useful for specific image communications applications such as facsimile, videotex, and audiographic conferencing. The intent is that the generic processes will be incorporated into the various CCITT Recommendations for terminal equipment for these applications.

This Specification aims to follow the procedures of CCITT and ISO/IEC JTC1 on "Presentation of CCITT | ISO/IEC Common Text," currently under elaboration by CCITT and ISO/IEC JTC1. The alignment of this Specification to the final version of the above joint drafting rules might result in minor editorial changes to this Specification. It is envisaged that the final CCITT Recommendation | International Standard shall fully conform with the joint drafting rules of CCITT and ISO/IEC JTC1.

Annexes A, B, C, and D are normative and thus form an integral part of this Specification. Annexes E, F, G, and H are informative and thus do not form an integral part of this Specification.

1 Scope

This CCITT Recommendation | International Standard is concerned with compliance tests
for the continuous-tone still image encoding processes, decoding processes, and
compressed data formats specified in Part 1.

This Specification

- specifies compliance tests for the Part 1 compressed data
  formats;

- specifies compliance tests for the Part 1 encoding processes;

- specifies compliance tests for the Part 1 decoding processes;

- specifies a method for constructing application-specific
  compliance tests;

- gives guidance and examples on how to implement these tests in
  practice.

This Specification specifies normative generic compliance tests for the Part 1 encoding
and decoding processes. These compliance tests are applicable to "stand-alone" generic
implementations of one or more of the encoding and decoding processes specified in
Part 1. Among the purposes of these tests is to ensure that generic encoder (and decoder)
implementations compute the discrete cosine transform (DCT) and quantization functions
with sufficient accuracy.

1

2 Normative references

ISO IS 5807 (1985), Information processing -- Documentation symbols and conventions for data, program and system flowcharts, program network charts and system resources charts.

3 Definitions, abbreviations, and symbols

3.1 Definitions and abbreviations

For the purposes of this International Standard, the following definitions apply.

arith.: An abbreviation for arithmetic coding.

(coding) process 1: Coding process with baseline sequential DCT, 8-bit sample precision.

(coding) process 2: Coding process with extended sequential DCT, Huffman coding, 8-bit sample precision.

(coding) process 3: Coding process with extended sequential DCT, arithmetic coding, 8-bit sample precision.

(coding) process 4: Coding process with extended sequential DCT, Huffman coding, 12-bit sample precision.

(coding) process 5: Coding process with extended sequential DCT, arithmetic coding, 12-bit sample precision.

(coding) process 6: Coding process with spectral selection only, Huffman coding, 8-bit sample precision.

(coding) process 7: Coding process with spectral selection only, arithmetic coding, 8-bit sample precision.

(coding) process 8: Coding process with spectral selection only, Huffman coding, 12-bit sample precision.

(coding) process 9: Coding process with spectral selection only, arithmetic coding, 12-bit sample precision.

(coding) process 10: Coding process with full progression, Huffman coding, 8-bit sample precision.

(coding) process 11: Coding process with full progression, arithmetic coding, 8-bit sample precision.

(coding) process 12: Coding process with full progression, Huffman coding, 12-bit sample precision.

3

(coding) process 13: Coding process with full progression, arithmetic coding, 12-bit sample precision.

(coding) process 14: Coding process with lossless, Huffman coding, 2- through 16-bit sample precision.

(coding) process 15: Coding process with lossless, arithmetic coding, 2- through 16-bit sample precision.

(coding) process 16: Coding process with extended sequential DCT, Huffman coding, 8-bit sample precision in hierarchical mode.

(coding) process 17: Coding process with extended sequential DCT, arithmetic coding, 8-bit sample precision in hierarchical mode.

(coding) process 18: Coding process with extended sequential DCT, Huffman coding, 12-bit sample precision in hierarchical mode.

(coding) process 19: Coding process with extended sequential DCT, arithmetic coding, 12-bit sample precision in hierarchical mode.

(coding) process 20: Coding process with spectral selection only, Huffman coding, 8-bit sample precision in hierarchical mode.

(coding) process 21: Coding process with spectral selection only, arithmetic coding, 8-bit sample precision in hierarchical mode.

(coding) process 22: Coding process with spectral selection only, Huffman coding, 12-bit sample precision in hierarchical mode.

(coding) process 23: Coding process with spectral selection only, arithmetic coding, 12-bit sample precision in hierarchical mode.

(coding) process 24: Coding process with full progression, Huffman coding, 8-bit sample precision in hierarchical mode.

(coding) process 25: Coding process with full progression, arithmetic coding, 8-bit sample precision in hierarchical mode.

(coding) process 26: Coding process with full progression, Huffman coding, 12-bit sample precision in hierarchical mode.

(coding) process 27: Coding process with full progression, arithmetic coding, 12-bit sample precision in hierarchical mode.

(coding) process 28: Coding process with lossless, Huffman coding, 2- through 16-bit sample precision in hierarchical mode.

(coding) process 29: Coding process with lossless, arithmetic coding, 2- through 16-bit sample precision in hierarchical mode.

compliance test: The procedures specified in this Part 2 of this Specification which determine whether or not an embodiment of an encoding process, compressed data stream, or decoding process complies with Part 1 of this Specification.

compressed image test data (stream): Compressed image data generated to test a particular coding process. (Distributed as part of the compliance test data.)

compressed image validation data (stream): Compressed image data generated for validation of a particular coding process. (Distributed as part of the compliance test data.)

compressed test data (stream): Either compressed image test data or table specification test data or both.

decoder reference test data: Quantized DCT coefficient data generated by the reference FDCT and reference quantizer from the reconstructed image data output by the reference decoder whose input is the compressed image test data to be used in the DCT-based decoder compliance tests. [After the test data has been generated, a description of the format needs to be inserted here.] (Distributed as part of the compliance test data.)

encoder reference test data: Quantized DCT coefficient data generated by the reference FDCT and reference quantizer from the source image test data to be used in the DCT-based encoder compliance tests. (Distributed as part of the compliance test data.)

generic: Applicable to a broad range of applications, i.e. application independent.

Huff.: An abbreviation for Huffman coding.

quantized coefficient validation data: Quantized DCT coefficient data generated from the source image validation test data to be used in the DCT-based encoder validation tests. (Distributed as part of the compliance test data.)

reference DCT-based decoder: An embodiment of the DCT-based decoding processes which generates the decoder reference test data. It consists of an entropy decoder, the reference dequantizer, and the reference IDCT.

reference DCT-based encoder: An embodiment of the DCT-based encoding processes which generated the DCT-based compressed image test data streams. It consisted of the reference FDCT, the reference quantizer, and an entropy encoder.

5

reference forward discrete cosine transform; reference FDCT: A double precision floating point embodiment of the FDCT described in A.3.3 of Part 1 of this Specification.

reference inverse discrete cosine transform; reference IDCT: A double precision floating point embodiment of the IDCT described in A.3.3 of Part 1 of this Specification.

reference quantizer: An embodiment of the quantization described in A.3.4 in Part 1 of this Specification.

source image test data: Source image data to be used as input to encoder compliance tests. (Distributed as part of the compliance test data.)

table specification test data (stream): Table specification data generated to test decoder compliance with abbreviated format compressed data. (Distributed as part of the compliance test data.)

3.2 Symbols

The symbols used in this Specification are listed below.

B_{ij}: quantization value at the ith row and jth column in the quantization tables defined in Annex B

DF: differential frame flag

E_{ij}: quantization value at the ith row and jth column in the quantization tables used in testing for greater accuracy defined in Annex E

F: the divisor used to generate E_{ij}> from B_{ij} as defined in E.1

FS: first scan in frame flag

G: guaranteed in compressed data

H-L: hierarchical lossless processes

H-S: hierarchical sequential DCT-based processes without final lossless scans

HP: hierarchical progression flag

LL: lossless processes

P(FULL): full progressive DCT-based processes with both spectral selection and successive approximation

6

P(SA): progressive DCT-based successive approximation processes

P(SS): progressive DCT-based spectral selection processes

RI: restart interval flag

S(B): baseline sequential DCT-based process

S(E): extended sequential DCT-based processes

O: optional in compressed data

7

4 General

The purpose of this clause is to give an informative overview of this Part 2 Specification and the principles underlying it. Another purpose is to introduce some of the terms which are defined in clause 3. (Terms defined in Part 1 clause 3 continue to apply in Part 2.)

Part 2 concerns compliance testing for embodiments of the elements specified in Part 1. For encoders and decoders - embodiments of the Part 1 encoding and decoding processes - this document makes a distinction between GENERIC embodiments and APPLICATION-SPECIFIC embodiments. For the former, compliance tests themselves are specified herein; for the latter, this document specifies a method for defining compliance tests. Compliance tests are also specified herein for compressed data streams - embodiments of the Part 1 compressed data formats.

> NOTE - Like many compliance tests, those specified herein for generic encoders and decoders are not exhaustive tests of their respective functional specifications. Therefore, passing these tests does not guarantee complete functional correctness. This observation has two implications: (1) the tests do not fully guarantee complete interoperability between independently- implemented encoders and decoders, and (2) the tests for embodiments of the DCT-based processes do not guarantee that encoders or decoders will have some well defined image-quality-producing capability. These limitations are discussed in more detail below.

4.1 Purpose of the compliance tests

The purpose of compliance tests is to provide designers, manufacturers, or users of a product with a set of procedures for determining whether the product meets a specified set of requirements with some confidence. In addition, the compliance tests specified herein are intended to achieve the following specific goals:

- increase the likelihood of compressed data interchange;

- decrease the likelihood that DCT-based encoders or decoders will yield reduced image quality as a result of computing the DCT or quantization procedures with insufficient accuracy;

- help implementors to meet the Part 1 requirements for encoders and decoders as fully as possible.

4.2 Compressed data compliance tests

The aim of the compliance tests specified in clause 5 is to determine whether a particular compressed image data stream or table-specification data stream meets the interchange

8

format or abbreviated format requirements specified in Part 1. These tests are performed on the compressed data itself.

4.3 Encoder and decoder compliance tests

This subclause summarizes the considerations which have led to the encoder and decoder compliance tests set out in Part 2.

4.3.1 Encoder versus decoder requirements

Part 1 imposes more requirements on decoders than on encoders. This difference is based on the philosophy that any encoder should be allowed to produce only compressed images with a limited range of parameter values, but that decoders must handle images with broad ranges of parameters in order to facilitate interchange. Specifically, a decoder is required to handle either (a) the full range and combination of the parameter values specified by its coding process (in which case it qualifies as a generic decoder), or (b) a subset of the same defined by some application (in which case it is an application-specific decoder - see 4.3.2.)

4.3.2 Generic versus application-specific decoders

Each coding process specified in Part 1 is defined for a fairly broad range of parameters. It is recognized, however, that many applications may require only a limited subset of these. For example, a simple picture database might use only grayscale images of fixed pixel dimensions.

Consequently, the committee which prepared this Specification has defined a distinction between generic and application-specific decoders. The former concept is important to facilitate interchange as applications become increasingly inter-connected, and for hardware or software decoder products which can be embedded within many different applications. The latter concept allows application-oriented standards bodies to define a subset of a Part 1 coding process as its requirements.

This distinction, along with the decoder requirements philosophy in 4.3.1, means that the compliance test for generic decoders should exercise, as much as possible, the full range and combination of the parameter values specified by its coding process. It also means that a compliance test for application-specific decoders should exercise only the combination and range specified by the application.

Although comprehensive in many ways, the compliance tests for generic decoders do not test the full allowed range of all parameters. Many parameters have larger allowed ranges than it is feasible to test. Also, for some parameters, e.g., Number of samples per line (X) and Number of lines (Y), it is not desirable to test their full allowed range since few applications require functionality over the entire range.

9

According to the encoder requirements philosophy, any encoder may operate on limited ranges of parameter values only, suggesting that encoders are by nature application-specific. Therefore, there is no concept of a generic encoder, and no defined encoder compliance test intended to exercise different parameter values. (The only generic aspect of encoder compliance concerns DCT accuracy, as explained in 4.3.3.)

4.3.3 Computational accuracy of DCT and quantization

In Part 1, the FDCT, quantizer, and IDCT are defined as ideal mathematical formulae. Because these formulae imply infinite precision, implementors must decide how to approximate them. Efficiency or cost considerations may encourage lower-accuracy approximations, but it is the combination of the DCT and the table-based method of quantization - which accommodates psychovisual thresholding - that gives the DCT-based processes their excellent image-quality-producing capability. This capability may be degraded if the DCT/quantization procedures are computed with insufficient accuracy. Therefore, this Specification provides a method of compliance testing aimed at discouraging such degradation.

Because there is no point in requiring that the FDCT be computed with greater accuracy than necessary for the subsequent quantization procedure, the compliance testing method for DCT-based encoders is concerned with the accuracy of the quantized DCT coefficients. (Basing the test on quantized coefficients also meets the practical constraint that, for product implementations, unquantized coefficients are typically not externally observable.) For symmetry, the method of decoder compliance testing imposes IDCT/dequantization accuracy requirements which are consistent with those imposed on the FDCT/quantization.

It is important to note that required accuracy is a function of the quantization tables used in these tests. A table with larger (coarser) quantization values will make for a less stringent test than one with smaller (finer) values. Therefore, passing the accuracy test means that the encoder or decoder is likely to perform comparably to an encoder or decoder with an ideal FDCT or IDCT, BUT ONLY WHEN USING THE SPECIFIC QUANTIZATION TABLE EMPLOYED IN THE TEST. An encoder which passes the test with a moderately coarse quantization table will not be guaranteed to perform as well with a finer quantization table as an ideal encoder.

For the generic DCT-based compliance tests specified herein, a set of quantization tables requiring moderate accuracy is specified. Encoders and decoders which achieve this accuracy will yield image quality sufficient for many applications, without incurring undue computational burden. Applications requiring greater or lesser accuracy may specify different quantization tables for application-specific compliance tests.

10

4.3.4 Summary: generic compliance test considerations

The compliance tests for generic decoders have been defined to exercise the full range and combination of parameter values specified by the coding process being tested.   The compliance tests for generic decoders have been designed so that decoders which satisfy the requirements of these tests are likely to be suitable for use within many different applications or for interchanging data between applications.

The generic compliance tests for DCT-based encoders and decoders define quantization tables requiring a level of computational accuracy which will yield image quality sufficient for many applications.

4.3.5 Procedures for constructing application-specific compliance tests

Application-specific compliance tests are used for testing compliance of application-specific decoders, i.e., decoders which implement a subset of a coding process, or for testing the accuracy of encoders and decoders for use in applications which have greater or lesser accuracy requirements than specified by the generic compliance tests. Application-specific compliance tests are constructed by applications standards bodies to satisfy the requirements of a particular application. This Specification contains the procedures for constructing application-specific compliance tests.

Two different procedures are defined for construction of application-specific compliance tests: one for DCT-based processes and one for lossless processes. Application-specific compliance tests for DCT-based processes may specify quantization tables which are selected according to the accuracy requirements of the application.

4.4 Availability of compliance test data

Standardized compliance test data is used to perform the encoder and decoder compliance tests.   There are two types of compliance test data which are used by the encoder compliance tests: source image test data and encoder reference test data.   Similarly, there are two types of compliance test data which are used by the decoder compliance tests: compressed test data and decoder reference test data.

The compliance test data for the encoder compliance tests and the generic decoder compliance tests are available from [ISO contact] to parties who wish to determine compliance of an encoder or decoder. [The WG10 committee anticipates that compliance test data will be available from ISO but at the current time is unable to provide details of the procedure for obtaining this data.] Information about compliance test data for application-specific compliance tests should be obtained from the standards body which maintains standards for the particular application area.

11

5 Compressed data format compliance testing

In order to determine compressed data format compliance, the test procedures in 5.1, 5.2 or 5.3 shall be performed.

There are separate tests for the following compressed data streams:

5a) Compressed image data encoded by non-hierarchical processes in interchange format (See 5.1.1)

5b) Compressed image data encoded by hierarchical processes in interchange format (See 5.1.2)

5c) Compressed image data encoded by non-hierarchical processes in abbreviated format (See 5.2.1)

5d) Compressed image data encoded by hierarchical processes in abbreviated format (See 5.2.2)

5e) Compressed data in abbreviated format for table specifications (See 5.3).

Twenty-nine coding processes are defined in each of the first paragraphs of Part 1 Annexes F, G, H, and J. They are assigned numbers in Part 2 Clause 3 (Definitions) as "(coding) process n" where n is an integer from 1 to 29.

Part 1 Annex B contains the syntax requirements for the compressed data. Part 1 B.1.3 and Figure B.1 give the conventions for the syntax figures. The markers are identified by the marker assignments in Part 1 Table B.1.

Tables 1, 3, and 4 give specific references to syntax requirements for markers. Markers and marker segments which are required in the compressed data are denoted "G". Those that may optionally be present in the compressed data are denoted "O ". A dash (-) indicates non-compliance if the particular marker or marker segment is present in the compressed data for that coding process.

If a marker is present, its parameters are required and not optional.

The Part 1 references in the left-most columns of Tables 1, 3 and 4 indicate where the syntax requirements for each marker segment are stated. There is no significance to the order of markers in the tables.

NOTES -

1) The tests are partial as they check mainly the syntactical correctness of

12

the data. Passing the test does not ensure that the compressed data comply with all the requirements of Part 1.

2) The flow charts do not use most values of the parameters. Future extensions may include more elaborate test procedures based on parameters' values.

3) There is no requirement in this Specification that any tester shall implement the procedures in precisely the manner specified by the flow charts in this Clause. It is necessary only that a tester implement the equivalent function specified in this Clause.

4) For simplicity of exposition, the buffer holding the compressed data is assumed to be large enough to contain the entire compressed data stream.

5) In any case that there is any conflict between this Clause and Part 1, Part 1 shall take precedence.

5.1 Interchange Compressed Image Data Format Syntax Compliance Tests

5.1.1 Non-hierarchical coding processes syntax compliance test

Figure 1 gives the non-hierarchical coding processes syntax compliance test main procedure.

```
    ( Non-hierarchical_syntax_test )
                 |
            ------------
            || Find_SOI ||
            ------------
                 |
            ---------------
            || Check_frame ||
            ---------------
                 |
                 |            no
        < All required >-----( Error )
          markers found?
              |yes
              |            no
          < Marker >-------( Error )
           order OK?
             |yes
             |
             |
          ( Done )
```

Figure 1. Non-hierarchical syntax test procedure

13

"All required makers found" means that all markers designated with G in the Table 1 column of the process under test were found. A missing required marker makes the compressed data under test non-compliant with the syntax. All other markers found should have an O in the relevant column. A marker found in the compressed data which has a (-) in the relevant column or is missing from the table makes the compressed data under test non-compliant with the syntax.

The high-level syntax in Part 1 B.2.1 and Part 1 Figure B.2 gives the required order to answer the "Marker order OK?" test for non-hierarchical coding processes.

Table 1. Marker syntax requirements for non-hierarchical coding processes

	Part 1			Process														
	Refer.	Fig.	Table	1	2	3	4	5	6	7	8	9	10	11	12	13	14	15
SOI	B.2.1	B.2		G	G	G	G	G	G	G	G	G	G	G	G	G	G	G
EOI	B.2.1	B.2		G	G	G	G	G	G	G	G	G	G	G	G	G	G	G
RST$_m$	B.2.1	B.2		O	O	O	O	O	O	O	O	O	O	O	O	O	O	O
SOS	B.2.3	B.4	B.3	G	G	G	G	G	G	G	G	G	G	G	G	G	G	G
DNL	B.2.5	B.12	B.10	O	O	O	O	O	O	O	O	O	O	O	O	O	O	O
Non-differential frames:																		
SOF$_0$	B.2.2	B.3	B.2	G	-	-	-	-	-	-	-	-	-	-	-	-	-	-
SOF$_1$	B.2.2	B.3	B.2	-	G	-	G	-	-	-	-	-	-	-	-	-	-	-
SOF$_2$	B.2.2	B.3	B.2	-	-	-	-	-	G	-	G	-	G	-	G	-	-	-
SOF$_3$	B.2.2	B.3	B.2	-	-	-	-	-	-	-	-	-	-	-	-	-	G	-
SOF$_9$	B.2.2	B.3	B.2	-	-	G	-	G	-	-	-	-	-	-	-	-	-	-
SOF$_{10}$	B.2.2	B.3	B.2	-	-	-	-	-	-	G	-	G	-	G	-	G	-	-
SOF$_{11}$	B.2.2	B.3	B.2	-	-	-	-	-	-	-	-	-	-	-	-	-	-	G
Tables/miscellaneous:																		
DQT	B.2.4.1	B.6	B.4	G	G	G	G	G	G	G	G	G	G	G	G	G	O	O
DHT	B.2.4.2	B.7	B.5	G	G	O	G	O	G	O	G	O	G	O	G	O	G	O
DAC	B.2.4.3	B.8	B.6	O	O	O	O	O	O	O	O	O	O	O	O	O	O	O
DRI	B.2.4.4	B.9	B.7	O	O	O	O	O	O	O	O	O	O	O	O	O	O	O
COM	B.2.4.5	B.10	B.8	O	O	O	O	O	O	O	O	O	O	O	O	O	O	O
APP$_n$	B.2.4.6	B.11	B.9	O	O	O	O	O	O	O	O	O	O	O	O	O	O	O

Table 2 gives the parameter column in Part 1 Tables B.2 through B.11 which should be used to determine the allowed range of parameter values in marker segments for non-hierarchical processes.

Table 2. Parameter column in Part 1 Annex B tables for non-hierarchical processes

	sequential DCT		progressive DCT	lossless
	baseline	extended		
Non-differential frames:				
SOF$_0$	G	-	-	-
SOF$_1$	-	G	-	-
SOF$_2$	-	-	G	-
SOF$_3$	-	-	-	G
SOF$_9$	-	G	-	-
SOF$_{10}$	-	-	G	-
SOF$_{11}$	-	-	-	G

14

5.1.2. Hierarchical coding processes syntax compliance test

Figure 2 gives the hierarchical coding processes syntax compliance test main procedure.

```
                    ( Hierarchical_syntax_test )
                                |
                        -------------
                        || Find SOI ||
                        -------------
                              |
                              |<-------------------------------------
                              |       no    --------------------    |
                          < DHP? >-----|| Check_tables/misc. ||---
                              | yes     --------------------
                              |
                    ---------------------------
                    | Check DHP parameters     |
                    | Position BP after segment |
                    | HP = 1                   |
                    ---------------------------
                              |
                        -------------
                        || Skip_fill ||
                        -------------
                              |
            ---------->|
        |         ---------------
        |         || Check_frame ||
        |         ---------------
        |       no    |
        --------< EOI?>
                    | yes
                    |
                    |           no
            < All required >-----( Error )
              markers found?
                    |yes
                    |           no
              < Marker >-------( Error )
                order OK?
                    |yes
                    |
                 ( Done )
```

Figure 2. Hierarchical syntax test procedure

The "Check DHP parameters" procedure is not specified here and is left to the tester. The tester should use the references given in Table 3 in the line containing DHP. In order to check that the parameters' values are in range, the appropriate column to be used can be found in Table 4.

BP is the pointer to the compressed data stream bytes. After checking parameters, it is positioned after the segment. HP is the hierarchical progression flag.

15

An EOI marker determines the end of the compressed data stream. If an EOI marker has not been found before BP points outside the compressed data, the compressed data stream under test is non-compliant.

The "All required makers found" test means that all markers designated with G in the column of the process of Table 3 for hierarchical processes were found. A missing required marker makes the compressed data under test non-compliant with the syntax. All other markers found should have an O in the relevant column. A marker found in the compressed data which has a (-) in the relevant column makes the compressed data under test non-compliant with the syntax.

The high-level syntax in Part 1 B.3.1 and Part 1 Figure B.13 gives the required order to answer the "Marker order OK?" test for hierarchical coding processes.

Table 3. Marker syntax requirements for hierarchical coding processes

	Part 1			Process													
	Refer.	Fig.	Table	16	17	18	19	20	21	22	23	24	25	26	27	28	29
SOI	B.3.1	B.13		G	G	G	G	G	G	G	G	G	G	G	G	G	G
EOI	B.3.1	B.13		G	G	G	G	G	G	G	G	G	G	G	G	G	G
RST_m	B.2.1	B.2		O	O	O	O	O	O	O	O	O	O	O	O	O	O
SOS	B.2.3	B.4	B.3	G	G	G	G	G	G	G	G	G	G	G	G	G	G
DNL	B.2.5	B.12	B.10	O	O	O	O	O	O	O	O	O	O	O	O	O	O
DHP	B.3.2	B.13	B.2	G	G	G	G	G	G	G	G	G	G	G	G	G	G
EXP	B.3.3	B.14	B.11	O	O	O	O	O	O	O	O	O	O	O	O	O	O
Non-differential frames:																	
SOF_0	B.2.2	B.3	B.2	O	O	-	-	O	O	-	-	O	O	-	-	-	-
SOF_1	B.2.2	B.3	B.2	O	-	G	-	O	-	O	-	O	-	O	-	-	-
SOF_2	B.2.2	B.3	B.2	-	-	-	-	O	-	O	-	O	-	O	-	-	-
SOF_3	B.2.2	B.3	B.2	-	-	-	-	-	-	-	-	-	-	-	-	G	-
SOF_9	B.2.2	B.3	B.2	-	O	-	G	-	O	-	O	-	O	-	O	-	-
SOF_{10}	B.2.2	B.3	B.2	-	-	-	-	-	O	-	O	-	O	-	O	-	-
SOF_{11}	B.2.2	B.3	B.2	-	-	-	-	-	-	-	-	-	-	-	-	-	G
Differential frames:																	
SOF_5	B.2.2	B.3	B.2	O	-	O	-	O	-	O	-	O	-	O	-	-	-
SOF_6	B.2.2	B.3	B.2	-	-	-	-	O	-	O	-	O	-	O	-	-	-
SOF_7	B.2.2	B.3	B.2	O	-	O	-	O	-	O	-	O	-	O	-	O	-
SOF_{13}	B.2.2	B.3	B.2	-	O	-	O	-	O	-	O	-	O	-	O	-	-
SOF_{14}	B.2.2	B.3	B.2	-	-	-	-	-	O	-	O	-	O	-	O	-	-
SOF_{15}	B.2.2	B.3	B.2	-	O	-	O	-	O	-	O	-	O	-	O	-	O
Tables/miscellaneous:																	
DQT	B.2.4.1	B.6	B.4	G	G	G	G	G	G	G	G	G	G	G	G	O	O
DHT	B.2.4.2	B.7	B.5	G	O	G	O	G	O	G	O	G	O	G	O	G	O
DAC	B.2.4.3	B.8	B.6	O	O	O	O	O	O	O	O	O	O	O	O	O	O
DRI	B.2.4.4	B.9	B.7	O	O	O	O	O	O	O	O	O	O	O	O	O	O
COM	B.2.4.5	B.10	B.8	O	O	O	O	O	O	O	O	O	O	O	O	O	O
APP_n	B.2.4.6	B.11	B.9	O	O	O	O	O	O	O	O	O	O	O	O	O	O

Table 4 gives the parameter column in Part 1 Tables B.2 through B.11 which should be used to determine the allowed range of parameter values in marker segments for hierarchical processes.

Table 4. Parameter column in Part 1 Annex B tables for hierarchical processes

	sequential DCT baseline	sequential DCT extended	progressive DCT	lossless
Non-differential frames:				
SOF_0	G	–	–	–
SOF_1	–	G	–	–
SOF_2	–	–	G	–
SOF_3	–	–	–	G
SOF_9	–	G	–	–
SOF_{10}	–	–	G	–
SOF_{11}	–	–	–	G
Differential frames:				
SOF_5	–	G	–	–
SOF_6	–	–	G	–
SOF_7	–	–	–	G
SOF_{13}	–	G	–	–
SOF_{14}	–	–	G	–
SOF_{15}	–	–	–	G

5.2 Abbreviated Compressed Data Format Syntax Requirements

5.2.1 Non-hierarchical coding processes syntax compliance test

The compliance testing for abbreviated format compressed image data syntax is the same as for the interchange format compressed image data given in 5.1.1 except that some or all of the table specifications may be omitted (see Part 1 B.4). If all of the tables are removed from a marker segment, the marker and its length parameter are also removed.

5.2.2 Hierarchical coding processes syntax compliance test

The compliance testing for abbreviated format compressed image data syntax is the same as for the interchange format compressed image data given in 5.1.2 except that some or all of the table specifications may be omitted (see Part 1 B.4). If all of the tables are removed from a marker segment, the marker and its length parameter are also removed.

17

5.3 Abbreviated format for table specification data syntax compliance test

Figure 3 gives the abbreviated format for table specification data syntax compliance test main procedure.

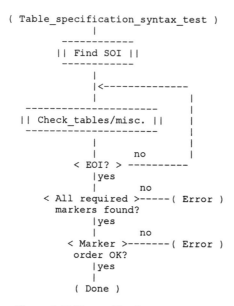

Figure 3. Table specification syntax test procedure

An EOI marker determines the end of the compressed data stream. If an EOI marker has not been found before BP points outside the compressed data, the compressed data stream under test is non-compliant.

The "All required makers found" test means that all markers designated with G in the column of the process of Table 5 for abbreviated format for table specifications were found. A missing required marker makes the compressed data under test non-compliant with the syntax. All other markers found should have an O in the relevant column. A marker found in the compressed data which has a (-) in the relevant column or is missing from Table 5 makes the compressed data under test non-compliant with the syntax.

The high-level syntax in Part 1 B.5 and Part 1 Figure B.15 gives the required order to answer the "Marker order OK?" test for abbreviated format for table-specification data.

18

Table 5. Marker syntax requirements for abbreviated format for table specification data

```
                        Part 1         |
            Marker Refer.    Fig.  Table|
                                        |
            SOI    B.5       B.15       |        G
            EOI    B.5       B.15       |        G
            Tables/miscellaneous:       |
            DQT    B.2.4.1 B.6    B.4   |        O
            DHT    B.2.4.2 B.7    B.5   |        O
            DAC    B.2.4.3 B.8    B.6   |        O
            DRI    B.2.4.4 B.9    B.7   |        O
            COM    B.2.4.5 B.10   B.8   |        O
            APPn   B.2.4.6 B.11   B.9   |        O
```

5.4 Additional procedures

Figure 4 gives the "Find_SOI" procedure in which the SOI marker is identified. This determines the start of the compressed data stream.

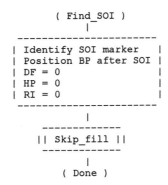

```
              ( Find_SOI )
                   |
        -----------------------
        | Identify SOI marker    |
        | Position BP after SOI  |
        | DF = 0                 |
        | HP = 0                 |
        | RI = 0                 |
        -----------------------
                   |
            -------------
          || Skip_fill ||
            -------------
                   |
              ( Done )
```

Figure 4. Procedure to find the SOI marker

The procedure of identifying the SOI marker skips any preceding fill bytes and may require information about where the compressed data starts which is outside of this Specification. A failure to find SOI at the start of the compressed data makes the compressed data under test non-compliant.

The hierarchical-progression flag (HP), the differential-frame flag (DF), and restart-interval flag (RI) are cleared. They will be set by a DHP marker, an EXP marker, and a DRI marker respectively. The HP and DF flags allow some procedures to be shared for both non-hierarchical and hierarchical processes.

Figure 5 shows the "Skip_fill" procedure.

```
         ( Skip_fill )
              |
              |              no
         < B = X'FF'? >-----( Error )
              | yes
              |<--------------
              |                      |
              |              ------------
              |              | BP = BP + 1 |
              |              ------------
              |         yes       |
         < B2 = X'FF'? >--------
              | no
              |
          ( Done )
```

Figure 5. Skip fill bytes procedure

First, the byte B pointed to by BP shall be an X'FF' byte. Then BP is incremented past any extra "fill" X'FF' bytes so that it will point to last X'FF' byte. Note that B2 is the byte next to B and is pointed to by BP+1.

Figure 6 gives the "Check_frame" procedure.

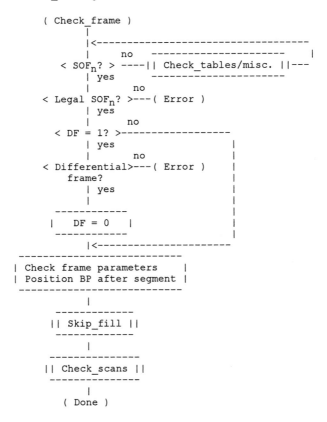

```
              ( Check_frame )
                    |
                    |<------------------------------------
                    |      no    --------------------           |
               < SOFn? > ----|| Check_tables/misc. ||---
                    | yes     --------------------
                    |              no
            < Legal SOFn? >---( Error )
                    | yes
                    |           no
              < DF = 1? >------------------
                    | yes                  |
                    |          no          |
            < Differential>---( Error )    |
                frame?                      |
                    | yes                   |
                    |                       |
              ------------                  |
              |   DF = 0   |                |
              ------------                  |
                    |<---------------------
        --------------------------
        | Check frame parameters   |
        | Position BP after segment |
        --------------------------
                    |
              -------------
              || Skip_fill ||
              -------------
                    |
              ---------------
              || Check_scans ||
              ---------------
                    |
                ( Done )
```

Figure 6. The procedure for checking a frame for all coding processes.

The "Legal SOFn?" test verifies that, in Table 1 for non-hierarchical processes, and Table 3 for hierarchical processes, the intersection of the column of the given process number and the line of the found SOF_n marker code, is designated by either G or O.

If the EXP marker has set the DF flag to 1 (in hierarchical processes only), a differential frame must follow.

The "Check frame parameters" procedure is not specified here and is left to the tester. The tester should use the references given in Table 1 in the line containing the found SOF_n for non-hierarchical coding processes. Table 2 gives the column which is used to determine allowed range of parameter values based on the SOF_n marker for non-hierarchical coding processes. The tester should use the references given in Table 2 in the line containing the found SOF_n for hierarchical coding processes. Table 4 gives the column which is

21

used to determine allowed range of parameter values based on the SOF_n marker for hierarchical coding processes.

Figure 7 gives the "Check_tables/misc." procedure for all processes.

```
          ( Check_tables/misc. )
                    |
                    |                   no
      < Legal tables/misc. >---( Error )
              marker?
                   | yes
                   |    no
              < DRI? >------------
                   | yes         |
         --------------------    |
         | BP = BP + 4      |    |
         | IF( (B+B2)=0 )   |    |
         |   THEN RI=0      |    |
         |   ELSE RI=1      |    |
         | BP = BP - 4      |    |
         --------------------    |
                   |<-------------
                   |
         --------------------------
         | Check marker parameters   |
         | Position BP after segment |
         --------------------------
                   |
              -------------
              || Skip_fill ||
              -------------
                   |
               ( Done )
```

Figure 7. Procedure to check the tables/miscellaneous markers

The "Legal tables/misc. marker?" test verifies that, in Table 1 for non-hierarchical processes, Table 3 for hierarchical processes, and Table 5 for abbreviated format for table specifications, under the tables/miscellaneous caption, the intersection of the column of the given process number and the line of the marker code to be checked, is designated by either G or O.

If the marker is DRI (Define Restart Interval), four is added to BP so that B points at the most significant byte and B2 at the least significant byte of the parameter Ri in order to use them to set the restart-interval flag RI. Then four is subtracted from BP to reposition BP at the DRI marker.

The "Check marker parameters" procedure is not specified here and is left to the tester. The tester should use the references given in Table 1 for non-hierarchical processes, Table 3 for hierarchical processes, and Table 5 for abbreviated format for table specifications, in the line containing the found table/misc. marker. In order to check that the parameters' values are in range, the appropriate column to be used can be found in Table 2 for non-hierarchical processes and Table 4 for hierarchical processes.

Figure 8 gives the "Check_scans" procedure for all coding processes.

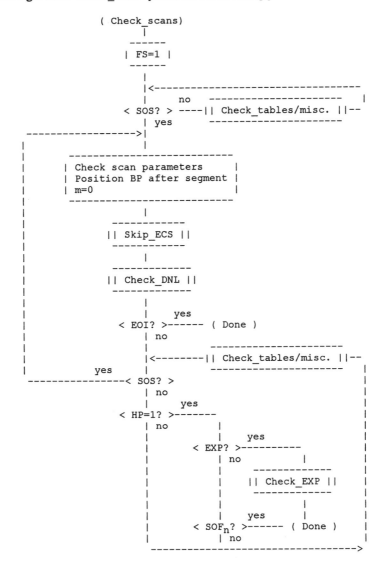

Figure 8. The procedure for checking scans

FS is the first-scan-in-frame flag needed for syntax checking of the DNL marker in the "Check_DNL" procedure.

23

The "Check scan parameters" procedure is not specified here and is left to the tester. The tester should use the references given in Table 1 for non-hierarchical processes and Table 3 for hierarchical processes on the line containing SOS. In order to check that the parameters' values are in range, the appropriate column to be used can be found in Table 2 for non-hierarchical processes and Table 4 for hierarchical processes.

An EOI marker determines the end of the compressed data stream. If an EOI marker has not been found before BP points outside the compressed data, the compressed data stream under test is non-compliant.

Figure 9 gives the "Skip_ECS" procedure.

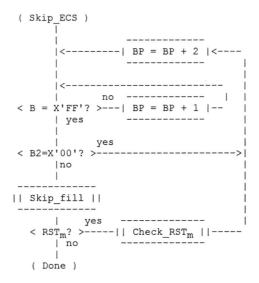

Figure 9. Skip over entropy-coded segment procedure

Figure 10 gives the "Check_DNL" procedure.

```
              ( Check_DNL )
                   |            no
                < DNL? >-------------------
                   | yes                    |
                   |           no           |
               < FS = 1? >-----( Error )    |
                   | yes                     |
         ---------------------------          |
        | Check DNL parameters     |          |
        | Position BP after segment |         |
         ---------------------------          |
                   |                           |
              -------------                    |
             || Skip_fill ||                   |
              -------------                    |
                   |<----------------------
                   |
                ------
               | FS=0 |
                ------
                   |
               ( Done )
```

Figure 10. Check DNL procedure

The "Check DNL parameters" procedure is not specified here and is left to the tester. The tester should use the references given in Table 1 for non-hierarchical processes, and in Table 3 for hierarchical processes, in the line containing DNL. In order to check that the parameters' values are in range, the appropriate column to be used can be found in Table 2 for non-hierarchical processes and Table 4 for hierarchical processes.

Figure 11 gives the "Check_EXP" procedure.

```
              ( Check_EXP )
                   |
         ---------------------------
        | Check EXP parameters     |
        | Position BP after segment |
         ---------------------------
                   |
              -------------
             || Skip_fill ||
              -------------
                   |
                ------
               | DF=1 |
                ------
                   |
               ( Done )
```

Figure 11. Check EXP procedure

25

The "Check EXP parameters" procedure is not specified here and is left to the tester. The tester should use the references given in Table 3 for hierarchical processes, in the line containing EXP. The parameters' values are independent of the process.

DF is set to check that the next SOF_n is a differential SOF marker.

Figure 12 gives the "Check_RST_m" procedure.

```
         ( Check_RSTm )
              |
              |       yes
        < RI = 0? >------( Error )
              | no
              |                  no
    < m = (7 AND RSTm)? >------( Error )
             |yes
             |
       -------------------
      | m = (m+1) AND 7   |
       -------------------
             |
          ( Done )
```

Figure 12. Check restart marker procedure

If the restart-interval flag (RI) is zero, RST_m markers are not allowed. The three least significant bits of the RST_m marker shall agree with the modulo counter m.

6 Encoder compliance tests

An encoder is considered compliant to an encoding process if it satisfies the requirements stated in Clause 6 in Part 1 of this Specification and satisfies the requirements on accuracy for the compliance tests defined for that process in Part 2 of this Specification.

6.1 Compliance tests for DCT-based encoders

In order to determine compliance of DCT-based encoders, the test procedure set forth in A.1.1 and A.1.2 shall be performed.   An encoder is found to be compliant if the resulting test data meet the requirements on accuracy specified in A.1.2.

6.2 Compliance tests for lossless encoders

No lossless encoder compliance tests are defined or required.

6.3 Availability of compliance test data

Source image test data and encoder reference test data are available from [ISO contact] to parties who wish to determine compliance of a DCT-based encoder. [The WG10 committee anticipates that compliance test data will be available from ISO but at the current time is unable to provide details of the procedure for obtaining this data.]

27

7 Decoder compliance tests

A decoder is considered compliant to an decoding process if it satisfies the requirements stated in Clause 7 in Part 1 of this Specification and satisfies the requirements on accuracy for the compliance tests defined for that process in Part 2 of this Specification.

7.1 Compliance tests for DCT-based decoders

In order to determine compliance of DCT-based decoders, the test procedure set forth in A.1.3 and A.1.4 shall be performed.   An decoder is found to be compliant if the resulting test data, for all the tests specified for a particular process, meet the requirements on accuracy specified in A.1.4.

7.2 Compliance tests for lossless decoders

In order to determine compliance of lossless decoders, the test procedure set forth in A.2.2 shall be performed.   A decoder is found to be compliant if the resulting test data, for all the tests specified for a particular process, exactly match the decoder reference test data.

7.3 Availability of compliance test data

Compressed image test data and decoder reference test data are available from [ISO contact] to parties who wish to determine compliance of a DCT-based decoder. [The WG10 committee anticipates that compliance test data will be available from ISO but at the current time is unable to provide details of the procedure for obtaining this data.]

7.4 Compliance tests for DCT-based sequential mode decoding processes (Tests A, B, C, D, E, and F)

A list of the compliance tests for processes which utilize the DCT-based sequential mode of operation follow below:

Process 1. Baseline DCT, 8-bit sample precision.
       Required tests: A, B.

Process 2. Extended sequential DCT, Huffman decoding, 8-bit sample precision.
       Required tests: A, B, C.

Process 3. Extended sequential DCT, arithmetic decoding, 8-bit sample precision.
       Required tests: A, B, D.

28

Process 4. Extended sequential DCT, Huffman decoding, 12-bit sample precision.
           Required tests: A, B, C, E.

Process 5. Extended sequential DCT, arithmetic decoding, 12-bit sample precision.
           Required tests: A, B, D, F.

Compliance for baseline decoders (Process 1) requires successful completion of two tests, tests A and B. Each test defines its own compressed compressed image test data structure. The testing procedure shall be repeated for each test using the specified compressed image test data as test input and the output data produced by each test must satisfy the requirements on accuracy for all DCT-based decoders.

The structure of the compressed image test data used by the baseline process tests (Tests A and B) are described below:

     Test A:
        Compressed image test data stream A1:
                Interchange format syntax
                4 components
                A single interleaved scan
                Restart interval = 1/2 block row - 1
     Test B:
        Compressed image test data stream B1:
                Abbreviated format syntax
                Huffman and quantization tables
                No entropy coded segments

        Compressed image test data stream B2:
                Abbreviated format syntax
                255 components non-interleaved

Test A specifies a compressed image test data stream which conforms to the syntax of the Interchange Format. Test B employs two compressed image test data streams (B1 and B2) which conform to the abbreviated format syntax. These two compressed image test data streams must be decoded in succession, with compressed image test data stream B2 immediately following the EOI marker of compressed image test data stream B1. The output test data produced after compressed image test data stream B2 is decoded must satisfy the requirements on accuracy for all DCT-based decoders.

All other tests defined for the DCT-based sequential processes (tests C, D, E and F) employ two compressed image test data stream as input: one interleaved and one non-interleaved. See Annex C for a specification of the compressed image test data streams utilized.

7.5 Compliance tests for DCT-based progressive mode decoding processes (Tests G, H, I, J, K, L, M, and N)

A list of the compliance tests for decoding processes which utilize the DCT-based progressive mode of operation follow below:

Process 6.   Spectral selection only, Huffman decoding, 8-bit sample precision.
            Required tests: A, B, C, G.

Process 7.   Spectral selection only, arithmetic decoding, 8-bit sample precision.
            Required tests: A, B, D, H.

Process 8.   Spectral selection only, Huffman decoding, 12-bit sample precision.
            Required tests: A, B, C, E, G, I.

Process 9.   Spectral selection only, arithmetic decoding, 12-bit sample precision.
            Required tests: A, B, D, F, H, J.

Process 10. Full progression, Huffman decoding, 8-bit sample precision.
            Required tests: A, B, C, G, K.

Process 11. Full progression, arithmetic decoding, 8-bit sample precision.
            Required tests: A, B, D, H, L.

Process 12. Full progression, Huffman decoding, 12-bit sample precision.
            Required tests: A, B, C, E, G, I, K, M.

Process 13. Full progression, arithmetic decoding, 12-bit sample precision.
            Required tests: A, B, D, F, H, J, L, N.

The tests defined for the DCT-based progressive processes each employ a single compressed image test data stream to be used as input. See Annex C for a specification of the compressed image test data utilized.

7.6 Compliance tests for lossless mode decoding processes (Tests O and P)

A list of the compliance tests for processes which utilize the lossless mode of operation follow below:

Process 14. Lossless, Huffman decoding, 2- through 16-bit sample precision.
            Required tests: O.

Process 15. Lossless, arithmetic decoding, 2- through 16-bit sample precision.
            Required tests: P.

30

The required tests of each lossless mode process each utilize two compressed image test data streams having different sample precision: 8 and 16 bits respectively. Also, the compressed image test data streams have different encoding order: interleaved and non-interleaved. Each of the two compressed image test data streams is tested separately. See Annex C for a specification of the compressed image test data utilized.

7.7 Compliance tests for hierarchical mode decoding processes

Hierarchical compliance tests utilize compressed image test data comprised of several stages as input. Every stage within a particular hierarchical compliance test employs one of the processes numbered 2-15, as modified for hierarchical mode.

Every hierarchical compliance test designates that a decoder shall pass additional tests which address functional subsets of the current test.

7.7.1 Compliance tests for hierarchical mode with DCT-based sequential decoding processes (Tests Q and R)

A list of the compliance tests for decoders which utilize DCT-based sequential processes in the hierarchical mode of operation follow below:

Process 16. Extended sequential DCT, Huffman decoding, 8-bit sample precision.
            Required tests: A, B, C, Q.

Process 17. Extended sequential DCT, arithmetic decoding, 8-bit sample precision.
            Required tests: A, B, D, R.

Process 18. Extended sequential DCT, Huffman decoding, 12-bit sample precision.
            Required tests: A, B, C, E, Q.

Process 19. Extended sequential DCT, arithmetic decoding, 12-bit sample precision.
            Required tests: A, B, D, F, R.

The tests which address DCT-based sequential configurations (Tests Q and R) utilize compressed image test data streams having 6 frames. See Annex C for a specification of the compressed image test data utilized.

Decoders capable of processing a final lossless stage shall also complete an additional test: Test S for Huffman decoders and test T for arithmetic decoders.

31

7.7.2 Compliance tests for hierarchical mode with DCT-based progressive processes (Tests Q and R)

A list of the compliance tests for decoders which utilize DCT-based progressive processes in the hierarchical mode of operation follow below:

Process 20. Spectral selection only, Huffman decoding, 8-bit sample precision.
             Required tests: A, B, C, G, Q.

Process 21. Spectral selection only, arithmetic decoding, 8-bit sample precision.
             Required tests: A, B, D, H, R.

Process 22. Spectral selection only, Huffman decoding, 12-bit sample precision.
             Required tests: A, B, C, E, G, I, Q.

Process 23. Spectral selection only, arithmetic decoding, 12-bit sample precision.
             Required tests: A, B, D, F, H, J, R.

Process 24. Full progression, Huffman decoding, 8-bit sample precision.
             Required tests: A, B, C, G, K, Q.

Process 25. Full progression, arithmetic decoding, 8-bit sample precision.
             Required tests: A, B, D, H, L, R.

Process 26. Full progression, Huffman decoding, 12-bit sample precision.
             Required tests: A, B, C, E, G, I, K, M, Q.

Process 27. Full progression, arithmetic decoding, 12-bit sample precision.
             Required tests: A, B, D, F, H, J, L, N, R.

Decoders capable of processing a final lossless stage shall also complete an additional test: test S for Huffman decoders and test T for arithmetic decoders.

7.7.3 Compliance tests for hierarchical mode with lossless decoding processes (Tests S and T)

A list of the compliance tests for decoders which utilize lossless processes in the hierarchical mode of operation follow below:

Process 28. Lossless, Huffman decoding, 2- through 16-bit sample precision.
             Required tests: O, S.

Process 29. Lossless, arithmetic decoding, 2- through 16-bit sample precision.
             Required tests: P, T.

The tests which address lossless configurations (tests S and T) utilize compressed image test data streams having 5 frames. See Annex C for a specification of the compressed image test data utilized.

7.8 Summary of decoder compliance test requirements

Table 6 summarizes the tests required to determine compliance for each decoding process.

Table 6. Decoder compliance test requirements for each process

```
                              Test
Process | A B C D E F G H I J K L M N O P Q R S T
--------|------------------------------------------
   1    | G G
   2    | G G G
   3    | G G   G
   4    | G G G   G
   5    | G G   G   G
   6    | G G G       G
   7    | G G   G       G
   8    | G G G   G   G   G
   9    | G G   G   G   G   G
  10    | G G G       G       G
  11    | G G   G       G       G
  12    | G G G   G   G   G   G
  13    | G G   G   G   G   G   G
  14    |                           G
  15    |                             G
  16    | G G G                     G   O
  17    | G G   G                     G   O
  18    | G G G   G                 G   O
  19    | G G   G   G                 G   O
  20    | G G G       G             G   O
  21    | G G   G       G             G   O
  22    | G G G   G   G   G         G   O
  23    | G G   G   G   G   G         G   O
  24    | G G G       G       G     G   O
  25    | G G   G       G       G     G   O
  26    | G G G   G   G   G   G     G   O
  27    | G G   G   G   G   G   G     G   O
  28    |                           G       G
  29    |                             G       G
```

Required tests are denoted 'G'. Optional additional tests are denoted 'O'.

33

Annex A

Procedures for determining generic encoder and decoder compliance

(This annex forms an integral part of this Recommendation | International Standard.)

The compliance test procedures defined within this specification require that output data sets generated by the device under test match reference data sets within the requirements on accuracy for the process being tested. Compliance test procedure for DCT-based processes are defined separately from the test procedure for lossless processes.

A.1 Compliance test procedures for DCT-based processes

This section describes the compliance tests for DCT-based processes. DCT-based process implementation accuracy is always assessed by comparison of quantized DCT data. Processes with 8-bit sample precision and the corresponding processes with 12-bit sample precision are tested using identical test procedures; processes with 12-bit sample precision are tested by left justifying the 8-bit source image data within the 12-bit samples. Left justification can be accomplished by multiplying the 8-bit samples by 16.

A.1.1 DCT-based encoder compliance test procedure - Introduction

The encoder compliance test procedure for DCT-based processes creates a compressed image data set from the source image test data. The compressed image data is then entropy decoded using a reference entropy decoder and the decoded quantized DCT coefficients are compared with the DCT coefficients of the encoder reference test data. If the uncoded quantized DCT coefficients generated by the encoder under test are directly accessible (as possibly in a software implementation), the entropy encoding and reference entropy decoding steps may be omitted. If the quantized DCT coefficients are not directly accessible, then it is the responsibility of the implementor to provide a reference entropy decoder which is compatible with the generated compressed image data so that the necessary test data can be obtained.

The difference between the encoder implementation's quantized DCT coefficients and the encoder reference test data must not exceed the requirements on accuracy contained in A.1.2. These requirements apply to processes with 8-bit input precision and to processes with 12-bit input precision. A block diagram of the encoder testing procedure is shown in Figure A.1.

A-1

```
                        ---------------   -------------
                        | DCT-based   |   | Reference |    +
Source test image -->|  Encoder    |-->| Entropy   |---> + --> Output test data
                        | Under Test  |   | Decoder   |    - ^
                        ---------------   -------------      |
                                                             |
                                      Encoder reference test data
```

Figure A.1 - Testing procedure for a DCT-based encoder

Standard data sets can be obtained from [ISO contact] which contain the source image test data and the encoder reference test data to be subtracted from the output data produced by the device under test. Two encoder reference test data sets are available: one for processes employing 8-bit precision and one for processes with 12-bit input precision.

A.1.2 Procedure for determining compliance of a DCT-based encoder

This procedure is used to determine whether a proposed implementation of a DCT-based encoder satisfies the requirements for compliance. The procedure is as follows:

1. With the supplied test image as input and using the quantization tables specified in Annex B, generate compressed image data using the encoder under test.

2. Decode the compressed image data using the reference entropy decoder to obtain the quantized transform coefficients for the encoder under test.

3. Subtract the decoded quantized DCT coefficients from the corresponding quantized DCT coefficients of the encoder reference test data supplied to obtain the error values. 8x8 blocks that were completed by extension, or blocks that were added to complete an MCU as defined in A.2.4 of Part 1 shall not be considered. The values of all absolute differences shall not exceed one.

A.1.3 DCT-based decoder compliance test procedure - Introduction

A DCT-based decoder is tested by first decoding the compressed image test data. The output image is then used as input to the reference FDCT and quantizer. The output of the reference FDCT and quantizer is then compared to the decoder reference test data. The reference FDCT and quantizer shall be constructed be the implementor according to the definitions in Clause 3.

The quantized coefficients produced from the output image of the decoder under test shall meet the requirement on accuracy given in A.1.4.

These requirements apply to processes with 8-bit output precision and to processes with 12-bit output precision. A block diagram of the decoder testing procedure is shown in Figure A.2.

A-2

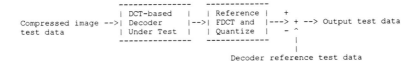

Figure A.2 - Testing procedure for a DCT-based decoder

The compressed image test data and the decoder reference test data are available from [ISO contact] to any parties who wish to determine compliance of a decoder. Two decoder reference test data sets are available: one for processes employing 8-bit output precision and one for processes with 12-bit output precision.

A.1.4 Procedure for determining compliance of a DCT-based decoder

This procedure is used to determine whether a proposed implementation of a decoder satisfies the requirements for compliance. The procedure is as follows:

1. Decode the supplied compressed image test data using the decoder under test.

2. Calculate the quantized DCT coefficients from the decoded output image according to the FDCT and quantization procedures defined in Annex A of Part 1 implemented with double precision floating point accuracy.

3. For each quantized coefficient, subtract the reference quantized coefficient in the decoder reference test data supplied.   8x8 blocks that were completed by extension, or blocks that were added to complete an MCU as defined in A.2.4 of Part 1 shall not be considered.   The values of all absolute differences shall not exceed one.

A.2 Compliance tests for lossless processes

This subclause describes the compliance test procedure for lossless processes. The lossless compliance tests require exact accuracy; output test data shall match the decoder reference test data with no differences.

A.2.1 Lossless encoder compliance test procedure

No lossless encoder compliance test procedure is defined.

A.2.2 Lossless decoder compliance test procedures

Lossless decoders are tested by decoding the compressed image test data produced by a reference encoder and comparing the output image with the image produced when the same compressed image data is decoded by a reference decoder. The output image

A-3

produced by the decoder under test shall exactly match the decoder reference test data(no differences). The same requirements apply to all lossless decoders regardless of the output precision. A block diagram of the general lossless testing procedure is shown in Figure A.3:

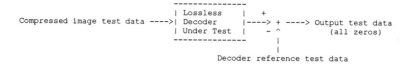

Figure A.3 - General testing procedure for a lossless decoder

The compressed image test data and the decoder reference test data are available from [ISO contact] to any parties who wish to determine compliance of a lossless decoder process.

Annex B

Quantization tables for generic compliance testing of DCT-based processes

(This annex forms an integral part of this Recommendation | International Standard.)

This annex specifies the quantization tables used in compliance tests for all DCT-based processes. These quantization tables are used for compliance testing of generic encoders and decoders.

B.1 Quantization tables

The quantization tables are defined in Tables B.1 - B.4. The compliance tests for those processes having 8-bit sample precision use the values in these tables as shown. The compliance tests for those processes having 12-bit sample precision use the table values multiplied by 4. The tables are presented in "orthogonal", not "zig-zag", representation.

The values of quantization table for component A are shown in Table B.1.

Table B.1   Quantization table for component A

8	6	5	8	12	20	26	30
6	6	7	10	13	29	30	28
7	7	8	12	20	29	35	28
7	9	11	15	26	44	40	31
9	11	19	28	34	55	52	39
12	18	28	32	41	52	57	46
25	32	39	44	52	61	60	51
36	46	48	49	56	50	52	50

The values of quantization table for component B are shown in Table B.2.

Table B.2   Quantization table for component B

9	9	12	24	50	50	50	50
9	11	13	33	50	50	50	50
12	13	28	50	50	50	50	50
24	33	50	50	50	50	50	50
50	50	50	50	50	50	50	50
50	50	50	50	50	50	50	50
50	50	50	50	50	50	50	50
50	50	50	50	50	50	50	50

B-1

The values of quantization table for component C are shown in Table B.3.

Table B.3 Quantization table for component C

16	17	18	19	20	21	22	23
17	18	19	20	21	22	23	24
18	19	20	21	22	23	24	25
19	20	21	22	23	24	25	26
20	21	22	23	24	25	26	27
21	22	23	24	25	26	27	28
22	23	24	25	26	27	28	29
23	24	25	26	27	28	29	30

The values of quantization table for component D are shown in Table B.4.

Table B.4 Quantization table for component D

16	16	19	22	26	27	29	34
16	16	22	24	27	29	34	37
19	22	26	27	29	34	34	38
22	22	26	27	29	34	37	40
22	26	27	29	32	35	40	48
26	27	29	32	35	40	48	58
26	27	29	34	38	46	56	69
27	29	35	38	46	56	69	83

B.2 Quantization table assignment for DCT-based compliance tests

For the DCT-based encoder compliance tests the quantization table assignment is shown in Table B.5 .

Table B.5 Encoder quantization table assignments

Component	Quantization Table defined in	Quantization Table Selector (Tqi)
A	Table B.1	0
B	Table B.2	1
C	Table B.3	2
D	Table B.4	3

Annex C

Compressed test data stream structure for decoder compliance tests

(This annex forms an integral part of this Recommendation | International Standard.)

Decoder compliance tests define specific compressed image data to be used as test input. Compliance tests employ reference test data which are to be compared with the test data generated by the device under test.

This Annex contains specifications of the structure of the compressed test data sets utilized by decoder compliance tests.

The compressed test data streams, A1, C1, C2, ... T2, define parameter values in the frame header as follows:

Sample precision (P)	= varies with test
Number of lines (Y)	= 257
Number of samples per line (X)	= 255
Number of components (N$_f$)	= 4
Component identifiers (C$_i$)	= 200 (component A)
	150 (component B)
	100 (component C)
	50 (component D)
Horizontal sampling factors (H$_i$)	= 1, 1, 3, 1
Vertical sampling factors (V$_i$)	= 1, 2, 1, 4
Quantization table selectors (T$_q$)	= 0, 1, 2, 3

Sample precision (P) utilized is either 8 or 12 for DCT-based processes, and is either 8 or 16 for lossless processes.

These compressed data streams describe an image which, after decoding, has components with the following dimensions:

Component A	85 samples x 65 lines
Component B	85 samples x 129 lines
Component C	255 samples x 65 lines
Component D	85 samples x 257 lines

These dimensions are derived from the number of lines (Y), number of samples per line (X), horizontal sampling factors (H$_i$), and the vertical sampling factors (V$_i$) parameters in the frame header.

C-1

The compressed test data stream, B2, defines parameter values in the frame header as follows:

Sample precision (P)   = 8
Number of lines (Y)   = 257
Number of samples per line (X)   = 255
Number of components (N$_f$)   = 255
Component identifiers (C$_i$)   = 254 (component A)
                                              253 (component B)
                                              252 (component C)
                                              251 (component D)
                                              250 (component A)
                                          (decrements by 1 for each scan)
                                              ...0 (component C)
Horizontal sampling factors (H$_i$)   = (1, 1, 3, 1) repeated for 255 components
Vertical sampling factors (V$_i$)   = (1, 2, 1, 4) repeated for 255 components
Quantization table selectors (T$_q$)   = (0, 1, 2, 3) repeated for 255 components

C.1 Non-Hierarchical decoder compliance tests

This section contains specifications of the structure of the compressed test data streams utilized by non-hierarchical decoder compliance tests.

C.1.1 Compressed test data stream structure for the baseline decoding process (Test A)

Compressed test data stream A1

SOI
COM   •   •
DQT $(P_q = 0)$ quant tables
DRI   restart interval $(R_i = 5)$
SOF$_0$ frame parameters $(P=8)$
DHT $(T_h = 0\text{-}1)$ Huffman tables
APP0   •   •
SOS   scan parameters $(N_s = N_f)$
entropy-coded   **data segment**   •   •
EOI

**(Interchange format
for compressed data)**

C-3

C.1.2 Compressed test data stream structure for the baseline decoding process (Test B)

Compressed test data stream B1

SOI
COM • •
DQT ($P_q = 0$) quant tables
DHT (Th = 0-1) Huffman tables
EOI

(Abbreviated format for table-specification data)

Compressed test data stream B2

SOI
COM • •
SOF0 frame parameters (P=8, N_f=255)
DRI restart interval ($R_i = 11$)
SOS scan parameters ($N_s = 1$)
entropy-coded data segment (0)
• • •
entropy-coded data segment (15)
DQT ($P_q = 0$) quant tables
DHT ($T_h = 0-1$) Huffman tables
DRI restart interval ($R_i = 0$)
entropy-coded data segment (16)
• • •
entropy-coded data segment (255)
EOI

(Abbreviated format for compressed image data)

C.1.3 Compressed test data structure for the extended sequential Huffman decoding process, 8-bit sample precision (Test C)

Compressed test data stream C1   **Compressed test data stream C2**

SOI
COM
•
•
DQT ($P_q = 0$) quant tables
DRI restart interval ($R_i = 10$)
SOF1 frame parameters ($P = 8$)
DHT ($T_h = 0\text{-}3$) Huffman tables
SOS scan parameters ($N_s = N_f$)
entropy-coded data segment • •
EOI

SOI
COM
•
•
DQT ($P_q = 0$) quant tables
SOF1 frame parameters ($P = 8$)
DHT ($T_h = 0\text{-}3$) Huffman tables
DRI restart interval ($R_i = 10$)
SOS scan parameters ($N_s = 1$)
entropy-coded data segment • •
• • (one scan for each component)
EOI

C-5

C.1.4 Compressed test data structure for the extended sequential arithmetic decoding process, 8-bit sample precision (Test D)

Compressed test data stream D1　　　**Compressed test data stream D2**

Compressed test data stream D1
SOI
COM • •
DQT $(P_q = 0)$ quant tables
DRI restart interval $(R_i = 10)$
SOF9 frame parameters $(P = 8)$
DAC $(T_a = 0\text{-}3)$ AC cond tables
SOS scan parameters $(N_s = N_f)$
entropy-coded data segment • •
EOI

Compressed test data stream D2
SOI
COM • •
DQT $(P_q = 0)$ quant tables
SOF9 frame parameters $(P = 8)$
DAC $(T_a = 0\text{-}3)$ AC cond tables
DRI restart interval $(R_i = 10)$
SOS scan parameters $(N_s = 1)$
entropy-coded data segment • •
• • (one scan for each component)
EOI

C.1.5 Compressed test data structure for the extended sequential Huffman decoding process, 12-bit sample precision (Test E)

Compressed test data stream E1   **Compressed test data stream E2**

SOI
COM • •
DQT $(P_q = 0)$ quant tables
DRI restart interval $(R_i = 10)$
SOF1 frame parameters $(P = 12)$
DHT $(T_h = 0\text{-}3)$ Huffman tables
SOS scan parameters $(N_s = N_f)$
entropy-coded data segment • • •
EOI

SOI
COM • •
DQT $(P_q = 1)$ quant tables
SOF1 frame parameters $(P = 12)$
DHT $(T_h = 0\text{-}3)$ Huffman tables
DRI restart interval $(R_i = 10)$
SOS scan parameters $(N_s = 1)$
entropy-coded data segment • • •
(one scan for each component)
EOI

C-7

C.1.6 Compressed test data structure for the extended sequential arithmetic decoding process, 12-bit sample precision (Test F)

Compressed test data stream F1   Compressed test data stream F2

SOI
COM
•
•
DQT ($P_q = 0$) quant tables
DRI restart interval ($R_i = 10$)
SOF9 frame parameters ($P = 12$)
DAC ($T_a = 0$-3) AC cond tables
SOS scan parameters ($N_s = N_f$)
entropy-coded data segment
•
•
EOI

SOI
COM
•
•
DQT ($P_q = 1$) quant tables
SOF9 frame parameters ($P = 12$)
DAC ($T_a = 0$-3) AC cond tables
DRI restart interval ($R_i = 10$)
SOS scan parameters ($N_s = 1$)
entropy-coded data segment
•
•
•
(one scan for each component)
EOI

C-8

C.1.7 Compressed test data structure for the progressive spectral selection Huffman decoding process, 8-bit sample precision (Test G)

Compressed test data stream G1

SOI
COM
•
•
DQT $(P_q = 0)$ quant tables
DRI   restart interval $(R_i = 10)$
SOF2   frame parameters $(P = 8)$
DHT $(T_h=0\text{-}3)$ Huffman tables
SOS   scan parameters $(N_s = 3,$ $S_s = S_e = 0,$ $A_h = A_l = 0\ )$
entropy-coded   **data segment**   •   •
DHT $(T_h=0\text{-}3)$ Huffman tables
SOS   scan parameters $(N_s = 1,$ $S_s = S_e = 0,$ $A_h = A_l = 0)$
entropy-coded   **data segment**   •   •
•   •   (total of 10 scans)
EOI

C-9

C.1.8 Compressed test data structure for the progressive spectral selection arithmetic decoding process, 8-bit sample precision (Test H)

Compressed test data stream H1

SOI
COM • •
DQT $(P_q = 0)$ quant tables
DRI restart interval $(R_i = 10)$
SOF10 frame parameters $(P = 8)$
DAC $(T_a = 0\text{-}3)$ AC cond tables
SOS scan parameters $(N_s = 3,$ $S_s = S_e = 0,$ $A_h = A_l = 0\,)$
entropy-coded data segment • •
DAC $(T_a = 0\text{-}3)$ AC cond tables
SOS scan parameters $(N_s = 1,$ $S_s = S_e = 0,$ $Ah = Al = 0)$
entropy-coded data segment • •
• • (total of 10 scans)
EOI

C.1.9 Compressed test data structure for the progressive spectral selection Huffman decoding process, 12-bit sample precision (Test I)

Compressed test data stream I1

SOI
COM   •   •
DQT $(P_q = 0)$ quant tables
DRI   restart interval $(R_i = 10)$
SOF2   frame parameters $(P = 12)$
DHT $(T_h = 0\text{-}3)$ Huffman tables
SOS   scan parameters $(N_s = 3,$ $S_s = S_e = 0,$ $A_h = A_l = 0\ )$
entropy-coded   **data segment**   •   •
DHT $(T_h = 0\text{-}3)$ Huffman tables
SOS   scan parameters $(N_s = 1,$ $S_s = S_e = 0,$ $A_h = A_l = 0)$
entropy-coded   **data segment**   •   •
•   •   (total of 10 scans)
EOI

C-11

C.1.10 Compressed test data structure for the progressive spectral selection arithmetic decoding process, 12-bit sample precision (Test J)

Compressed test data stream J1

SOI
COM
•
•
DQT $(P_q = 0)$ quant tables
DRI   restart interval $(R_i = 10)$
SOF10   frame parameters $(P = 12)$
DAC $(T_a = 0\text{-}3)$ AC cond tables
SOS   scan parameters $(N_s = 3,$ $S_s = S_e = 0,$ $A_h = A_l = 0$)
entropy-coded   **data segment**   •   •
DAC $(T_a = 0\text{-}3)$ AC cond tables
SOS   scan parameters $(N_s = 1,$ $S_s = S_e = 0,$ $Ah = Al = 0)$
entropy-coded   **data segment**   •   •
•   •
(total of 10 scans)
EOI

C.1.11 Compressed test data structure for the full progressive Huffman decoding process, 8-bit sample precision (Test K)

Compressed test data stream K1

SOI
COM • •
DQT ($P_q = 0$) quant tables
DRI restart interval ($R_i = 10$)
SOF2 frame parameters ($P = 8$)
DHT ($T_h = 0\text{-}3$) Huffman tables
SOS scan parameters ($N_s = 1$, $S_s = S_e = 0$, $A_h = 0, A_l = 1$)
entropy-coded data segment • •
DHT ($T_h = 0\text{-}3$) Huffman tables
SOS scan parameters ($N_s = 3$, $S_s = S_e = 0$, $A_h = 0, A_l = 1$)
entropy-coded data segment • •
• • (total of 15 scans)
EOI

C-13

C.1.12 Compressed test data structure for the full progressive arithmetic decoding process, 8-bit sample precision (Test L)

Compressed test data stream L1

SOI
COM
•
•
DQT
$(P_q = 0)$
quant tables
DRI
restart interval
$(R_i = 10)$
SOF10
frame parameters
$(P = 8)$
DAC
$(T_a = 0\text{-}3)$
AC cond tables
SOS
scan parameters
$(N_s = 1,$
$S_s = S_e = 0,$
$A_h = 0, A_l = 1)$
entropy-coded
data segment
•
•
DAC
$(T_a = 0\text{-}3)$
AC cond tables
SOS
scan parameters
$(N_s = 3,$
$S_s = S_e = 0,$
$A_h = 0, A_l = 1)$
entropy-coded
data segment
•
•
•
•
(total of 15 scans)
EOI

C-14

C.1.13 Compressed test data structure for the full progressive Huffman decoding process, 12-bit sample precision (Test M)

Compressed test data stream M1

SOI
COM • •
DQT $(P_q = 0)$ quant tables
DRI restart interval $(R_i = 10)$
SOF2 frame parameters $(P = 12)$
DHT $(T_h = 0\text{-}3)$ Huffman tables
SOS scan parameters $(N_s = 1,$ $S_s = S_e = 0,$ $A_h = 0, A_l = 1)$
entropy-coded data segment • •
DHT $(T_h = 0\text{-}3)$ Huffman tables
SOS scan parameters $(N_s = 3,$ $S_s = S_e = 0,$ $A_h = 0, A_l = 1)$
entropy-coded data segment • •
• • (total of 15 scans)
EOI

C-15

C.1.14 Compressed test data structure for the full progressive arithmetic decoding process, 12-bit sample precision (Test N)

Compressed test data stream N1

SOI
COM • •
DQT $(P_q = 0)$ quant tables
DRI restart interval $(R_i = 10)$
SOF10 frame parameters $(P = 12)$
DAC $(T_a = 0\text{-}3)$ AC cond tables
SOS scan parameters $(N_s = 1,$ $S_s = S_e = 0,$ $A_h = 0, A_l = 1)$
entropy-coded data segment • •
DAC $(T_a = 0\text{-}3)$ AC cond tables
SOS scan parameters $(N_s = 3,$ $S_s = S_e = 0,$ $A_h = 0, A_l = 1)$
entropy-coded data segment • •
• • (total of 15 scans)
EOI

C-16

C.1.15 Compressed test data structure for the lossless Huffman decoding process, 2-through 16-bit sample precision (Test O)

Compressed test data stream O1   Compressed test data stream O2

SOI
COM
•
•
DRI
restart interval
($R_i = 84$)
SOF3
frame parameters
(P=8)
DHT
($T_h = 0\text{-}3$)
Huffman tables
SOS
scan parameters
($N_s = N_f$)
entropy-coded
data segment
•
•
EOI

SOI
COM
•
•
SOF3
frame parameters
(P=16)
DHT
($T_h = 0\text{-}3$)
Huffman tables
DRI
restart interval
($R_i = 11$)
SOS
scan parameters
($N_s = 1$)
entropy-coded
data segment
•
•
DRI
restart interval
($R_i = 0$)
•
•
(one scan for each
component)
EOI

C.1.16 Compressed test data structure for the lossless arithmetic decoding process, 2- through 16-bit sample precision (Test P)

Compressed test data stream P1	Compressed test data stream P2

<table>
<tr><td align="center">SOI</td></tr>
<tr><td align="center">COM
•
•</td></tr>
<tr><td align="center">DRI
restart interval
($R_i = 84$)</td></tr>
<tr><td align="center">SOF11
frame parameters
(P=8)</td></tr>
<tr><td align="center">DAC
($T_a = 0\text{-}3$)
AC cond tables</td></tr>
<tr><td align="center">SOS
scan parameters
($N_s = N_f$)</td></tr>
<tr><td align="center">entropy-coded
data segment
•
•</td></tr>
<tr><td align="center">EOI</td></tr>
</table>

<table>
<tr><td align="center">SOI</td></tr>
<tr><td align="center">COM
•
•</td></tr>
<tr><td align="center">SOF11
frame parameters
(P=16)</td></tr>
<tr><td align="center">DAC
($T_a = 0\text{-}3$)
AC cond tables</td></tr>
<tr><td align="center">DRI
restart interval
($R_i = 11$)</td></tr>
<tr><td align="center">SOS
scan parameters
($N_s = 1$)</td></tr>
<tr><td align="center">entropy-coded
data segment
•
•</td></tr>
<tr><td align="center">DRI
restart interval
($R_i = 0$)
•
•</td></tr>
<tr><td align="center">(one scan for each
component)</td></tr>
<tr><td align="center">EOI</td></tr>
</table>

C.2 Hierarchical decoder compliance tests

This section contains specifications of the structure of the compressed test data streams utilized by hierarchical decoder compliance tests.

C.2.1 Compressed test data structure for hierarchical DCT-based sequential Huffman decoding processes (Test Q)

Compressed test data stream Q1

SOI
COM   •   •
DHP   hierarchical   parameters
DQT   quant tables
SOF1   frame parameters
DHT   (T_h=0-3)   Huffman tables
SOS   scan parameters   ($N_s = N_f$)
entropy-coded   **data segment**   •   •
EXP   expand 2:1,2:1
SOF5   frame parameters
DHT   (T_h=0-3)   Huffman tables
SOS   scan parameters   ($N_s = N_f$)
entropy-coded   **data segment**   •   •
•   •   (repeat for total of 4   differential frames)
EOI

Compressed test data stream Q2

SOI
COM   •   •
DHP   hierarchical   parameters
DQT   quant tables
SOF1   frame parameters
DHT   (T_h=0-3)   Huffman tables
SOS   scan parameters   ($N_s = 1$)
entropy-coded   **data segment**   •   •
•   •   (one scan for each   component)
EXP   expand 2:1,2:1
SOF5   frame parameters
•   •   (repeat for total of 4   differential frames)
EOI

C-19

C.2.2 Compressed test data structure for hierarchical DCT-based sequential arithmetic decoding processes (Test R)

Compressed test data stream R1   **Compressed test data stream R2**

Compressed test data stream R1	Compressed test data stream R2
SOI	**SOI**
COM · ·	**COM** · ·
DHP hierarchical parameters	**DHP** hierarchical parameters
DQT quant tables	**DQT** quant tables
SOF9 frame parameters	**SOF9** frame parameters
DAC (T_a = 0-3) AC cond tables	**DAC** (T_a = 0-3) AC cond tables
SOS scan parameters ($N_s = N_f$)	**SOS** scan parameters ($N_s = 1$)
entropy-coded data segment · ·	**entropy-coded data segment** · ·
EXP expand 2:1,2:1	· (one scan for each component)
SOF13 frame parameters	**EXP** expand 2:1,2:1
DAC (T_a = 0-3) AC cond tables	**SOF13** frame parameters
SOS scan parameters ($N_s = N_f$)	· · (repeat for total of 4 differential frames)
entropy-coded data segment · ·	**EOI**
· · (repeat for total of 4 differential frames)	
EOI	

C.2.3 Compressed test data structure for the hierarchical lossless Huffman decoding processes (Test S)

Compressed test data stream S1

SOI
COM • •
DHP hierarchical parameters
SOF3 frame parameters (P=8)
DHT (T_h=0-3) Huffman tables
SOS scan parameters (N_s = N_f)
entropy-coded data segment • •
EXP expand 2:1,2:1
SOF7 frame parameters
DHT (T_h=0-3) Huffman tables
SOS scan parameters (N_s = N_f)
entropy-coded data segment • •
• • (repeat for total of 4 differential frames)
EOI

Compressed test data stream S2

SOI
COM • •
DHP hierarchical parameters
SOF3 frame parameters (P=16)
DHT (T_h=0-3) Huffman tables
SOS scan parameters (N_s = 1)
entropy-coded data segment • •
• • (one scan for each component)
EXP expand 2:1,2:1
SOF7 frame parameters
DHT (T_h=0-3) Huffman tables
SOS scan parameters (N_s = 1)
entropy-coded data segment • •
• • (one scan for each component
• • (repeat for total of 4 differential frames)
EOI

C-21

C.2.4 Compressed test data structure for the hierarchical lossless arithmetic decoding processes (Test T)

Compressed test data stream T1   **Compressed test data stream T2**

Compressed test data stream T1	Compressed test data stream T2
SOI	SOI
COM • •	COM • •
DHP hierarchical parameters	DHP hierarchical parameters
SOF11 frame parameters (P=8)	SOF11 frame parameters (P=16)
DAC $(T_a=0\text{-}3)$ AC cond tables	DAC $(T_a = 0\text{-}3)$ AC cond tables
SOS scan parameters $(N_s = N_f)$	SOS scan parameters $(N_s = 1)$
entropy-coded data segment • •	entropy-coded data segment • •
EXP expand 2:1,2:1	• • (one scan for each component)
SOF15 frame parameters	EXP expand 2:1,2:1
DAC $(T_a = 0\text{-}3)$ AC cond tables	SOF15 frame parameters
SOS scan parameters $(N_s = N_f)$	DAC $(T_a = 0\text{-}3)$ AC cond tables
entropy-coded data segment • •	SOS scan parameters $(N_s = 1)$
• • (repeat for total of 4 differential frames)	entropy-coded data segment • •
EOI	• • (one scan for each component • • (repeat for total of 4 differential frames)
	EOI

Annex D

Construction of application-specific compliance tests

(This annex does not form an integral part of this Recommendation | International Standard.)

Many application-specific encoders and decoders are incapable of operating over the full range of parameter values specified in Part 1 for a particular process. Also, the requirements on accuracy defined in Annex A in combination with the quantization tables specified in Annex B are not a guarantee of satisfactory image quality for all application environments. Therefore, each application may construct application-specific compliance tests which constrain the range of parameter values or specify different quantization tables.

D.1 Procedure for construction of application-specific compliance tests for DCT-based processes

The parameters which may be constrained for DCT-based processes are as follows:

> Number of samples per line (X)
> Number of lines (Y)
> Number of image components in frame (N_f)
> Horizontal sampling factor (H_i)
> Vertical sampling factor (V_i)
> Number of image components in scan (N_s)

The following parameters may be constrained to the range [0, (maximum number of tables)-1]:

> Quantization table identifier (T_q)
> Huffman table identifier (T_h)
> AC conditioning table identifier (T_b)
> Quantization table selector (Tq_i)
> DC entropy coding table selector (Td_i)
> AC entropy coding table selector (Ta_i)

The values chosen for quantization depend completely on the image quality requirements of the application.

D-1

The following procedure shall be followed for construction of application-specific compliance tests for DCT-based processes.

1. Specify desired number of components, image dimensions and sampling factors.

2. Create test image from the provided source image test data. Start by filling the first component, proceed from left to right, top to bottom, and end with the last component. If necessary, replicate the provided source image test data.

3. Compute encoder reference test data by applying double precision floating point FDCT and quantize using application-specific quantization tables.

4. Produce compressed test data stream for decoding process by encoding quantized DCT coefficients using the entropy encoder.

5. Compute decoder reference test data by applying double precision floating point IDCT and inverse quantizer to the encoder reference test data. Clip the resulting output data to the range of the sample precision ([0,255] for 8-bit precision and [0,4095] for 12-bit precision). Apply the FDCT and quantizer used in step 2 to the clipped output data to produce the decoder reference test data.

NOTE: The entropy encoder which is needed to generate the compressed test data stream for application-specific compliance tests shall be developed and validated by the creators of the application-specific compressed test data.

D.2 Procedure for construction of application-specific compliance tests for lossless processes

The parameters which may be constrained for lossless processes are as follows:

Number of samples per line (X)
Number of lines (Y)
Number of image components in frame (N_f)
Horizontal sampling factor (H_i)
Vertical sampling factor (V_i)
Number of image components in scan (Ns)

D-2

The following parameters may be constrained to the range [0, (maximum number of tables)-1]:

> Huffman table identifier (T_h)
> AC conditioning table identifier (T_b)
> DC entropy coding table selector (T_{di})
> AC entropy coding table selector (T_{ai})

The following procedure shall be followed for construction of application-specific compliance tests for lossless processes.

1. Specify desired number of components, image dimensions and sampling factors.

2. Create test image from the provided source image test data. Start by filling the first component, proceed from left to right, top to bottom, and end with the last component. If necessary, replicate the provided source image test data.

3. Compute encoder reference test data by applying the reference lossless encoder. Encoder reference test data is used as the compressed test data stream.

4. Compute decoder reference test data by applying the reference lossless decoder to the compressed test data stream.

NOTE: The reference lossless encoder and reference lossless decoder which are needed to generate the test data for application-specific compliance tests shall be developed and validated by the creators of the application-specific test data.

D.3 Procedure for determining compliance of application-specific encoders and decoders

The procedure for application-specific compliance testing follows the procedure defined in Annex A, using the reference test data and the compressed test data stream generated by the procedure defined in D.1 and D.2. The output test data shall satisfy the requirements on accuracy defined in Annex A except the value of absolute differences. The values of all absolute differences shall not exceed a level set by the application.

Annex E Compliance test data for testing of greater computational accuracy

(This annex does not form an integral part of this Recommendation | International Standard.)

Greater computational accuracy of an encoder's FDCT/quantizer, or of a decoder's IDCT, can be tested by first changing the quantization tables shown in Annex B, and then following the Annex A procedures for testing generic DCT-based encoders or decoders. In addition to the compliance test data supplied for compliance testing of generic encoders and decoders, test data which utilizes different quantization tables are provided to assist implementors in measurement of greater computational accuracy. This annex specifies the structure of the compressed image data and quantization tables utilized in this additional test data.

E.1 Quantization table scale factors

The quantization tables used in the test data provided for the measurement of greater computational accuracy are derived by dividing the value of quantization tables in Annex B by a scale factor, F , where F is a power of 2.

Quantization table values for a given value of F are computed from the following formula:

$$E_{ij} = \lceil B_{ij} / F \rceil \text{ for i=0-7 ; for j=0-7 ;}$$

where $\lceil\ \rceil$ represents the ceiling function: rounding up to the next larger integer,

E_{ij} is the quantization table value used in testing for greater accuracy,

B_{ij} is the quantization table value defined in Annex B,

F   is the scale factor.

E-1

Table E.1 contains the resulting values of the quantization table for Component A (see Annex B) for F = 2.

Table E.1 - Quantization Table for Component A for F=2

4	3	3	4	6	10	13	15
3	3	4	5	7	15	15	14
4	4	4	6	10	15	18	14
4	5	6	8	13	22	20	16
5	6	10	14	17	28	26	20
6	9	14	16	21	26	29	23
13	16	20	22	26	31	30	26
18	23	24	25	28	25	26	25

E.2 Tests for measurement of greater computational accuracy

Test data is provided for six tests. The test designations and the values of F utilized are shown below:

Test	F
A2	2
A3	4
A4	8
A5	16
A6	32
A7	64

E.2.1 Encoder Tests

The encoder tests for measurement of greater computational accuracy are performed by following the procedure set forth in Annex A.1.1 and A.1.2. The test is successfully completed if the generated test data satisfies the requirements on accuracy specified in A.1.2.

E.2.2 Decoder Tests

The decoder tests for measurement of greater computational accuracy are performed by following the procedure set forth in Annex A.1.3 and A.1.4. The test is successfully completed if the generated test data satisfies the requirements on accuracy specified in A.1.4.

The decoder tests use test data with structure identical to that used by test A. See C.1.1 for a detailed specification of the compressed test data stream structure.

E-2

Annex F

Specification of supported parameter ranges

(This annex does not form an integral part of this Recommendation | International Standard.)

Many application-specific decoding devices are incapable of processing compressed images over the full range of parameter values specified by the particular decoding process. In addition, the generic compliance tests do not test many of the parameters over their entire allowable range of values.

For these reasons, it is strongly recommended that, for certain parameters, implementors publish the allowable parameter ranges for each decoding process supported. A table similar to table F.1 should be used for this purpose.

It is recommended that the DCT accuracy level be one for most applications (see A.1.4). For those applications requiring greater precision, lossless coding should be considered. It is recognized that certain applications may have a DCT accuracy level greater than one.

Table F.1 Parameter ranges supported

Parameter	Size (bits)	sequential DCT baseline	extended	progressive DCT	lossless
Sample precision (P)	8				
Number of lines (Y)	16				
Number of samples per line (X)	16				
Number of components in frame (Nf)	8				
Horizontal sampling factors (Hi)	4				
Vertical sampling factors (Vi)	4				
Quantization table selectors (Tqi)	8				
Number of components in scan (Ns)	8				
Other information					

F-1

Annex G

Test data for validation of implementations

(This annex does not form an integral part of this Recommendation | International Standard.)

This section contains a description of image source data, intermediate results of various procedures, compressed data streams, and final reconstructed images which may be used to validate implementations of the various processes specified by this Specification. The tests described in this annex are informative and do not replace or modify the compliance test requirements.

G.1 Test data description

Referring to Figure 4 of Part 1 of this Specification, test data for validation of encoder processes have the following elements :

1.	source image data
2.	unquantized FDCT coefficients
3.	quantized DCT coefficients
4.	compressed data streams.

Referring to Figure 6 of Part 1, test data for validation of decoder processes includes items 3 and 4 above, and also

5.	dequantized DCT coefficients
6.	reconstructed image data

For hierarchical encoder and decoder validation tests, additional intermediate reconstructed image data before and after upsampling are included. In addition, down sampling source image data are given. For lossless processes DCT-related data is not required.

G.2 Source image data

Six test image data sets are provided. For three precisions, 8-bit, 12-bit, and 16-bit, test sets X and Y are provided.

Test set X is a three component image with 8 bits per sample which was extracted from a CCIR Rec. 601-1 resolution image. For higher precisions, the low order bits are random. This image is 128 x 128 samples with three components having horizontal sampling factors of 2:1:1. This test set has both smooth areas and high detail areas.

Test set Y is a four component image, and is the same test data as used for compliance

G-1

testing. Test set Y uses a wider range of sampling factors and image dimensions which are not multiples of eight.

G.3 Compressed data streams

Validation tests are defined for representative examples of all of the modes of operation described in Part 1 of this Specification. These tests are listed in Table G.1. Note that two compressed data streams are supplied for each validation test, corresponding to the two test image data sets. The quantization tables used in the DCT-based modes are the example tables given in Annex K of Part 1 of this Specification. The modes of operation listed in Table G.1 are:

S(B)	- baseline sequential DCT
S(E)	- extended sequential DCT
P(SS)	- progressive DCT with spectral selection
P(SA)	- progressive DCT with successive approximation
P(FULL)	- progressive DCT with both spectral selection and successive approximation.
LL	- lossless
H-S	- hierarchical using the sequential DCT mode. This test does not include a final lossless correction.
H-L	- hierarchical using lossless mode. The final output is lossless. The first differential frame uses a point transform.

Table G.1   Validation compressed data streams

Validation Test	JPEG Mode	P (bits)	Entropy Coding	Coding Tables	N_f	Interleave	DRI
1 a	S(B)	8	Huff.	sample	max	no	no
1 b	S(B)	8	Huff.	sample	max	yes	no
1 c	S(B)	8	Huff.	custom	max	no	yes
1 d	S(B)	8	Huff.	custom	max	yes	yes
2 a	S(E)	8	arith.	default	max	no	yes
2 b	S(E)	8	arith.	default	max	yes	yes
3 a	P(SS)	8	Huff.	custom	max	no	yes
3 b	P(SS)	8	Huff.	custom	max	yes (DC)	yes
3 c	P(SA)	8	Huff.	custom	max	yes (DC)	yes
3 d	P(Full)	8	Huff.	custom	max	yes (DC)	yes
4 a	P(SS)	8	arith.	custom	max	no	yes
4 b	P(SS)	8	arith.	custom	max	yes (DC)	yes
4 c	P(SA)	8	arith.	custom	max	yes (DC)	yes
4 d	P(Full)	8	arith.	custom	max	yes (DC)	yes
5 a	LL	8	Huff.	custom	1	--------	yes
5 b	LL	8	Huff.	custom	max	yes	yes
5 c	LL	12	Huff.	custom	1	--------	yes
5 d	LL	12	Huff.	custom	max	yes	yes
5 e	LL	16	Huff.	custom	1	--------	yes
5 f	LL	16	Huff.	custom	max	yes	yes
6 a	LL	8	arith.	custom	1	--------	yes
6 b	LL	8	arith.	custom	max	yes	yes
6 c	LL	12	arith.	custom	1	--------	yes
6 d	LL	12	arith.	custom	max	yes	yes
6 e	LL	16	arith.	custom	1	--------	yes
6 f	LL	16	arith.	custom	max	yes	yes
7	S(E)	12	Huff.	custom	1	--------	yes
8	S(E)	12	arith.	custom	1	--------	yes
9	H-S	8	Huff.	custom	max	yes	yes
10	H-S	8	arith.	custom	max	yes	yes
11	H-L	8	Huff.	custom	max	yes	yes
12	H-L	8	arith.	custom	max	yes	yes

The sample Huffman tables referred to in the baseline sequential validation tests (tests 1a and 1b) are given in Annex K of Part 1 of this Specification.

In tests 2a and 2b default conditioning is used. However, this default conditioning is explicitly specified in the compressed data stream by means of a DAC marker segment.

Annex H

Examples and guidelines

(This annex does not form an integral part of this Recommendation | International Standard.)

This annex provides examples of test data and other guidelines.

H.1 Examples of application-specific compliance tests

This section contains two examples of application-specific compliance tests which utilize constrained parameter sets.

The first example utilizes a test image with three components, having samples per line and number of lines which are a multiple of 8, simplified sampling ratios and compressed image data encoded using the baseline process.

The second example utilizes a test image with one component, having samples per line and number of lines which are a multiple of 8, simplified sampling ratios and compressed image data encoded using the baseline process.

The parameter values used in these examples are shown in Tables H.1 and H.2.

Table H.1 Parameter values used in example 1

Parameter	Size (bits)	sequential DCT baseline process constrained to:
Sample precision (P)	8	8
Number of lines (Y)	16	256
Number of samples per line (X)	16	256
Number of components in frame (Nf)	8	3
Horizontal sampling factors (Hi)	4	2, 1, 1
Vertical sampling factors (Vi)	4	2, 1, 1
Quantization table selectors (Tqi)	8	0, 1, 1
Number of components in scan (Ns)	8	3

H-1

Table H.2 Parameter values used in example 2

Parameter	Size (bits)	sequential DCT baseline process constrained to:
Sample precision (P)	8	8
Number of lines (Y)	16	256
Number of samples per line (X)	16	256
Number of components in frame (Nf)	8	1
Horizontal sampling factors (Hi)	4	1
Vertical sampling factors (Vi)	4	1
Quantization table selectors (Tqi)	8	0
Number of components in scan (Ns)	8	1

REFERENCES

1   A. K. Jain. *Fundamentals of Digital Image Processing.* Englewood Cliffs, NJ: Prentice Hall (1989).

2   R. W. G. Hunt. *Measuring Colour.* New York: Halsted Press (1987).

3   R. W. G. Hunt. *The Reproduction of Colour in Photography, Printing, & Television.* England: Fountain Press (1987).

4   H. Peng (IBM). Private communication (Mar. 1991).

5   F. I. Van Nes and M. A. Bouman. Spatial Modulation Transfer in the Human Eye. *J. of the Opt. Soc. of Am.* 57(3):401-6 (Mar. 1967).

6   L. A. Olzak and J. P. Thomas. Seeing Spatial Patterns. *In* K. R. Boff, L. Kaufman, and J. P. Thomas (eds.), *Handbook of Perception and Human Performance.* pp. 1-56. New York: John Wiley (1986).

7   S. J. Daly, M. Rabbani, and C. Chen. Digital Image Compression and Transmission System Visually Weighted Transform Coefficients. U.S. Pat. No. 4,780,761 (Oct. 25, 1988).

8   J. O. Limb. Distortion Criteria of the Human Viewer. *IEEE Trans. on Systems, Man, and Cybernetics.* SMC-9(12):778-93 (Dec. 1979).

9   K. T. Mullen. The Contrast Sensitivity of Human Colour Vision to Red-Green and Blue-Yellow Chromatic Gratings. *J. Physiol.* 359:381-400 (1985).

10   D. B. Judd and G. Wyszecki. *Color in Business, Science and Industry.* pp. 282-7. New York: John Wiley and Sons (1975).

11   R. L. De Valois and K. K. De Valois. Spatial Vision. *In* D. E. Broadbent, J. L. McGaugh, N. J. Mackintosh, M. I. Posner, E. Tulving, and L. Weiskrantz (eds.), *Oxford Psychology Series.* vol. 14. p. 78. New York: Oxford University Press (1988).

12   N. Ahmed, T. Natarajan, and K. R. Rao. Discrete Cosine Transform. *IEEE Trans. Computers.* C-23:90-3 (Jan. 1974).

13   P. A. Wintz. Transform Picture Coding. *Proc. of the IEEE.* 60(7):809-20 (Jul. 1972).

14   H. Lohscheller. Subjectively Adapted Image Communication System. *IEEE Trans. on Commun.* COM-32(12):1316-22 (Dec. 1984).

15   H. A. Peterson, H. Peng, J. H. Morgan, and W. B. Pennebaker. Quantization of color image components in the DCT domain. *Proc. SPIE.* 1453:210-22 (1991).

16   A. J. Ahumada Jr. and H. A. Peterson. Luminance-Model-Based DCT Quantization for Color Image Compression. *Proc. SPIE.* 1666: (Feb. 1992).

17   E. Linzer and E. Feig. New DCT and Scaled-DCT Algorithms for Fused Multiply/Add Architectures. *ICASSP.* 2201-4 (1991).

18   K. R. Rao and P. Yip. *Discrete Cosine Transform.* New York: Academic Press (1990).

19   J. W. Cooley and J. W. Tukey. An algorithm for the machine calculation of complex Fourier series. *Math. Comp.* 19:297-301 (1965).

20   P. Duhamel and M. Vetterli. Fast Fourier Transforms: A Tutorial Review and a State of the Art. *Signal Processing.* 19:259-99 (1990).

21   W. Chen, C. H. Smith, and S. C. Fralick. A Fast Computational Algorithm for the Discrete Cosine Transform. *IEEE Trans. on Commun.* COM-25(9):1004-9 (Sept. 1977).

22   E. Feig and E. Linzer. Discrete Cosine Transform Algorithms for Image Data Compression. *Proceedings Electronic Imaging '90 East*, pp. 84-7. Boston, MA (Oct. 29-Nov. 1, 1990).

23   A. Ligtenberg and M. Vetterli. A discrete Fourier-cosine transform chip. *IEEE J. on Selected Areas in Commun.* SAC-4:49-61 (Jan. 1986).

24   A. Ligtenberg, R. H. Wright, and J. H. O'Neill. A VLSI Orthogonal Transform Chip for Real-Time Image Compression. *Visual Communication & Image Process.* II (Oct. 1987).

25   M. Vetterli and H. Nussbaumer. Simple FFT and DCT Algorithms with Reduced Number of Operations. *Signal Processing.* 6:267-78 (1984).

26   R. M. Haralick. A Storage Efficient Way to Implement the Discrete Cosine Transform. *IEEE Trans. Computers.* C-25:764-5 (Jul. 1976).

27   B. D. Tseng and W. C. Miller. On Computing the Discrete Cosine Transform. *IEEE Trans. on Computers.* C-27(10):966-8 (Oct. 1978).

28   S. Winograd. On Computing the Discrete Fourier Transform. *Mathematics of Computation.* 23(141):175-99 (Jan. 1978).

29   H. F. Silverman. An Introduction to Programming the Winograd Fourier Transform Algorithm (WFTA). *IEEE Trans. on Acoustics, Speech, and Signal Proc.* ASSP-25(2):152-65 (Apr. 1977).

30   Y. Arai, T. Agui, and M. Nakajima. A Fast DCT-SQ Scheme for Images. *Trans. of the IEICE.* E 71(11):1095 (Nov. 1988).

31   M. Vetterli. Trade-offs in the Computation of Mono- and Multi-dimensional DCT's. Columbia University, Center for Telecom. Res. *Tech. Rep.* (Jun. 1988).

32   J. Granata, M. Conner, and R. Tolimieri. The Tensor Product: A Mathematical Programming Language for FFTs and other Fast DSP Operations. *IEEE SP Magazine.* 40-8 (Jan. 1992).

33   M. An, I. Gertner, M. Rofheart, and R. Tolimieri. Discrete Fast Fourier Transform Algorithms: A Tutorial Survey. *In* P. W. Hawkes (ed.), *Advances in Electronics and Electron Physics.* vol. 80. pp. 2-67. New York: Academic Press, Inc. (1991).

34   M. Rabbani and P. W. Jones. *Digital Image Compression Techniques.* Bellingham, WA: SPIE Optical Engineering Press (1991).

35   D. A. Huffman. A Method for the Construction of Minimum-Redundancy Codes. *Proc. IRE.* 40(9):1098-101 (Sept. 1952).

36   D. S. Thornton and K. L. Anderson. A BNF for the JPEG Image Compression Header Structure. *Proceedings Electronic Imaging '90 East*, pp. 96-9. Boston, MA (Oct. 29-Nov. 1, 1990).

37   A. V. Aho, R. Sethi, and J. D. Ullman. *Compilers: Principles, Techniques, and Tools.* Addison-Wesley Publishing Company (1988).

38   C. E. Shannon. *The Mathematical Theory of Communication.* Illinois: The University of Illinois Press (1949).

39   P. Elias. *In* N. Abramson, *Information Theory and Coding.* New York: McGraw-Hill Book Company (1963).

40   W. B. Pennebaker, J. L. Mitchell, G. L. Langdon, and R. B. Arps. An overview of the basic principles of the Q-coder adaptive binary arithmetic coder. *IBM J. Res. Develop.* 32(6):717-26 (Nov. 1988).

41   J. L. Mitchell and W. B. Pennebaker. Optimal hardware and software arithmetic coding procedures for the Q-Coder. *IBM J. Res. Develop.* 32(6):727-36 (Nov. 1988).

42   W. B. Pennebaker and J. L. Mitchell. Probability estimation for the Q-Coder. *IBM J. Res. Develop.* 32(6):737-52 (Nov. 1988).

43   J. L. Mitchell and W. B. Pennebaker. Software implementations of the Q-Coder. *IBM J. Res. Develop.* 32(6):753-74 (Nov. 1988).

44   R. B. Arps, T. K. Truong, D. J. Lu, R. C. Pasco, and T. D. Friedman. A multi-purpose VLSI chip for adaptive data compression of bi-level images. *IBM J. Res. Develop.* 32(6):775-95 (Nov. 1988).

45   F. Ono, M. Yoshida, T. Kimura, and S. Kino. Subtraction-type Arithmetic Coding with MPS/LPS Conditional Exchange. *Annual Spring Conference of IECED, Japan.* D-288 (1990).

46   C. Chamzas and D. Duttweiler. Probability estimation in arithmetic coders. In preparation (1992).

47   G. G. Langdon and J. J. Rissanen. Compression of Black-White Images with Arithmetic Coding. *IEEE Trans. on Commun.* 29(6):858-67 (Jun. 1981).

48   G. G. Langdon and J. J. Rissanen. A Simple General Binary Source Code. *IEEE Trans. Info. Theory.* 28(5):800-3 (Sept. 1982).

49   C. B. Jones. An Efficient Coding System for Long Source Sequences. *IEEE Trans. Info. Theory.* IT-27(3):280-91 (May 1981).

50   J. J. Rissanen. Generalized Kraft Inequality and Arithmetic Coding. *IBM J. Res. Develop.* 20:198-203 (May 1976).

51   R. Pasco. Source Coding Algorithms for Fast Data Compression. Ph.D. Thesis. Stanford University. Stanford, CA 94305 (1976).

52   F. Rubin. Arithmetic Stream Coding Using Fixed Precision Registers. *IEEE Trans. Info. Theory.* 25(6):672-5 (Nov. 1979).

53   J. J. Rissanen and G. G. Langdon. Arithmetic Coding. *IBM J. Res. Develop.* 23(2):149-62 (Mar. 1979).

54   G. G. Langdon. An Introduction to Arithmetic Coding. *IBM J. Res. Develop.* 28(2):135-49 (Mar. 1984).

55   F. Ono, S. Kino, M. Yoshida, and T. Kimura. Bi-level image coding with Melcode - Comparison of block type code and arithmetic type code. *Proc. of Globecom '89*, pp. 225-60. (Nov. 1989).

56   S. W. Golomb. Run-Length Encodings. *IEEE Trans. Info. Theory.* IT-12:399-401 (Jul. 1966).

57   J. G. Cleary, I. H. Witten, and R. M. Neal. Arithmetic Coding for Data Compression. *Commun. of the ACM.* 30(6):520-40 (Jun. 1987).

58   G. G. Langdon. Method for Carry-Over Control in a FIFO Arithmetic Code String. *IBM TDB*. 23(1):310-12 (Jun. 1980).
59   C. Chamzas (AT&T). X3L3 presentation (1989).
60   R. C. Pasco and W. B. Pennebaker. Overlap of Marker Codes with the Final Bits of an Arithmetic Code String. ISO/IEC JTC1/SC2/WG8 JBIG-N160 (Sept. 8, 1989).
61   J. J. Rissanen (IBM). Private communication (1985).
62   D. R. Helman, G. G. Langdon, N. Martin, and S. J. P. Todd. Statistics Collection for Compression Coding with Randomizing Feature. *IBM TDB*. 24(10):4919 (Mar. 1982).
63   W. B. Pennebaker. A Basis for a QM-coder. X3L2.8 Oral presentation (Aug. 14, 1989).
64   P. Flajolet. Approximate Counting: A Detailed Analysis. *BIT*. 25:113-34 (1985).
65   A. G. Tescher. Transform Image Coding. *Aerospace Corp. Res. Rep.* SAMSO-TR-78-127. pp. 1-110 (May 1978).
66   G. Goertzel. An algorithm for the evaluation of finite trigonometric series. *Am. Math. Monthly*. 65(1):34-5 (Jan. 1958).
67   G. Wallace, R. Vivian, and H. Poulsen. Subjective testing results for still picture compression algorithms for international standardization. *Proc. of Globecom '88*, pp. 1022-7. (1988).
68   G. G. Langdon, Jr. Compression of Multilevel Signals. U.S. Pat. No. 4,749,983 (Jun. 7, 1988).
69   G. Goertzel and J. L. Mitchell. Symmetrical Optimized Adaptive Data Compression/ Transfer/Decompression System. U.S. Pat. No. 4,633,490 (Dec. 30, 1986).
70   J. L. Mitchell and W. B. Pennebaker. Arithmetic Coding Data Compression/De-Compression by Selectively Employed, Diverse Arithmetic Coding Encoders and Decoders. U.S. Pat. No. 4,891,643 (Jan. 2, 1990).
71   W. B. Pennebaker and J. L. Mitchell. Probability Adaptation for Arithmetic Coders. U.S. Pat. No. 4,935,882 (Jun. 19, 1990).
72   G. G. Langdon, J. L. Mitchell, W. B. Pennebaker, and J. J. Rissanen. Arithmetic Coding Encoder and Decoder System. U.S. Pat. No. 4,905,297 (Feb. 27, 1990).
73   R. B. Arps (IBM). Private communication (1990).
74   G. G. Langdon (IBM). Private communication (1985).
75   J. L. Mitchell and K. L. Anderson. Speedup Mode. ISO/IEC JTC1/SC2/WG8 JPEG-241 (Jan. 6, 1989).
76   H. Jeffreys. *Theory of Probability*. London: Oxford University Press (1961).
77   A. Agresti. *Categorical Data Analysis*. New York: John Wiley and Sons (1990).
78   R. N. Williams. *Adaptive Data Compression*. Boston: Kluwer Academic Publishers (1991).
79   W. R. Equitz (IBM). Private communication (Aug. 1992).
80   J. L. Mitchell and W. B. Pennebaker. Evolving JPEG color data compression standard. *In* M. Nier and M. E. Courtot (eds.), *Standards for Electronic Imaging Systems*. vol. CR37. pp. 68-97. Washington: SPIE Optical Engineering Press (1991).
81   B. Niss. Prediction of AC Coefficients from the DC Values. ISO/IEC JTC1/SC2/WG8 N745 (May 1988).
82   S. Wu and A. Gersho. Improved Decoder for Transform Coding with Application to the JPEG Baseline System. *IEEE Trans. on Commun.* 40(2):251-4 (Feb. 1992).
83   J. C. King (Adobe). Vendor/application worksheet (Jun. 9, 1992).
84   Corporate contact: Sohail Kahn, Marketing Director, AT&T Microelectronics, 555 Union Blvd., MS 2C-271, Allentown, PA, U.S.A. 18103. P: (215)439-6309; F: (215)778-4593.
85   G. T. Warner (AT&T). Vendor/application worksheet (Jun. 5, 1992).
86   J. O. Jensen (AGI). Vendor/application worksheet (Jun. 5, 1992).
87   S. Hite (AutoView). Vendor/application worksheet (Jul. 10, 1992).
88   E. Rutins. Vendor/application worksheet (Jul. 1992).
89   G. Nishite (Cal DMV). Vendor/application worksheet (Jul. 27, 1992).
90   Corporate contact: Rick Rasmussen, Vice President of Marketing, C-Cube Microsystems, Inc., 1778 McCarthy Blvd., Milpitas, CA, U.S.A. 95035. P: (408)944-6300; F: (408)944-6314.
91   J. Anderson (C-Cube). Vendor/application worksheet (Aug. 3, 1992).
92   Corporate contact: Inez Conover, Vice President of Sales & Marketing, Data Link, 620 Herndon Parkway, Suite 120, Herndon, VA, U.S.A. 22070. P: (703)318-7300; F: (703)318-7309.
93   T. Nock (Data Link). Vendor/application worksheet (Aug. 11, 1992).
94   Corporate contact: Tom Plunkett, Chief Operating Officer, Discovery Technologies, 2405 Trade Centre Drive, Longmont, CO, U.S.A. 80503. P: (303)651-6500; F: (303)651-7545.
95   T. Plunkett (DTI). Vendor/application worksheet (Jul. 6, 1992).
96   L. Sprengelmeyer. Vendor/application worksheet (Aug. 3, 1992).
97   T. Kinsman (Eastman Kodak Company). Vendor/application worksheet (Aug. 6, 1992).
98   T. Kinsman (Eastman Kodak Company). Vendor/application worksheet (Aug. 24, 1992).
99   M. H. Woehrmann (Handmade Software). Vendor/application worksheet (Jul. 24, 1992).
100   L. Kelly-Mahaffey (IBM). Vendor/application worksheet (Jun. 12, 1992).
101   I. R. Finlay (IBM). Vendor/application worksheet (Aug. 10, 1992).
102   R. Hawks (Identix). Vendor/application worksheet (Jun. 5, 1992).
103   T. Williams (IIT). Vendor/application worksheet (Jun. 5, 1992).
104   T. G. Lane (IJG). Vendor/application worksheet (Jun. 19, 1992).
105   Corporate contact: Gregory Kisor, President, Information Technologies Research, Inc., 3520 W. Hallandale Beach Blvd., Pembroke Park, FL, U.S.A. 33023. P: (305)962-9961; F: (305)962-6546.
106   G. Kisor (ITR). Vendor/application worksheet (Jun. 2, 1992).
107   W. C. Lewis (Lewis Siwell). Vendor/application worksheet (Jun. 17, 1992).

108 L. Blauner (LSI). Vendor/application worksheet (May 22, 1992).
109 P. H. Ang, P. A. Ruetz, and D. Auld. Video Compression Makes Big Gains. *IEEE Spectrum*. 28(10):16-9 (Oct. 1991).
110 L. Bijnagle (Moore Data). Vendor/application worksheet (Jul. 14, 1992).
111 V. Andelin (NBSI). Vendor/application worksheet (Aug. 24, 1992).
112 H. Ibaraki (NEL). Vendor/application worksheet (Aug. 7, 1992).
113 A. Bassan (Optibase). Vendor/application worksheet (May 12, 1992).
114 A. Bassan (Optibase). Vendor/application worksheet (Jul. 8, 1992).
115 Corporate contact: Dr. James S. Tyler, President, Optivision, Inc., 4009 Miranda Avenue, Palo Alto, CA, U.S.A. 94306. P: (415)855-0200; F: (415)855-0222.
116 B. Johnston (Optivision). Vendor/application worksheet (Aug. 21, 1992).
117 K. Löser (PKI). Vendor/application worksheet (Jul. 27, 1992).
118 A. Mathur (PRISM). Vendor/application worksheet (Aug. 24, 1992).
119 Corporate contact: Marty Hollander, Vice President, Marketing, Storm Technology, Inc., 1861 Landings Drive, Mountain View, CA, U.S.A. 94043. P: (415)691-6670; F: (415)691-9825.
120 A. M. Gunning (STORM). Vendor/application worksheet (Jun. 29, 1992).
121 Corporate contact: Leonard G. Roberts, Vice President of Sales & Marketing, Telephoto Communications Inc., 11722-D Sorrento Valley Road, San Diego, CA, U.S.A. 92121-1084. P: (619)452-0903; F: (619)792-0075.
122 A. B. Barnhart (Telephoto). Vendor/application worksheet (Jun. 2, 1992).
123 G. Cruickshank (Tribune Solutions). Vendor/application worksheet (Jul. 2, 1992).
124 Corporate contact: Dick Moeller, Chairman of the Board & Chief Executive Officer, VideoTelecom Corp., 1901 W. Braker Lane, Austin, TX, U.S.A. 78758. P: (512)834-2700; F: (512)834-3792.
125 D. Hein (VTC). Vendor/application worksheet (May 21, 1992).
126 Corporate contact: Don Dehaan, Vice President of Sales, XImage Corporation, 1050 North Fifth Street, San Jose, CA, U.S.A. 95112. P: (408)288-8800; F: (408)993-1050.
127 P. Johnson (XImage). Vendor/application worksheet (Jul. 13, 1992).
128 Corporate contact: Howard Gordon, President, Xing Technology Corporation, 456 Carpenter Canyon, Arroyo Grande, CA, U.S.A. 93420. P: (805)473-0145; F: (805)473-0147.
129 H. Gordon (Xing). Vendor/application worksheet (May 26, 1992).
130 Corporate contact: Michael Nell, Vice President of Marketing, Zoran Corporation, 1705 Wyatt Dr., Santa Clara, CA, U.S.A. 95054. P: (408)986-1314; F: (408)986-1240.
131 S. E. Brook (Zoran Corporation). Vendor/application worksheet (Jul. 24, 1992).
132 J. Yoshida. Fuji flashes digital still camera. *Electronic Engineering Times*. (Sept. 9, 1991).
133 M. Kenner (3M). Vendor/application worksheet (May 27, 1992).
134 ETSI Project Team 22. Re-using Matrices and Tables. ISO/IEC JTC1/SC2/WG8 JPEG-515 (Feb. 27, 1990).
135 I. Sebestyen. Use of JPEG in ETSI-Photovideotex (CCITT). ISO/IEC JTC1/SC2/WG10 JPEG-736 (Mar. 29, 1991).
136 E. Hamilton (C-Cube). Vendor/application worksheet (Aug. 21, 1992).
137 A. C. Wilson. A new image for image processing. *Mil. & Aerosp. Electronics*. 3(6):17-9 (Aug. 1992).
138 (Aldus Corporation). TIFF: Tag Image File Format Revision 6.0. (Apr. 6, 1992).
139 G. D. Wallenstein. *Setting Global Telecommunication Standards: the Stakes, the Players, and the Process*. Boston: Artech House (1990).
140 A. Macpherson. *International Telecommunication Standards Organization*. Boston: Artech House (1990).
141 G. D. Wallenstein. *Setting Global Telecommunication Standards: the Stakes, the Players, and the Process*. p. 85. Boston: Artech House (1990).
142 G. P. Hudson. Photovideotex Image Compression Algorithms - Towards International Standardisation. *Esprit '87 Conference*. Brussels (Sept. 1987).
143 Japan. Flowcharts of Candidate Algorithms for Standardization. ISO/IEC JTC1/SC2/WG8 N446 (Mar. 1987).
144 M. Postl. Quadtree Extension of Block Truncation coding. ISO/IEC JTC1/SC2/WG8 N449 (Mar. 1987).
145 ESPRIT-PICA. Adaptive Discrete Cosine Transform for Still Image Communication on ISDN. ISO/IEC JTC1/SC2/WG8 N453 Rev. 1 (Mar. 1987).
146 D. Tricker on behalf of the ESPIRT-PICA Predictive Coding. Progressive Recursive Binary Nesting. ISO/IEC JTC1/SC2/WG8 N461 (Mar. 1987).
147 Germany. A Hybrid Algorithm for the Coding of Colour Images. ISO/IEC JTC1/SC2/WG8 N462 Rev. 1 (Mar. 1987).
148 G. K. Wallace. DCT with Low Block-to-Block Distortion. ISO/IEC JTC1/SC2/WG8 N464 (Mar. 1987).
149 AT&T Bell Laboratories. Block List Transform (BLT) Coding of Images. ISO/IEC JTC1/SC2/WG8 N465 (Mar. 1987).
150 Germany. Flowchart of the HPC Coding Algorithm. ISO/IEC JTC1/SC2/WG8 N469 (Mar. 1987).
151 J. L. Mitchell, W. B. Pennebaker, and C. F. Touchton. DPCM using Adaptive Binary Arithmetic Coding. ISO/IEC JTC1/SC2/WG8 N471 (Mar. 1987).

152   J. L. Mitchell, W. B. Pennebaker, C. A. Gonzales, and C. F. Touchton. Progress Towards a Color Image Data Compression Standard. *Proceedings Electronic Imaging '89 West*, pp. 189-94. Pasadena, CA (Apr. 10-13, 1989).

153   ESPRIT-PICA. Adaptive Discrete Cosine Transform Coding Scheme for Still Image Communication on ISDN. ISO/IEC JTC1/SC2/WG8 N502 Rev. 1 (Jun. 1987).

154   C. A. Gonzales, J. L. Mitchell, W. B. Pennebaker, and C. F. Touchton. DPCM Using Adaptive Binary Arithmetic Coding. ISO/IEC JTC1/SC2/WG8 N509 (Jun. 1987).

155   K. Ogura. Generalized Block Truncation Coding (GBTC). ISO/IEC JTC1/SC2/WG8 N510 (Jun. 1987).

156   KDD. Progressive Coding Scheme (PCS). ISO/IEC JTC1/SC2/WG8 N506 (Jun. 1987).

157   C. F. Touchton, W. B. Pennebaker, J. L. Mitchell, and C. A. Gonzales. A Predictive Image Codec using Adaptive Binary Arithmetic Coding. ISO/IEC JTC1/SC2/WG8 N639 (Jan. 1988).

158   ISO/ADCTG. Adaptive Discrete Cosine Transform Coding Scheme for Still Image Telecommunication Services. ISO/IEC JTC1/SC2/WG8 N640 (Jan. 1988).

159   Japan. Block Separated Component Progressive Coding (BSPC). ISO/IEC JTC1/SC2/WG8 N641 (Jun. 1987).

160   A. Leger, J. L. Mitchell, and Y. Yamazaki. Still Picture Compression Algorithms Evaluated for International Standardization. *Proc. of Globecom '88*, pp. 1028-1032. (Nov. 1988).

161   J. L. Mitchell. CCITT/ISO Photographic Image Data Compression Standard Functionality and Goals. ISO/IEC JTC1/SC2/WG8 N766 Rev. 0 (Jan. 1988).

162   J. L. Mitchell. CCITT/ISO Photographic Image Data Compression Standard Functionality and Goals. ISO/IEC JTC1/SC2/WG8 N766 Rev. 1 (May 1988).

163   G. Wallace. Addendum to N766 Functionality Statement. ISO/IEC JTC1/SC2/WG8 N766 Annex 2 (May 1988).

164   B. Niss. Prediction of AC coefficients from DC values revisited. ISO/IEC JTC1/SC2/WG8 N147 (Jun. 1988).

165   A. Ligtenberg and G. K. Wallace. The Emerging JPEG Compression Standard for Continuous Tone Images: An Overview. *Proc. of PCS '90*, pp. 6.1.1-4. Cambridge, MA (Mar. 26-28, 1990).

166   A. Leger and T. Omachi. Sequential Coding in the Emerging JPEG Compression Standard. *Proc. of PCS '90*, pp. 6.2.1-3. Cambridge, MA (Mar. 26-28, 1990).

167   W. B. Pennebaker, J. Vaaben, and J. L. Mitchell. Progressive Coding in the Emerging JPEG Compression Standard. *Proc. of PCS '90*, pp. 6.3.1-3. Cambridge, MA (Mar. 26-28, 1990).

168   U.S. National Body. Resolutions. X3L3 SD-17/91-09 (Sept. 27, 1991).

169   I. R. Finlay. Study of CD 10918-2 Method of IDCT Conformance Test. ISO/IEC JTC1/SC29/WG10 N9 (Mar. 1992).

170   ISO CD 11544 - Progressive Bi-level Image Compression. ISO/IEC JTC1/SC2/WG9 S1R4.1 (Sept. 16, 1991).

171   D. Le Gall. MPEG: A Video Compression Standard for Multimedia Applications. *Commun. of the ACM*. 34(4):46-58 (Apr. 1991).

172   K. R. McConnell, D. Bodson, and R. Schaphorst. *FAX: Digital Facsimile Technology and Applications*. Boston: Artech House (1992).

173   (CCITT). Video Codec for Audiovisual Services at p × 64 kbit/s. International Telegraph and Telephone Consultative Committee Recommendation H.261. CDM XV-R 37-E (Aug. 1990).

174   M. Liou. Overview of the px64 kbit/s Video Coding Standard. *Commun. of the ACM*. 34(4):59-63 (Apr. 1991).

175   R. B. Arps. Binary Image Compression. *In* W. K. Pratt (ed.), *Image Transmission Techniques*. pp. 219-76. San Francisco: Academic Press Inc. (1979).

176   K. Toyokawa. System for Compressing Bi-Level Data. U.S. Pat. No. 4,901,363 (Feb. 13, 1990).

177   K. L. Anderson, W. B. Pennebaker, and J. L. Mitchell. Compression of Binary Halftones. *IBM Res. Rep.* RC 14733. pp. 1-11 (Jun. 1989).

178   C. Chamzas and D. Duttweiler. Matching arithmetic coding models to dithering periods. In preparation (1992).

179   D. Sheinwald and R. C. Pasco. Deriving Deterministic Prediction Rules from Reduction Schemes. Accepted for publication (1993).

180   B. Astle. A Proposal for MPEG Video Report. ISO/IEC JTC1/SC2/WG11 MPEG 91/74 (Apr. 15, 1991).

181   C. A. Gonzales (IBM). Private communication (1990).

182   E. Hamilton (C-Cube). Private communication (Sept. 18, 1992).

183   W. B. Pennebaker. Adaptive Quantization within the JPEG Sequential Mode. Patent application (May 1991).

184   B. Niss. A simple algorithm for exact coding of images. ISO/IEC JTC1/SC2/WG8 JPEG-244 (Jan. 1989).

185   W. B. Pennebaker and J. L. Mitchell. Probability Adaptation for Arithmetic Coders. U.S. Pat. No. 5,099,440 (Mar. 24, 1992).

186   J. A. Osborne and C. Selfiert. Narrow Bandwidth Signal Transmission. U.S. Pat. No. 4,665,436 (May 12, 1987).

187   C. A. Gonzales, J. L. Mitchell, and W. B. Pennebaker. Adaptive Graylevel Image Compression System. U.S. Pat. No. 4,725,885 (Feb. 16, 1988).

188   W. B. Pennebaker and J. L. Mitchell. *JPEG Still Image Data Compression Standard*. New York: Van Nostrand Reinhold (1993).

189   (ISO/IEC). Directives—Part 3: Drafting and presentation of International Standards.

INDEX

abbreviated format   99, 351, 354, 381, 401, 562, 567-9, 579
   compliance tests   567-9
   compressed image data   99, 401, 562, 567
      example   124-5
   table specification data   99, 117, 401, 562, 568-9
      example   124
AC coefficient   33-4, 344, 354, 378
AC prediction   38-9, 261-5, 534-6
adaptive coding   73
   arithmetic   73, 75-7, 354
   Huffman   73-5
   single pass   73-7
   two pass   75
adaptor   73-7
ADCT   302, 306
   refining   306-7
additive color   16
aliasing   10-2, 181
alphabet   74, 143
anti-aliasing   16
application environments   342-3, 354, 357
application specific
   compliance tests   551, 558-60, 613-15, 625-6
application standards   6, 292-3, 559
   ANPA/IPTC   292
   ETSI   292
   NITF   293
applications   5, 267-93
   additions to JPEG syntax, *see* syntax
   APPn markers   107, 397-8
   bank-check processing   292, 325
   bulletin board systems   272-3
   digital cameras   277, 290
   driver licences   273, 283, 285
   insurance   272
   medical   21, 275, 281
   mug shots   279, 281-2, 289
   publishing   269-70, 285-8, 291
   real estate   283
   teleconferencing   288

arithmetic coding   5, 140-3, 345, 348, 409-30, 543
   see also QM-coder
   binary arithmetic coding   75, 142, 150, 409-11
   carry-over   143, 410-1
   code stream conventions   220-3, 410-11
   code stream termination   142
   conditioning   76, 146-7, 359, 410
   Elias coding   142
   interval subdivision   140-1, 410
   principles   140-3, 409-10, 425
   probability estimation   143, 156-7, 233-52, 411
   renormalization   142, 358, 410-1
   symbol ordering   142, 220-3, 410
arithmetic coding conditioning   394-5
   AC conditioning   213, 459
   DC conditioning   208-9, 457-8
   parameters   208, 213, 394-5, 458-9
   restrictions   118, 395, 449, 469, 477
   specification   117-8, 345, 394
   table identifier   395
arithmetic coding statistical models   203-18, 449-60, 469-74
   decision trees
      coding a value   206, 452-6, 471-4
      coding AC coefficients   210, 450-2
      coding later stage   215
   hierarchical mode spatial
   corrections   216, 501
   initial conditions   161, 204, 458-9, 494-5
      see also Initenc, Initdec
   lossless mode   216-8, 494-5
      2-D conditioning   216-8, 494
   progressive coding of AC   214, 488-90
      correction bits   215, 489-90
      EOBx   215, 489
      first-stage   214, 488
      later-stage   214-6, 488-90
   progressive coding of DC   214, 488
      first stage   214, 488
      later stage   214, 488

sequential coding of AC 210-3, 456-9, 470-4
  conditioning 213, 458-9
  decision sequence 211, 450-6
  decision tree 206, 210
  EOB decision 210-1, 450-1, 459, 470-1
  example 213
  magnitude categories 211, 454-6
  zigzag sequence 210, 450
sequential coding of DC 206-10, 457-8, 469-74
  conditioning 206, 208-10, 457-8
  decision sequence 207, 450, 452-6
  decision tree 206
  DPCM difference coding 206-10
  magnitude categories 207, 454-6
  statistics area 209, 458
aspect ratio 13

Backus-Naur form, *see* **BNF**
backwards compatibility 333
band, *see* spectral band
baseline 6, 86, 169, 343, 348, 353-4, 443, 462
  compliance tests 578, 593-4
  compressed data examples 121-2
  sequential DCT system 81-6, 443-8, 462-8
basis functions 31-3
Bayesian estimation 150, 156, 233-5, 247-8
bi-level images 5, 341
  see also JBIG
binary arithmetic coding, *see* arithmetic coding
bit plane coding 325
black/white images, *see* bi-level images
Block List Transform Coding 302
block truncation coding 72, 302
blocking artifacts 38, 181-2
  removal of 261-5, 534-6
blocks 38, 344, 360, 375
BNF 127-34
  for JPEG 130-4
brightness 18
byte stuffing 105, 150, 160, 354, 448, 480

CCITT 1, 292, 295, 297-8
  ODA 292, 339
  recommendations 7
    H.261 319, 546
    T.4 317
    T.6 318
    T.81 292, 311, 335, 337-543
    T.83 545, 547-626
  SGVIII 292, 548
chrominance 18, 24
  quantization 37, 503
CMY 16
CMYK 17
code book 73
code stream conventions for QM-coder
  bottom of interval 151, 221
    carry propagation 150, 158
  matching code streams 221-4
  top of interval 221
    borrow propagation 222-3
    initialization 223-4
coding models 355
  see also encoder models
  see also decoder models
  see also JPEG coding models

coding process 355
  see also process
color coordinates 17
color spaces 17
colorblind 20
compliance testing 41, 309, 311-3, 339, 367-9, 545-626
  application specific 551, 561, 613-15
  DCT accuracy 560, 585-7, 617-9
  decoder 561, 578-88, 615, 618
  encoder 561, 577, 615, 618
  syntax 562-76
  see also quantization tables
component identifier 112
  see also frame header parameters
components 17, 343, 349
compressed data 109-10, 355
  bit order 110, 381, 408, 430
  byte alignment 105, 110, 448
  byte order 110, 381, 430
  format 99, 381-403
compressed image data 99, 355
compression measures 78-9
  bits/pixel 78-9
  compression ratio 78-9
compression performance 5, 253-9
compression systems 65-6
conditional probabilities 146-7, 410
context 76, 147, 319, 355, 411
  index 147, 150, 411
contrast sensitivity function 24-5
  chrominance 24-5, 37
  diagonal 24
  luminance 24-5, 37, 179
  suprathreshold contrast response 179
control procedures 98, 431-40
  hierarchical 98, 498-9, 501-2
  non-hierarchical 98, 431-40
correlation 70, 72
  between components 114, 190
  DCT 29, 72
  in images 70
  in statistical modelling 143
  KLH 29
cosine basis functions 31-3, 38, 355-7

data ordering 99, 372
  see also interleaved
  see also non-interleaved
data organization 97-134
data units 101, 355, 372, 435, 461, 479, 491
DC coefficient 33-4, 344, 355
DCT 5, 29-64, 356, 541
  coefficients 344, 355
  compliance tests 560, 577-8, 585-7
  computational complexity 39
  dequantization 35-6, 71, 345, 355, 376, 461
  fast IDCT 52-3
  fast 1-D 41-52
    based on DFT 41, 49-50
    complexity 42-3
    flowgraphs 45, 52
    rotations in 44-8
    scaled 52
    simple example 42-3
  fast 2-D 53-64
    complexity 53
    flowgraphs 54-63
    other approaches 63-4
    scaled 53-63

FDCT   33, 39-40, 344, 356, 375-6
forward, *see* FDCT
Goertzel   178-9
IDCT   33, 39-40, 345, 357, 375-6, 462
   clamping   111, 462
inverse, *see* IDCT
mathematical definition   39-41, 375-6,
   560
precision   111, 442, 455
quantization   29, 34-8, 71, 344, 376, 442
   1-D   30-3
   2-D   33-4
DCT coefficient   33-4, 355
DCT decomposition   30-4
decision tree   157, 204-5
   compact notation   157, 204-5
   see also arithmetic coding statistical
   models
decoder models   65-74, 169-88
   DCT   70-1, 170, 175, 178
   DPCM   71-2, 172
decoder requirements   369, 559
dequantization   35-6, 71, 376, 461
   display adjusted decoding   266, 536-7
descriptors   65-6, 170, 203
DFT   41
   16-point   50-1
differential, *see* hierarchical coding models
differential pulse code modulation, *see*
DPCM
digital halftoning   15
dimensions   100, 349, 371, 387
discrete cosine transform   5, 29-64
   see also DCT
discrete Fourier transform, *see* DFT
distortion   67, 118, 531-2
DPCM   69-70, 169

encoder models   65-72, 169-88
   DCT   69-71, 172-8, 442
      AC   442
      DC   442
   DPCM   71-2, 171-2
encoder requirements   368, 559
end of image, *see* marker EOI
end of spectral selection   115
entropy   135-47
   definition   136
entropy coding   345, 348
   concepts   135-47
   examples   137-47
   major functional blocks   73-7
entropy decoder   67, 345
entropy encoder   65-6, 344-6
entropy-coded segments   97-9, 105-6, 356,
381, 383, 574
   stuffed zero byte   105, 150, 160, 448
entropy-coding table selection   114, 389
entropy-coding tables   66
   see also Huffman tables, arithmetic coding
errors
   recovery from   120, 230-1, 475
   transmission   67

facsimile   5, 318, 339
fast software for QM-coder   225-9
   conditional exchange   229
   decoder software   228-9
      parallel operation   228-9
   encoder software   225-8
      condition codes   225

negative probabilities   226
probability estimation   225-6
register mapping   225
FDCT, *see* DCT
file formats   291-3
fill bytes   105, 130-1, 381, 436, 570
filtering   12
   downsampling   12, 96, 347, 356, 499,
   531
   high-pass   12
   low-pass   12, 531
   upsampling   12, 96, 347, 359, 499
fixed-length codes   137
fixed probability   212, 214-6, 226, 458, 488,
490
frame   98, 347, 352, 356, 571
   component specification   111-2
   control procedure   431-2, 437
   header   98, 106, 110-3, 356, 386-8, 403

gamma correction   20, 26
generic embodiments   558-61, 585
Group 3   317-8
Group 4   317-8

halftoning   15
hierarchical coding   79, 348, 532-4
hierarchical coding models   79, 93-6, 185-8,
499-502
   differential DCT-based   106, 185-6,
   497-502
   differential lossless   106, 187-8, 497,
   499-502
   downsampling   96, 187-8, 499, 531
   modulo 65536   187, 379, 502
   point transform   96, 116, 121, 187, 378,
   500, 502
   upsampling   96, 187-8, 499-500
      bilinear interpolation   121, 499-500
hierarchical mode   81-2, 93-6, 346-8, 353,
356, 497-502
   compliance tests   565-7, 609-12
   differential correction   106, 500
   header   120-1, 399-400
   lossless   96, 106, 499-500, 502
histogram
   DCT coefficient probabilities.   173
   differences   70
   intensities   69
hue   18
Huffman codes   138-40
   bit order   110, 466
Huffman coding   138-40, 345, 348, 357, 543
   coding tree   139
   in MPEG   328-9
   JPEG Huffman coding   189-201
      see also Huffman statistical models
   principles   138-40
   restart   467, 475
   termination of code stream   105-6, 383
Huffman statistical models   189-98
   baseline   443-8, 462-8
   hierarchical mode spatial
   corrections   198
   lossless coding statistical model   198,
   493
   progressive coding of AC   480-8
      additional bit   198, 485
      correction bits   197-8, 485-7
      end-of-band   194, 480
      EOB run-length codes   194-5, 481
      first-stage   194-6, 480-4

first-stage example 196
later-stage 196-8, 484-8
later-stage example 198
2-D code table 195, 481
progressive coding of DC 194, 480
  correction bits 196-7
  first stage 194-6, 480
  first-stage example 195
  later stage 196-8, 480
  later-stage example 197
sequential coding of AC 192-4, 444-8, 463-8
  additional bits 191, 445
  end-of-block, see EOB
  EOB 192, 442, 444, 461
  example 194
  magnitude categories 193, 445, 460, 500
  run length/amplitude 192-3, 444-5, 464
  zigzag sequence 190, 442
  ZRL 192-3, 444
  2-D code table 192-3, 445, 460
sequential coding of DC 190-2, 443-4, 462
  appended bits 191, 443
  DPCM difference coding 190-2, 378, 443
  example 191
  magnitude categories 190-1, 443, 460
Huffman table 357, 444
  baseline 443, 446
  custom 75, 235, 307, 504
  fixed 75, 307
  restrictions 117, 393, 443, 445-6, 449, 462, 468, 477
  typical 509-17
Huffman table generation 198-201, 405-8
  Generate_code_table 200-1, 407
  Generate_size_table 200-1, 406
  generating decoder tables 201, 464-5
    Decoder_tables 201, 465
  generating encoder tables 201, 407-8
    Order_codes 201, 408
Huffman table specification 117, 345, 405-8
  creating a specification 199-200, 503-8
    Adjust_BITS 199-200, 507
    Code_size 199-200, 505
    Count_BITS 199-200, 506
    Sort_input 199-200, 508
  examples 121-2, 127, 517-8
human visual system 23-7, 542

IDCT, see DCT
IEC 1, 295, 298-9, 339-40
illumination 20
image 352
  binary 15-6
  color 16-8, 355
  color-palette 15, 21
  concepts 9-22
  continuous-tone 15, 343
  digital 10, 356
  grayscale 4, 15, 343, 356
  limited bits/pixel 16
  sequences 21
  vocabulary 9-22
image compression models 67-73
information 136-7

relation to probability 136
inheritance of tables 119, 392, 394
interchange format 99, 107, 118, 342, 351, 357, 381, 579
  compliance tests 562-7
  examples 127
  requirements 367
interframe 21
interleaved data 81, 99, 101-5, 113, 349, 357, 373-4, 388-90, 579
  in scan 113, 479
  number of components 114, 388-9
internal representation 100, 372
  nonrectangular sampling 100
  rectangular arrays 12, 100, 371
International Electrotechnical Commission see IEC
International Organization for Standardization see ISO
International Telegraph and Telephone Consultative Committee, see CCITT
ISO 1, 292, 295-7
ISO standards 6
  10918-1 309-11, 337-543, 545
  10918-2 311-3, 545, 547-626
  11172 317
  11544 317
ISO/IEC JTC1/SC29/WG10, see JPEG
ISO/IEC JTC1/SC29/WG11, see MPEG
ISO/IEC JTC1/SC29/WG9, see JBIG

JBIG 5, 149, 234, 244, 246-52, 317-25
  see also QM-coder
  adaptive template 320
  bit plane coding 325
  compression performance 323-4
  progressive mode 320-2
  sequential mode 322-3
  template models 318-23
JFIF 292-3
Joint Bi-level Image Experts Group, see JBIG
Joint Photographic Experts Group, see JPEG
JPEG 339, 541
  DIS Part 1 5, 335, 337-543
  draft DIS Part 2 5, 547-626
  functional requirements 303-6
  future directions 331-3
  goals 302, 312-6, 558
  history 5, 301-16, 339, 548
  joint coordination 299, 339, 548
JPEG coding models
  see sequential DCT-based models
  see progressive DCT-based models
  see lossless models
JPEG DIS Part 1 5, 335-543, 548
  Annex E 98, 431-40
  Annex J 98, 121, 497-502
  Annex K 6, 38, 188, 503-37
  Annex L 6, 539-40
  Annex M 6, 541-3
  Clause 3 6, 335, 342-53
  Clause 4 (definitions) 4, 6, 335, 354-66
  patents 6, 336, 539-40
  symbols 360-66
  table of contents 337
JPEG DIS Part 2 5, 545-626
  Clause 3 (definitions) 553-7
  table of contents 547
JPEG entropy coding
  see Huffman coding
  see QM-coder

JPEG extensions 331-3, 563
JTC1 299

KLH transform 29, 35

level of difficulty 4-5
  easy ○ 4
  hard ● 5
  intermediate ◑ 5
level shift 30, 171, 357, 375, 442, 462
  see also zero shift
linear color transformations 18
lossless coding model 182, 184-5, 491-6
  first line 93, 492-3
  point transform 115-6, 378, 493, 531-4
  prediction 182-5, 346, 379, 492-3
    at edges 184, 492
    calculation 184, 493
    initialization. 120, 492
    modulo calculation 185, 379, 493,
      496, 502
  prediction selection 115, 492
  switched prediction 332-3
lossless compression 10, 67, 343, 357
lossless mode 70, 78, 81-2, 92-3, 344, 346,
  353, 357
  compliance tests 563-4, 578, 580, 587-8,
    607-8
  compression performance 257-9, 314
  DPCM 81, 92-3
  examples of compressed data 127
lossy 4, 67, 78, 343, 357
luminance 18, 24
  quantization 37, 503
luminance-chrominance representation 18

marker 98, 105-21, 351, 357, 381, 403
  definitions 105-9, 381-3
  DHP 107-8, 120-1, 399-400, 498, 565,
    569
  DNL 107-9, 165, 398-9, 423, 432, 434,
    438, 468, 564-6, 573, 575
  EOI 107-9, 385, 431, 437, 564-9
  EXP 107-8, 121, 400, 498-9, 569, 573,
    575
  JPG 107-8, 382
  JPGn 107-8, 383
  marker code 105
  non-SOFn 107-9
  RES 107, 109, 383
  RSTm 107, 109, 385, 432-3, 438, 564-6,
    576
  SOFn 106, 109, 111, 386, 431-2, 436,
    564-7, 571, 573
  SOF0 106, 109, 111, 386, 564-7, 571,
    573
  SOI 107, 109, 385, 431, 436-7, 563-9
  SOS 107, 109, 113, 387, 564-6, 573
  syntax requirements 564, 566, 569
  tables/misc. 107, 117-21, 390-8, 431-2,
    436, 564-73
    APPn 107, 397-8
    COM 107, 396-7
    DAC 107, 117-8, 394-5, 458-9, 495
    DHT 107-8, 117, 392-4, 405
    DQT 107-8, 113, 118-9, 391-2
    DRI 107-8, 119-20, 396, 438, 569,
      572
    TEM 107, 109, 309, 383, 420
marker segments 97-9, 105-21, 351, 358,
381, 383
Markov

sources 146
  see also probability estimation
MCU 101-5, 351, 358, 373
  cautionary note 104
  control procedure 434-5, 440
  examples 101-5
  in final interval 432
  incomplete 101, 374, 586-7
  limit on number of data units 116, 333,
    389
  prediction
    DC 103-4, 355, 442, 461
    lossless 104
  quantization 103
  relation to sampling factors 101, 389
MELCODE 150, 308
Minimax coder 150, 156, 233
minimum coded units, *see* MCU
model tables 66
  see also quantization tables
modes of operation 5, 81-96, 346, 358
  domain of applicability 531-4
  see also hierarchical mode
  see also lossless mode
  see also progressive DCT-based mode
  see also sequential DCT-based mode
Motion Picture Experts Group, *see* MPEG
motion sequences 5
MPEG 5, 266, 317, 325-9
  adaptive quantization 266, 329
  group-of-pictures 326-8
  Huffman coding 328-9
  macroblocks 327-9
  pictures 326
  sequence 326
  slices 326

new work item 333
non-interleaved data 99, 101-3, 349, 373,
388, 579
nondifferential, *see* frame header
nonlinearity 20, 26
number of
  components in frame 111-2, 386-8
  components in scan 113, 388-90
  data units in MCU 116-7
  lines 111-2, 358, 386-8
  samples per line 111-2, 386-8

optimal procedures for QM-coder 219-25
  conditional exchange 220, 229
  hardware 219-23
    parallel operations 220, 228-9
  software 219-23
    parallel decoder 228-9
    serial operations 220
orthogonal waveforms 32-3
orthonormal 40

parameters 358, 381
  constrained 560, 613-5, 619, 625-6
  compliance tests 591
PCM 68-9
pel 13
pixel 13
point transform 358
  in scan header 115-6
  see also hierarchical coding
  see also lossless coding
  see also progressive DCT-based mode
precision 10, 13, 25, 111, 346, 358, 372

DCT-based modes 111, 348, 387-8, 578-80
   12-bit 111, 190-5, 207, 211, 459-60, 475, 477, 497, 578-82, 589
   hierarchical mode 111, 497, 499-500, 581-2
   lossless mode 111, 387-8, 492, 580-2
   sample 10, 13-6
    see also frame header
prediction 69
   initialization 171, 184, 442, 492-3
   see also AC prediction
   see also lossless coding model
   see also MCU
predictive coding 72
probability estimation 5, 74, 156-7, 225-6, 233-52
   approximate counting 156, 233
   Bayesian estimation 233-6, 247-8
   coding inefficiency 236
     mixed context 247
     single context 245-6
   conditional exchange 234, 241-2, 246, 411, 426-7
   granularity 235, 249
   initial learning 247-8, 415
   Markov chain modelling 236-47
     approximate 238-40
     mixed contexts 240, 243-4
     single context 240-3
   other estimation tables 249-52
     JPEG-FA table 250
   QM-coder state machine 235, 237, 415-7
     nontransient states 238-9
     transient states 237
   QM-coder table 239, 416-7
   Robustness 249
process 78, 342-4, 353, 355, 368-9, 441, 447, 491, 497, 553-5, 578-83
progressive DCT-based coding
models 173-82
   distortion at low bit-rates 179-83
   IDCT for progressive decoders 178-9
   spectral selection 174, 179-82
   successive approximation 175, 179-82
     fat zero 178, 537
     first-stage 175-7
     later-stage 177-8
progressive DCT-based mode 81-2, 86-92, 256-7, 346, 358, 477-90
   compliance tests 563-4, 580, 585-7, 599-606
   compression performance 256-7
   examples of compressed data 125-6
   full progression 92, 126, 182, 256, 356, 477
   number of components 112
   scan restrictions 86, 115, 479-80
   spectral selection 86-9, 359, 477-80
     end of 115, 388-90, 479
     start of 115, 388-90, 479
   successive approximation 86, 90-3, 359, 477-80
     point transform 115-6, 378, 477, 479-80, 483, 537
   syntax 106, 113-7, 386, 388-390
Progressive Recursive Binary Nesting 302
pulse code modulation, see PCM
Px64, see CCITT H.261
pyramidal coding 78-9, 347-8

Q-coder 149, 249, 308
QM-coder 149-67
   approximate multiplication 150, 153, 411
   Byte_in 165-7, 429
   Byte_out 158-60, 420
   carry bit 159, 354, 411
   carry-over 150, 411, 419-21
   Code_D(S) 157-8
   Code_0(S) 151, 157-8, 204, 226, 412-13
   Code_1(S) 151, 157-8, 204, 227, 412-13
   code stream convention 150-2, 220-3, 410-1
   conditional exchange 150, 234, 355, 411, 413-4, 426-7
     decoder 164-5, 425-7
     encoder 154-5, 414-5
   Decode(S) 164-5, 204, 425-7
   Flush 151, 162-3, 204, 423-4, 449
     Pacman 162, 165
     termination 151
   Initdec 166, 204, 430, 469
   Initenc 151, 204, 422, 449
   initialization 204
     decoder 166, 430
     encoder 160-2, 422
   input 165-7
   integer representation 153-4, 410-11
   interval subdivision 151-4, 410-11
   output 158-61
     latency 230, 357, 420
     pure 150, 222-3, 230, 234
   probability estimation 150, 156-7, 355, 411, 415-8
     approximate counting 156, 233
     coding inefficiency
     fast attack 161, 248-9
     initial learning 161, 233, 247-8, 415
     renormalization-driven 156, 417
     state machine 156, 358, 416-7
   register conventions
     decoder 163-4, 228, 425-6
     encoder 159-60, 411-12
   Renorm_d 166, 428-9
   Renorm_e 159, 419
   renormalization 150
     decoder 164-6, 428-9
     encoder 152-3, 419-22
   restart 120, 475
   resynchronization of decoders 230-1, 469, 475
   spacer bits 159-60, 411
   speedup mode 231-2
   stack count 160, 419-22
   stacked data 160
   symbol ordering 150-2, 220-2
     converting between 224-5
   termination of code stream 162-3, 423
     Pacman 162
   test sequences 518-30
   versus Q-coder 229-30, 249
quantization 29, 34-8, 344, 376, 442
   adaptive 264, 266, 329, 331-2
     approximate 264-6, 332
     see also MPEG
   precision 118-9, 392
   rounding 36, 376
   tables 37, 344, 358, 391-2, 461
     compliance test 560, 589-90, 617-8
     example 37, 503

identifier 119, 392
precision 118, 391-2
selection 112, 387
specification 118-9, 391-2
uniform step 36, 359, 376
values 35, 37, 170, 358
visibility threshold 37, 503, 531
reconstructed images 2-4, 65, 83-5, 87-91, 94-5, 356
clamping 111, 462
reference FDCT 40-1, 555-6, 586-7
reference IDCT 40-1, 555-6
reference test data 555, 561, 585-8
reflectance 26-7
resolution 10
restart interval 98, 119, 351, 359, 442, 475, 491, 579
control procedure 433-4, 438-9
example 126
final 119
restart marker 98, 106, 120, 351, 359
RGB 16
run-length 148, 359, 444

samples 10, 359, 344
sample precision 111
sampling aperture 10
sampling factor 100-4, 112-3, 357, 359, 371
example 101-4, 113
see also frame parameters
sampling grid 12
sampling interval 10
saturation 18
scan 81, 86, 98, 347, 352, 359, 573
component specification 114
control procedure 432-3, 438
header 98, 113-7, 359, 388-90, 403
segments 351
see also entropy-coded segments
see also marker segments
sequential DCT-based coding models 71-2, 169-73, 442
AC coding model 172-3, 442
DC coding model 70, 171-2, 442
zigzag ordering 171-3, 442
sequential DCT-based mode 81-6, 346, 359, 441-75
baseline system 86, 343, 353, 441-8, 461-8
compliance tests 563-4, 578-9, 585-7
compression performance 254-6
examples of compressed data 121-3
extended systems 86, 343, 353, 449-60, 468-75, 595-8
skew 237
skew coder 149
source image 65, 356, 371
spatial masking 25
spectral band 86, 115, 347, 477, 479
spectral selection, *see* progressive DCT-based mode
standards committees 5, 295-9, 301
ANPA/IPTC 292
ETSI 292
IPTC 339
nomenclature 336
see also CCITT
see also IEC

see also ISO
see also JBIG
see also JPEG
see also MPEG
start of frame, *see* marker SOFn
start of image, *see* marker SOI
start of scan, *see* marker SOS
start of spectral selection 115
statistical models 65
see also arithmetic coding statistical models
see also Huffman statistical models
tutorial 143-7, 456
statistics areas 147, 209, 212, 216-7, 359
statistics bins 147, 204, 359, 458-9
stuffed bytes 105, 150, 160
sub-band coding 73
subsampling 12, 181
subtractive color 16
successive approximation
bit position high 115-6
bit position low 115-6
see also progressive DCT-based mode
symbol ordering for QM-coder 220-3, 410
converting between 224-5
inverting code streams 224-5
LPS above MPS 220-3
MPS above LPS 220-3
symbols 65-6, 170
syntax 97-134
application-specific additions 97, 100
in JPEG DIS Part 1 383-401
see also compliance tests

tables/misc., *see also* marker
technical specifications 307-9
tool kit 81
transcoding 77, 359, 449
transform coding 72
trichromatic theory 16
tristimulus values 20

uniform perceptual spaces 20, 25
CIELAB 20, 26
CIELUV 20, 26

validation tests 308, 621-3
source image data 556, 621
variable-length codes 137-43
expansion 143
vector quantization 72-3, 302
vector quantization enhanced decoding 264
vendors 5, 267-91
visible distortion 23, 78, 503
blocking artifacts 38, 181, 534
suppression of 182, 261-5, 534-6
visually-weighted quantization 29, 34-8, 71, 542, 560

Weber's Law 26
Worksheet 267-9

YCbCr 19
YIQ 19
YUV 18

zero-shift 38, 115
see also level shift
zigzag sequence 33-4, 171-3, 190, 210, 345, 359, 378, 442, 445, 479